Microscopy and Microanalysis for Lithium-Ion Batteries

The past three decades have witnessed the great success of lithium-ion batteries, especially in the areas of 3C products, electrical vehicles, and smart grid applications. However, further optimization of the energy/power density, coulombic efficiency, cycle life, charge speed, and environmental adaptability are still needed. To address these issues, a thorough understanding of the reaction inside a battery or dynamic evolution of each component is required. *Microscopy and Microanalysis for Lithium-Ion Batteries* discusses advanced analytical techniques that offer the capability of resolving the structure and chemistry at an atomic resolution to further drive lithium-ion battery research and development.

- Provides comprehensive techniques that probe the fundamentals of Li-ion batteries.
- Covers the basic principles of the techniques involved as well as its application in battery research.
- Describes details of experimental setups and procedure for successful experiments.

This reference is aimed at researchers, engineers, and scientists studying lithium-ion batteries including chemical, materials, and electrical engineers, as well as chemists and physicists.

Microscopy and Microanalysis for Lithium-Ion Batteries

Edited by
Cai Shen

CRC Press
Taylor & Francis Group
Boca Raton London New York

CRC Press is an imprint of the
Taylor & Francis Group, an **informa** business

First edition published 2023
by CRC Press
6000 Broken Sound Parkway NW, Suite 300, Boca Raton, FL 33487-2742

and by CRC Press
4 Park Square, Milton Park, Abingdon, Oxon, OX14 4RN

CRC Press is an imprint of Taylor & Francis Group, LLC

ISBN: 978-1-032-28952-6 (hbk)
ISBN: 978-1-032-28954-0 (pbk)
ISBN: 978-1-003-29929-5 (ebk)

DOI: 10.1201/9781003299295

Typeset in Times
by codeMantra

Contents

Preface

The 2019 Nobel Prize in Chemistry was awarded to three scientists for their contributions to the development of lithium-ion batteries (LIBs). John B. Goodenough, M. Stanley Whittingham, and Akira Yoshino share the prize for their work on these rechargeable devices, which are now widely used for portable electronics, electric vehicles, and large-scale energy storage systems.

LIBs have made a great contribution to the transition from fossil fuel dependence to carbon neutral society. However, with the increasing demand for lithium batteries, the bottleneck of lithium battery technology is gradually exposed. At present, the technical bottleneck of batteries is mainly in five aspects: energy density, cycle life, safety, charging speed, and temperature tolerance. For example, the energy density of LIBs currently is in the range of 150–300 Wh/kg, with only a few manufacturers such as Tesla achieving 300 Wh/kg. However, the energy density in gasoline can reach 2,000 Wh/kg, almost ten times that of LIBs. In other words, it would take more than 400 kg of LIBs to produce the same energy as a 60-L fuel tank. In addition to energy density, several other factors, such as cycle life, are important to the widespread use of LIBs. To solve these problems, it is clear that joint efforts of scientists and engineers around the world are needed.

LIB is a delicate system which mainly includes anode, electrolyte, separator, and cathode. The design of high-performance LIBs depends on careful preparation and optimization of all these important components. To achieve this goal, the assistance of advanced characterization techniques is essential. The history of human science and technology is also the history of the development of instruments. Advanced characterization techniques can greatly promote the development of new materials and new instruments, and the birth of new instruments in turn to promote the development of science and technology.

In the process of LIB development, we hope to obtain a lot of useful information to help us to analyze the materials and devices to understand their performance in various aspects. At present, electrochemical measurement and characterization techniques are commonly used in the research of LIB materials and devices. Among them, electrochemical test mainly includes (i) charge and discharge test, which mainly depends on battery charge and discharge performance and rate; (ii) cyclic voltammetry (CV), which mainly depends on the reversibility of battery charge and discharge, peak current, and peak position; and (iii) electrochemical impedance spectroscopy (EIS) AC impedance test, which depends on the resistance and polarization of the battery.

Apart from the electrochemical test, the other commonly used characterization methods can be roughly divided into the following six parts: (i) composition characterization (ICP, SIMS, XPS, STXM), (ii) morphology characterization (SEM, TEM, AFM), (iii) crystal structure characterization (XRD, NEXAFS, ND, NMR, TEM), (iv) characterization of material functional groups (Raman, IR), (v) observation of ion transport in materials (ND, NMR, AFM), and (vi) micromechanical and electrical properties of materials (AFM).

It is my great pleasure and honor to invite those experts who are working in the frontier of LIB research to publish this book. The contents of this book include Chapter 1: Lithium-ion batteries; Chapter 2: Electron microscopy for advanced battery research; Chapter 3: Characterizing the localized electrochemical phenomena in Li-ion batteries by using SPM-based techniques ; Chapter 4: Atom probe tomography; Chapter 5: X-ray diffraction; Chapter 6: ICP-based techniques for LIBs characterization; Chapter 7: Secondary ion mass spectrometry; Chapter 8: Nuclear magnetic resonance microscopy: atom to micrometer; Chapter 9: Differential electrochemical mass spectrometry; Chapter 10: Thermal analysis of Li-ion batteries; Chapter 11: Electrochemical impedance spectroscopy; and Chapter 12: Synchrotron X-ray and neutron techniques.

The book does not cover every aspect of characterization techniques because of limited space. That does not mean those not included are not important. As mentioned above, to solve the bottleneck of LIBs, we need to relay on every single tool that can help to understand the fundamental and practical aspect of LIBs. We hope to include as many and more comprehensive chapters as possible in future editions.

The publication of this book was encouraged and supported by Chunsheng Wang and Hong Li who both played important roles during my academic research on LIBs. I would also like to thank the referees who contributed to the review of this book. They are Anyang Cui, Arun Devaraj, Alexander M. Korsunsky, Bingxin Huang, Fudong Han, Guiming Zhong, Haifeng Dai, Jun Lu, Jian Wang, Jiangwei Wang, Jiangyong Wang, Kun Liang, Lidan Xing, Mark E. Orazem, Manickam Minakshi, Qiuan Huang, Riqiang Fu, Sergio Brutti, Simon Wiemers-Meyer, Wenbo Yue, Xiangwen Gao, Xiaoji Xu, Xingyue Yong, Yanbin Shen, Yihan Zhu, and Yingke Zhou.

Finally, I would like to thank Allison Shatkin from Taylor & Francis Group for her continuous support to publish this book.

Cai Shen, PhD, FRSC
University of Nottingham Ningbo China

Acknowledgments

Cai Shen would like to thank financial support from the National Natural Science Foundation of China (Grant Nos. 22175192 and U2032126).

Editor

Cai Shen earned a Ph.D. in Chemistry from the University of St Andrews, UK, in 2008. Sequentially, he continued his research at the University of Maryland, Heidelberg University, and Aarhus University. He was an Associate Professor and then Professor at the Ningbo institute of Materials Technology and Engineering, Chinese Academy of Sciences during 2013–2022. He joined the University of Nottingham Ningbo China in February 2022. He has published more than 100 papers in peer-reviewed journals including *JACS*, *ACS Nano*, *Nano Energy*, *Small Methods*, and *Energy Storage Materials*. He is on the editorial board for a number of journals including *Journal of Microscopy*, *Materials* and *Energies*. His research interest mainly focuses on lithium-ion batteries and application of scanning probe microscopy. He was admitted as a Fellow of the Royal Society of Chemistry (FRSC) in 2022.

Contributors

N. Angulakshmi
Department of Materials Engineering and Convergence Technology
Gyeongsang National University
Jinju, South Korea

Yimeng Chen
CAMECA Instrument Inc.
Fitchburg, Wisconsin, USA

Limin Guo
Laboratory of Advanced Spectro-Electrochemistry and Lithium-Ion Batteries
Dalian Institute of Chemical Physics
Chinese Academy of Sciences
Dalian, China
and
College of Environment and Chemical Engineering
Dalian University
Dalian, Liaoning, China

Ankur Jain
Department of Mechanical and Aerospace Engineering
University of Texas at Arlington
Arlington, Texas, USA

Murugavel Kathiresan
CSIR-Central Electrochemical Research Institute
Karaikudi, India

Jie Liu
Center for High Pressure Science & Technology Advanced Research (HPSTAR)
Beijing, China

Xiaosong Liu
National Synchrotron Radiation Laboratory
University of Science and Technology of China
Hefei, Anhui, China

Chenjie Lou
Center for High Pressure Science & Technology Advanced Research (HPSTAR)
Beijing, China

Paweł Piotr Michałowski
Łukasiewicz Research Network – Institute of Microelectronics and Photonics
Warsaw, Poland

Sascha Nowak
MEET Battery Research Center
University of Münster
Münster, Germany

Kei Ono
School of Vehicle and Mobility
Tsinghua University
Beijing, China

Long Pang
Laboratory of Advanced Spectro-Electrochemistry and Lithium-Ion Batteries
Dalian Institute of Chemical Physics
Chinese Academy of Sciences
Dalian, China

Mohammad Parhizi
Underwriters Laboratories
Houston, Texas, USA

Zhangquan Peng
Laboratory of Advanced Spectro-Electrochemistry and Lithium-Ion Batteries
Dalian Institute of Chemical Physics
Chinese Academy of Sciences
Dalian, China
and
School of Applied Physics and Materials
Wuyi University
Jiangmen, China

A. Manuel Stephan
CSIR-Central Electrochemical Research Institute
Karaikudi, India

Mingxue Tang
Center for High Pressure Science & Technology Advanced Research (HPSTAR)
Beijing, China

Shangshang Wang
School of Vehicle and Mobility
Tsinghua University
Beijing, China

Xuefeng Wang
Institute of Physics
Chinese Academy of Sciences
Beijing, China

M. Stanley Whittingham
Binghamton University
State University of New York
Vestal, New York, USA

Fengxia Xin
Binghamton University
State University of New York
Vestal, New York, USA

Zhi Yang
Laboratory of Advanced Spectro-Electrochemistry and Lithium-Ion Batteries
Dalian Institute of Chemical Physics
Chinese Academy of Sciences
Dalian, China

Pengfei Yu
Shanghai Institute of Microsystem and Information Technology
Chinese Academy of Sciences
Shanghai, China

Kaiyang Zeng
Department of Mechanical Engineering
National University of Singapore
Singapore, Singapore

Jianbo Zhang
School of Vehicle and Mobility
Tsinghua University
Beijing, China

Yelong Zhang
Laboratory of Advanced Spectro-Electrochemistry and Lithium-Ion Batteries
Dalian Institute of Chemical Physics
Chinese Academy of Sciences
Dalian, China
and
School of Applied Physics and Materials
Wuyi University
Jiangmen, China

Zhiwei Zhao
Laboratory of Advanced Spectro-Electrochemistry and Lithium-Ion Batteries
Dalian Institute of Chemical Physics
Chinese Academy of Sciences
Dalian, China

Jiacheng Zhu
Institute of Physics
Chinese Academy of Sciences
Beijing, China

1 Lithium-Ion Batteries

Fengxia Xin and M. Stanley Whittingham
Binghamton University, State University of New York

CONTENTS

1.1 INTRODUCTION

Minimizing CO_2 emissions and developing clean, stable and renewable energy technologies, i.e., solar, tide and wind, are imperative for our society. In various energy conversion and storage systems, electrochemical energy storage satisfies the requirements

DOI: 10.1201/9781003299295-1

of lightweight, high energy/power density and long life to convert intermittent renewable energy into controllable and continuous power. Lithium-ion batteries (LIBs) are considered as promising energy storage and supply systems, where Li^+ ions act as charge carriers.[1] Since the first successful demonstration of LIBs in the 1970s and the successful launch of commercial LIBs in 1990s, LIBs dominate the energy storage industry, ranging from personal electronic devices to electrical vehicles.[2,3]

LIBs are widely used in portable electronic devices and hybrid/full electric vehicles (EVs).[4] LIBs power the revolution in portable electronics with an increase in energy density from 80 Wh/kg of the first Sony's commercialized LIB to ~300 Wh/kg of Tesla's batteries. The pursuit of high specific energy density is a key research and development direction for LIBs in hybrid/full EVs. If EVs replace the majority of gasoline-powered transportation, LIBs will significantly reduce greenhouse gas emissions. The high energy efficiency of LIBs may also allow their use in different types of electric grid applications, improving the quality of energy harvested from solar, wind, tide and other renewable sources, and contributing to an energy-sustainable economy.

Most of the recent research is focused on improving the performance of LIBs, which depends on the properties and characteristics of the component materials such as cathode, anode, electrolyte and separator, as well as cell engineering and system integration.

This chapter provides an overview of LIBs, including their origin, history, structure and working mechanism. It should be noted that the principle and chemical mechanism are also suitable for other metal-ion rechargeable batteries beyond LIBs. Then, the progress and challenges of different cathode and anode materials for LIBs are presented. Moreover, different kinds of electrolytes are also discussed. In the final section, challenges, opportunities and future developments in LIBs and other metal-ion batteries are discussed.

1.2 ORIGIN OF Li-ION BATTERIES

The chemical element Li was discovered in 1817 by Johan August Arfwedson by analyzing the mineral petalite ($LiAlSi_4O_{10}$) and was first reported by Jöns Jakob Berzelius in 1818.[5,6] In 1821, Li was successfully isolated through the electrolysis of Li oxide (Li_2O) by William Thomas Brande.[7] Lewis and Keyes began to study the electrochemical properties of Li metal in 1913 through Li/Hg intermetallics.[8] The relative potentials between Li, Li/Hg intermetallic and a calomel (Hg_2Cl_2) electrode in a propylamine solution of LiI were measured. The potential of Li metal was 3.3044 V (vs. calomel) and described as "the highest electrode potential hitherto measured". This statement still holds today and sets the foundation of all modern efforts in quest of a battery based on Li metal or Li derivatives.

The pursuit of high-energy Li-based batteries started in 1950s. In 1958, as water and air had to be avoided when using Li because of its reactive nature, William S. Harris examined the solubility and conductivity of salts in various non-aqueous electrolytes – including cyclic esters (ethylene carbonate, propylene carbonate, γ-butyrolactone and γ-valerolactone), molten salts and inorganic Li salt ($LiClO_4$) – dissolved in propylene carbonate.[9] He observed the formation of a passivation layer that was capable of preventing a direct chemical reaction between Li and electrolyte while still allowing ionic transport, which led to studies on the stability of LIBs. These studies also increased interest in the commercialization of primary LIBs.

Since the late 1960s, non-aqueous 3V Li-ion primary batteries have been available in the market with different cathodes, including Li-sulfur dioxide batteries,[10] Li-carbon monofluoride batteries,[11] Li-manganese oxide batteries,[12] and Li-halogen batteries.[13,14] Simultaneously, advances in the understanding of Li-ion primary batteries with different materials gave birth to rechargeable (secondary) LIBs. A broad description of key discoveries and technical achievements that eventually led to the birth of LIBs can also be seen in some review papers.[15,16]

1.3 HISTORY OF LITHIUM-ION BATTERIES

1.3.1 Brief History

Extensive research on rechargeable LIBs operating at room temperature began in early 1970s after the discovery of intercalation reactions in energy storage.[17] In chemistry, intercalation is the reversible insertion of an ion or a molecule into a crystalline lattice without any significant change in the lattice structure, except for minor expansions or contractions. Initially, researchers have created superconductors by modifying the carrier density of chalcogenides through intercalating ions or organic molecules into the structure of host material.[18] M. Stanley Whittingham demonstrated the first rechargeable LIBs using layered TiS_2 as a cathode, Li metal as an anode and $LiClO_4$ in dioxolane as an electrolyte. Following the demonstration of TiS_2-based LIBs, different metal chalcogenides were investigated by various research groups as electrode materials for LIBs.[19] In fact, most of the LIBs utilize intercalation reactions at both electrodes for energy storage, indicating that the base technology has not changed in almost 40 years. Furthermore, a Li-aluminum (Al) alloy was used as an anode,[20] which was formed by placing a sheet of Li on the top of Al. In the final stage of battery construction, prior to cell sealing, the electrolyte was added, which enabled the reaction between Li and Al sheets to form Li-Al alloy.

In 1980, John Goodenough first proposed the utilization of layered Li cobalt oxide ($LiCoO_2$) as high-energy and high-voltage cathode material.[21] In 1983, Goodenough also identified manganese spinel ($LiMn_2O_4$) as a low-cost cathode material.[22] However, the lack of safe anode materials limited the application of layered oxide cathodes in LIBs. In 1987, Yohsino et al. filed a patent and built a prototype cell using a carbonaceous anode and $LiCoO_2$ cathode.[23] Both carbon anode and $LiCoO_2$ cathode are stable in air, which is highly beneficial for engineering and manufacturing. Sony succeeded in the commercialization of first rechargeable LIBs based on a carbon anode (petroleum coke) and $LiCoO_2$ cathode in the early 1990s. The as-designed battery demonstrated an open circuit voltage of >3.6 V and an energy density of ~150 Wh/kg. Since then, much research has been and is still ongoing in improving the performance of electrode materials.

1.3.2 Basic Structure of Lithium-Ion Batteries

Though various types of electrode materials, electrolytes and separators have been explored, the basic design of LIBs is still the same as proposed by M. Stanley Whittingham in 1976.[2] Typically, LIBs are mainly composed of cathode and anode materials, current collectors, electrolytes and separators. To coat the cathode and anode materials, which are often called active materials on current collectors, slurries are often first made. The slurries are prepared by mixing altogether a liquid solvent, a polymeric binder, conductive additive and active materials. The cathode and anode are isolated from each other by a separator. The separator provides electrical insulation between the electrodes, while allowing the transport of Li^+ ions in the electrolyte phase through the separator. In principle, the electrolyte should be ionically conductive and electronically insulating; however, the actual behavior of electrolytes is much more complicated. Intercalation/deintercalation is the most classic charge storage mechanism in rechargeable batteries.[24] In the case of graphite anode and $LiCoO_2$ cathode, as shown in Figure 1.1, Li^+ ions are deintercalated from the layered $LiCoO_2$ host during the charging process, passed across the electrolyte and intercalated between graphite layers.[25] This process is reversed during the discharge process. The electrochemical reactions during the charge/discharge process can be briefly described as follows:

$$\text{Cathode}: LiCoO_2 \leftrightarrow Li_{1-x}CoO_2 + xLi^+ + xe^-$$

$$\text{Anode}: 6C + xLi^+ + xe^- \leftrightarrow Li_xC_6$$

The size of LIBs depends on the application and power specifications. LIBs come in different forms such as coin, cylindrical, prismatic and pouch cells (Figure 1.2).[26] Coin cells are mainly used in non-rechargeable portable devices, including medical implants, watches, hearing aids, car keys and memory backups. Typical applications

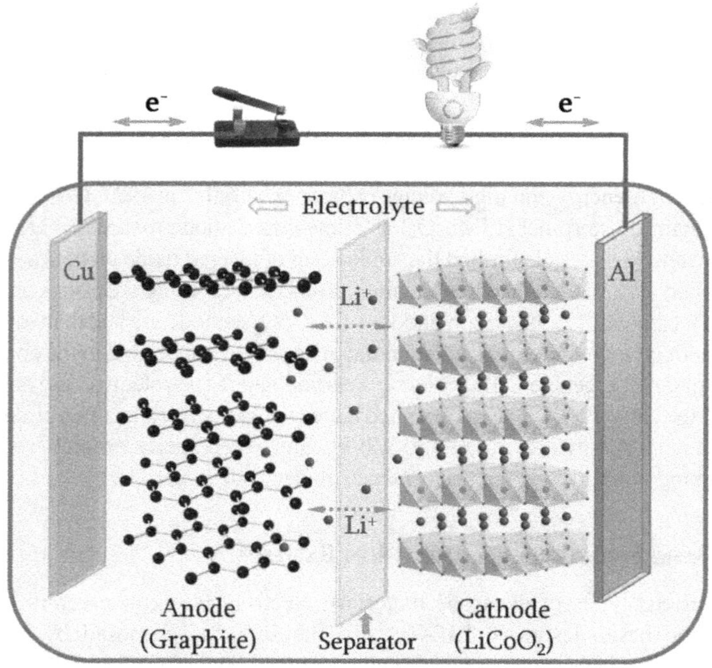

FIGURE 1.1 Schematic of the configuration of LIBs.[25] (Reproduced with permission of Elsevier.)

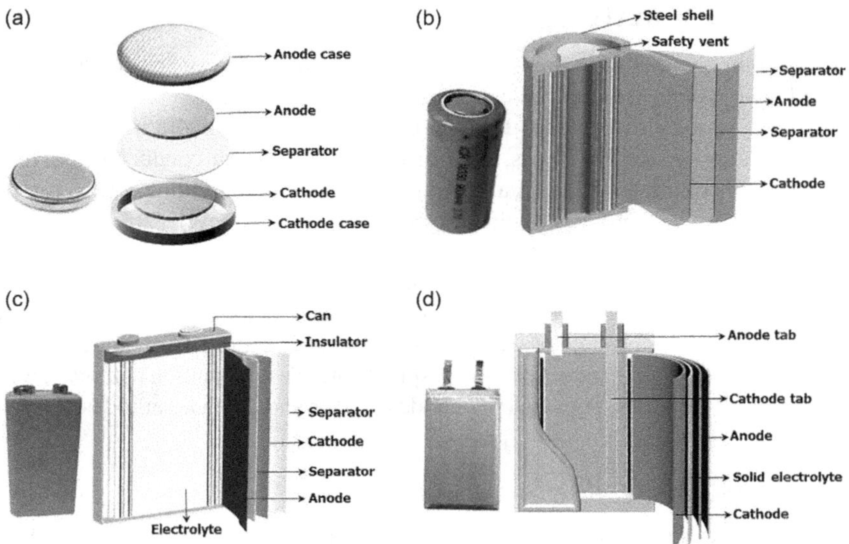

FIGURE 1.2 Schematic illustration of typical rechargeable battery configurations: (a) coin, (b) cylindrical, (c) prismatic and (d) pouch shapes.[26] (Reproduced with permission under the terms of the Creative Commons CC BY license.)

for cylindrical cells are power tools, medical instruments, laptops and e-bikes. The prismatic cells are predominantly found in cellular phones, tablets and low-profile laptops. The pouch cell offers a simple, flexible and lightweight solution to battery design.

1.3.3 Beyond Lithium-Ion Batteries

Though the new battery technologies may not be able to directly compete with LIBs in market share due to their unique characteristics in terms of performance, cost and size, a couple of promising battery systems are being developed for future applications. In line to develop batteries with a high energy density, Li-sulfur (Li-S) and Li-air (Li-O_2) have garnered widespread research attention due to their highest theoretical energy density. As non-lithium metals are much more abundant than Li, non-lithium metal-ion batteries have also emerged and shown great promise in energy storage applications. For instance, sodium-ion batteries (SIBs) are one of the most promising "beyond-Li" energy storage technologies because most of the LIBs' *know-how* can be directly used for the development of SIBs. Similarly, magnesium-ion batteries (MIBs) exhibit the advantage of enhanced safety by avoiding the formation of dendrites and a comparable energy density to LIBs at a significantly low cost. Furthermore, zinc-ion batteries (ZIBs) can be considered as one of the most promising alternatives for grid-energy storage. Lastly, aluminum-ion batteries (AIBs), with a three-electron transfer reaction, can theoretically achieve three times higher energy density than LIBs. Because of abundant aluminum resource and low cost, AIBs have become one of the most promising next-generation battery systems.

1.4 CATHODE MATERIALS FOR LITHIUM-ION BATTERIES

As a key component of LIBs, cathode materials with high energy density, long cyclic life and high safety are highly desired.[27] Compared with graphite anode, the relatively low capacity of cathodes has become a bottleneck for the improvement in the energy density of LIBs. Thus, a lot of research has been conducted to explore advanced cathode materials with a high energy density. In addition, other key performance parameters of LIBs, including rate capability and cyclic life, are critical, which are determined by the intrinsic chemistry of cathode materials. For example, the rate capability is highly related to the electronic and ionic conductivities of cathode materials, whereas the cyclic life is strongly influenced by both the composition of cathode materials and state of charge. Cathode materials for LIBs can be classified into several groups, i.e., layered cathodes, spinel cathodes, polyanion cathodes, disordered rock-salt cathodes, conversion cathodes, sulfur and oxygen cathodes.

1.4.1 Layered Cathodes

Layered cathodes are characterized by a specific structure, consisting of alternating layers of Li^+ ion and a metal compound. The layered structure is the earliest form of intercalation compounds for cathode materials in LIBs. Metal chalcogenides were first studied as a possible intercalating cathode.[28] Titanium disulfide (TiS_2), a hexagonal closed-packed structure with Li intercalation sites, is widely studied due to its high energy density (~130 Wh/kg or 280 Wh/L) and extended cyclic life.[2,27] Figure 1.3 shows the layered structure of $LiTiS_2$ compound, where Li^+ ions can intercalate between the sheets of transition metal sulfide.[27] $LiTiS_2$ intercalation compound is an ideal example of intercalation, showing complete solubility of Li for all values of x, ranging from 0 to 1. Figure 1.3 also shows the charge/discharge curve of $LiTiS_2$ at the current density of 10 mA/cm^2.[29]

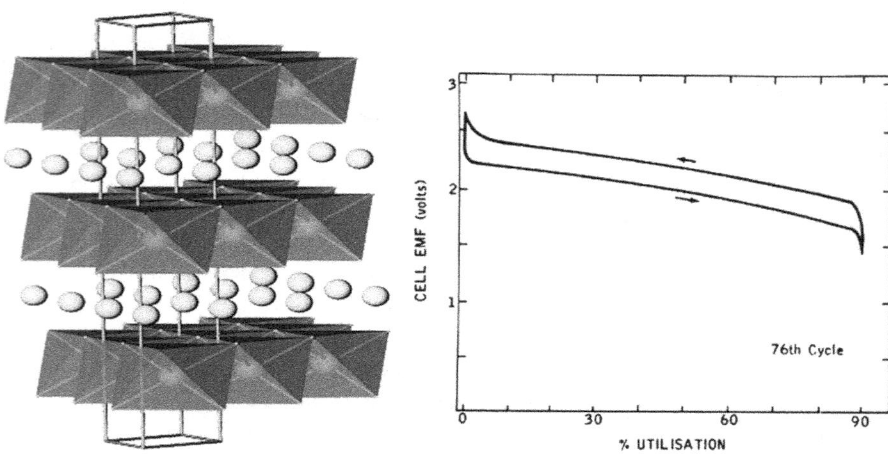

FIGURE 1.3 (Left) Layered structure of $LiTiS_2$.[27] (Reproduced with permission of American Chemical Society.) (Right) Charge/discharge curve of Li/TiS_2 at the current density of 10 mA/cm^2.[29] (Reproduced with permission of Elsevier.)

Vanadium pentoxide (V_2O_5) and molybdenum trioxide (MoO_3) are two of the earliest studied layered transition metal oxides (TMOs) for rechargeable lithium batteries. The layered TMOs, with the same structure as layered dichalcogenides, were most extensively studied due to their high voltage and high energy density.

The layered lithium cobalt oxide ($LiCoO_2$, LCO) is the first widely investigated and commercially successful form of layered TMO cathode. LCO was recognized and introduced by John Goodenough[21] and commercialized by Sony, combining with the carbonaceous anode to fabricate LIBs.[30,31] The monovalent Li^+ and trivalent Co^{3+} ions are ordered on the alternate (111) planes of the rock salt structure with a cubic close-packed array of oxide ions. The Co and Li, located on octahedral sites, occupy alternating layers and form a hexagonal symmetry. LCO displays a relatively high theoretical specific capacity of 274 mAh/g and a high theoretical volumetric capacity of 1,363 mAh/cm^3. However, when more than half of Li^+ ions ($x > 0.5$) is extracted from the layered crystal lattice, a series of irreversible structural transformations happen in $Li_{1-x}CoO_2$, leading to an irreversible capacity loss. In general, elemental doping, surface coating and co-modification can effectively improve structural stability and promote cycling stability. Nowadays, the practical specific capacity of commercial LCO can reach 185 mAh/g, corresponding to a charge cut-off voltage of 4.5 V.

$LiNiO_2$ (LNO) possesses the same crystal structure with $LiCoO_2$ and a similar theoretical specific capacity of 275 mAh/g. Eagerly considered by Jeff Dahn et al., the relatively high energy density and low cost drove the research on LNO.[32,33] Pure LNO is difficult to synthesize and most reports suggest excess Ni as in $Li_{1-x}Ni_{1+x}O_2$, where Ni^{2+} ions tend to substitute Li^+ sites during synthesis and even delithiation processes.[34] In fact, such *Li/Ni mixing* is a notorious intrinsic issue in high-nickel layered oxides. Moreover, LNO also suffers from detrimental structural transitions, blocking Li-ion diffusion pathways and resulting in poor cyclic performance and shows inferior thermal stability as low Li content appears to be unstable due to the highly effective equilibrium oxygen partial pressure.[35] Combined Ni and Co, researchers found $LiNi_{1-y}Co_yO_2$ as an effective way to reduce cationic disorder and increase thermal stability.[36] In addition, compared with LNO, the $LiNi_{1-y}Co_yO_2$ phase is easily prepared without the necessity of using oxygen atmosphere. Furthermore, $LiMnO_2$ (LMO) is studied because Mn is much cheaper and less toxic compared to Co or Ni. The practical use of pure $LiMnO_2$ cathode in LIBs is hindered due to the cooperative John–Teller effect associated with high-spin Mn^{3+}, which caused severe structural degradation and rapid capacity fading. Combined Ni and Mn, one interesting material is $LiNi_{0.5}Mn_{0.5}O_2$, which renders a stable capacity of 160 mAh/g in the voltage range of 2.5 V–4.3 V.[37–39]

To solve the thermal instability of LCO and LNO, layered ternary TMOs, such as $LiNi_{1-x-y}Co_xMn_yO_2$ (NMC) and $LiNi_{1-x-y}Co_xAl_yO_2$ (NCA), are the most practical candidates for the widespread applications in terms of reversible capacity, rate capability and cost-effectiveness. NMC can be considered as a solid solution of $LiNiO_2$, $LiCoO_2$ and $LiMnO_2$. Figure 1.4 presents the phase diagrams of three individual lithiated oxides, i.e., $LiNiO_2$, $LiCoO_2$ and $LiMnO_2$, with various compositions of Ni, Co and Mn.[40] In the structure of ternary metal oxides, each metallic element performs its function and delivers great synergy. The introduction of Mn in NMC and Al in NCA can maintain structural stability and lower the cost. Ni allows high Li extraction and

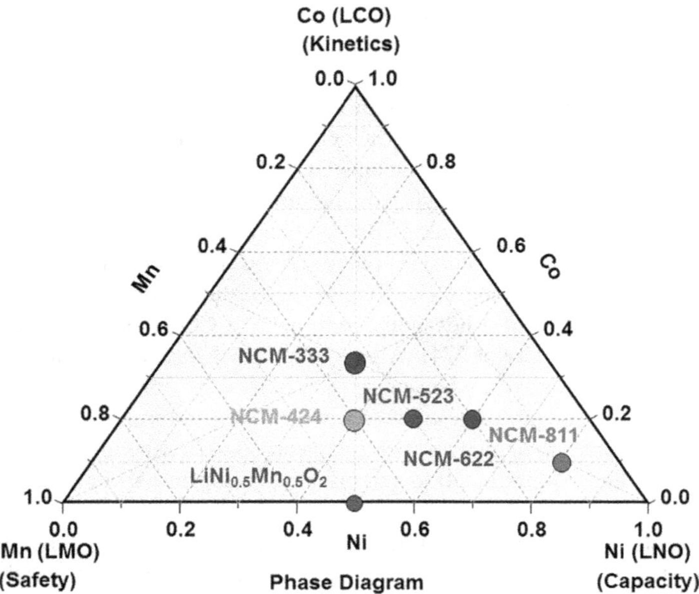

FIGURE 1.4 Phase diagram of the ternary system generated from LNO, LCO and LMO with some representative compositions shown.[52] (Reproduced with permission of American Chemical Society.)

provides a high capacity, while Co prevents Ni from occupying Li sites and, thus, guarantees the high reversible capacity. The research on NMC cathode began with low Ni contents by different groups.[41–44] In 2001, Ohzuku and Makimura[43] synthesized and characterized $LiNi_{0.33}Mn_{0.33}Co_{0.33}O_2$ (NMC111), whereas Jeff Dahn and co-workers[44] investigated $LiNi_xCo_{1-2x}Mn_xO_2$ ($x = 0.25$ and 0.375). Hereafter, a series of NMC was developed and commercialized, including NMC422, NMC532, NMC622 and NMC 811. The increase in Ni content enhances the charge storage capacity. For example, NMC811 exhibits a reversible capacity of 200 mAh/g, whereas NMC111 exhibits a far lower capacity of only 150 mAh/g.[45,46] However, the capacity fading and thermal instability are aggravated due to the unstable material properties of high-nickel NMC. The instability of Ni^{4+} ions toward the liquid organic electrolyte and moisture traces increases the reactivity of cathode materials. Surface coating and partial elemental substitution are the most common and effective methods to overcome the performance degradation. Surface coating can reduce the reactions between electrolyte and cathode materials, while elemental substitution can affect the electronic structure to stabilize the material and prevent structural collapse during Li removal. For example, our group has synthesized niobium (Nb)-coated and -substituted NMC 811 to improve the electrochemical performance. Nb coating stabilized the surface, decreased the first cycle loss and improved the rate capability, whereas Nb substitution improved capacity retention during extended cycling by stabilizing the lattice structure.[47–50]

Similar to NMC, NCA can also be considered as a solid solution of $LiNiO_2$, $LiCoO_2$ and $LiAlO_2$. The most popular formulation of NCA is $LiNi_{0.8}Co_{0.15}Al_{0.05}O_2$, showing a comparable specific capacity of 200 mAh/g to NMC811.[51] It should be noted that NCA incorporates a relatively low level of the third metal (Al: 5%–10%), which is much lower than Mn in NMC (10%–40%). This is because the use of higher levels of Al (>10%) results in severe capacity decay and poor Li^+ ion diffusion through the structure.

Li-rich layered oxides, especially Li-rich Mn-based cathode materials ($xLi_2MnO_3 \cdot (1-x)LiMO_2$, $0<x<1$, M = Ni, Co, Mn, named as LMR), are the most promising next-generation cathode materials. Compared with Ni-rich NMC and NCA, LMR cathodes are cheaper and exhibit higher specific capacity. The crystal structure of LMR cathode is generally composed of a $LiMO_2$ component, with rhombohedral R-3m structure, and a monoclinic Li_2MO_3 component with C2/m structure, as shown in Figure 1.5.[53] The layered-structured Li_2MnO_3 phase acts as a Li-ion reservoir and structural stabilizer. Different from Li-O-TM configurations in traditional $LiMO_2$ materials, an O^{2-} ion in Li_2MnO_3 is surrounded by four Li and two Mn, representing a different local environment from Li-O-Li configuration. LMR can deliver an initial specific discharge capacity of around ≥250 mAh/g.

As discussed in the above sections, cathodes contain transition metal ions, i.e., Mn, Ni, and Co. The dissolution of transition metal ions from cathodes into electrolytes and final deposition on the anode is inevitable, especially under elevated temperature and high cell voltage.[54,55] The transition metal ions dissolution is one of the major causes of capacity and power fade in LIBs.

The dissolutions of transition metal ions are acid-base reactions, as the trace levels of HF found in the electrolytes. These reactions lead to the formation of electronically and ionically insulating LiF on the cathode's surface and cause transition metal ions dissolution. In addition, such acidic species might result from under elevated temperature and high cell voltage. To address and eliminate the issue of transition metal ion dissolution, two strategies have been widely used. One is to stabilize the

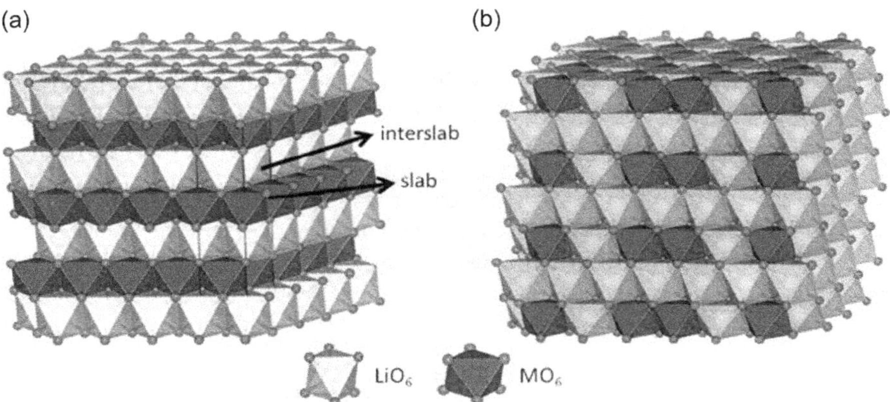

FIGURE 1.5 Crystal structure of (a) rhombohedral $LiMO_2$ phase (space group: R-3m, M = Ni, Co, Mn) and (b) monoclinic Li_2MO_3 phase (space group: C2/m).[53] (Reproduced with permission of Elsevier.)

structure of cathode materials: (i) Surface coating can provide a chemically inert layer for their protection; (ii) cationic and anodic substitution at the bulk level has also been well practiced to tune the crystal chemistry toward more stable cathode materials. The other is to add electrolyte additives to scavenge the water and/or HF impurities in the electrolyte.

1.4.2 Spinel-Structured Cathode Materials

Spinel-structured cathodes provide an opportunity to eliminate Co. They are cheap and environment friendly, as well as render excellent rate performance due to the presence of three-dimensional (3D) Li^+ diffusion channels. The most widely researched spinel cathode is $LiMn_2O_4$.[22] The practical specific capacity of $LiMn_2O_4$ is about 120 mAh/g and the theoretical specific capacity is 148 mAh/g. $LiMn_2O_4$ possesses a potential of around 4 V (vs. Li/Li^+), exhibits excellent cyclability and retains the capacity at high rates. Hence, LMO is a promising candidate for high-power LIBs. The main issues of this spinel structure come from the Jahn–Teller effect and the dissolution of Mn^{3+} in the electrolyte, leading to irreversible structural degradation and capacity decay. To improve the electrochemical properties and remedy the shortcomings, other elements are introduced into the Mn-based spinel framework to form $LiM_{0.5-x}Mn_{1.5+x}O_4$ (M = Al, Ti, Cr, Fe, Co, Ni, Cu and Zn). For instance, Ni-substituted spinel material, denoted as $LiNi_{0.5}Mn_{1.5}O_4$, is considered as a promising high-voltage spinel cathode, as it operates on Ni^{4+}/Ni^{2+} double redox at ~4.7 V (vs. Li/Li^+) with a theoretical capacity of 147 mAh/g and a practical capacity of 125 mAh/g.[56–60]

1.4.3 Polyanion Cathodes

Polyanion cathodes are composed of a 3D network of transition metal polyhedra and polyanionic groups. Polyanion cathodes possess high thermal stability due to covalently bonded oxygen atoms. The most extensively investigated polyanion cathodes are phosphates ($LiMPO_4$, M = Fe, Mn, Co or Ni) with ordered olivine structure. $LiFePO_4$ (LFP) is the most widely studied and fully developed material among olivine phosphates.[61] As shown in Figure 1.6, during the charging process, LFP changes into $FePO_4$ due to delithiation, whereas a reversible transformation from $FePO_4$ to LFP occurs due to lithiation during the discharge process.[62] LFP exhibits a practical specific capacity of 120–160 mAh/g (theoretical capacity: 170 mAh/g), a smooth operating voltage of 3.4 V (vs. Li/Li^+) and long cyclic life. However, low intrinsic electronic conductivity and slow 1D Li diffusion rate limit the rate performance. Therefore, LFP is usually coated with carbon or mixed with other conductive agents to enhance the rate capability. In addition to phosphates, silicates (Li_2MSiO_4)), fluorophosphates ($LiVPO_4F$) and fluorosulfates ($LiMSO_4F$) have been explored as polyanion cathodes for LIBs.

1.4.4 Disordered Rock-Salt Cathodes

Disordered rock-salt structured cathodes, especially cation-disordered rock-salt (DRX), are a new type of high-energy-density cathode material.[63,64] 3D host structure in DRXs is considerably more stable, as random cation distribution eliminates

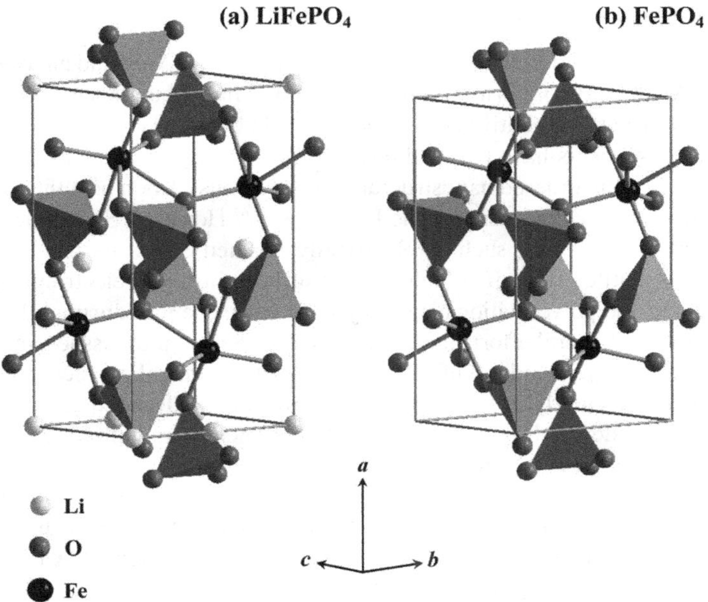

FIGURE 1.6 Crystal structures of (a) $LiFePO_4$ and (b) $FePO_4$ (PO_4 environments of phosphorus have been represented as tetrahedra).[62] (Reproduced with permission of American Chemical Society.)

the changes in interlayer spacing. DRX-based oxides and oxyfluorides often consist of earth-abundant elements, along with Co-free chemistry, wide compositional space and large charge storage capacities. Some DRXs are capable of delivering reversible capacity and energy densities well over 300 mAh/g and 1,000 Wh/kg, respectively. The electronic conductivity of the metal oxides limits the rate performance of cathode materials, which can be improved by various strategies.[65] One should also note that carbon coatings can improve the Li diffusion coefficient of V-based, Mn-based and Co-based DRS oxides.

1.4.5 Conversion Cathode Materials

Unlike the intercalation-based cathodes, the chemical bonds in conversion cathodes are repeatedly broken and reformed during electrochemical reactions. Moreover, conversion cathodes can store multiple Li^+ ions per metal center because of multi-electron conversion reactions, resulting in three to five times higher capacity than conventional cathodes. Among several types of conversion materials, metal fluorides are the most promising due to their high theoretical potential (3.55 V [vs. Li/Li^+], CuF_2) and gravimetric capacity (713 mAh/g, FeF_3).[66] However, transition metal fluorides suffer from poor ionic and electronic conductivities due to a large bandgap induced by the ionic character of the transition metal–fluoride bond. A considerable performance improvement has been achieved by the utilization of various carbon composites, cation and anion substitution and nanostructuring.

1.4.6 Sulfur and Oxygen

Sulfur (S) delivers a theoretical specific capacity of 1,675 mAh/g and an average voltage of 2.15V (*vs.* Li/Li^+). Furthermore, low cost and high abundance also endow the S with a significant competitive advantage. It should be noted that S is a Li-free cathode. Ji and co-workers have realized a reversible capacity of up to 1,320 mAh/g by introducing a highly ordered nanostructured mesoporous carbon host into the S cathode, which aroused enthusiasm for the Li-S battery.[67] However, S cathode possesses some inherent drawbacks, such as electrically isolated S and its reaction product Li_2S, serious shuttle effect of soluble Li polysulfides in liquid electrolyte and electrode cracking and pulverization caused by volumetric changes during lithiation/delithiation process. Several efforts have been devoted to solve these issues, including (i) sulfur hosts, interlayers and modified separators with chemical/physical confinement function; (ii) molecularly designed active materials; (iii) advanced electrolytes with reduced polysulfide solubility; (iv) catalysts/redox mediators with kinetic accelerating functions; and (v) anode protection.[68]

Oxygen is a gaseous active material that can be acquired from the air, which can remarkably lower the total weight of working batteries. Moreover, the conversion reaction between O_2 and Li can deliver a high charge storage capacity.[69] A typical Li-air battery can achieve the highest gravimetric energy density with a theoretical energy density of 3,500 Wh/kg and a practical energy density of 950 Wh/kg. However, electrolyte instability, degradation of air electrode and instability of Li anode lead to poor cyclability and low energy efficiency of the system. Upgrading electrolytes, protecting metallic Li anode and designing catalysts can be adopted to resolve these issues.[70]

1.5 ANODE MATERIALS FOR LITHIUM-ION BATTERIES

Another effective approach to increase the energy density of LIBs is to pursue high-capacity anode materials. Based on the electrochemical lithiation/delithiation mechanism, anodes used in LIBs are broadly classified into three main categories: intercalation anodes, alloy anodes and conversion anodes.

1.5.1 Intercalation Anodes

1.5.1.1 Carbon-Based Materials

Carbon-based materials, including natural/synthetic graphite and soft/hard carbon, represent the most viable candidates for LIBs because of their low cost, abundance, low delithiation potential (*vs.* Li/Li^+), high Li diffusivity, high electrical conductivity and low volumetric changes during the lithiation/delithiation process. It is worth emphasizing that the carbon-based anode enabled LIBs to become commercially viable about 40 years ago and still carbon is the most desirable choice as an anode material. For instance, the conventional graphite anode has merits of 3D structural stability, moderate energy density, theoretical gravimetric capacity of 372 mAh/g, theoretical volumetric capacity of 735 mAh/cm^3 and low cost.[71] To date, graphite is the dominant anode material for commercial LIBs. During the charging process, Li^+ ions from the electrolyte penetrate the carbon and form a Li/carbon intercalation compound, i.e., Li_xC, which is a reversible reaction, storing 1 Li atom per 6C atoms. One should note that the gravimetric capacity of carbon is higher than most of the cathode materials, but the volumetric capacity of commercial graphite is still low, ranging from 330 to 430 mAh/cm^3. In addition, nanostructured carbon-based anode materials, such as 1D nanotubes, nanowires, nanofibers, 2D graphene and porous carbon-based anode, have been developed to increase the energy and power densities of LIBs.[72]

1.5.1.2 Insertion-Type Transition Metal Oxide Anodes

The most classic insertion-type TMO anodes are based on titanium, e.g., $Li_4Ti_5O_{12}$ (LTO) and TiO_2, where the redox center is reduced to Ti^{3+} upon lithiation and reoxidized to Ti^{4+} when alkali metal cations are subsequently de-inserted. The spinel-structured LTO, which can store up to three Li^+ ions per unit, undergoes a reversible two-phase reaction from spinel to rock-salt structure, resulting in a theoretical capacity of 175 mAh/g, which corresponds to a chemical formula of $Li_7Ti_5O_{12}$. The structures of both $Li_4Ti_5O_{12}$ and $Li_7Ti_5O_{12}$ are illustrated in Figure 1.7.[73] LTO exhibits low volumetric changes (~0.2%) and is considered as a zero-strain material, which results in superior cyclic performance. Furthermore, the high delithiation potential (1.55V *vs.* Li/Li$^+$) can prevent the growth of Li dendrites and guarantee the safety of LTO-based LIBs. Also, the Li-ion diffusion coefficient is one order of magnitude higher than the graphite, which ensures excellent rate capability. However, LTO has an inherently poor electronic conductivity, which hinders the rate performance. Two different approaches are typically adopted to improve the rate capability: accelerating charge-transfer reaction by enhancing ionic diffusion and electronic conductivity via surface modification or ionic doping, and reducing Li^+ diffusion distance in the bulk phase by designing a nanostructured LTO anode.[74] Moreover, various structures of TiO_2 have been explored as anode materials for LIBs. Owing to the characteristic parallel channels along the [010] direction, TiO_2-B nanotubes or nanowires are considered as one of the most appealing structures for the incorporation and diffusion of Li^+ ions.[75]

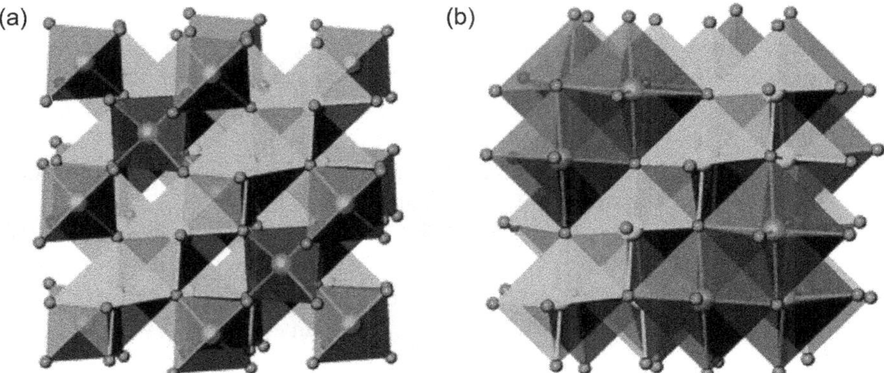

FIGURE 1.7 (a) Spinel structured $Li_4Ti_5O_{12}$, where the blue-colored tetrahedra represent Li and green-colored octahedra represent disordered Li and Ti; and (b) rock-salt structured $Li_7Ti_5O_{12}$, where blue-colored octahedra represent Li and green-colored octahedra represent disordered Li and Ti.[73] (Reproduced with permission of American Chemical Society.)

1.5.2 Alloying Anodes

Since A. Dey demonstrated that Li metal could electrochemically react with other metals at room temperature in an organic electrolyte,[76] alloying reactions between Li and different metallic or semi-metallic elements and compounds are being extensively investigated. The group IV elements, especially Si and Sn, are the main focus for the alloying reactions owing to the high-capacity Li-rich binary alloys and low working potential (except graphite). The theoretical capacity of Si and Sn is 3,579 and 994 mAh/g, respectively, indicating promise for next-generation LIBs.

1.5.2.1 Si and Si-Based Compounds

The reaction mechanism of a pure Si anode with Li can be explained as follows[77]:

$$\text{During discharge}: \text{Si (crystalline)} + x\text{Li} + xe^- \rightarrow \text{Li}_x\text{Si (amorphous)}$$

$$+ (3.75 - x)\text{Li}^+ + (3.75 - x)e^- \rightarrow \text{Li}_{15}\text{Si}_4\text{(crystalline)}$$

$$\text{During charge}: \text{Li}_{15}\text{Si}_4\text{(crystalline)} \rightarrow \text{Si(amorphous)} + y\text{Li}^+ + ye^- + \text{Li}_{15}\text{Si}_4\text{(residual)}$$

The given two-phase reaction is observed in all stages of the first discharge and completely vanishes after the complete formation of a binary alloy ($Li_{15}Si_4$). Hence, only a single-phase reaction is observed in subsequent cycles. However, drastic volumetric expansion (300%–400%) and huge stress generation are accompanied by the lithiation/delithiation process of Si, causing a series of severe destructive consequences: (i) structural integrity of electrode is deteriorated due to gradually enhanced pulverization during repeated discharge/charge processes, (ii) disconnection between the electrode and current collector happens due to the presence of interfacial stress, and (iii) continuous consumption of Li^+ ions occurs during the continuous formation-breaking-reformation process of solid electrolyte interface (SEI) layer (Figure 1.8).[78] Several design strategies have been developed to develop nanostructured Si anodes and overcome these issues because (i) nanostructured Si anodes can survive the battery operation without mechanical fracture,[79,80] (ii) a combination of nanostructured Si with nanostructured C, such as a typical yolk-shell or wire-in-tube structure, can ensure efficient transportation of electrons and ions during battery operations,[81] and (iii) covalently bonded nanostructured Si with C can reduce the accidental disconnection caused by the separation of active material and current collector.

In addition, Si monoxide (SiO) is also considered as an anode candidate for LIBs. The absolute volume change of SiO is smaller than Si during the discharge/charge process. Moreover, several oxygen-containing compounds formed by Li insertion are expected to play a matrix role in the alleviation of volumetric changes.

1.5.2.2 Tin (Sn) and Sn-Based Compounds

In 1997, Fujifilm Celltec Co. Ltd., (Japan) announced its Stalion battery using tin-based amorphous oxide as an anode. The material was composed of Sn-O as an active center for Li insertion and other glass-forming elements, which could provide a gravimetric capacity of >600 mAh/g,[82] encouraging more researchers to engage in

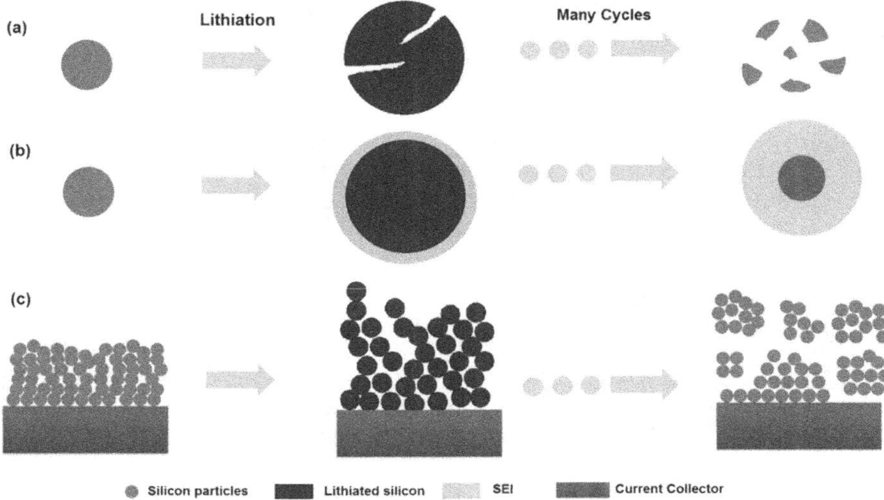

FIGURE 1.8 Three different failure mechanisms of Si electrode: (a) electrode pulverization, (b) continuous breaking and re-growth of the SEI layer and (c) collapse of the entire electrode.[78] (Reprinted with permission of Elsevier.)

Sn-based materials. The basic reaction mechanisms of Sn-based oxide materials can be given as follows:

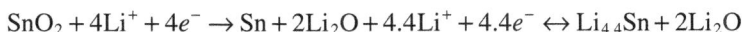

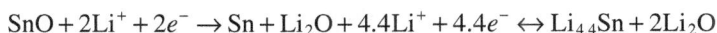

$$SnO_2 + 4Li^+ + 4e^- \rightarrow Sn + 2Li_2O + 4.4Li^+ + 4.4e^- \leftrightarrow Li_{4.4}Sn + 2Li_2O$$

$$SnO + 2Li^+ + 2e^- \rightarrow Sn + Li_2O + 4.4Li^+ + 4.4e^- \leftrightarrow Li_{4.4}Sn + 2Li_2O$$

The volumetric change during the transition from Sn to fully lithiated Sn ($Li_{4.4}Sn$) is around 260%, which is the main reason for the drastic capacity decay of Sn anode.[83] During the charge/discharge process, one Li-Sn phase grows at the expense of another, and the difference in lattice parameters and structure between different Li-Sn alloys causes the accumulation and release of stress/strain, leading to particles cracking, SEI reformation, loss of connection with current collector and electrode failure, resulting in a rapid deterioration of cyclic performance. A lot of efforts have been made to overcome these issues, including the design and exploration of different Sn-based compounds with variable sizes, shapes and porosities.

Over the years, various Sn-based oxides or oxide glasses were reported with high capacities, excellent cyclic stability and superior rate performance. However, the initial coulombic efficiency should be improved. For example, Dahn et al. have studied the electrochemical performance of Sn-based compounds and structural changes during the reaction of Li with different tin oxide glasses, demonstrating a significant irreversible capacity loss (200–700 mAh/g) during the first cycle.[84,85]

Instead of using Sn oxide as an inactive dispersant, an alternative approach is to start with Sn-based alloy. In 2005, Sony released their Nexelion 14430-sized battery with nanostructured Sn-Co-C-based anode. Sony claimed that the volumetric capacity could be increased by more than 30% over conventional LIBs.[86] In 2011, Sony announced another iteration of the Nexelion cell (18650-sized battery) with a capacity of 3.5 Ah and a volumetric energy density of 723 Wh/L in the voltage range of 2.0V–4.3V.[87] Furthermore, encouraged by the success of Sony's Sn-Co-C anode, the research was focused on Sn-transition metal alloys, such as Sn-Cu, Sn-Ni, Sn-Fe and Sn-Co alloys.[88–91] Table 1.1[92] summarizes lithiated density, gravimetric and volumetric capacities of Sn, Sn_5Fe, Sn_2Fe, Sn_2Co, Ni_3Sn_4 and Cu_6Sn_5 for the formation of $Li_{4.4}Sn + M$ compared with Li metal and graphite.

1.5.3 Conversion Anodes

Conversion is a reversible electrochemical reaction (often called a displacement reaction), where the transition metal compound (MX_y, X = P, S, O, F or Cl) is electrochemically destroyed and subsequently reduced as a metal (M^0). When utilized as anodes in LIBs, the electrochemical reaction mechanism of transition metal compounds, such as oxides, phosphides, sulfides and nitrides (M_xN_y; M = Fe, Co, Cu, Mn, Ni and N = O, P, S and N), is the reduction (oxidation) of transition metal along with the composition (decomposition) of Li compounds (Li_xN_y; N = O, P, S and N). The electrochemical conversions reactions can be described as follows:

$$M_xN_y + ze^- + zLi^+ \leftrightarrow xM + yLi_zN_y\ ,$$

here M = Fe, Co, Cu, Mn, Ni and N = O, P, S and N.

Conversion reaction-based TMOs, including Fe_2O_3, Co_3O_4, MnO, CuO and NiO, are typical anodes for LIBs. Advantageously, these compounds exhibit significantly low volumetric changes upon charge/discharge and exhibit high capacities (around 1,000 mAh/g for Fe_2O_3) owing to the multi-electron transfer reaction per transition metal during the conversion of TMO into transition metal and Li oxide. However, the

TABLE 1.1
Lithiated Density, Gravimetric and Volumetric Capacities of Sn, Sn_5Fe, Sn_2Fe, Sn_2Co, Ni_3Sn_4 and Cu_6Sn_5 Alloys[92]

Alloy	Lithiated Density	Gravimetric Capacity (mAh/g)	Volumetric Capacity (Ah/L)	Volume Expansion (%)
Li	0.534	3,860	2,062	Infinite
Graphite	2.16	372	804	10
Sn	2.58	993	2,562	258
Sn_5Fe	2.71	929	2,517	242
Sn_2Fe	2.89	804	2,324	250
Sn_2Co	2.92	796	2,323	265
Sn_4Ni_3	3.08	724	2,231	234
Sn_5Cu_6	3.40	604	2,056	174

Source: Reproduced with permission under a Creative Commons Attribution 4.0 International License.

industrialization of conversion-reaction-based TMO anodes is hindered by their poor electrical conductivity and poor cyclic performance.

Similar to alloy anodes, the conversion anodes also have the issues of material pulverization at the particle-level, unstable SEI layer, and morphological and volumetric changes at the electrode level. In order to cycle the conversion oxides, a nano-design strategy is needed for the interconversion of multiple solid phases.[93,94]

1.5.4 Metallic Li Anode

Apart from the aforementioned metals and compounds, metallic Li can also be used as an anode material. In fact, metallic Li was used in the infancy of Li battery research, including the first viable LIB pioneered by Stanley Whittingham at Exxon in 1970s. Li metal is the ultimate anode choice because of its highest theoretical capacity (3,860 mAh/g or 2,061 mAh/cm^3), low density (0.59 g/cm^3) and lowest electrochemical potential (−3.04V *vs.* the standard hydrogen electrode).[3,95–97] Furthermore, a Li metal anode is indispensable for Li-S and Li-air systems, both of which are being intensively studied for next-generation energy storage applications.[68] Unlike the intercalation/deintercalation mechanisms in graphite anode, a conversion reaction between metallic Li and Li-ions occurs in the case of Li metal anode.[96]

Once the anode is replaced by Li metal, the specific energy of LIBs with Li-containing TMO as cathodes can increase from 280 to ~440 Wh/kg, while the Li-S and Li-air systems can further boost the specific energy to ~650 and ~950 Wh/kg, respectively. The volumetric energy density of Li-air batteries approaches that of gasoline.[97] However, uncontrollable Li dendrites growth leads to a short lifespan and catastrophic safety hazards, restricting the practical applications Li anode. A few effective strategies have been adopted to overcome these challenges, including electrolyte modification, introducing a protective layer, nanostructured anodes and membrane modification. Although experimental results are promising, it is still a long way to go before it can be applied for practical application in batteries due to the reactive nature of Li metal.

1.6 ELECTROLYTES

Electrolytes are ubiquitous and indispensable in all electrochemical devices. The role of electrolytes is to serve as a medium for transferring charges between cathode and anode. The electrolyte is in close contact with other components, including cathode, anode and separator. The interfaces, mainly the interfaces between electrolytes and electrodes, often dictate the performance of LIBs. Thus, the electrolyte must demonstrate stability against both cathode and anode surfaces.[98] The desired electrolyte for LIBs should meet the following requirements: high ionic conductivity, electrochemically stable in a wide potential range, chemically stable, thermally stable, cost-effective, simple preparation, low toxicity and environmental friendliness. Furthermore, the electrochemical operating window of electrolytes should be modified to develop high-voltage cathode and low-voltage anode materials.[99] Given the importance of electrode–electrolyte interphases to the battery performance, the electrode/electrolyte interfaces, namely, solid–electrolyte interphase (SEI) and cathode–electrolyte interphase (CEI), which are formed via electrochemical decomposition of the electrolytes at the anode/electrolyte and cathode/electrolyte interfaces, respectively, will be briefly introduced first.

1.6.1 Electrode/Electrolyte Interfaces

In general, the electrode–electrolyte interphase can be considered as a thin film covering the electrode particles due to the decomposition reactions of electrolyte, preserving the active material from latter degradative mechanisms.

The anode–electrolyte interphase, which is called solid–electrolyte interphase (SEI), influences the insertion of Li^+ from the solvated phase into the solid phase and constitutes the rate-limiting step for lithiation of most electrode materials.[100] The electrochemical working potentials of anode materials are below the reduction potential of organic carbonates commonly employed in Li-based battery electrolytes (around ~1V *vs.* Li/Li^+). During battery charging, electrochemical reduction of electrolyte occurs and produces a passivating SEI layer on the anode surface. The observed decomposition reactions of the initial SEI components coupled with the observed changes to the SEI composition upon aging lead to the following proposed mechanism for the evolution of the SEI (Figure 1.9).[101] SEI is a Li-ion conductor but an electronic insulator, leading to the termination of SEI growth at a certain thickness. A stable SEI layer allows for high coulombic efficiency and long-term stability of anodes due to the surface passivation effect. However, owing to repetitive large volumetric changes during lithiation and delithiation, the electrode/electrolyte interface moves and changes significantly, making it extremely challenging to maintain a stable SEI for high-capacity electrode materials. In other cell chemistries, such as graphite and Si anode electrodes, the SEI formation processes only involve the electrochemically reductive decomposition of the electrolyte on the electrode because of the chemical stability of graphite and Si. The formation of SEI on Li consists

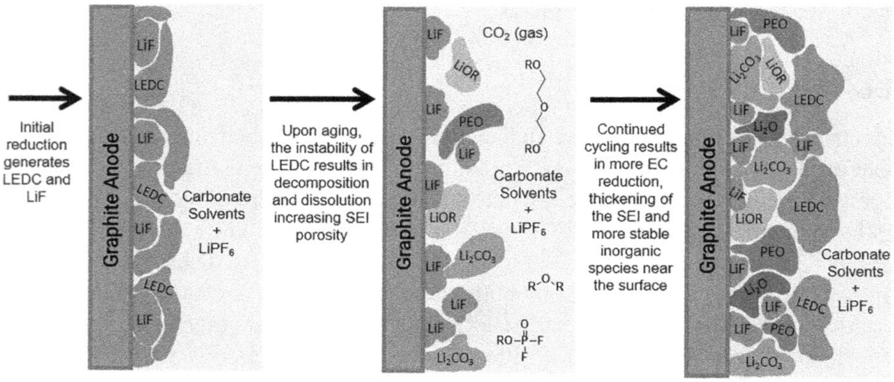

FIGURE 1.9 Schematic figure of the initial SEI formed on the graphite anode, the effect of acid-mediated thermal decomposition reactions on the structure of the SEI, and the further reduction of electrolyte leading to the thickening of the SEI.[101] (Reprinted with permission of Elsevier.)

of chemical and electrochemical reactions. Hence, though SEI formed on Li metal exhibits similar functions to SEI formed on graphite and Si electrodes, the fundamental distinctions among both types of SEIs should not be neglected.[102]

Unlike the SEI, CEI has been less thoroughly studied, as the operating potentials of most cathodes do not deviate much from the thermodynamic stability window of the commercial carbonate electrolytes.[103] With the development of high-voltage LIBs, the electrolyte will oxidize and decompose on the surface of the cathode as the cutoff voltage beyond the oxidation stability of electrolytes.[103] The understanding of CEI becomes increasingly important due to the requirement of high-voltage operation. However, the precise structure and composition of the CEI remain under debate, and how its structure and chemistry impact battery safety are yet to be fully understood. Exchange reactions and nucleophilic reaction mechanisms are often proposed for CEI formation of different cathode materials.

1.6.2 Organic Electrolytes

Organic electrolytes or non-aqueous electrolytes, which commonly consist of a Li salt solved in an organic solvent or solvent mixture, are the most commonly used in LIBs due to their high ionic conductivity and superior wettability toward both porous separator and electrodes. When the first Li intercalation cathode, i.e., TiS_2, was invented, ether-based electrolyte ($LiClO_4$ in dioxolane) was used because the operating potential of TiS_2 is moderate (<3.0V *vs.* Li/Li^+), which is well within the stability limit of ethers.[28] When layered TMOs were used as high-voltage cathode materials, the electrolytes were shifted from ether-based to ester-based due to the good oxidative stability toward the cathode and excellent capability to dissolve a variety of Li salts to provide good ionic conductivity. Among all ester-based electrolytes, propylene carbonate (PC) and ethylene carbonate (EC) are two popular solvents. The first-generation LIBs, which were commercialized by Sony in 1990 using amorphous carbon as an anode, utilized PC-based electrolytes. On the other hand, EC-based electrolytes were developed when graphite was used as a replacement for amorphous carbon. Furthermore, $LiPF_6$ is widely used as a Li salt in organic electrolytes because of its well-balanced properties, which are suitable for different battery environments. Finally, $LiPF_6$ dissolved in a mixture of EC and a linear carbonate selected from dimethyl carbonate (DMC), diethylene carbonate (DEC) or ethyl methyl carbonate (EMC) is used in the majority of commercial LIBs. In addition, electrolyte additives, which can effectively improve the performance and cyclic life of LIBs by directing SEI formation, increasing ionic conductivity of the electrolyte, increasing thermal stability of $LiPF_6$ and protecting electrodes against dissolution and overcharge, have been extensively used in electrolyte formulations.[104–106] Electrolyte additives are usually inorganic compounds and organic compounds with various functional groups, such as unsaturated carbon bonds, sulfur-containing components, halogen-containing components and other components. Generally speaking, for anodes, electrolyte additives are reduced at higher potentials than the electrolyte solvents and passivate the electrode's surface to prevent further reduction of the electrolyte solvents; for cathodes, electrolyte additives are expected to oxidize prior to the solvents and cover the electrode surface to prevent the oxidative decomposition of the electrolytes. Considering the different type of electrode materials, the choice of electrolyte additives may change. For example, fluorine- and phosphorus-containing electrolyte additives are promising agents able to improve the performance of high-voltage cathode materials, while butyl sultone can improve the low electronic conductivity of $LiFePO_4$ and its rate capability at low temperature.[107,108]

1.6.3 Aqueous Electrolytes

Instead of flammable organic electrolytes, the utilization of aqueous electrolytes with high ionic conductivities is promising from the viewpoints of cost, environment and safety.[109] Therefore, lithium nitrate aqueous electrolyte was first used by Dahn's group to construct an aqueous LIB by employing VO_2 anode and $LiMn_2O_4$ cathode in 5 M $LiNO_3$ electrolyte with an average operating voltage of 1.5V and an energy density of ~55 Wh/kg.[110] However, the possible side reactions of H_2 or O_2 evolution and dissolution/side reactions of the electrode material with water or dissolved O_2 often lead to low coulombic efficiency and poor cyclic performance. In addition, various metal oxides cathodes, such as $LiCoO_2$, $LiMn_2O_4$ or NMC, exhibit pH sensitivity in aqueous electrolytes.

Recently, "water-in-salt" electrolytes are developed by adding high concentrations of different salts in water. Wang's group first demonstrated that the anodic and cathodic limit of water decomposition can be increased by reducing the electrochemical activity of water by manipulating Li^+ ion solvation structure with a suitable anion ($(N(SO_2CF_3)^{2-}$ or $TFSI^-$).[111] The formation of an interface between electrode and electrolyte during the first charge process extended the electrochemical window to ~3.0V. It should be noted that the use of highly concentrated salts in water-in-salt electrolytes hinders practical applications. Hence, research efforts should be devoted to reduce electrolyte costs without sacrificing the unique properties of SEI forming ability and increasing water activity.

1.6.4 Ionic Liquids

Ionic liquids (ILs), which are composed of cations and anions of salts without any solvents, have low melting points and good ionic conductivities. Different cations, e.g., imidazolium, quaternary ammonium, pyrrolidinium and piperidinium, combined with anions, e.g., PF_6^-, BF_4^- and bis(trifluoromethanesulfonyl)imide ($TFSI^-$), deliver different properties for electrolyte applications. Recently, room-temperature ILs have attracted much more attention due to their wide electrochemical operating range, non-volatility and improved thermal stability. However, they also suffer from inferior power performance compared to commercial organic electrolytes. Hence, there is a lot of work to be completed before commercialization.

1.6.5 Solid-State Electrolytes

Traditional LIBs have critical safety issues because of the utilization of highly flammable organic liquid electrolytes with low thermal stability and low flashing point, which are easy to cause fire accidents and explosions. Therefore, solid-state electrolytes

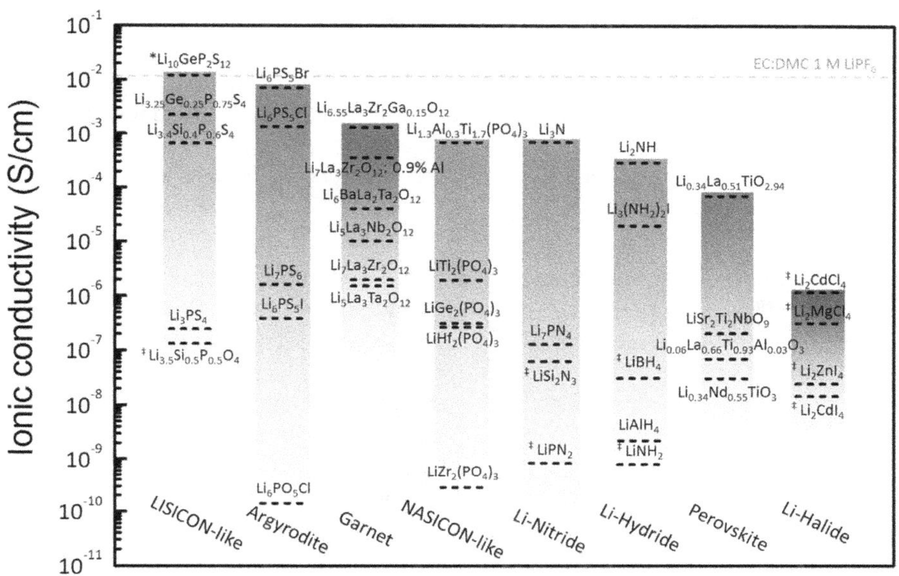

FIGURE 1.10 Li ionic conductivity of selected inorganic solid electrolytes and a typical organic liquid electrolyte.[115] (Reprinted with permission of American Chemical Society.)

(SSEs) are being developed to pursue next-generation energy storage devices with high energy density and improved safety. Compared with liquid electrolytes, SSEs are nonvolatile, nonflammable and possess high thermal stability, which make them suitable for a broad operating temperature range. SSEs provide a physical barrier layer to separate cathode and anode electrodes and prevent thermal runaways under high temperatures or impacts.[112] In addition, SSEs can realize the successful utilization of Li metal anodes by effectively suppressing the formation of Li dendrites.[113] To meet commercial requirements, high ionic conductivity, favorable mechanical properties and outstanding interfacial stability with electrodes are the most fundamental requirements for solid electrolytes.[114] Therefore, several types of SSEs have been developed, including inorganic solid electrolytes (sodium superionic conductor-type, perovskite-type, garnet-type and sulfide-type), polymer and composite solid electrolytes and thin-film solid electrolytes.[115,116] Figure 1.10 compares the Li ionic conductivity of selected inorganic solid electrolytes with a typical organic liquid electrolyte.[115] A number of structural families, such as LISICON-like, argyrodite and garnet can achieve high ionic conductivities in the range of 10^{-2} to 10^{-3} S/cm at room temperature and the Li^+ conductivity within each structural family can vary greatly by up to 5–6 orders of magnitude.

1.7 SUMMARY AND OUTLOOK

Electrochemical energy storage systems and technologies are in continuous development owing to the worldwide demand to overcome the current energy issues and satisfy the requirements of an energy-sustainable society. LIBs are an excellent option for energy storage in a wide range of energy levels from portable electronics through transportation vehicles to loading-leveling stationary storage. In this chapter, we presented a brief introduction to Li-ion batteries, as well as beyond-Li batteries, and summarized the development of cathode, anode and electrolyte materials for LIBs.

The future development of LIBs and other rechargeable batteries should consider the following aspects:

i. High-energy-density cathode materials with high capacity and high voltage should be further developed by exploring novel materials and expanding our fundamental understanding of the structure–composition–property–performance relationships. It is worth emphasizing that the pursuit of high-capacity cathode materials is the most effective approach to improve the energy density of batteries, while high-voltage cathode materials can decrease the complexity of LIB modules. For anode materials, the mixing of multiple materials to pursue complementary properties should be a direction.
ii. Fast charging or extremely fast charging (XFC) is a key driver and a long-term strategic goal in enhancing the consumer recharge experience. According to the United States Advanced Battery Consortium (USABC), fast charging is to obtain 80% of state of charge of battery within 15 minutes, which means the battery pack can be charged to 80% of SOC at 4C rate or higher. While the maximum charge rate of most commercial LIBs is limited to 3C, the fast charging cathode and anode materials should be explored.
iii. A complete understanding of the relationship between composition and/or structural evolution and battery performance is still limited. The past decade has seen an explosion of innovative experimental and modeling methods to investigate complex materials at multiple scales. However, advanced characterization techniques (i.e., in situ and operando tools, cryo-electron microscopy) and theoretical/computational analyses are still required for the further development of battery materials.
iv. Both high- and low-temperature environments severely affect the battery capacity and service life. In a high-temperature environment, LIBs may produce thermal runaway, resulting in short-circuiting, combustion, explosion and other safety problems. On the other hand, Li dendrites may appear in LIBs at low temperatures, causing short-circuiting, failure to start and other operational faults. Hence, an efficient battery thermal management system (BTMS) is of utmost significance for maximizing the lifetime of battery modules/packs.
v. Sustainable recycling technology of LIBs is partially established; however, recycling technology is far from maturity and improving the recycling technology of LIBs is a continuous effort. It must be firmly convinced that the combined efforts of the government, end users, battery manufacturers and recyclers should be given more attention.

REFERENCES

[1] M. Li, J. Lu, Z. Chen and K. Amine. 2018. 30 years of lithium-ion batteries. *Adv. Mater.* 30 (33):1800561.

[2] M. S. Whittingham. 1976. Electrical energy storage and intercalation chemistry. *Science* 192 (4244):1126–1127.

[3] J.-M. Tarascon and M. Armand. 2001. Issues and challenges facing rechargeable lithium batteries. *Nature* 414:359–367.

[4] J. Deng, C. Bae, A. Denlinger and T. Miller. 2020. Electric vehicles batteries: requirements and challenges. *Joule* 4 (3):511–515.

[5] A. Arfwedson. 1818. Untersuchung einiger bei der Eisen-Grube von Utö vorkommenden Fossilien und von einem darin gefundenen neuen feuerfesten Alkali. *J. Chem. Phys* 22:93–117.

[6] J. J. Berzelius. 1817. Ein neues mineralisches Alkali und ein neues Metall. *J. Chem. Phys* 21:44–48.

[7] W. T. Brande. 1821. *A manual of chemistry.* 2nd ed. Vol. 1: London: J. Murray.

[8] G. N. Lewis and F. G. Keyes. 1913. The potential of the lithium electrode. *J. Am. Chem. Soc.* 35 (4):340–344.

[9] W. S. Harris. 1958. Electrochemical studies in cyclic esters, University of California, Berkeley.

[10] W. F. Meyers and J. W. Simmons. 1969. U.S. Pat., 3,423,242.

[11] N. Watanabe and M. Fukuda. 1970. U.S. Pat., 3,536,532.

[12] H. Ikeda, T. Saito and H. Tamura. 1977. Lithium-manganese dioxide cell. *Denki Kagaku oyobi Kogyo Butsuri Kagaku* 45:314–318.

[13] J. R. Moser. 1972. U.S. Pat., 3,660,163.

[14] A. A. Schneider and J. R. Moser. 1972. U.S. Pat., 3,674,562.

[15] M. Winter, B. Barnett and K. Xu. 2018. Before Li ion batteries. *Chem. Rev.* 118 (23):11433–11456.

[16] M. V. Reddy, A. Mauger, C. M. Julien, A. Paolella and K. Zaghib. 2020. Brief history of early lithium-battery development. *Materials* 13 (8):1884.

[17] M. S. Whittingham. 1976. The role of ternary phases in cathode reactions. *J. Electrochem. Soc.* 123 (3):315–320.

[18] F. R. Gamble, J. H. Osiecki, M. Cais, R. Pisharody, F. J. DiSalvo and T. H. Geballe. 1971. Intercalation complexes of Lewis bases and layered sulfides: a large class of new superconductors. *Science* 174 (4008):493–497.

[19] D. W. Murphy and F. A. Trumbore. 1976. The chemistry of TiS_3 and $NbSe_3$ cathodes. *J. Electrochem. Soc.* 123 (7):960.

[20] B. M. L. Rao, R. W. Francis and H. A. Christopher. 1977. Lithium-aluminum electrode. *J. Electrochem. Soc.* 124 (10):1490–1492.

[21] K. Mizushima, P. Jones, P. Wiseman and J. B. Goodenough. 1980. Li_xCoO_2 $0<x\leq 1$: a new cathode material for batteries of high energy density. *Mater. Res. Bull.* 15 (6):783–789.

[22] M. Thackeray, W. David, P. Bruce and J. Goodenough. 1983. Lithium insertion into manganese spinels. *Mater. Res. Bull.* 18 (4):461–472.

[23] A. Yoshino, K. Sanechika and T. Nakajima. 1987. U.S. Pat., 4,668,595.

[24] M. Winter, J. O. Besenhard, M. E. Spahr and P. Novák. 1998. Insertion electrode materials for rechargeable lithium batteries. *Adv. Mater.* 10 (10):725–763.

[25] C. Liu, Z. G. Neale and G. Cao. 2016. Understanding electrochemical potentials of cathode materials in rechargeable batteries. *Mater. Today* 19 (2):109–123.

[26] Y. Liang, C.-Z. Zhao, H. Yuan, Y. Chen, W. Zhang, J.-Q. Huang, D. Yu, Y. Liu, M.-M. Titirici, Y.-L. Chueh, H. Yu and Q. Zhang. 2019. A review of rechargeable batteries for portable electronic devices. *InfoMat* 1 (1):6–32.

[27] M. S. Whittingham. 2004. Lithium batteries and cathode materials. *Chem. Rev.* 104: 4271–4301.

[28] M. S. Whittingham. 1977. U.S. Pat., 4,009,052.

[29] M. S. Whittingham. 1978. Chemistry of intercalation compounds: metal guests in chalcogenide hosts. *Prog. Solid State Chem.* 12 (1):41–99.

[30] T. Nagaura. 1990. Lithium ion rechargeable battery. *Prog. Batteries Sol. Cells* 9:209.

[31] K. Ozawa. 1994. Lithium-ion rechargeable batteries with $LiCoO_2$ and carbon electrodes: the $LiCoO_2$/C system. *Solid State Ion.* 69 (3):212–221.

[32] J. R. Dahn, U. von Sacken and C. A. Michal. 1990. Structure and electrochemistry of $Li_{1\pm y}NiO_2$ and a new Li_2NiO_2 phase with the $Ni(OH)_2$ structure. *Solid State Ion.* 44 (1):87–97.

[33] J. R. Dahn, U. von Sacken, M. W. Juzkow and H. Al-Janaby. 1991. Rechargeable $LiNiO_2$/ carbon cells. *J. Electrochem. Soc.* 138 (8):2207–2211.

[34] A. Rougier, P. Gravereau and C. Delmas. 1996. Optimization of the composition of the $Li_{1-z}Ni_{1+z}O_2$ electrode materials: structural, magnetic, and electrochemical studies. *J. Electrochem. Soc.* 143 (4):1168.

[35] H. Arai, S. Okada, Y. Sakurai and J.-I. Yamaki. 1998. Thermal behavior of $Li_{1-y}NiO_2$ and the decomposition mechanism. *Solid State Ion.* 109 (3–4):295–302.

[36] P. Kalyani and N. Kalaiselvi. 2005. Various aspects of $LiNiO_2$ chemistry: a review. *Sci. Technol. Adv. Mater.* 6 (6):689.

[37] T. Ohzuku, A. Ueda, M. Nagayama, Y. Iwakoshi and H. Komori. 1993. Comparative study of $LiCoO_2$, $LiNi_{12}Co_{12}O_2$ and $LiNiO_2$ for 4 volt secondary lithium cells. *Electrochim. Acta* 38 (9):1159–1167.

[38] T. Ohzuku and Y. Makimura. 2001. Layered lithium insertion material of $LiNi_{1/2}Mn_{1/2}O_2$: a possible alternative to $LiCoO_2$ for advanced lithium-ion batteries. *Chem. Lett.* 30 (8):744–745.

[39] K. Kang. 2006. Electrodes with high power and high capacity for rechargeable lithium batteries. *Science* 311 (5763):977–980.

[40] F. A. Susai, H. Sclar, Y. Shilina, T. R. Penki, R. Raman, S. Maddukuri, S. Maiti, I. C. Halalay, S. Luski, B. Markovsky and D. Aurbach. 2018. Horizons for Li-ion batteries relevant to electro-mobility: high-specific-energy cathodes and chemically active separators. *Adv. Mater.* 30 (41):1801348.

[41] Z. Liu, A. Yu and J. Y. Lee. 1999. Synthesis and characterization of $LiNi_{1-x-y}Co_{x-}Mn_yO_2$ as the cathode materials of secondary lithium batteries. *J. Power Sources* 81:416–419.

[42] M. Yoshio, H. Noguchi, J.-I. Itoh, M. Okada and T. Mouri. 2000. Preparation and properties of $LiCo_yMn_xNi_{1-x-y}O_2$ as a cathode for lithium ion batteries. *J. Power Sources* 90 (2):176–181.

[43] T. Ohzuku and Y. Makimura. 2001. Layered lithium insertion material of $LiCo_{1/3}Ni_{1/3}Mn_{1/3}O_2$ for lithium-ion batteries. *Chem. Lett.* 30 (7):642–643.

[44] Z. Lu, D. MacNeil and J. Dahn. 2001. Layered cathode materials Li $[Ni_xLi_{(1/3-2x/3)}Mn_{(2/3-x/3)}]O_2$ for lithium-ion batteries. *Electrochem. Solid-State Lett.* 4 (11):A191.

[45] M.-H. Kim, H.-S. Shin, D. Shin and Y.-K. Sun. 2006. Synthesis and electrochemical properties of $Li[Ni_{0.8}Co_{0.1}Mn_{0.1}]O_2$ and $Li[Ni_{0.8}Co_{0.2}]O_2$ via co-precipitation. *J. Power Sources* 159 (2):1328–1333.

[46] H.-J. Noh, S. Youn, C. S. Yoon and Y.-K. Sun. 2013. Comparison of the structural and electrochemical properties of layered $Li[Ni_xCo_yMn_z]O_2$ (x=1/3, 0.5, 0.6, 0.7, 0.8 and 0.85) cathode material for lithium-ion batteries. *J. Power Sources* 233:121–130.

[47] F. Xin, H. Zhou, X. Chen, M. Zuba, N. Chernova, G. Zhou and M. S. Whittingham. 2019. Li-Nb-O coating/substitution enhances the electrochemical performance of the $LiNi_{0.8}Mn_{0.1}Co_{0.1}O_2$ (NMC 811) cathode. *ACS Appl. Mat. Interfaces* 11 (38):34889–34894.

[48] F. Xin, H. Zhou, J. Bai, F. Wang and M. S. Whittingham. 2021. Conditioning the surface and bulk of high-nickel cathodes with a Nb coating: an in situ X-ray study. *J. Phys. Chem. Lett.* 12 (33):7908–7913.

[49] F. Xin, H. Zhou, Y. Zong, M. Zuba, Y. Chen, N. A. Chernova, J. Bai, B. Pei, A. Goel, J. Rana, F. Wang, K. An, F. J. L. Piper, G. Zhou and M. S. Whittingham. 2021. What is the role of Nb in Nickel-Rich layered oxide cathodes for lithium-ion batteries? *ACS Energy Lett.* 6:1377–1382.

[50] F. Xin, A. Goel, X. Chen, H. Zhou, J. Bai, S. Liu, F. Wang, G. Zhou and M. S. Whittingham. 2022. Electrochemical characterization and microstructure evolution of Ni-rich layered cathode materials by niobium coating/substitution. *Chem. Mater.* 34 (17):7858–7866.

[51] Y.-Q. Lai, M. Xu, Z.-A. Zhang, C.-H. Gao, P. Wang and Z.-Y. Yu. 2016. Optimized structure stability and electrochemical performance of $LiNi_{0.8}Co_{0.15}Al_{0.05}O_2$ by sputtering nanoscale ZnO film. *J. Power Sources* 309:20–26.

[52] A. Chakraborty, S. Kunnikuruvan, S. Kumar, B. Markovsky, D. Aurbach, M. Dixit and D. T. Major. 2020. Layered cathode materials for lithium-ion batteries: review of computational studies on $LiNi_{1-x-y}Co_xMn_yO_2$ and $LiNi_{1-x-y}Co_xAl_yO_2$. *Chem. Mater.* 32 (3):915–952.

[53] D. Tang, D. Liu, Y. Liu, Z. Yang and L. Chen. 2014. Investigation on the electrochemical activation process of $Li_{1.20}Ni_{0.32}Co_{0.004}Mn_{0.476}O_2$. *Prog. Nat. Sci.: Mater. Int.* 24 (4):388–396.

[54] C. Zhan, T. Wu, J. Lu and K. Amine. 2018. Dissolution, migration, and deposition of transition metal ions in Li-ion batteries exemplified by Mn-based cathodes-a critical review. *Energy Environ. Sci.* 11 (2):243–257.

[55] W. Li. 2020. Review-an unpredictable hazard in lithium-ion batteries from transition metal ions: dissolution from cathodes, deposition on anodes and elimination strategies. *J. Electrochem. Soc.* 167 (9):090514.

[56] Q. Zhong, A. Bonakdarpour, M. Zhang, Y. Gao and J. R. Dahn. 1997. Synthesis and electrochemistry of $LiNi_xMn_{2-x}O_4$. *J. Electrochem. Soc.* 144 (1):205–213.

[57] K. Amine, H. Tukamoto, H. Yasuda and Y. Fujita. 1997. Preparation and electrochemical investigation of $LiMn_{2-x}Me_xO_4$ (Me: Ni, Fe, and x=0.5, 1) cathode materials for secondary lithium batteries. *J. Power Sources* 68 (2):604–608.

[58] J. Xiao, X. Chen, P. V. Sushko, M. L. Sushko, L. Kovarik, J. Feng, Z. Deng, J. Zheng, G. L. Graff, Z. Nie, D. Choi, J. Liu, J.-G. Zhang and M. S. Whittingham. 2012. High-performance $LiNi_{0.5}Mn_{1.5}O_4$ spinel controlled by Mn^{3+} concentration and site disorder. *Adv. Mater.* 24 (16):2109–2116.

[59] G. Liang, V. K. Peterson, K. W. See, Z. Guo and W. K. Pang. 2020. Developing high-voltage spinel $LiNi_{0.5}Mn_{1.5}O_4$ cathodes for high-energy-density lithium-ion batteries: current achievements and future prospects. *J. Mater. Chem. A* 8 (31):15373–15398.

[60] R. Santhanam and B. Rambabu. 2010. Research progress in high voltage spinel $LiNi_{0.5}Mn_{1.5}O_4$ material. *J. Power Sources* 195 (17):5442–5451.

[61] A. K. Padhi, K. S. Nanjundaswamy and J. B. Goodenough. 1997. Phospho-olivines as positive-electrode materials for rechargeable lithium batteries. *J. Electrochem. Soc.* 144 (4):1188–1194.

[62] L. Castro, R. Dedryvère, M. El Khalifi, P. E. Lippens, J. Bréger, C. Tessier and D. Gonbeau. 2010. The spin-polarized electronic structure of $LiFePO_4$ and $FePO_4$ evidenced by in-lab XPS. *J. Phys. Chem. C* 114 (41):17995–18000.

[63] J. Lee, A. Urban, X. Li, D. Su, G. Hautier and G. Ceder. 2014. Unlocking the potential of cation-disordered oxides for rechargeable lithium batteries. *Science* 343 (6170):519–522.

[64] R. J. Clément, Z. Lun and G. Ceder. 2020. Cation-disordered rock-salt transition metal oxides and oxyfluorides for high energy lithium-ion cathodes. *Energy Environ. Sci.* 13 (2):345–373.

[65] R. Wang, X. Li, L. Liu, J. Lee, D.-H. Seo, S.-H. Bo, A. Urban and G. Ceder. 2015. A disordered rock-salt Li-excess cathode material with high capacity and substantial oxygen redox activity: $Li_{1.25}Nb_{0.25}Mn_{0.5}O_2$. *Electrochem. Commun.* 60:70–73.
[66] F. Wu and G. Yushin. 2017. Conversion cathodes for rechargeable lithium and lithium-ion batteries. *Energy Environ. Sci.* 10 (2):435–459.
[67] X. Ji, K. T. Lee and L. F. Nazar. 2009. A highly ordered nanostructured carbon-sulphur cathode for lithium-sulphur batteries. *Nat. Mater.* 8 (6):500–506.
[68] P. G. Bruce, S. A. Freunberger, L. J. Hardwick and J. M. Tarascon. 2012. Li-O_2 and Li-S batteries with high energy storage. *Nat. Mater.* 11 (1):19–29.
[69] H. D. Lim, B. Lee, Y. Bae, H. Park, Y. Ko, H. Kim, J. Kim and K. Kang. 2017. Reaction chemistry in rechargeable Li-O_2 batteries. *Chem. Soc. Rev.* 46 (10):2873–2888.
[70] G. Girishkumar, B. McCloskey, A. C. Luntz, S. Swanson and W. Wilcke. 2010. Lithium–air battery: promise and challenges. *J. Phys. Chem. Lett.* 1 (14):2193–2203.
[71] J. R. Dahn, T. Zheng, Y. Liu and J. S. Xue. 1995. Mechanisms for lithium insertion in carbonaceous materials. *Science* 270 (5236):590–593.
[72] L. Ji, Z. Lin, M. Alcoutlabi and X. Zhang. 2011. Recent developments in nanostructured anode materials for rechargeable lithium-ion batteries. *Energy Environ. Sci.* 4 (8):2682.
[73] E. M. Sorensen, S. J. Barry, H.-K. Jung, J. M. Rondinelli, J. T. Vaughey and K. R. Poeppelmeier. 2006. Three-dimensionally ordered macroporous $Li_4Ti_5O_{12}$: effect of wall structure on electrochemical properties. *Chem. Mater.* 18 (2):482–489.
[74] G.-N. Zhu, Y.-G. Wang and Y.-Y. Xia. 2012. Ti-based compounds as anode materials for Li-ion batteries. *Energy Environ. Sci.* 5 (5):6652.
[75] A. R. Armstrong, G. Armstrong, J. Canales, R. García and P. G. Bruce. 2005. Lithium-ion intercalation into TiO_2-B nanowires. *Adv. Mater.* 17 (7):862–865.
[76] A. Dey. 1971. Electrochemical alloying of lithium in organic electrolytes. *J. Electrochem. Soc.* 118 (10):1547–1549.
[77] C. M. Park, J. H. Kim, H. Kim and H. J. Sohn. 2010. Li-alloy based anode materials for Li secondary batteries. *Chem. Soc. Rev.* 39 (8):3115–3141.
[78] H. Wu and Y. Cui. 2012. Designing nanostructured Si anodes for high energy lithium ion batteries. *Nano Today* 7 (5):414–429.
[79] U. Kasavajjula, C. Wang and A. J. Appleby. 2007. Nano- and bulk-silicon-based insertion anodes for lithium-ion secondary cells. *J. Power Sources* 163 (2):1003–1039.
[80] C. K. Chan, H. Peng, G. Liu, K. McIlwrath, X. F. Zhang, R. A. Huggins and Y. Cui. 2008. High-performance lithium battery anodes using silicon nanowires. *Nat. Nanotechnol.* 3 (1):31–35.
[81] N. Liu, Z. Lu, J. Zhao, M. T. McDowell, H. W. Lee, W. Zhao and Y. Cui. 2014. A pomegranate-inspired nanoscale design for large-volume-change lithium battery anodes. *Nat. Nanotechnol.* 9 (3):187–192.
[82] Y. Idota, T. Kubota, A. Matsufuji, Y. Maekawa and T. Miyasaka. 1997. Tin-based amorphous oxide: a high-capacity lithium-ion-storage material. *Science* 276 (5317):1395–1397.
[83] M. Winter and J. O. Besenhard. 1999. Electrochemical lithiation of tin and tin-based intermetallics and composites. *Electrochim. Acta* 45 (1–2):31–50.
[84] I. A. Courtney and J. Dahn. 1997. Electrochemical and in situ X-ray diffraction studies of the reaction of lithium with tin oxide composites. *J. Electrochem. Soc.* 144 (6):2045–2052.
[85] I. A. Courtney and J. Dahn. 1997. Key factors controlling the reversibility of the reaction of lithium with SnO_2 and Sn_2BPO_6 glass. *J. Electrochem. Soc.* 144 (9):2943–2948.
[86] SONY's new Nexelion hybrid lithium ion batteries to have thirty-percent more capacity than conventional offering, https://www.sony.com/en/SonyInfo/News/Press/200502/05-006E/ (accessed Feb. 24, 2022).

[87] ソニー、ノートPC市場向けに「スズ系アモルファス負極」を採用した 3.5Ahの高容量リチウムイオン二次電池"Nexelion(ネクセリオン)"を開発, https://www.sony.com/ja/SonyInfo/News/Press/201107/11-078/ (accessed Feb. 24, 2022).
[88] X.-L. Wang, M. Feygenson, H. Chen, C.-H. Lin, W. Ku, J. Bai, M. C. Aronson, T. A. Tyson and W.-Q. Han. 2011. Nanospheres of a new intermetallic $FeSn_5$ phase: synthesis, magnetic properties and anode performance in li-ion batteries. *J. Am. Chem. Soc.* 133 (29):11213–11219.
[89] F. Xin, X. Wang, J. Bai, W. Wen, H. Tian, C. Wang and W. Han. 2015. A lithiation/delithiation mechanism of monodispersed MSn_5(M = Fe, Co and FeCo) nanospheres. *J. Mater. Chem. A* 3 (13):7170–7178.
[90] Z. Dong, Q. Wang, R. Zhang, N. A. Chernova, F. Omenya, D. Ji and M. S. Whittingham. 2019. Reaction mechanism of the Sn_2Fe anode in lithium-ion batteries. *ACS Omega* 4 (27):22345–22355.
[91] F. Xin, H. Zhou, Q. Yin, Y. Shi, F. Omenya, G. Zhou and M. S. Whittingham. 2019. Nanocrystal conversion-assisted design of Sn–Fe alloy with a core–shell structure as high-performance anodes for lithium-ion batteries. *ACS Omega* 4 (3):4888–4895.
[92] F. Xin and M. S. Whittingham. 2020. Challenges and development of tin-based anode with high volumetric capacity for li-ion batteries. *Electrochem. Energy Rev.* 3 (4):643–655.
[93] P. Poizot, S. Laruelle, S. Grugeon, L. Dupont and J. Tarascon. 2000. Nano-sized transition-metal oxides as negative-electrode materials for lithium-ion batteries. *Nature* 407 (6803):496–499.
[94] P.-L. Taberna, S. Mitra, P. Poizot, P. Simon and J.-M. Tarascon. 2006. High rate capabilities Fe_3O_4-based Cu nano-architectured electrodes for lithium-ion battery applications. *Nat. Mater.* 5 (7):567–573.
[95] M. S. Whittingham. 2012. History, evolution, and future status of energy storage. *Proc. IEEE* 100:1518–1534.
[96] W. Xu, J. Wang, F. Ding, X. Chen, E. Nasybulin, Y. Zhang and J.-G. Zhang. 2014. Lithium metal anodes for rechargeable batteries. *Energy Environ. Sci.* 7 (2):513–537.
[97] D. Lin, Y. Liu and Y. Cui. 2017. Reviving the lithium metal anode for high-energy batteries. *Nat. Nanotechnol.* 12 (3):194–206.
[98] K. Xu. 2004. Nonaqueous liquid electrolytes for lithium-based rechargeable batteries. *Chem. Rev.* 104 (10):4303–4418.
[99] J. B. Goodenough and Y. Kim. 2009. Challenges for rechargeable li batteries. *Chem. Mater.* 22 (3):587–603.
[100] E. Peled and S. Menkin. 2017. Review-SEI: past, present and future. *J. Electrochem. Soc.* 164 (7):A1703–A1719.
[101] S. K. Heiskanen, J. Kim and B. L. Lucht. 2019. Generation and evolution of the solid electrolyte interphase of lithium-ion batteries. *Joule* 3 (10):2322–2333.
[102] H. Wu, H. Jia, C. Wang, J. G. Zhang and W. Xu. 2020. Recent progress in understanding solid electrolyte interphase on lithium metal anodes. *Adv. Energy Mater.* 11 (5):2003092.
[103] K. Xu. 2014. Electrolytes and interphases in li-ion batteries and beyond. *Chem. Rev.* 114 (23):11503–11618.
[104] S. S. Zhang. 2006. A review on electrolyte additives for lithium-ion batteries. *J. Power Sources* 162 (2):1379–1394.
[105] M. R. Palacin. 2009. Recent advances in rechargeable battery materials: a chemist's perspective. *Chem. Soc. Rev.* 38 (9):2565–2575.
[106] A. M. Haregewoin, A. S. Wotango and B.-J. Hwang. 2016. Electrolyte additives for lithium ion battery electrodes: progress and perspectives. *Energy Environ. Sci.* 9 (6):1955–1988.

[107] Z. Zhang, L. Hu, H. Wu, W. Weng, M. Koh, P. C. Redfern, L. A. Curtiss and K. Amine. 2013. Fluorinated electrolytes for 5 V lithium-ion battery chemistry. *Energy Environ. Sci.* 6 (6):1806.

[108] L. Liao, T. Fang, X. Zhou, Y. Gao, X. Cheng, L. Zhang and G. Yin. 2014. Enhancement of low-temperature performance of $LiFePO_4$ electrode by butyl sultone as electrolyte additive. *Solid State Ion.* 254:27–31.

[109] C. Yang, J. Chen, T. Qing, X. Fan, W. Sun, A. von Cresce, M. S. Ding, O. Borodin, J. Vatamanu, M. A. Schroeder, N. Eidson, C. Wang and K. Xu. 2017. 4.0 V aqueous li-ion batteries. *Joule* 1 (1):122–132.

[110] W. Li, J. R. Dahn and D. S. Wainwright. 1994. Rechargeable lithium batteries with aqueous electrolytes. *Science* 264:1115–1118.

[111] L. M. Suo, O. Borodin, T. Gao, M. Olguin, J. Ho, X. L. Fan, C. Luo, C. S. Wang and K. Xu. 2015. "Water-in-salt" electrolyte enables high-voltage aqueous lithium-ion chemistries. *Science* 350 (6263):938–943.

[112] G. G. Eshetu, S. Grugeon, S. Laruelle, S. Boyanov, A. Lecocq, J.-P. Bertrand and G. Marlair. 2013. In-depth safety-focused analysis of solvents used in electrolytes for large scale lithium ion batteries. *Phys. Chem. Chem. Phys.* 15 (23):9145–9155.

[113] Q. Zhang, K. Liu, F. Ding and X. Liu. 2017. Recent advances in solid polymer electrolytes for lithium batteries. *Nano Res.* 10 (12):4139–4174.

[114] J. W. Fergus. 2010. Ceramic and polymeric solid electrolytes for lithium-ion batteries. *J. Power Sources* 195 (15):4554–4569.

[115] J. C. Bachman, S. Muy, A. Grimaud, H.-H. Chang, N. Pour, S. F. Lux, O. Paschos, F. Maglia, S. Lupart, P. Lamp, L. Giordano and Y. Shao-Horn. 2016. Inorganic solid-state electrolytes for lithium batteries: mechanisms and properties governing ion conduction. *Chem. Rev.* 116 (1):140–162.

[116] A. Manthiram, X. W. Yu and S. F. Wang. 2017. Lithium battery chemistries enabled by solid-state electrolytes. *Nat. Rev. Mater.* 2 (4):16103.

2 Electron Microscopy for Advanced Battery Research

Jiacheng Zhu and Xuefeng Wang
Institute of Physics, Chinese Academy of Sciences

CONTENTS

DOI: 10.1201/9781003299295-2

2.1 BASIC PRINCIPLE OF ELECTRON MICROSCOPY (EM)

2.1.1 Interaction between Electron and Specimen

When a beam of electrons is irradiated on a sample, the electrons interact with nuclei and electron cloud of atoms in the structure, thus changing the amplitude, phase position, momentum, and energy of the electrons to varying degrees. In this process, various characteristic signals are generated, from which we could obtain information about the microscopic morphology, structure, and composition of the materials.

Compared with neutrons and X-rays, which interact with nuclei and electrons, respectively, incident electrons could have Coulomb interaction with both. Therefore, the scattering cross-section of atoms to electrons is much larger than that of neutrons and X-rays, which means that electrons can be used to probe finer structures at nanoscale.

Electron scattering can be divided into elastic and inelastic scattering, according to whether the electron energy changes after scattering. **Elastic scattering** is mainly caused by the deflection of incident electrons by the positively charged nuclei and is the physical basis for electron diffraction and EM. The smaller the distance between incident electrons and a nucleus is, the larger its scattering angle will be. Electrons with a scattering angle greater than 90° are called **elastic backscattered electrons**. However, extra-nuclear electrons repel the incident electrons and produce reverse deflection, mainly causing **inelastic scattering** of the incident electrons, which is the basis of **electron energy loss spectrum**. If those electrons can still bounce off the surface of the sample after multiple scattering, inelastic backscattered electrons are formed. Based on the above circumstances, the backscattered electrons come from a depth range of hundreds of nanometers on the specimen surface, and its yield increases with the increase of atomic number. Therefore, it can be used to analyze morphology and display atomic number contrast.

Under the attack from the incident electron beam, the outer electrons of atoms are bombarded out and hence leave the surface of the sample to become free electrons in the vacuum, which are called **secondary electrons**. Generally emitted from the depth range of 5–10 nm on the surface layer, the secondary electron has low energy, as a result of which it is quite sensitive to the surface morphology (the principle of secondary electron image), and its yield has no dependence on atomic number.

If the thickness of a sample is large enough, some electrons will be depleted of all the kinetic energy through multiple inelastic scattering after being injected into the sample, and finally absorbed by the sample to become **absorbed electrons**. The contrast of the absorbed electron image is opposite to that of the secondary electron and backscattered electron image, so the absorbed electron could also produce the atomic number contrast.

If the sample is thin enough, the high-energy electron beam can penetrate this thin sample and become **transmitted electrons**. The mass thickness, composition, and crystalline structure of the sample's micro-area determine the signal characteristics of transmitted electrons. By collecting transmitted electrons and elastic scattering electrons to image on the phosphor screen, the atomic/nanoscale structure can be observed.

The high-energy incident electrons can also bombard out the atom's inner electron to make the atom into an excited state. At this time, the outer electron will transition to this inner layer in order to bring the atom back to the ground state, with the excess energy radiated in the form of X-rays at a specific wavelength, and that is called **characteristic X-ray**. If the energy released after the transition of the outer electron to the inner layer causes another outer electron to be excited and emitted, a nonradiative transition is then formed, and the newly excited electron is called **Auger electron**.

When the incident electron beam bombards the material, electrons are deflected and rapidly decelerated due to the Coulomb force of the nuclei, with the lost energy released in the form of electromagnetic radiation. Hence, continuous-wavelength X-rays, also known as **bremsstrahlung X-rays**, are generated as a result of the continuous change of electron velocity.

If the semiconductor is bombarded by an electron beam, the valence band electrons will be excited to the conduction band and leave holes in the valence band. When the excited electrons re-jump back to the valence band and recombine with the holes, visible light will be emitted directly, which is called **cathode fluorescence**.

At the same time, the irradiation of high-energy electron beam often causes irreversible damage to the samples, changing the original chemical composition, morphology, and crystal structure of the samples. From the perspective of mechanism, radiation damage mainly includes knock-on displacement, radiolysis, charging damage, and heating damage.[1] Atomic displacement is an important feature of radiation damage, which can cause defects in the material; the coordinated displacement of multiple atoms can induce amorphization, phase transition, diffusion, and segregation of the material (Figure 2.1).[1]

2.1.2 EM System: Guns, Lens, Aberrations, and Resolutions

2.1.2.1 Electron Guns

The core part of an EM is its electro-optical system, in which the electron gun, deflection coil, electromagnetic lens, diaphragm components, and sample stage are important modules for both scanning electron microscope (SEM) and transmission electron microscope (TEM). The EM uses an electron gun as the light source, which is located at the top of the whole device. In order to obtain high-quality EM images,

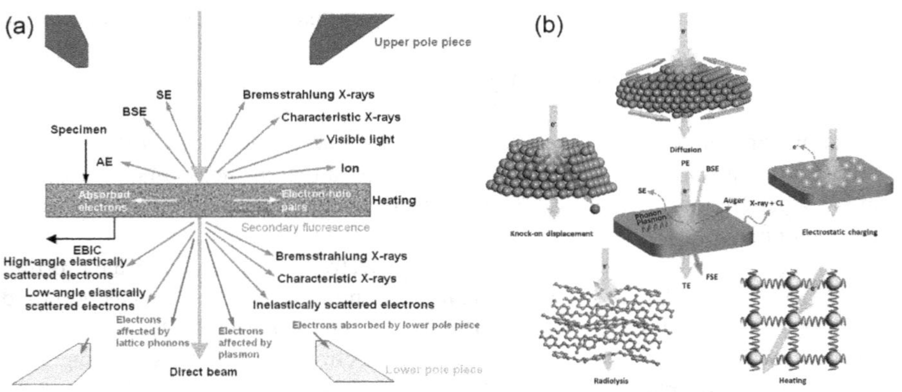

FIGURE 2.1 (a) Interaction between electron and sample. (b) Mechanism of electron radiation damage.[1]

the electron beam emitted by the electron gun should have high brightness and small beam spot diameter.

Usually, electron guns can be divided into two types: thermal-electron emission and field emission. The thermal-electron emission gun includes hairpin tungsten filament/monocrystal lanthanum hexaboride filament, an anode, and a grid cap. The cathode filament could be energized and heated to a high temperature in vacuum to emit electrons, forming a directional high-speed electron flow under the accelerating voltage. For a long time, most traditional SEM used hairpin tungsten filament, performance of which is worse than that with lanthanum hexaboride filament. Higher resolution can be achieved by the field-emission electron gun, which uses monocrystal tungsten as the tip material with a tip curvature radius of about 100 nm.[2] This tip is called the emitter. When a strong electric field is applied, the electrons inside the metal will be emitted to form field emission. The field-emission electron gun can be divided into two types: thermal field emission gun (also named Schottky field emission gun) and cold field emission gun. The cold field electron gun uses the (310) crystal plane of tungsten as the emitter, which does not need heating and has a high resolution. However, residual gas molecules will be adsorbed on the emitter when working at room temperature, which requires timing flash process; the thermal field electron gun performs field emission at high temperature without ion adsorption so that we can obtain stable emission current while its energy divergence is higher than that of the cold field gun.

2.1.2.2 The Lens of EM

The electron beam emitted by the electron gun needs to be concentrated through a condenser into an electron beam spot with a diameter small enough in order to focus on the sample. The resolution of EM is determined by the resolving power of the objective lens in the process of sample detection by the electron beam. In TEM, for example, the imaging system usually consists of an objective lens, an intermediate lens, and a projection lens. The objective lens, located between the astigmatic coil of the objective lens and the diffraction lens holders, can change the contrast and eliminate astigmatism for optimal resolution; the intermediate lens and the projection lens are located between the diffraction lens holders and the view-camera room to further magnify the image from the objective lens. The lenses mentioned above all use electromagnetic lens. Like the effect of glass convex lens to converge the light beam, the electromagnetic lens uses a rotating space magnetic field symmetrical to the axis of the mirror barrel to bend the motion of electrons to the axis by Lorentz force in order to form a focus, and the strong magnetic field is generated by a very stable direct excitation current through the coil with pole boots.

2.1.2.3 The Aberrations in the EM

In practical optical systems, the position obtained by focusing non-paraxial light rays is not consistent with that of focusing paraxial light rays, and thus the deviation from the ideal condition of Gaussian optics is called **aberration**. The light source used by EM is electron beam, the wavelength of which is smaller than that of visible light, resulting in higher resolution. However, aberration is still present in EMs, which is mainly divided into geometric aberration and chromatic aberration. Geometric aberration includes spherical aberration, comet, distortion, and astigmatism. Spherical aberration is caused by the fact that the far-axis electrons are refracted differently from the paraxial electrons as they pass through the lens. After the parallel electron beam passes through the lens, the rays are focused on an axial length rather than a point on the axis, which results in diffuse circular spot in the image plane. The spherical aberration of electromagnetic lenses cannot be completely eliminated because the lenses used are non-ideal. In advanced TEM, spherical aberration correctors are used to improve resolution, and in SEM, spherical aberration is reduced by reducing aperture angle and other methods. When the object point is on the outer edge of the lens center axis, it can make the electron beam and the center axis tilted at an angle, which destroys the symmetry of the lens and produces coma. Coma means the formation of fuzzy image points extending in a certain direction like comets on the imaging plane. However, since the sample point does not shift too much in the EM generally, the influence of the system comet can be ignored. Furthermore, because of the inhomogeneity of the focal length of the electromagnetic lens, the imaging point in the state of underfocus or overfocus forms elliptic beam spots which are 90° interleaving each other; thus the imaging point is blurred due to the large beam spots in the normal focus position. This phenomenon is called astigmatism, which can be divided into residual astigmatism and secondary astigmatism. It is commonly found in condensing lens, objective lens, and other diffraction lenses, and can be adjusted by astigmatism eliminators. Finally, if the acceleration voltage provided by the equipment is unstable, the energy of accelerating electrons will be different; the electron beams with different wavelengths will focus on different focal points, resulting in blur imaging. This phenomenon is called chromatic aberration, which can be effectively alleviated by stabilizing the acceleration voltage and excitation current.

2.1.2.4 The Resolutions of EM Imaging

Once all the aberrations have been minimized, the best spatial resolution of EM will be achieved. In an optical system, resolution is a measure of the capability to separate the images of two adjacent object points. The limit of resolution usually adopts Rayleigh criterion: When the center of one Airy spot coincides with the first dark ring of the other one, it is just enough to distinguish these two image points. Besides, Airy spots are focused points of light formed through circular apertures in a diffractive confined system. Then according to Abbe diffraction limit formula, we could learn that the smaller the wavelength of the incident electron beam is, the higher the resolution of the EM will be. Its resolving power is also related to the numerical aperture (NA) of the electromagnetic lens. Therefore, the resolution of EM can be estimated based on Abbe diffraction limit formula (Equation 3.1).

$$R = \frac{1.22\lambda f}{D} = \frac{0.61\lambda}{\mu \sin \alpha} = \frac{0.61\lambda}{\text{NA}} \tag{3.1}$$

where λ is the wavelength of the electron beam, f is the focal length of the electromagnetic lens, D is the diameter of the electromagnetic lens, μ is the refractive index of the medium between the object and objective lens, α is the half-angle of the light cone entering the objective lens, and $\mu \sin\alpha$ is called the numerical aperture (NA).

When a high voltage is applied to accelerate the electron beam in TEM, its resolution can reach the level of 0.1 nm.

2.1.3 General Information from EMs

Various information of the materials can be obtained by EM through the interaction between electron beams and the sample such as morphology, crystal structure, defects, elemental identification and distribution, and electronic structure. Since the wavelength of the electron beam can be adjusted arbitrarily by varying the acceleration voltage, structural information at different scales can be achieved, especially at the atomic scale. Nowadays, the electron microscopes, such as SEM, TEM, high-resolution transmission electron microscope (HRTEM), scanning transmission electron microscope (STEM), and aberration-corrected transmission electron microscope (ACTEM), have been widely used in physics (study of crystal structure and defects, and surface physical phenomena), chemistry (study on the structure and morphology of inorganic/organic composite materials, additives, catalysts, and the microscopic chemical reaction mechanism of materials), biology (the study of molecular biology, genetic engineering, synthetic proteins, and microorganisms such as bacteria and viruses), etc.

EM has the unique advantage in imaging. Other techniques such as X-ray diffraction (XRD), Raman, and Infrared spectroscopy (IR) could only give us the average information about the chemical composition of the sample materials, which is not accurate enough for the analysis of battery reaction mechanism. While combining with selected area electron diffraction (SAED) and energy dispersive spectrometer (EDS), EM technique could give the structural and compositional information of microregions, and show us about the micro-defect distribution of the electrode materials. What's more, the utilization of in-situ EMs could show extensive information about the electrochemical and mechanical properties of the battery materials in a realistic operating environment.

2.1.3.1 Information from SEM

By collecting backscattered electrons and secondary electrons generated on the surface, SEM can be used to explore the morphology of solid materials. Characteristic structures and morphologies can be directly visualized and discerned, including one-dimensional (1D) fibers, tubes, two-dimensional (2D) layers, and three-dimensional (3D) features such as spinel, spheres, cubes, and other regular and irregular shapes. Pores, surface roughness, and contaminants can also be distinguished from SEM imaging. Coupled with slicing technology, 3D morphology can be constructed, uncovering the features inside the bulk.

2.1.3.2 Information from TEM

Using scattered and transmitted electrons created by high-energy electron beams when penetrating thin samples, TEM can recognize the microstructure of samples at the nano/atomic scale, including morphology, composition, lattice structure, dislocation, atom displacements, etc. Advanced aberration-corrected STEM enables to observe the light atoms such as Li, and capture their location, movement, and interaction with other atoms during different conditions.

2.1.3.3 Information from Electron Diffraction

Similar to X-ray, the elastic scattered electrons interacting with samples could produce strong characteristic diffraction patterns which are closely related to the crystal structure. By adjusting the electric current of the intermediate lens of the TEM, the object plane of the intermediate lens could be removed to the rear focal plane of the objective lens, and then the TEM can be converted into diffraction mode with the selected aperture placed in the image plane of the objective lens. The electron diffraction patterns of crystal structures in specific regions of the materials can be obtained, which is called selected area electron diffraction (SAED). It is widely used to determine the phases, defects, and orientation. Single crystal, polycrystals, and amorphous materials display distinctive diffraction patterns due to their corresponding atom arrangement. Compared with X-ray and neutron diffraction, electron diffraction outperforms in analyzing the microstructure in local domains with strong signals.

2.1.3.4 Information from EDS

Through analyzing the energy of characteristic X-ray, the type of element can be determined by EDS. On the other hand, the strength of characteristic X-ray is related to the content of the element. Coupled with SEM or TEM, EDS is applied to identify the species, content, and distribution of elements in the samples, which is especially useful to discern the interface, interdiffusion, coating, and heterogenetic phases.

2.1.3.5 Information from EELS

When the incident electron beam interacts with the specimen, it will lose energy due to inelastic scattering. The energy loss of electrons could involve phonon excitation, plasma excitation, interband transition, inner layer electron excitation, secondary electron excitation, and bremsstrahlung. The partial energy lost value of the electrons is the eigenvalue of corresponding element in the sample. By collecting the transmitted electron signal and displaying its intensity according to its energy loss, we could get electron energy loss spectroscopy (EELS), which could be employed to analyze element types (especially for light elements) and valence states at a high energy resolution. Different elements, same element with varied oxidation states, and local structures can be told from EELS, which is of great significance to analyze the fine structure information in local domains.

2.2 SCANNING ELECTRON MICROSCOPY

2.2.1 General Information from SEM for Battery Materials and Interfaces

SEM uses focused electron beam to scan and image the sample surface point by point. The signal can be from secondary electrons, backscattered electrons, or/and absorption electrons (Figure 2.2a). By collecting, amplifying, and displaying these signals, the surface tomography and roughness of the sample can be obtained. SEM equipped with EDS accessories can further analyze the type and distribution of elements, and semi-quantitatively determines their content. Therefore, SEM is widely used to study the morphology, porosity, and tortuosity of the battery anode, cathode, separators, solid electrolyte materials and their interfaces. Using field-emission SEM, Xu et al.[3] found that the synthesized ZnS nanomaterials show spherical morphology with a diameter of ~300 nm (Figure 2.2b), which are uniformly covered by the insulating Li_2S/Li_2S_2 (Figure 2.2c) when they were used as a catalyst for S cathode and thus greatly improve the cycling reversibility of the Li-S battery. Chen et al.[4] doped $LiNi_{0.8}Co_{0.15}Al_{0.05}O_2$ (NCA) cathode material by boric acid and found that the size of secondary particles increases with the increase in boron content. After 100 cycles at a high temperature, cracks and thick solid-electrolyte interphase (SEI) layers appear on the bare NCA particles (Figure 2.2d) while the NCA particles doped with 1.5 mol% boric acid still maintain original morphology and have thin SEI layers (Figure 2.2e). Gao et al.[5] constructed an interphase, consisting of organic elastomeric salts (LiO-$(CH_2O)_n$-Li) and inorganic nanoparticle salts (LiF, -NSO_2-Li, Li_2O), to protect the $Li_{10}GeP_2S_{12}$ (LGPS) solid electrolyte from reaction with Li metal anode. Cross-section image by SEM demonstrates that this interphase is helpful to eliminate the cracks formed in the solid electrolyte (Figure 2.2f and g).

2.2.2 In-Situ/Operando SEM: Heating and Biasing

For ex-situ observation, the samples are harvested from the cycled batteries, which may be metastable or unstable to decompose easily. Then they are subjected to a series of processes prior to observation, including washing, drying, loading onto the stages, and transferring into the EM. During the above period, structural reconstruction/relaxation, contamination, and damage are potentially occurred, which will lead to some artifacts. To avoid such issues, in-situ experiments are suggested, which allows to observe the morphology evolution of samples during battery operation.[6] Therefore, it is expected to provide more reliable information than the ex-situ scenario.

The advantages of SEM are the high spatial resolution (<50 nm), large operating chamber for the setup, environmental function, and can combine EDS, focused ion

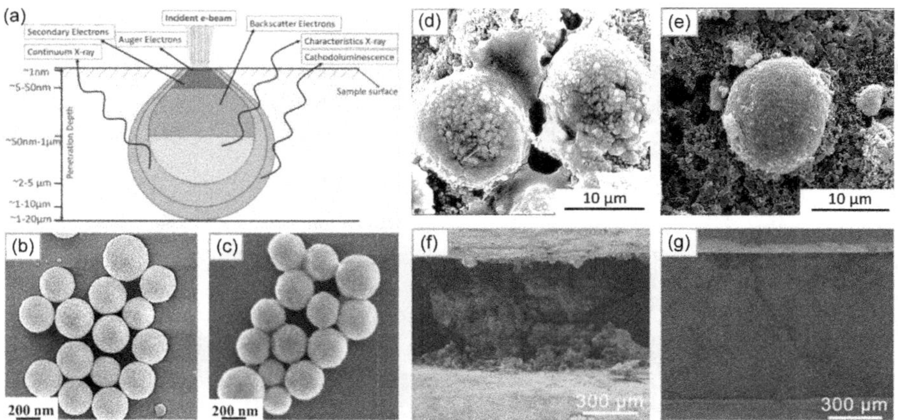

FIGURE 2.2 (a) The emission range of the sample's electronic signal. (b) FESEM image of ZnS nanospheres. (c) FESEM image of ZnS nanospheres after sulfur loading. (d and e) SEM image of (d) pristine NCA cathode and (e) 1.5% B-doped NCA cathode after 100 cycles at 2.8–4.3V and 55°C. (f and g) Side-view SEM images of the cycled Li|LGPS|Li cells using (f) bare Li and (g) nanocomposite-stabilized Li electrodes, respectively.

beam (FIB), backscattering, or some other technologies.[7] For in-situ SEM technology, a small battery is assembled inside the instrument for electrochemical cycling, using polymer solid electrolyte, ionic liquid, or conventional organic electrolyte sealed with windows.[6,8] Thus, this technique plays an irreplaceable role in studying the morphological and structural evolution of electrode materials, especially in revealing crack-propagated mechanical failure in Ni-rich Li-($Ni_xCo_yMn_z$)O_2 cathode nanoparticles during nanocell operating.[9] Except for studying cathodes, Chen et al.[10] constructed an in-situ SEM system based on ionic liquid to observe the lithiation/delithiation of SnO_2 anode material (Figure 2.3a and b). Ionic liquids with low vapor pressure are suitable as electrolytes in the high-vacuum chamber of SEM.[11] In their experiment, they captured the volume expansion of single SnO_2 particle after lithiation forming Li_2O and Li_xSn and observed mechanical induced cracks on the electrode particles during the cycling process (Figure 2.3c and d). Chen et al.[12] observed that the silicon (Si) particles were subjected to a volume expansion of 300% after lithiation, and the particles could not fully recover to their original form after delithiation with partial Li trapping in the Si anode. During repeated lithiation and delithiation, the anisotropy volume expansion leads to irreversible structural changes and the pulverization of Si particles.

Considering that ionic liquid is not widely adopted in conventional lithium-ion batteries (LIBs), Rong et al.[13] used SiN_x film as the window material to encapsulate liquid organic electrolyte, which can simulate the real operating environment of a battery with liquid electrolyte. This setup includes a top silicon chip with a SiN_x membrane viewing window and a bottom chip made of quartz. Two Cu current collectors are installed on the top chip with a microgap lying in the center of the viewing window (Figure 2.4a and b). The electrode material is deposited on the Cu current collector by physical deposition. The electrolyte used in the battery is 1.0M lithium bis(trifluoromethane sulfonyl)imide (LiTFSI) in 1,3-dioxolane (DOL)

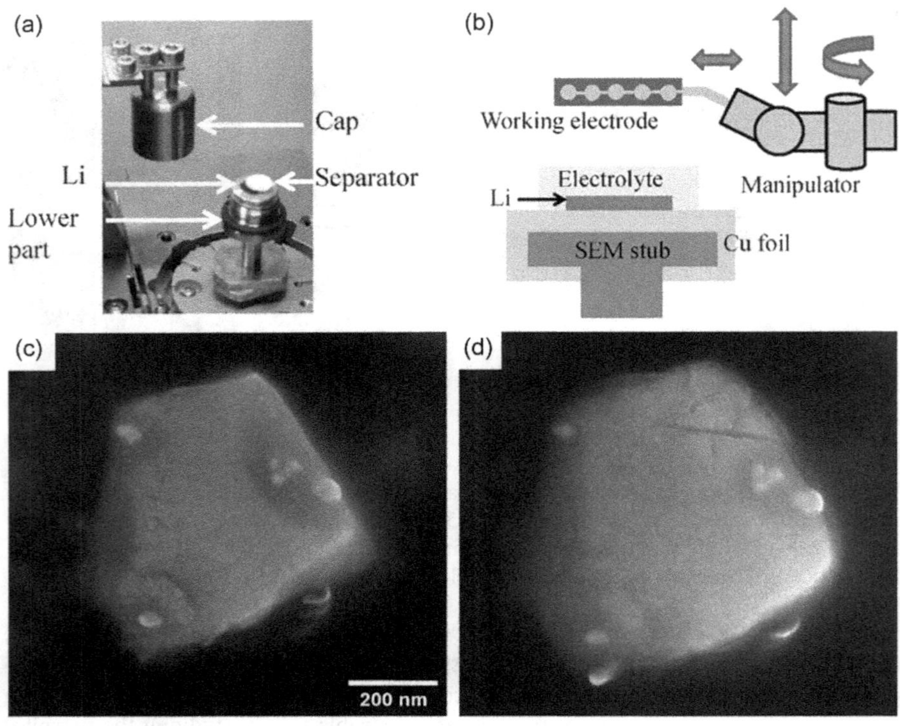

FIGURE 2.3 (a) Home-built transfer system (opened) containing the lower part of the cell. (b) Inside the SEM, the electrode under investigation is placed onto the separator. (c and d) Single SnO_2 particle before (c) and after (d) the first lithium insertion.

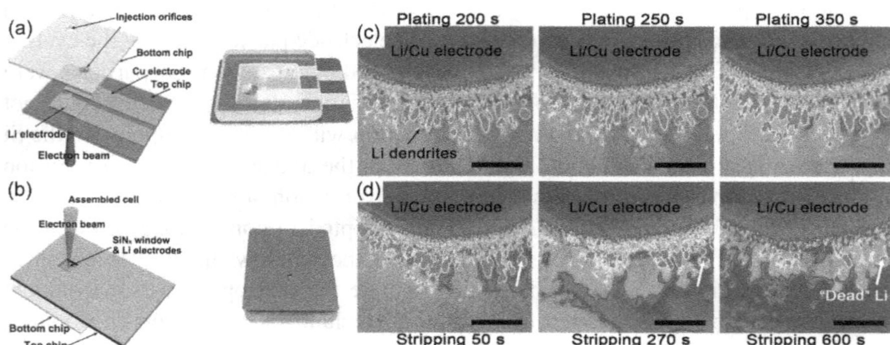

FIGURE 2.4 (a and b) Schemes and photos of an in situ SEM-electrochemical liquid cell: (a) bottom and (b) top views. (c and d) A time lapse series of SEM images of the processes of (c) lithium plating and (d) stripping under $0.15\,mA/cm^2$ on the Li/Cu electrode using the LiTFSI/DOL/DME electrolyte with the additive of $LiNO_3$ (1 wt%).

and 1,2-dimethoxyethane (DME) (volume ratio 1:1) with the additives of $LiNO_3$ or Li_2S_8. With this setup, they observed the real-time growth and dissolution of Li dendrites and detected the formation of dead Li (Figure 2.4d). The result shows that the electrolyte additive has an important effect on the growth rate and mechanism of lithium dendrite. Adding $LiNO_3$ and lithium polysulfide simultaneously in the ether-based electrolyte can inhibit the dendrite growth and result in smooth surface during lithium deposition.

Compared with liquid electrolyte, solid electrolyte has intrinsic merits to be used in the in-situ battery. Solid-state battery is also a hot topic as a next-generation high-energy and safe battery. Nagao et al.[14] built an all-solid-state Li-metal batteries based on sulfide solid electrolyte and used in-situ SEM to monitor the Li depositing and stripping process on it (Figure 2.5a and b). They visualized varied morphology of Li deposition dependent on the current densities. When the current density is less than 0.01 mA/cm^2, the deposition and dissolution of Li are mostly reversible without

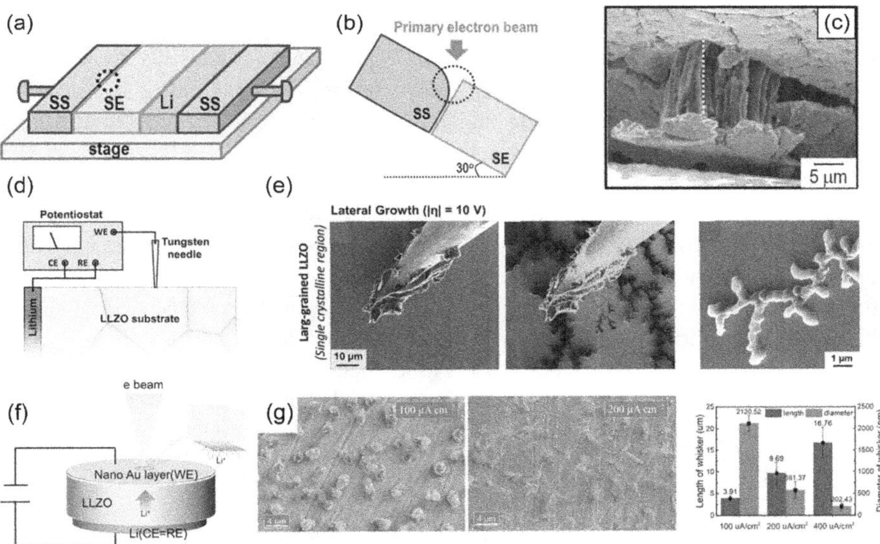

FIGURE 2.5 (a) Schematic of the all-solid-state cell for the in-situ SEM observation (b) with the tilt of the stage at an angle of 30°. (c) SEM images of Li deposition in the interface between the solid electrolyte $80Li_2S$-$20P_2S_5$ and stainless-steel (SS) at 5 mA/cm^2 for 10 minutes after the short circuit. (d) Schematic experimental setup for the in-situ measurements, which could measure the ionic conductivity of LLZO single particle using FIB manipulator. The macroscopic Li counter electrode (CE) was inherently also used as reference electrode (RE). The Li microelectrode was in-situ generated by plating on a tungsten needle and connected as working electrode (WE). (e) SEM images of lateral, dendritic growth along the inorganic solid electrolyte surface before and after plating at high negative overpotentials. (f) Schematic diagram of the experimental setup for electrochemical measurements. A thin gold layer was deposited on the finely polished garnet substrate as working electrode (WE). The Li counter electrode (CE) was welded at high temperature to eliminate the interface contributions to the overall impedance. (g) Effects of different current densities on the morphology of lithium deposition, the length, and diameter of Li whiskers.

any obvious dendritic growth; when the current density is larger than 1 mA/cm^2, the stress caused by uneven Li deposition induces large cracks in the sulfide electrolyte (Figure 2.5c). They also pointed out that inhibiting the growth of Li dendrites along the boundary of electrolyte is the key to solve the short circuit of the battery. In addition, in a FIB-SEM, Krauskopf et al.[15] applied a bias voltage between the tungsten needle (FIB manipulator) and Li metal electrode with $Li_{6.25}Al_{0.25}La_3Zr_2O_{12}$ (LLZO) solid electrolyte (Figure 2.5d) and observed the Li growth on the LLZO. When −10V overpotential is applied, Li deposition adopts a transverse growth (Figure 2.5e), which induces electrolyte cracks and electrochemical-mechanical fractures, thus changing the actual local current density. On the other hand, when the manipulator is in contact with a single LLZO particle, the bulk ionic conductivity of LLZO can be measured, which is 5.0×10^{-4} S/cm, avoiding the influence from the grain boundary impedance (Figure 2.5d). Rather than applying a bias voltage, Wang et al.[16] directly induced Li deposition through electron beam exposure on $Li_{6.4}La_3Zr_{1.4}Ta_{0.6}O_{12}$ (LLZTO) solid electrolyte (Figure 2.5f). The electron beam will converge on the surface of electrically insulated LLZTO and form a local overpotential, so the Li deposition occurs without the participation of the manipulator (Figure 2.5g). The results demonstrated that Li shows a higher nucleation tendency and deposition kinetics in the defect region of polycrystalline electrolyte, suggesting that the active site of Li deposition is closely related to the defect species in the solid electrolyte.

Additionally, in-situ SEM has also been applied to study a complex system with gas involved reaction, such as in Li-O_2 battery. By purging partial oxygen into an environmental SEM (ESEM), Zheng et al.[17] constructed an all-solid-state Li-O_2 battery with the aligned carbon nanotube (CNT) as the cathode of oxygen carrier, Li metal as the anode, and the native oxide layer Li_2O on the surface of Li metal as the electrolyte (Figure 2.6a). Li ions and oxygen molecules react to form lithium peroxide (Li_2O_2) on the surface of CNT and then decompose during discharging and charging processes (Figure 2.6b). They found that change of current density

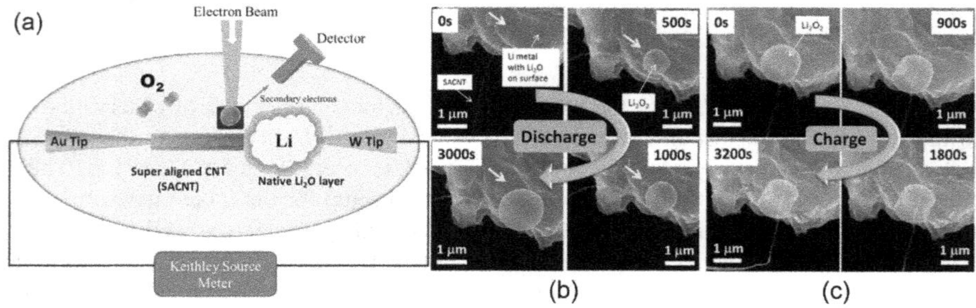

FIGURE 2.6 (a) A schematic view of microscale all-solid-state Li-O_2 battery assembled in the ESEM chamber. (b and c) Discharge and charge processes of the Li-O_2 battery. (b) Images captured at 0, 500, 1,000, and 3,000 s show the growth process of a spherical particle, which can grow up to 1.5 μm. A bias of −3V was applied on SACNT vs. Li metal to initiate the discharge process. (c) Images captured at 0, 900, 1,800, and 3,200 s show the decomposition process of the spherical particle. 8V was applied on SACNT vs. Li metal to initiate the charge process.

and surface chemistry of CNT leads to the formation of Li_2O_2 with varied morphology, such as sphere, conformal film, and red-blood-cell-like shape. In addition, the decomposition of Li_2O_2 particles starts from the surface of the particles rather than its contact site with CNT during charging, indicating the sufficient electronic and ionic conductivity of Li_2O_2 for charge transfer.

2.2.3 Advanced SEM Integrated with Other Techniques, Such as Raman and SIMS

Besides further increasing the spatial resolution of SEM, another development direction of SEM is to integrate other techniques into "all-in-one system" with extended functions. Currently, Raman and secondary ion mass spectrometry (SIMS) have been successfully integrated into SEM as a complementary tool to each other. For example, Raman spectroscopy is sensitive to the chemical composition, crystallinity, and phase of the sample through analyzing the signal of Raman frequency shift after the incident laser is scattered by the molecular structure of the specimen, which is a fast, simple, qualitative and quantitative method for analyzing molecular vibration and rotation information. SEM coupled with Raman enables to increase the spatial resolution of Raman and endows SEM capability to probe the molecular vibration. We can use SEM to find the accurate location of interest, and measure the microscopic composition of the selected point through Raman. With the Raman-SEM integrated technique, Hintennach et al.[18] was able to distinguish the exfoliated graphite particle from the intact graphite in the electrode after multiple cycles based on their Raman fingerprint features (E band for exfoliated vs. G band for intact graphite, Figure 2.7a) caused by lithiation. Their distribution can be easily identified based on Raman mapping (Figure 2.7b), which is hard to realize for conventional SEM-EDS. These results demonstrate the strength of integrated techniques.

Besides, there are some pieces of equipment combining SEM with SIMS. It is well known that SIMS is a technique sensitive to precise analysis of surface composition. SIMS bombards the sample surface with high-energy primary ion beams to sputter out molecules and atoms from the surface of the material, forming charged secondary ions. These secondary ions are then collected by a mass analyzer and their mass to charge ratios are analyzed in order to obtain information about the chemical composition of the near-surface atomic layer of the specimen. The advanced time-of-flight SIMS (TOF-SIMS) could measure ion mass by analyzing the time it takes for secondary ion to fly from the sample surface to the detector, which can largely enhance the analysis sensitivity. Moreover, FIB-SEM system integrated with TOF-SIMS can simultaneously achieve 3D chemical characterization of solid materials, high ion mass resolution, and high spatial resolution of ion distribution imaging. Bessette et al.[19] used FIB-SEM unified with TOF-SIMS to quantify the spatial concentration of Li ions in the $LiNi_{0.5}Mn_{0.3}Co_{0.2}O_2$ (NMC532) cathode material during electrochemical cycles. This work overcame the deficiency of traditional EDS in detecting light elements and trace elements, and can accurately detect the low Li content in the cathode particles. They imaged the ion spatial distribution maps of Li and Mn elements in the single pristine and cycled NMC particles (Figure 2.7c and d). After removing

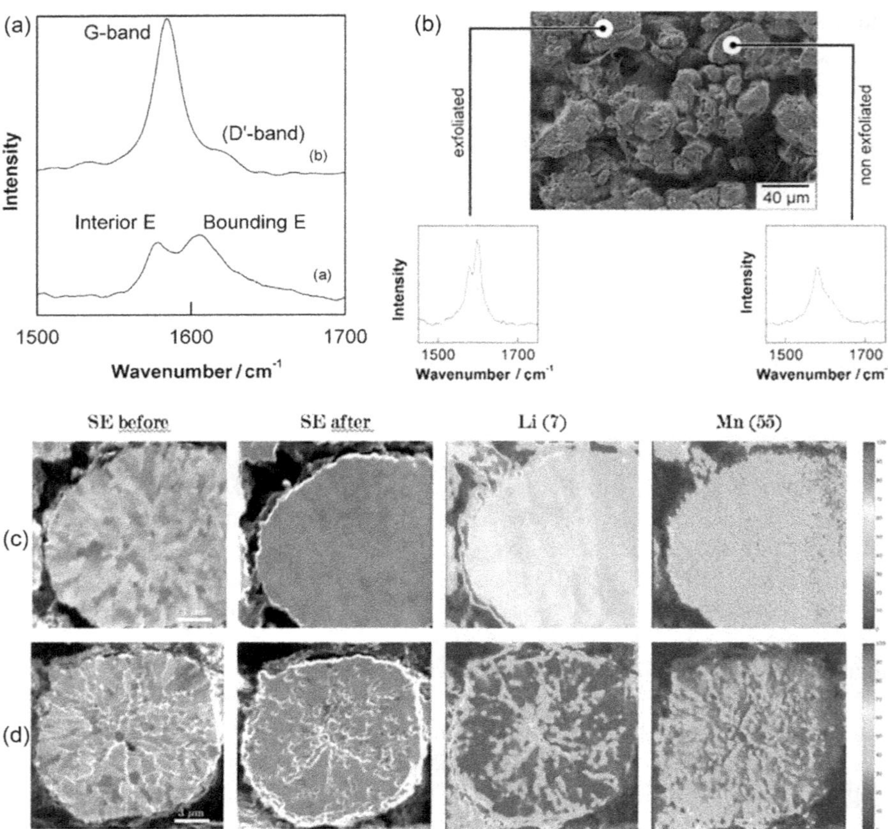

FIGURE 2.7 (a and b) Exemplary Raman spectra of two arbitrarily selected points on a graphite electrode. The electrode was electrochemically cycled up to very positive potentials (2.0V–5.5V vs. Li/Li^+, 20 cycles, 1 M $LiClO_4$ in EC/DMC 1:1). (a) The E band (exfoliation band) detected at the exfoliated parts of the graphite electrode and (b) the G band (with the weak shoulder of the D′ band) seen at the intact parts of the electrode. (c and d) FIB-SEM images before and after analysis and ion distributions of Li, Mn of pristine (c) and cycled (d) cathode. The ions maps represent the cumulative data over 100 frames of analysis and normalized according to the maximum intensity in the species for each sample.

the baneful effects caused by high brightness and edge effects generally present in high topography, they found out that Li ions are trapped in the primary grains of the secondary particles rather than at the grain boundaries between particles.

2.3 FOCUSED ION BEAM

2.3.1 FIB for Cross-Section Imaging

Different from the accelerated electrons generated by the electron gun in the electron-optical system of SEM, FIB uses the ion gun to accelerate the ion beam generated by the ion source (Ga, He, Ne ion source), and then interacts with the sample after focusing. Since the mass of ions is much larger than that of electrons, ion beams can not only perform imaging like electron beams, but also sputter atoms on solid surfaces. By carefully controlling the energy and strength of the ion beam, it is possible to carry on nanomachining precisely to produce tiny parts or remove unwanted materials, making FIB a versatile tool for micro/nano processing. Using FIB to cut sections at specific locations of the electrode material, the bulk and interface structure can be directly observed. Golozar et al.[20] used in-situ SEM and FIB to probe the growth of Li metal in a solid battery with $LiFePO_4$ cathode and solid PEO-based polymer electrolyte. Due to the edge effect and absence of pressure, Li dendrites formed at the anode edge, which were then lifted out by nanomanipulator and milled using a Ga ion source in FIB (Figure 2.8a). The dendrite exhibits a hollow nanostructure with the wall thickness of around 100 nm (Figure 2.8b), which mainly consists of the residual SEI layer because of the dissolution of inner Li. Porous electrolyte structure was also present in the region far from the anode, indicating that dendrites penetrate into the polymer electrolyte layer and cause electrolyte cracking (Figure 2.8c). Furthermore, Golozar et al.[21] found the porosity and the depletion of lithium in the vicinity of the isle. The cross-section analysis combined with EDS (Figure 2.8d) shows the high-concentration distribution of C, F, and N around the isle, and S away from the isle due to the gradual decomposition of LiTFSI forming Li_3N, Li_2S, LiF, LiC_xF_y, Li_xCNF_3, and Li_ySO_x in succession. Among them, LiF and other components have low ionic conductivity, which may hinder the dissolution of Li in the isle and form dead Li, resulting in irreversible capacity attenuation.

2.3.2 FIB for 3D Morphology

Application of FIB technology has greatly promoted the development of 3D reconstruction analysis. FIB layer-by-layer slicing and SEM imaging can be alternately controlled by software, thus obtaining 3D reconstruction. Combining FIB 3D reconstruction technology with EDS enables researchers to characterize the internal morphology and element distribution of materials in three dimensions. FIB and electron back-scattered diffraction (EBSD) can also be combined to characterize the orientation, morphology, size, and distribution of grains in polycrystalline materials.

Liu et al.[22] used FIB-SEM to achieve the 3D images of a $LiCoO_2$ electrode with clear contrast among the three phases: $LiCoO_2$ particles, carbonaceous phases (carbon and binder), and the electrolyte. To improve the contrast, a low-viscosity silicone resin was applied as a filling material enabling full infiltration of low-porosity electrodes and provides good contrast among phases, the latter of which allows for automatic segmentation. A typical 2D cross-section from the 3D data set is shown in Figure 2.9c, where obvious signals from Co, C, and Si elements are detected by EDS, representing the distribution of the $LiCoO_2$ (blue), CB (green), and silicone resin (red), respectively. From the 3D reconstruction, microstructural parameters such as phase volume fraction, surface area density, and feature size distribution can be extracted, and the electrolyte connectivity, tortuosity, and tortuosity distribution (tortuosity along the planar section within the electrode) can be determined. It is apparent that the active $LiCoO_2$ material occupies most of the reconstructed volume (77 v%), while the carbon phases (10.3 v%) and electrolyte (12.7 v%) have a relatively low volume fraction.

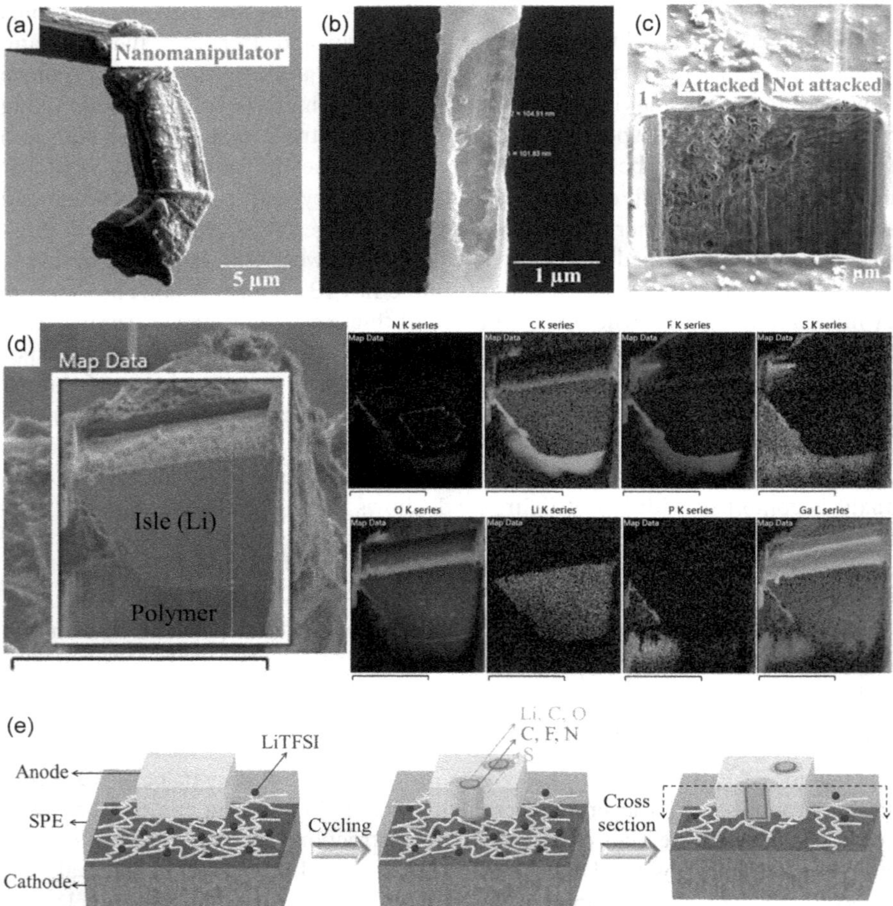

FIGURE 2.8 (a and b) SEM images showing dendrite on the edge of the anode and thickness of the wall of a dendrite. (c) High-magnification SEM image of FIB-milled region on the SPE far from anode after battery cycling showing two areas of attacked and not attacked. (d) SEM image showing the cross-section of one isle, and EDS mapping of this FIB-milled isle showing a high concentration of N, C, and F surrounding the isle. (e) Schematic of the battery during cycling showing the appearance of the isles and the FIB-milled cross-section.

2.3.3 FIB for Preparing Thin Samples for TEM

Thin specimen is requested for TEM observation, which is usually less than 100 nm. In the past, samples can be prepared by manually grinding, electropolishing, mechanical polishing, ion sputtering, and other means, but these processes are always cumbersome, and difficult to accurately select the specific micro/nano region of the interest. In this regard, FIB slicing technology has all kinds of convenience that the above methods do not have. This technique can be used to directly obtain nanoscale lamella from the sample for TEM with minimal damage to the sample, and it is the

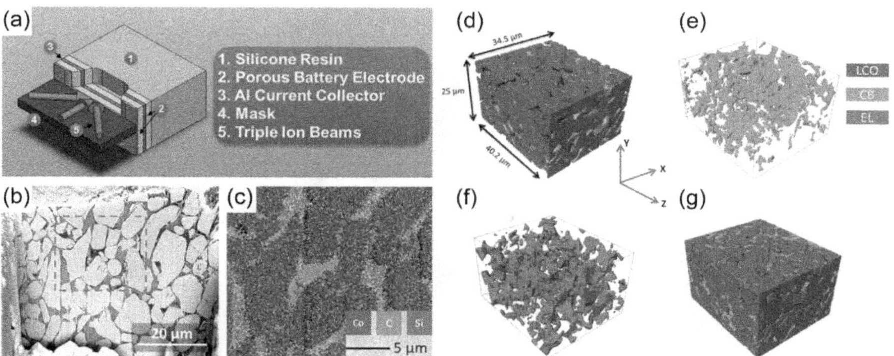

FIGURE 2.9 (a) Schematic view of the silicone resin-infiltrated $LiCoO_2$ electrode surface milled by a triple ion-beam cutter. (b) Cross-sectional SEM image of the fresh $LiCoO_2$ cathode sample after FIB milling. For the region corresponding to the red-dashed rectangle region in (b), (c) shows EDS chemical mapping. Three-dimensional reconstruction renderings of the electrode showing (d) $LiCoO_2$, (e) carbon and binder, (f) electrolyte, and (g) all three phases superimposed.

first choice for samples with solid-solid interfaces buried beneath in other solid materials.[23] Typical procedures include (i) deposit protective layer on the surface of samples, (ii) cross-section milling, (iii) J-cutting, (iv) lifting out, and (v) mounting on grid (Figure 2.10). Pt is usually deposited as a protective layer and as a binder to bridge the sample and substance at room temperature (25°C), which, however, cannot be realized at liquid nitrogen temperature during cryogenic FIB (cryo-FIB). Alternatively, a smart way is to use the redeposition of sputtered material instead of Pt deposition as a connection. As shown in Figure 2.10b, at −180°C, several rectangular milling patterns are drawn at the junction of tungsten probe and lamella top surface. A 10 pA ion beam current is then used to mill through the patterned region, where the superfluous splattered substance will redeposit at the surrounding region and connect lamella with the tungsten probe. The lamella can then be lifted out by the tungsten probe without any Pt deposition. As the lamella is in contact with the Cu grid post, the same method is applied again where several rectangular milling patterns are milled through to let reposition connect the lamella with the Cu grid post. By using this destructive method, Cheng et al. obtained the lamella lift-out fully under cryogenic conditions, successfully maintained the morphology of Li metal, and preserved the Li/lithium phosphorous oxynitride (LiPON) interphase for TEM characterizations.[24]

2.3.4 Cryogenic FIB for Beam-Sensitive Samples

As we know, ion milling in FIB uses strong high-energy ion beams to collide with the sample and transfer kinetic energy, ejecting atoms from the sample, which is destructive and can cause severe damage to the samples, especially for the fragile materials containing organic species and light elements. Due to the low melting temperature, low density, poor thermal conductivity, and low shear modulus of Li metal, it is very sensitive to Ga ion implantation. Local evaporation of lithium metal is very likely to

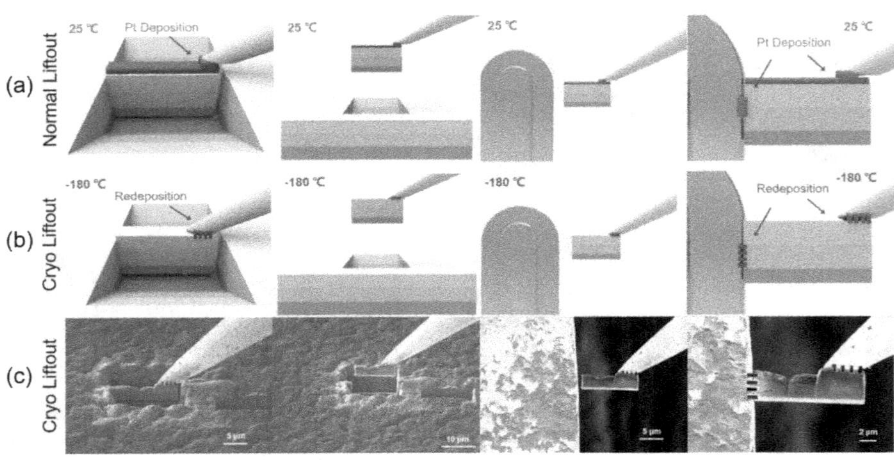

FIGURE 2.10 Schematic procedures for FIB to prepare thin lamella for TEM imaging.[24] During TEM sample preparation in FIB, the lamella needs to be connected with the tungsten probe for lift-out process. (a) Conventional method is depositing Pt as the connection at room temperature (25°C). (b and c) At −180°C, a 10 pA ion beam current is used to mill through the patterned region, where the sputtered materials will redeposit at the surrounding region and connect lamella with the tungsten probe.

occur, resulting in the redeposition of high-melting-point components in the SEI and other parts. When 5 nA ion beam current is used for the cross-section milling and for cleaning at room temperature, a dense commercial lithium foil will be subjected to be loose and porous, and is implanted with a large number of impurities containing Ga ions and O ions (Figure 2.11).[25] If milled at room temperature and cleaned at cryogenic temperature, part of the damaged layer could be removed, but Li metal is still destroyed; only when both the milling and cleaning processes are performed at −170°C, the as-prepared Li foil can keep its original dense structure. Therefore, cryo-FIB can be employed to get the cross-section of beam-sensitive materials for SEM imaging or a thin lamella for TEM observing.

Lee et al.[25] used cryo-FIB method as a quantitative tool to analyze bulk structures of electrochemical deposition of lithium, including stacking density of Li metal, size, and distribution of voids. They performed progressive continuous milling on a large range of current collector planes and simultaneously collected a series of high-resolution SEM cross-section images, which were reconstructed using Amira-Avizo software (Figure 2.12a–f). They compared the morphology of Li deposits in 1.0 M $LiPF_6$ EC-EMC, 4.6 M LiFSI-DME (SSEE), and 4.6 M LiFSI + 2.3 M LiTFSI DME (BSEE) electrolyte. It turns out that the Li metal in $LiPF_6$ electrolyte grows in dendrites and eventually forms a large network of interconnected voids. In the SSEE and BSEE electrolyte, the Li deposition of less than 4 μm thickness is highly dense with a smooth surface and less Li dendrites, and thus has higher Cu/Li coverage rate (more than 85%) and less excess Li/electrolyte surface area according to the statistical analysis (Figure 2.12g), but some voids are still present in the Li layer and at the Li/Cu interface. This work provides a clear picture of 3D microstructure of Li deposits.

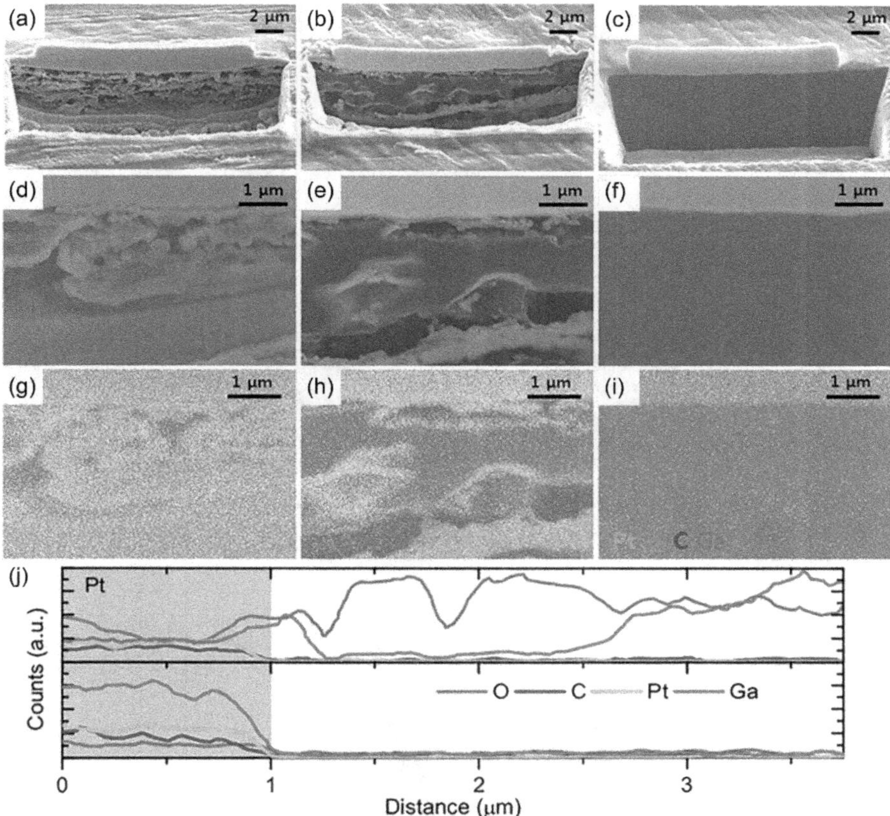

FIGURE 2.11 SEM images and EDS elemental mapping of cross-sections of commercial Li metal foil. (a, d, g) Cross-sectioned and cleaned at room temperature, (b, e, h) cross-sectioned at room temperature and cleaned at cryogenic temperature, and (c, f, i) cross-sectioned and cleaned at cryogenic temperature. (j) Quantitative elemental line scans through room-temperature (top) and cryogenic-temperature (bottom) cross-sections.

Cryo-FIB enables to transfer liquid–solid interface into solid–solid interface, thus making it possible to characterize the liquid–solid interface by EM. Zachman et al.[26] investigated the growth of Li dendrites in liquid electrolyte by cryo-STEM technology. Plunge-frozen was used to fast solidify the liquid electrolyte; thus, the liquid–solid interface between the glassy liquid electrolyte and Li metal remains in its original state at cryogenic temperature. Each successive cross-section was imaged (Figure 2.13c and d), revealing two distinct deposit morphologies, which are referred to as type I and type II dendrites. Type I dendrites are roughly 5 µm across with low curvature, whereas type II dendrites are generally hundreds of nanometers thick and tortuous (Figure 2.13e). EDS results demonstrate that a thick SEI layer of 300–500 nm forms outside the Li metal (Figure 2.13f and g), suggesting that more lithium irreversibly migrates to the SEI layer than we used to think. However, if the sample is prepared by conventional means for TEM observation, the soft extension

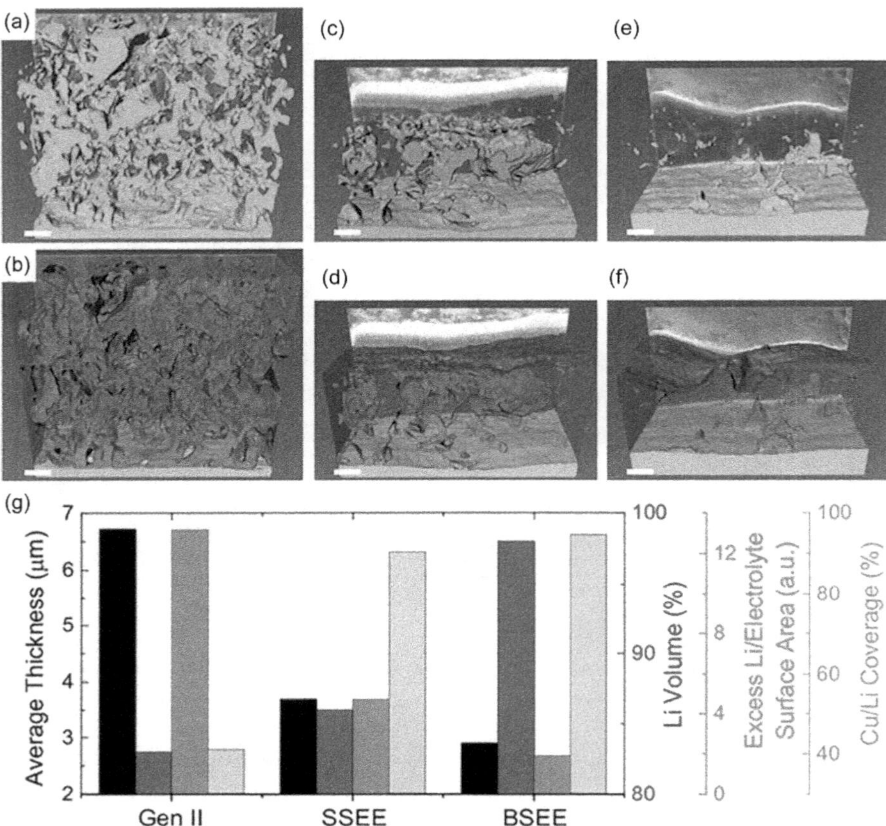

FIGURE 2.12 3D reconstruction of voids (blue) and bulk Li metal (burgundy) with 1 μm scale bar of first cycle electrochemically deposited Li in (a and b) 1.0 M $LiPF_6$ EC: EMC, (c and d) 4.6 M LiFSI-DME (SSEE), and (e and f) 4.6 M LiFSI + 2.3 M LiTFSI in DME (BSEE) along with (g) statistical analysis.

of the SEI will be removed during washing and drying processes, and the remaining SEI will be observed as a tight and dense thin layer of tens of nanometers thick.

Additionally, Jaiser et al.[27] observed the drying and film-forming process after electrode slurry coating through ex situ cryo-broad ion beam (BIB)-SEM technology. In the drying process after the preparation of electrode film, cryo-preservation is applied for wet film, and the film is cryo-transferred to BIB preparation of cross-section (Figure 2.14). The uniform shrinkage behavior of the electrode film in the thickness direction can be observed by cryo-SEM. Meanwhile, the liquid to solid mass ratio X_{NMP} decreases with the NMP evaporating. The distribution of graphite particles, liquid phase, and voids is visible, where the graphite particles are always uniformly distributed on the film during the whole process. In contrast, the binder is brought to the surface during solvent diffusing from the bulk to the surface and evaporating, resulting in a higher content of binder on the surface than in the bulk, exhibiting a gradient distribution of binder in the dry electrode film. All these works

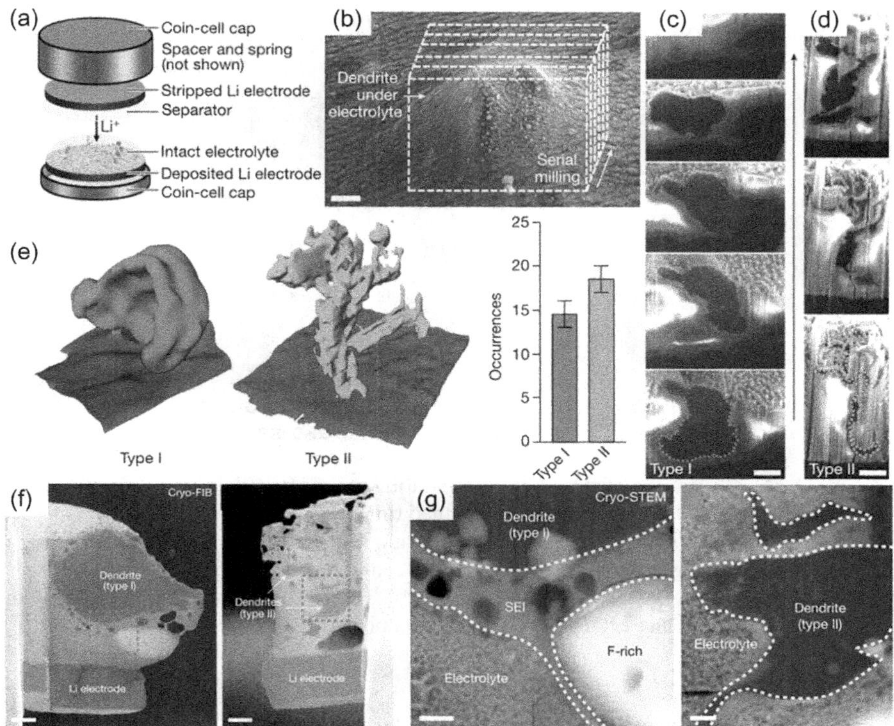

FIGURE 2.13 Characterization of dendrite morphologies by cryo-FIB and cryo-TEM. (a) Coin-cell arrangement used. (b) Raised regions in the electrolyte frozen on opened coin-cell electrodes reveal buried dendrite locations. (c and d) Two distinct dendrite morphologies, referred to as type I (c) and type II (d). (e) 3D reconstructions of the dendrite structures highlight the morphological differences. Roughly equal numbers of the two morphologies were present across many coin cells. (f) Electron-transparent cryo-FIB lift-out lamellae of type I and type II dendrites. (g) HAADF cryo-STEM imaging reveals an extended SEI layer on the type I dendrite, but not on the type II dendrite.

demonstrate the advantages of cryo-FIB in studying the internal cross-sections of beam-sensitive materials.

2.4 TRANSMISSION ELECTRON MICROSCOPY (TEM)

2.4.1 Introduction of High-Resolution TEM, STEM, Diffraction, and EELS

Compared with SEM, TEM has higher spatial resolution due to its higher energy and shorter wavelength. Thanks to small diffraction angle of electrons, observation of diffraction patterns in micro-areas is feasible, which is useful for studying the crystal structure of microcrystals, surface, and films. Ordinary TEM forms bright and dark field images based on the mass thickness contrast and diffraction contrast,

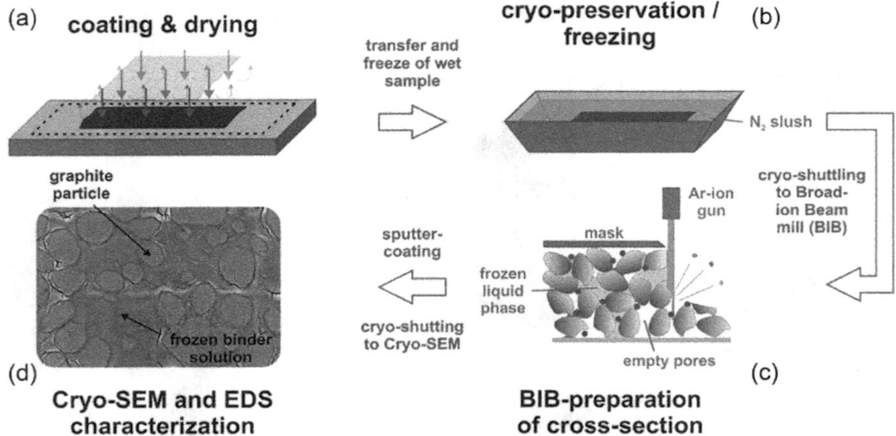

FIGURE 2.14 Illustration of the sample preparation procedure. (a) Initially, wet electrode films are dried convectively; (b) after a predefined time interval, the coated foil is cryo-preserved through submersion into a slushy nitrogen bath. From this point on, sample processing and transfer is exclusively conducted under cryo-conditions; (c) a smooth cross-section is prepared through broad ion beam (BIB) slope-cutting. (d) Subsequent to sputter-coating with tungsten, the sample is characterized by cryo-SEM and cryo-EDS.

respectively.[28] The mass thickness contrast is from the difference of scattering and absorption of incident electrons caused by the variance of density and thickness of various parts of the materials; the diffraction contrast is caused by the difference in the degree to which each part of the crystal plane meets the Bragg diffraction condition and the different amplitude of the structure, which can only be generated in the crystalline structure. Both belong to amplitude contrast.

2.4.1.1 High-Resolution TEM

Besides bright and dark field images, HRTEM can also form phase contrast image, which is the interference image based on the phase difference between the transmission beam and multiple diffraction beam involved in imaging. The more diffraction beams are involved in imaging, the richer and finer the information of sample structure will be. Lattice fringes reflecting the periodicity of crystal lattice are formed. The resolution of HRTEM can reach the atomic level of angstrom scale, enabling to obtain the 1D lattice fringes, 2D lattice images, and atomic structure images. At the same time, HRTEM can acquire electron diffraction pattern from different orientations of crystal and combine it with phase contrast images. Above all, HRTEM is widely used to observe the fine structure, atomic arrangement, twins, dislocation, etc. in the materials.[29] Under the weak phase object approximation of a sufficiently thin sample, the crystalline structure can be accurately reflected by adjusting to the Scherzer underfocus condition for imaging.

2.4.1.2 Scanning TEM

STEM scans point by point on the thin specimens with a small-sized, focused electron probe, and is often used in conjunction with the high-angle annular dark-field detector (HAADF) and annular bright-field detector (ABF) to form atomic-resolution images. HAADF receives inelastically scattered electrons or thermal diffuse scattering at high angles using an annular dark-field detector, and its image contrast is approximately proportional to the square of the atomic number Z, which provides an approximate method for identifying atomic species. With a spherical aberration corrector, sub-angstrom resolution can be achieved, enabling HAADF to accurately identify the position of atoms and atomic columns. To effectively visualize the light atoms, such as Li, ABF is utilized as a complementary method which preferentially receives the ring-shaped circumference (e.g., 12–24 mrad) of the direct (transmitted)-beam disk using an ABF detector. Therefore, a combination of HAADF and ABF-STEM provides a direct and robust method to visualize the light and heavy elements simultaneously at atomic resolution.

2.4.1.3 Electron Energy Loss Spectroscopy

EELS is a technique to analyze the energy loss of inelastic scattered electrons produced from excitation and ionization caused by incident electrons on the surface of materials. By analyzing the position of energy loss, the composition information of elements can be obtained. EELS directly analyzes the results of inelastic scattering interaction between incident electrons and the sample rather than the secondary process, so it has high detection efficiency and can analyze the element composition, chemical bonds, and electronic structure of the thin sample micro-area. Compared to EDS method, EELS has a better resolution of light elements. The application of EELS in TEM equipment allows a spatial resolution of 10^{-10}m scale and can be used to measure film thickness.

2.4.2 Atomic Structure: Bulk, Surface Construction, Coating, Doping, and Phase Transitions

2.4.2.1 Bulk Structure

Layered (Figure 2.15a) and spinel (Figure 2.15b) structure can be discerned from STEM images even with the same composition $LiCoO_2$, where layers consisting of Co atoms are clearly visible in the former while Co atoms occupy the both the 16d octahedral sites and 8a tetrahedral sites in the latter.[30,31] The varied atom arrangement will lead to discrepancy in Li^+ diffusion pathways and energy barrier. Defects such as grain boundaries (Figure 2.15c),[32] dislocations, and stacking faults are always observed in the structure due to its imperfect nature, which provides more variations in structural stability and Li^+ migration. To directly visualize the Li ions, Gu et al. applied ABF-STEM technique and observed that the remaining Li ions in partially delithiated $LiFePO_4$ preferably occupy every second layer along the *b* axis (Figure 2.15d–f).[33,34] This kind of staging phenomenon was also found in some layered intercalation compounds, such as Na^+ intercalating into MoS_2.[35] Besides

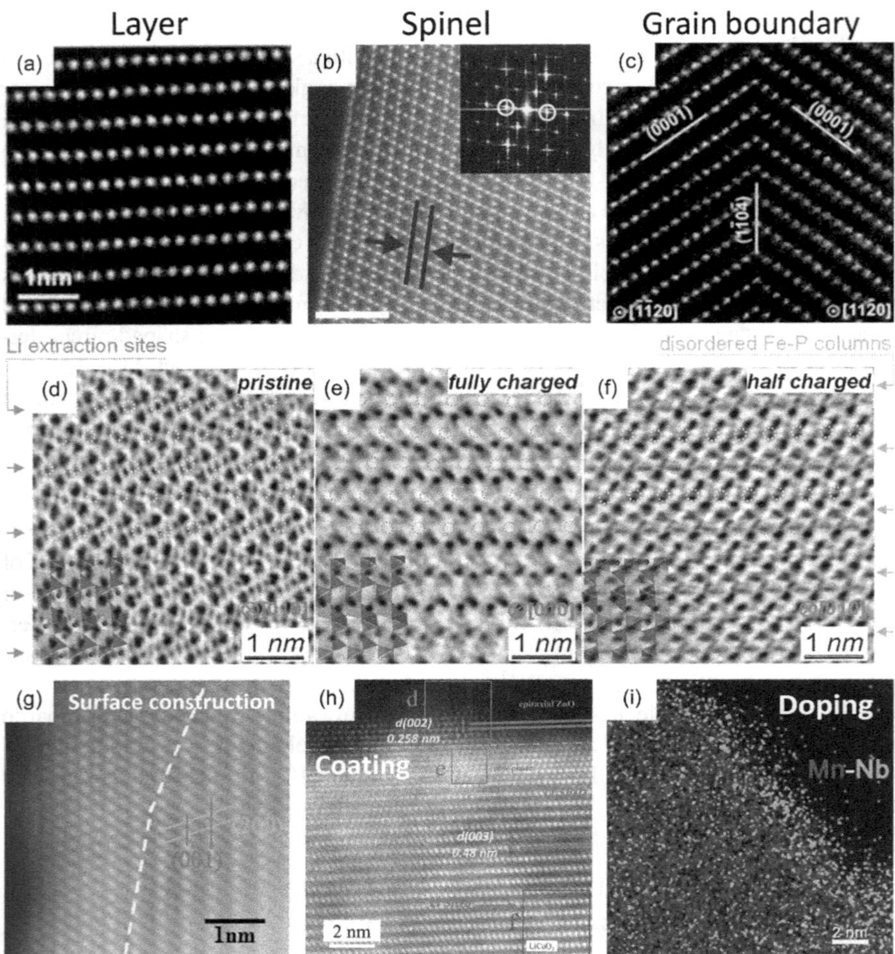

FIGURE 2.15 (a) HAADF-STEM micrograph of layered-type $LiCoO_2$. (b) Aberration-corrected HAADF-STEM image of spinel-type $LiCoO_2$ viewed down the [01-1] zone axis. (c) HAADF-STEM micrograph of a twin boundary in polycrystalline $LiCoO_2$. (d–f) ABF micrographs showing Li ions of partially delithiated $LiFePO_4$ at every other row. (d) Pristine material with the atomic structure of $LiFePO_4$ shown as inset. (e) Fully charged state with the atomic structure of $FePO_4$ shown for comparison. (f) Half-charged state showing the Li staging. (g) High-resolution HAADF-STEM image showing the presence of electrochemical-cycling-induced surface reconstructed layer on the $Li[Li_{0.2}Ni_{0.2}Mn_{0.6}]O_2$ particle surface. (h) HAADF-STEM image of $LiCoO_2$@ AlZnO particle. (i) The EDS mapping of Mn and Nb for the corresponding HAADF image of Li, Mn-rich layered oxide doped by Nb.

intercalation reaction, conversion reaction leads to the atoms rearrangement forming new phases and clusters such as FeF_2 and NiO.[36–38] These reaction mechanisms can be vividly revealed and pictured by the STEM and HRTEM.

2.4.2.2 Surface Construction

Surface reconstruction is prevailing in Li-rich oxides, where layered structure is transitioned into spinel or rock-salt structure (Figure 2.15g).[39] This is caused by the migration of transition metals (TMs) from the original TM layers to the Li ion layers: EDS results demonstrate that part of Ni is concentrated on the surface.[40] The migration is accelerated during charging process, when Li ions are successively extracted and leave more vacancies in the structure. As a consequence, the thickness of surface reconstruction region is increasing. EELS spectra show that the oxygen prepeak is gradually faded, which is associated with the transition of electrons from the 1*s* core state to unoccupied 2*p* states hybridized with 3*d* states in TM and suggests the formation of oxygen vacancy and the reduction of TMs.[41] Severe surface reconstruction will block the Li^+ diffusion and degrade the cycling performance.[42] It is worth noting that the electron beams for STEM imaging can also induce the formation of the oxygen vacancy and surface reconstruction, which will lead to artifacts or overestimated results.

2.4.2.3 Coating Methods

Coating a battery material with a thin layer has been widely used to modify the surface properties of battery materials and shown effective to improve the electrochemical performance of batteries. It is important to ensure a uniform coating, which requires TEM to check. Guo et al.[43] fabricated the SnO_2 and carbon-coated SnO_2 (SnO_2/C) hollow microspheres through hydrothermal reactions as high-capacity anode materials for LIBs. In the HRTEM images, the lattice fringes of the uncoated SnO_2 are extended to the edge of the particles, while there is an amorphous carbon layer on the surface of the SnO_2 particle. Carbon coating not only enhances the conductivity of SnO_2 but also facilitates reaction kinetics, forming SnO rather than Sn when it is fully charged. Therefore, the SnO_2/C particles have an improved cycling reversibility with a higher coulombic efficiency and specific capacity. To enhance the stability of $LiCoO_2$ at a high voltage of 4.6V, Cheng et al. coated $LiCoO_2$ particles with an AlZnO layer.[44] STEM-EDS demonstrates that the coating is complete and induces formation of disordered rock-salt phase on the surface (Figure 2.15h). This modified surface not only restrains the interfacial side reactions with electrolytes but also enhances the structural stability and lattice oxygen, achieving stable cycling of $LiCoO_2$ at 4.6V.

2.4.2.4 Doping Methods

Doping is another important means to tune the bulk/surface properties and electrochemical performance of battery materials. Due to the trace content of doped elements, HRTEM is essential to confirm its presence and occupancy in the structure. Liu et al. applied surface doping of niobium (Nb) in a Li-rich Mn-based layered oxide $Li_{1.2}Mn_{0.54}Ni_{0.13}Co_{0.13}O_2$.[45] STEM images show that the doped ions are in the Li layer near the oxide surface (Figure 2.15i); they bind the slabs via the strong Nb–O bonds and "inactivate" the surface oxygen, enhancing the structural stability of Li-rich oxide. The specific capacity of the modified oxide reaches 320 mAh/g in the initial cycle, 94.5% of which remains after 100 cycles. Zhang et al. achieved stable cycling of $LiCoO_2$ at 4.6 V (versus Li/Li^+) through trace Ti-Mg-Al co-doping.[46] Using STEM and other spectroscopic techniques, the incorporation of Mg and Al into the $LiCoO_2$

lattice is confirmed, which inhibits the undesired phase transition at voltages above 4.5 V. Even in trace amounts, Ti segregates significantly at grain boundaries and on the surface, modifying the microstructure of the particles while stabilizing the surface oxygen at high voltages.

2.4.3 In-Situ/Operando TEM: Biasing, Mechanical, and Heating

Although researchers have been able to combine multiple characterization methods to analyze the complex microstructure and electrochemical behavior of electrode materials at different scales, it is difficult to analyze the dynamic changes occurring inside the battery during operation without the appearance of in-situ characterization technology. The information of structural transformation and electrochemical reaction process of the electrode materials in the battery cycle can be predicted by comparing the data of the electrode of the initial state and after several cycles. However, during the sample transferring, some electrode materials that are sensitive to air may be damaged or contaminated, leading to artifacts and misleading results. In this regard, in-situ TEM[47] can directly observe the microstructure evolution of cathode, anode,[48,49] and their interfaces at the atomic level under the fields of force, heat, and bias.

Since higher vacuum and tinier space are required for TEM when compared with SEM, it is more challenging to build a suitable setup for in-situ TEM, developments of which are shown in Figure 2.16.[50] The early in-situ TEM technique for battery adopted an open-cell nanobattery, which used nanowire or nanoparticle electrode materials in contact with involatile solid electrolyte or ionic liquid electrolyte (Figure 2.16a).

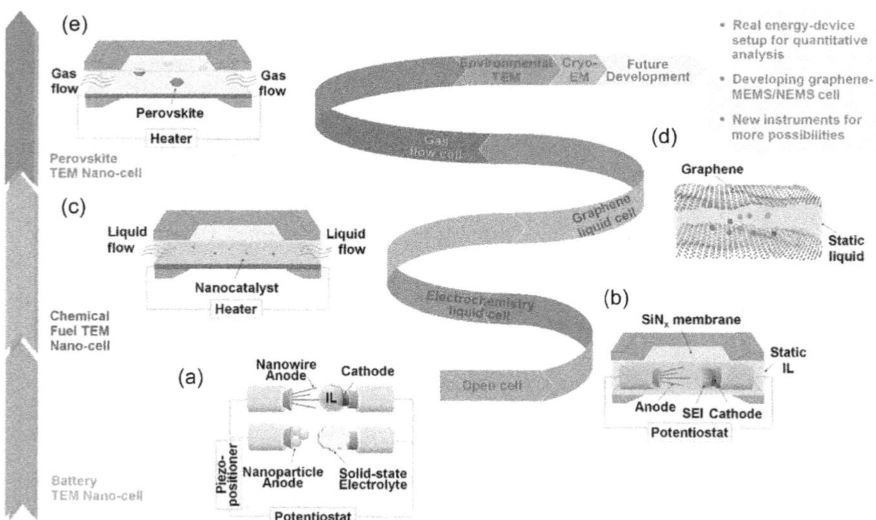

FIGURE 2.16 Development path of in-situ TEM nanocells and their applications for batteries. (a) Open-cell setup for LIB investigation. (b) Electrochemical liquid-cell setup for LIB investigation. (c) Electrochemical liquid-cell setup for fuel cell investigation. (d) Graphene liquid cell. (e) Gas flow cell for PSC investigation.

The native oxide (Li_2O) on the surface of Li metal is preferred to use as the electrolytes as well as LiPON and lithium lanthanum zirconium oxide (LLZO) in all-solid-state batteries (ASSBs).[51,52] Since the battery performance is highly dependent on the electrolyte chemistry, a closed cell with sealed windows was invented to encapsulate the conventional volatile organic electrolyte (Figure 2.16b).[53] The real-time electrochemical process can be monitored through the window (usually amorphous silicon nitride (SiN_x) layer) with an expense of reducing imaging resolution. To minimize the negative effect of the window, a graphene-bubble-based liquid nano-cell system was developed in the latest generation of in-situ TEM (Figure 2.16d), which can significantly improve the spatial resolution of the image even under the electron beam with low acceleration voltage.[54] With the constant innovation and development of in-situ TEM technology, we can probe the evolution of both the solid–liquid interfaces and the gas–solid/liquid interfaces.[17,55] Additionally, the in-situ technology has already been utilized to study the electrochemical reaction not merely in LIBs, but in potassium-ion (KIBs) and sodium-ion batteries (NIBs) as well.[56–60] In these new types of batteries, for instance, it could be used to visually monitor the volume change of sulfurized polyacrylonitrile (a low-cost anode candidate for KIBs)[56] or microporous carbon nanofiber/small-molecule sulfur composite (PCNF/S, a cathode material for K-S batteries)[57] during potassiation and depotassiation, and could also be employed to observe the first sodiation process of tin nanoparticles electrode in a NIB,[58] which evolves from crystalline Sn to amorphous Na_xSn ($x \sim 0.5$), finally to $Na_{15}Sn_4$ alloy. Through the in-situ TEM, the underlying linkage of electro-chemo-mechanics can be revealed, providing a battery working principle at the nano/atomic scale.

By applying bias voltage on the electrochemical nanocell, in-situ TEM can be used to explore the structural evolution of anode and cathode materials during charging and discharging. Huang et al.[61] assembled a nano-device inside the TEM with a bulk $LiCoO_2$ cathode, an ionic liquid electrolyte, and a single SnO_2 nanowire anode. Combining TEM images (Figure 2.17a) with SAED patterns (Figure 2.17b–e), they found out that lithiation of crystalline SnO_2 nanowire first forms amorphous Li_2O and Sn, and then transitions to crystalline tetragonal Sn and hexagonal $Li_{13}Sn_5$ (Figure 2.17e), nanograins of which are dispersed in amorphous substance (Figure 2.17f). An interface is present between the reacted and unreacted region, and rich in dislocation generated by local stress (Figure 2.17a). Similar behavior was observed with a Si nanowire but using ionic liquid as electrolyte (Figure 2.17g).[62] Crystalline Si undergoes amorphization after lithiation. Based on the real-time image record, the reaction rate is determinate to be 1.7 nm/s. In the HRTEM images, nanowire exhibits a core–shell structure with a distinct amorphous/crystalline interface (ACI) where the shell has been lithiated (Figure 2.17h).[63] Monitoring the dynamic lithiation process of monocrystalline Si at atomic scale reveals a ledge-mediated solid-state amorphization mechanism that Li^+ ions squeeze into the large open space between the two neighboring (111) bilayers of Si, resulting in the peeling off of the top layer and dissolution of Si atoms into the Li_xSi phase.

He et al.[64] applied in-situ STEM and electron diffraction technology to probe the morphology and structural change of magnetite (Fe_3O_4) particles during lithiation (Figure 2.18). Intercalation reaction occurs at first when the spinel-type Fe_3O_4 gradually becomes the Li-inserted rock-salt phase $Li_xFe_3O_4$, which shows a shallower

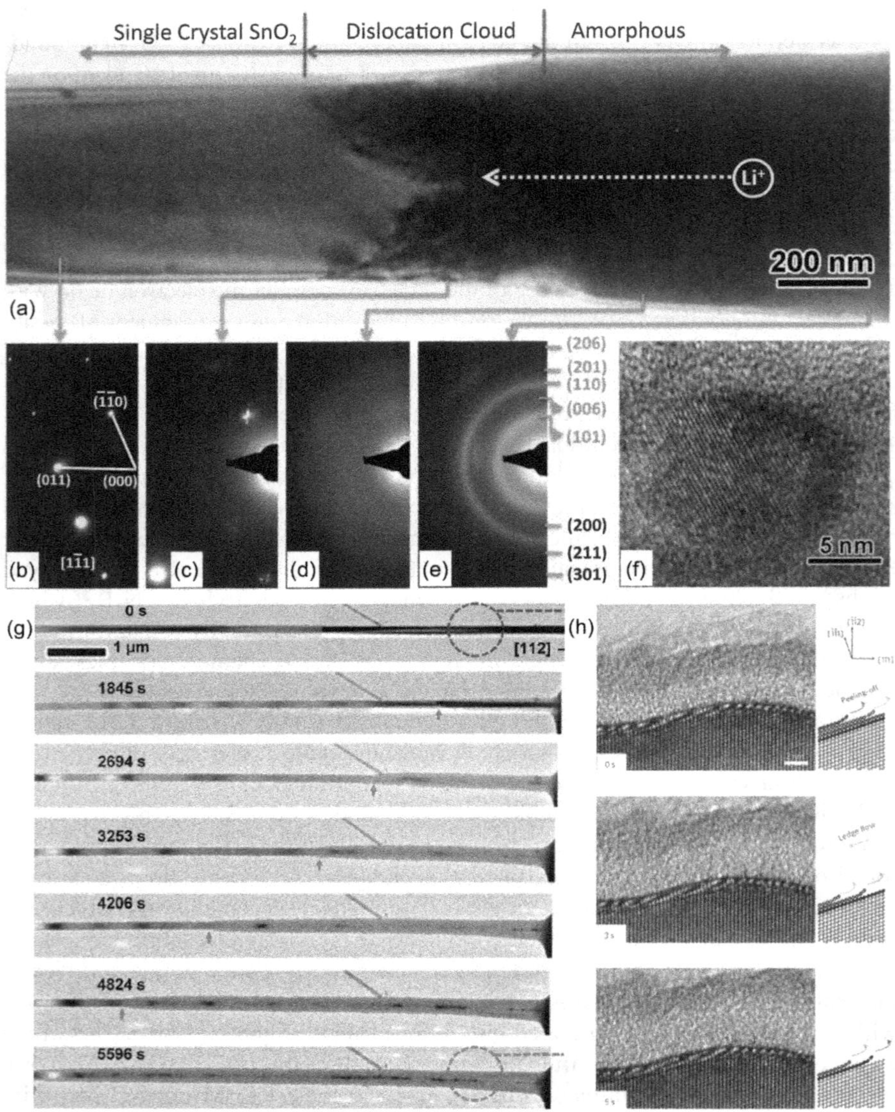

FIGURE 2.17 (a) TEM micrograph of the SnO_2 nanowire anode during charging at –3.5V against the $LiCoO_2$ cathode. (b–e) Electron diffraction patterns from the different sections of the nanowire. (g) Time-lapse TEM image series showing microstructural evolution of an intrinsic silicon nanowire during lithiation (pristine straight Si nanowire with a uniform diameter of 140 nm). The potential applied was –4V against the $LiCoO_2$ cathode. (h) HRTEM image sequences showing the lithiation process by lateral ledge flow (marked by colored arrows) in the amorphous/crystalline interface and simultaneous peeling-off of the {111} planes over 5 s.

contrast than that of Fe_3O_4 (Figure 2.18a). In the subsequent reaction, $Li_xFe_3O_4$ converts into Fe nanoparticles and Li_2O as demonstrated by real-time electron diffraction (Figure 2.18b). It is noted that conversion reaction starts before the intercalation reaction finishes probably due to the slow diffusion rate of Li^+ in $Li_xFe_3O_4$ leading to high concentration of Li^+ on the particle surface. According to the different contrast of Fe_3O_4 (red), $Li_xFe_3O_4$ (blue), Li_2O and Fe phases (green), their distribution is determined (Figure 2.18a). Both intercalation and conversion reactions follow a "shrinking-core" mode that reaction starts from the outer surface then diffuses into the inner region, and conversion tends to start through the outer facets.

Through using in-situ low-voltage TEM with spherical and chromatic aberration correction, Kuhne et al.[65] detected the Li intercalation into bilayer graphene. The device inside the TEM consists of a bilayer graphene supported by a Si_3N_4-covered Si substrate and a solid polymer electrolyte connecting the graphene to a metallic counter electrode on the right, thus forming a nanocell (Figure 2.19a). There is a hole in the center of Si_3N_4 membrane, allowing electrons pass through and visualize the bilayer graphene. In real-time TEM images (Figure 2.19b), a different crystalline lattice (Li lattice) appeared and extended in the area of original graphene crystal lattice during lithiation, which was also highlighted by the von Hann-filtered Fourier transform of the TEM image. After removing the graphene lattice and the moiré effects, the new phase is clearly shown (Figure 2.19c), which is found to be hexagonal Li metal rather than conventional cubic Li (Figure 2.19d). The presence of this superdense phase will lead to larger extra lithium storage capacity, far exceeding that expected from formation of regular LiC_6.

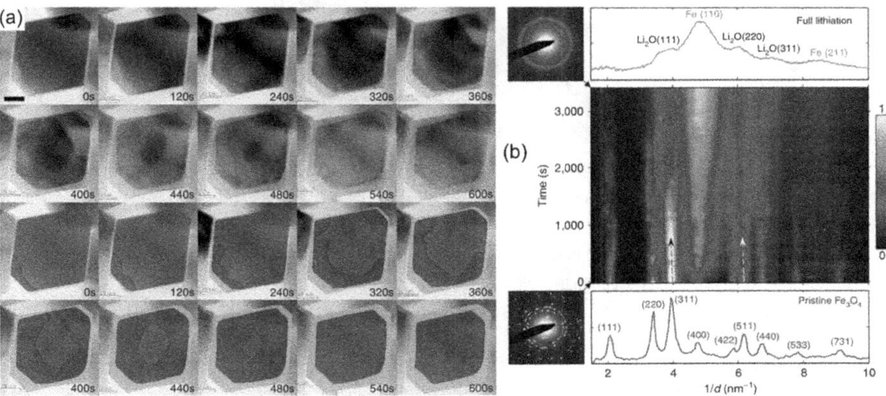

FIGURE 2.18 (a) In-situ BF-STEM image series showing phase evolution during lithiation. The overlaid false colors indicate different phases: pristine Fe_3O_4 (red), Li-inserted $Li_xFe_3O_4$ (blue), and Fe+Li_2O composite after conversion (green). (b) Phase evolution probed by in-situ electron diffraction. Electron diffraction intensity profile (the color map in the middle) as a function of reaction time during in-situ lithiation of Fe_3O_4 nanoparticles. The SAED patterns and corresponding radially integrated intensity profiles obtained at pristine (0 s) and fully lithiated (3,400 s) states are shown below and above the color map, respectively.

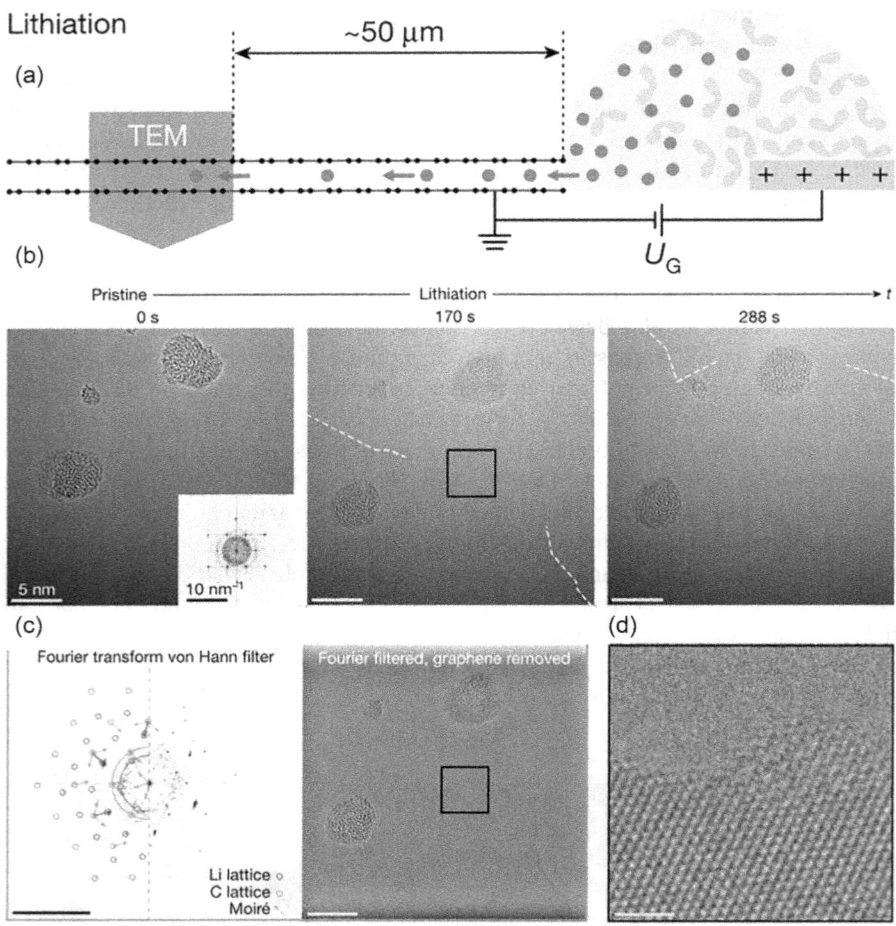

FIGURE 2.19 (a) Schematic side view of the device during in-situ TEM: the device during lithiation (U_G=5V at the counter electrode). About 50 μm away from its electrolyte-covered end, bilayer graphene is partially suspended over a hole in the Si_3N_4, supported by the Si substrate, allowing TEM investigations. (b) TEM images showing the propagating front (dashed white line) of a Li crystal forming inside bilayer graphene during lithiation. The images are acquired on the same sample area at consecutive times. (c) von Hann-filtered Fourier transform and Fourier-filtered version of the image in the middle of (b). Signals from bilayer graphene (Li) are highlighted in cyan (green); the origins of moiré artifacts are highlighted in bold magenta. (d) Magnified detail from the boxed areas in (c).

Besides applying bias, atomic force microscopy (AFM) can also be integrated into in-situ TEM to study the mechanical property of electrode materials. Zhang et al.[66] used in-situ environmental TEM to observe the growth process of an individual Li whisker. A CNT was used as the current collector and partial CO_2 was filled in to form Li_2CO_3 on the surface of Li metal as a solid electrolyte (Figure 2.20a). By applying an overpotential to the AFM cantilever (Figure 2.20b), Li metal first grows spherically,

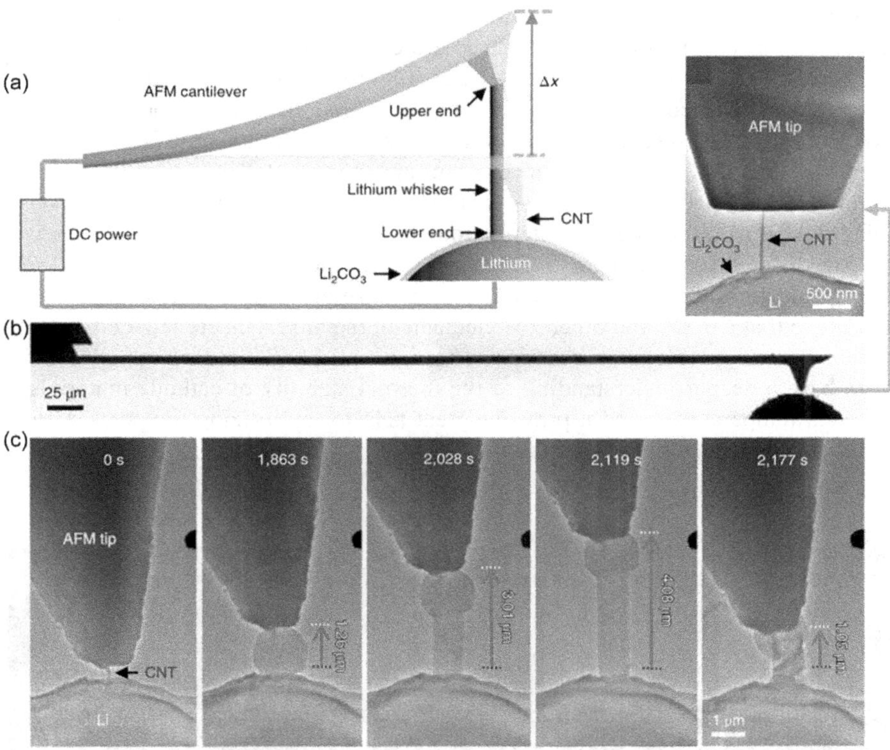

FIGURE 2.20 In-situ AFM-ETEM characterization of stress generation during Li whisker growth. (a) Schematic of the AFM-ETEM setup used for observation and measurement of Li whisker growth. (b) TEM images showing an AFM cantilever approaching the counter electrode of Li metal and a CNT attached to a flattened AFM tip. (c) Time-lapse TEM images of Li whisker growth.

and then adopts the root growth mode, presenting a mushroom shape (Figure 2.20c). With the increase of dendrite length, the tip of AFM cantilever is pushed up, thus the mechanical strength of Li metal can be measured (Figure 2.20c). Therefore, this technology enables to not only analyze the growth rate of Li dendrites at different periods, but also compare the stress intensity of Li whiskers with different crystal plane orientations, diameters, and deposition potentials. Surprisingly, the yield strength of Li dendrites reaches as high as 244 MPa and its growth stress up to 130 MPa. This study provides a quantitative standard for the selection and design of SEI with proper mechanical strength, which is conducive to inhibit the Li dendrite growth.

Additionally, in-situ heating TEM can be used to investigate the thermal stability of electrode materials. Hwang et al.[67] used in-situ TEM to monitor the thermal decomposition and phase transformation of $LiNi_{0.8}Mn_{0.1}Co_{0.1}O_2$ (NMC811) cathode material under heating. By means of electron diffraction and bright-field imaging to characterize the crystal structure, the original NMC811 materials do not have any obvious structural changes when heated from room temperature to 400°C

(Figure 2.21a–c). However, when NMC811 is charged to 4.3V, obvious phase transition from layer/spinel to rock-salt structure can be observed during heating with the porosity arising due to oxygen release (Figure 2.21d–g). According to SAED patterns, the charged NMC811 has strong reflections indexed as the [0001] zone axis from the layered structure accompanied by blurred spots of the [111] zone axis from the spinel structure at room temperature (Figure 2.21e), which is different from the SAED patterns indexed as [111] zone axis of rock-salt structure at 200°C (Figure 2.21g). And in the characterization of EELS (Figure 2.21h), the decrease of ΔE (the difference in the energy loss values at the highest point of pre-edge peak and main peak) for the oxygen K-edge and the increase of the L_3/L_2 intensity ratios for TM prove that Co, Ni, and other TM elements in the materials are reduced, accompanied by oxygen evolution reaction. The application of this technology enables us to have a deeper understanding of the thermal stability of cathode materials at different temperatures, and it provides great help for optimizing the proportion of metal elements in cathode materials.

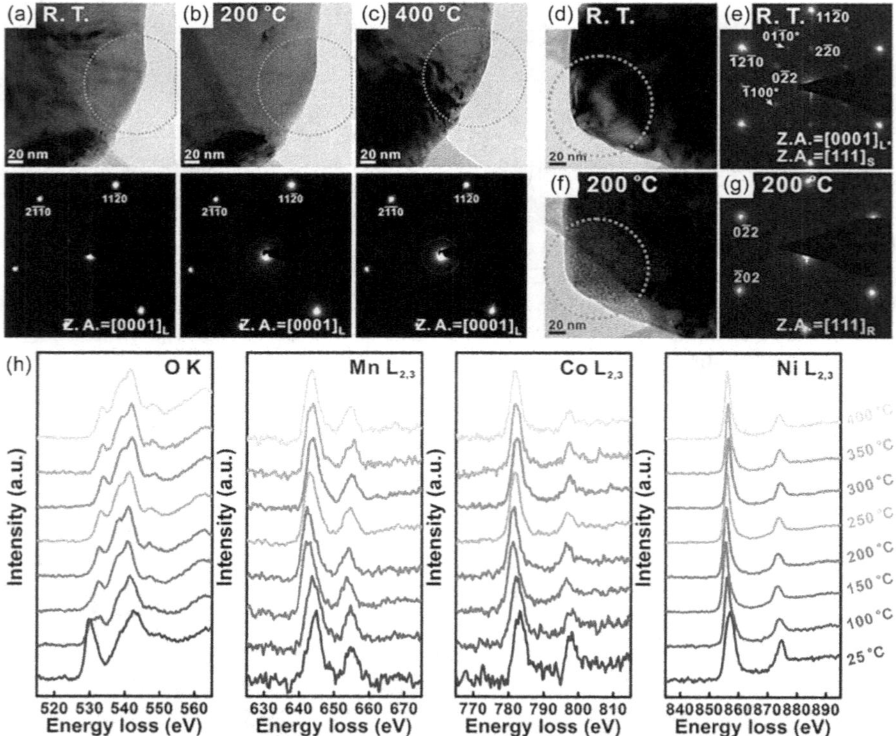

FIGURE 2.21 (a–c) BF images and SAED patterns from pristine NMC811 at room temperature, 200°C, and 400°C, respectively. (d-g) BF image and SAED pattern of the charged NMC811 at room temperature (d and e) and at 200°C (f and g). (h) Modification in the oxygen K-edge and the Mn, Co, and Ni $L_{2,3}$-edge EELS spectra of charged NMC811 with increasing temperature.

2.4.4 Cryo-TEM for Beam-Sensitive Samples: Li Metal, Solid Electrolyte Interphase (SEI), and Interfaces in Solid Batteries

When characterizing the battery materials by EMs, it is inevitable to encounter various irradiation-sensitive materials, especially for those containing light elements or at nanoscale. For examples, after obtaining a thin film of solid electrolyte LiPON by FIB, light spot trace was left on the thin film by electron beams.[24] Even for the oxide cathode materials, high-energy electron beams can cause migration of TM ions and irreversible decomposition, forming Li or O vacancies (Figure 2.22a–f)[68,69]. Worse more, when we characterize the Li metal by conventional TEM, the Li dendrites will bend, melt, and partially sublimate under the irradiation even at small dose during observation (Figure 2.22g–i).[70] These irradiation damages can be divided into primary damage (including ionization, chemical bond breaking, generation of free radicals and secondary electrons) and secondary damage (chemical reaction). On this regard, cryo-TEM is helpful to minimize the beam damage by low-temperature protection and low-dose imaging.[71] Under the same electron dose, the Li metal is stable under cryo-TEM, which provides a powerful method to uncover the structure mystery of the beam-sensitive materials.[72]

As an ideal anode material, Li metal has attracted tremendous attention but has the problems of dendritic growth and low coulombic efficiency, which hinders its practical use. Therefore, it is necessary to explore the micro/nano-structure of Li deposition under different growth conditions, and to find out the underlying correlation with the electrochemical performance.[73] Wang et al.[70], for the first time, applied cryo-TEM technology to reveal the nanostructure of electrochemically deposited lithium (EDLi) and the SEI on the surface while minimizing the beam damage to samples. By applying a reverse voltage in the coin cell, Li metal is directly deposited on the TEM grid inside the cell (Figure 2.23a), and the state of Li deposition can be adjusted by changing the current density and reaction time. In addition, they carried out cryo-XRD experiments on Li metal to prove that cooling down to −180°C had negligible effect on the crystalline structure of the Li metal. Then they were surprised to find that nucleation-dominated EDLi (at $0.5\,mA/cm^2$ for 5 minutes) is amorphous, while crystalline LiF nanograins are present in unevenly distributed SEI (Figure 2.23b). This conclusion was further supported by EELS analysis. Moreover, by using TEM combined with XPS, organic components in SEI can be effectively recognized. Further analysis showed that the EDLi grown in various electrolytes with different additives displays distinctive surface properties. This work not only proves the capability of cryo-TEM technique in the characterization of beam-sensitive materials, but also provides in-depth understanding of the evolution of EDLi.

At the same time, Li et al.[74] also adopted cryo-TEM technology to observe atomic structure of Li metal and its surface. Cu TEM grid with Li deposits were loaded onto the cryo-transfer holder in the liquid nitrogen and transferred to TEM at −170°C for observation. With high-resolution imaging, they could identify that Li dendrites grow in the form of single-crystal nanowire along the <111> (preferred), <110>, or <211> crystal orientations of BCC structure in carbonate-based electrolytes, where the Li atoms arrangement is clearly discerned (Figure 2.23c and d). This work greatly promoted the application of cryo-TEM for battery materials.

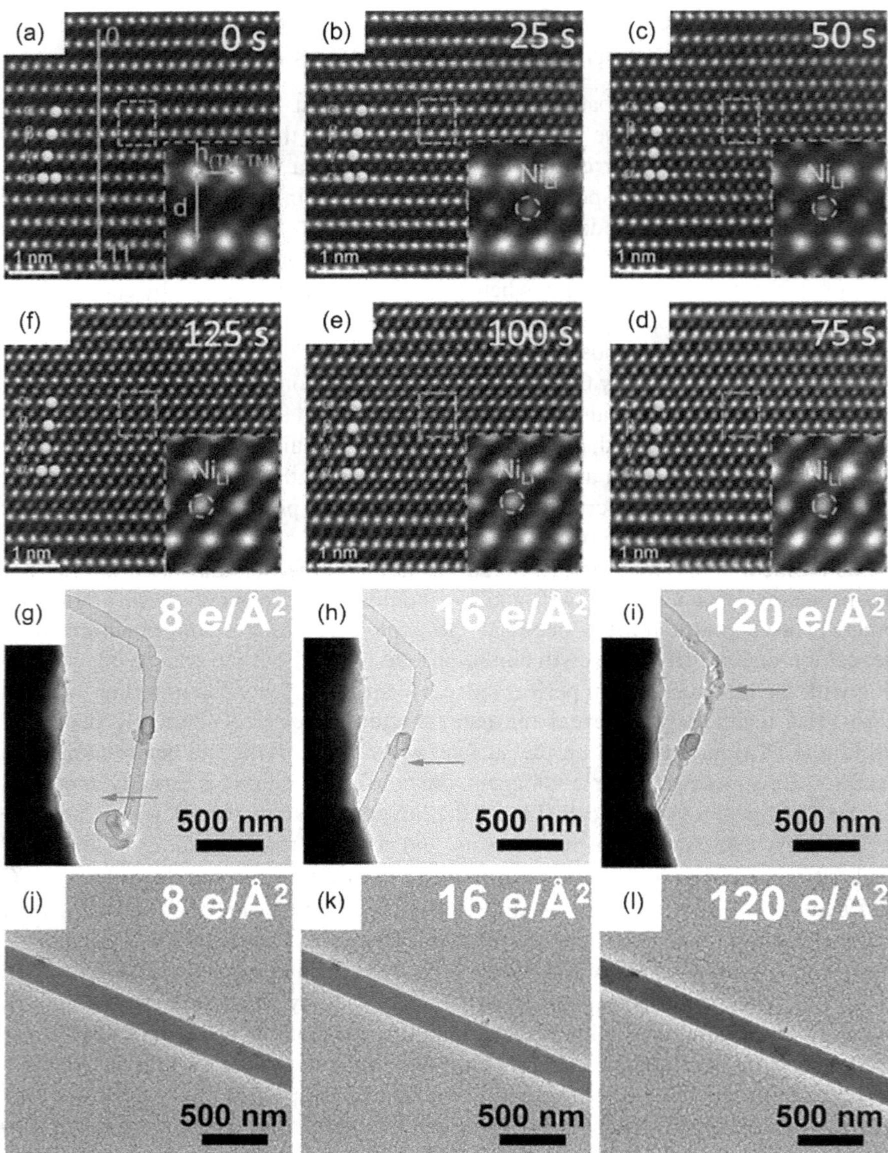

FIGURE 2.22 (a–f) Selected continuous HAADF-STEM images showing the gradual migration of TM ions and Li-Ni mixed arrangement of NMC811 materials under electron beam irradiation. The insets are magnifying images. (g–l) TEM images of the electrochemically deposited Li as a function of the electron irradiation dose at 300 K (panels g–i at 19,000× magnification) and 100 K (panels j–l at 19,000× magnification).

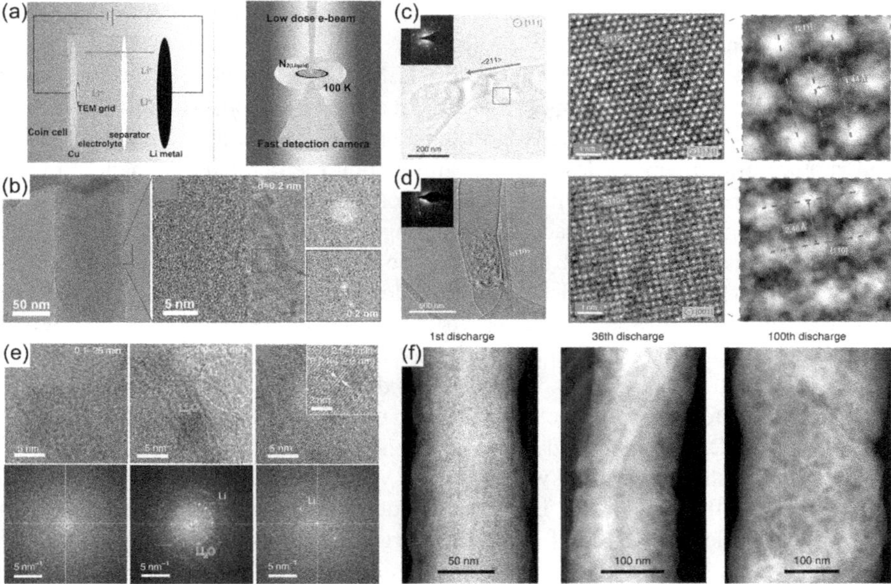

FIGURE 2.23 (a) Schematic cell configuration and schematic cryo-TEM imaging of the EDLi for TEM observation. (b) Cryo-TEM image and its regional zoomed-in image with the bulk and surface FFT result of the EDLi using conventional carbonate electrolyte. (c and d) TEM images of Li metal dendrites growing along the <211> (c, left) and <110> (d, left) directions (Inset: corresponding SAED pattern showing that the dendrites are single-crystalline). (c and d, middle) Magnified images of boxed regions in red and blue from the left images, showing the Li metal lattice at high resolution. (c and d, right) Atomic-resolution TEM images resolving individual Li atoms along the [111] and [001] zone axes. (e) Cryo-TEM images and their corresponding FFT patterns of the Li deposits at 0.1 mA/cm^2 for 25 minutes (left), 1.0 mA/cm^2 for 2.5 minutes (middle), and 2.5 mA/cm^2 for 1.0 minutes (right). And the images with the Li metal lattice and the characteristic bright diffraction spots highlighted by the red arrows are from the (110) plane of Li metal. (f) Cryo-STEM-HAADF images of the Si nanowires at different cycle numbers: first delithiated (left), 36th delithiated (middle), and 100th delithiated (right). With the progression of the cycling, Si becomes porous.

Subsequently, Wang et al.[75] continued to investigate the evolution of Li nanostructures during the nucleation and growth process at different deposition rate and time (Figure 2.23e). A disorder–order phase transition occurs during Li deposition, and the critical size for phase transition is about 5 nm. Through statistical analysis, more amorphous Li was formed under low current density (e.g., 0.1 mA/cm^2). The crystallization of Li requires a certain time to agglomerate sufficient atoms (more than 700 atoms), growth mechanism of which was further simulated by reactive molecular dynamics. The different nucleation states of Li metal determine its final growth mode. In combination with SEM and cryo-TEM images of the Li metal under different conditions, amorphous Li is found more favorable to form larger Li chunks. Compared to crystalline Li, glassy Li performs better in electrochemical reversibility. This work correlates the crystallinity of the Li metal to the nanostructures and performance, which is of great significance to enlighten the structural design of Li metal to achieve high-performance Li metal anode.

Besides Li metal, He et al.[76] used cryo-STEM-HAADF imaging combined with EDS elemental mapping to obtain the 3D structure of silicon (Si) and SEI during cycling with the help of advanced algorithms. Si nanowires underwent pulverization and rupture after many cycles and SEI grew along the percolation channel of the nanovoids (Figure 2.23f), resulting in plenty of fine particles encapsulated by SEI layer and losing electrical contact, thus forming dead Si. The gradual growth of the SEI toward the interior of Si particles can cause the capacity loss of Si anode.

The SEI is the most important and the least understood solid electrolyte in rechargeable Li batteries.[77] Using the cryo-TEM, Li et al.[74] pointed out that different SEI structures can be formed on the surface of Li metal under different electrolytes. A Mosaic structure of SEI was formed on Li dendrites in the ethylene carbonate-diethyl carbonate (EC-DEC) electrolyte, where crystalline nanograins such as Li_2CO_3 and Li_2O are embedded in the amorphous organic species (Figure 2.24a). A multilayered structure of SEI was generated with fluoroethylene carbonate (FEC) additive where the outer layer is composed of Li_2O and other inorganic species while the inner layer consists of amorphous organic components in contact with Li metal (Figure 2.24b). Although these two models have been proposed to describe the structure of SEI, its composition, nanostructure, and ionic conductivity mechanism (through bulk-phase conduction of particles or grain boundary conduction) of ideal SEI are still unknown.[78] Compared with XPS, which has been widely used to explore the SEI composition and acquire average surface information, cryo-TEM can provide spatial distribution of various SEI components. Yuan et al.[79] observed the microscopic composition and morphology of SEI visually through cryo-TEM (Figure 2.24c and d). The crystalline inorganic particles in the SEI are stacked with each other, but the stacking mode and compactness are varied in different SEIs, resulting in different mechanical properties of SEI. The SEI layer formed in the $LiPF_6$ EC/DEC electrolyte have the inorganic species such as LiF, Li_2O, and the organic species such as –COOLi, ROCOOR precipitated and randomly embedded in it, forming a mosaic microphase (Figure 2.24c). In addition, a multilayered structure with an inner organic-rich layer and outer inorganic-rich region was observed in the 1 M LiTFSI DOL/DME + 2 wt% $LiNO_3$ electrolyte (Figure 2.24d). They found that the composition and distribution of the organic part in SEI layers seriously affect CE of Li deposition, and SEI rich in dense inorganic components is beneficial to preventing the growth of Li dendrites.

Zhang et al.[80] analyzed the evolution of SEI nanostructures on Si particles during cycling by cryo-TEM (Figure 2.24e–l). Without FEC, the dominant inorganic component of the SEI is Li_2O. According to the EDS results, the thickness of SEI layer is estimated, which is thicker (40 nm) than that formed in FEC-containing electrolyte (25 nm). However, SEI became thinner during the charging process because the Li_2O reacts with Si to form lithium silicate (Li_xSiO_y), which reflects the intrinsic instability of SEI. LiF particles appeared obviously in SEI after FEC was added into the electrolyte. Thus, more LiF particles were retained on the Si surface after multiple cycling, which helps to keep a thin, dense, and stable SEI structure and maintain an equilibrium interplay between SEI and Li_xSi.

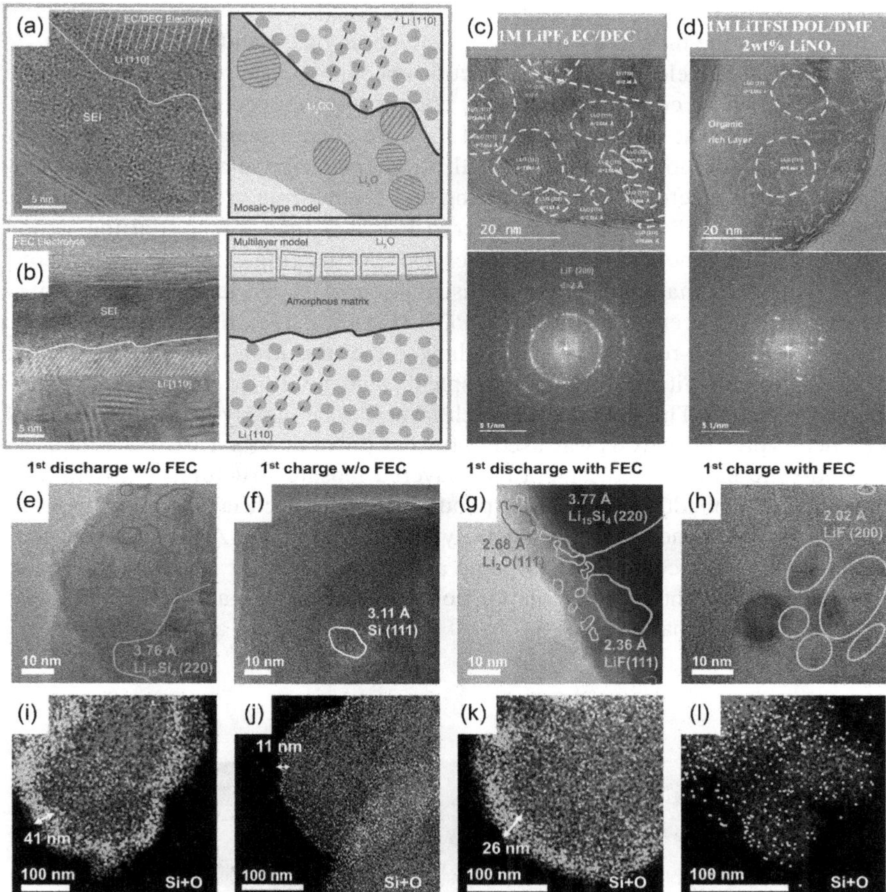

FIGURE 2.24 (a) Atomic-resolution image and schematic revealing the observed mosaic-type structure formed on Li dendrites in EC-DEC electrolyte. The lattice spacings of small crystalline grains of inorganic material dispersed throughout the amorphous film can be matched to Li carbonate (Li_2CO_3, orange circles) and Li oxide (Li_2O, red circles). (b) Atomic-resolution image and schematic of the multilayered SEI formed on Li dendrites in FEC electrolyte. (c and d) Cryogenic TEM images and FFT images of SEI layer formed in $LiPF_6$ EC/DEC and LiTFSI DOL/DME 2wt% $LiNO_3$. (e–h) HRTEM of Si anodes after first lithiation (e) and first delithiation (f) in the FEC-free electrolyte, after first discharge (g), and first charge (h) in the FEC-containing electrolyte. (i–l) EDS mapping of Si anodes after first discharge (i) and first charge (j) in the FEC-free electrolyte, after first discharge (k), and first charge (l) in FEC-containing electrolyte.

Furthermore, cryo-TEM technology can also be used to analyze a variety of complex interface structures caused by the different chemical/physical/mechanical properties of multiple solid compositions in ASSBs, including void, chemical/electrochemical reactions, grain boundary, solid-solid direct contact and other kinds of interaction and existence.[81] The difficulties in the analysis of ASSBs lie in the fact that the interfaces are buried in the bulk phase and difficult to peel off, with chemical and electrochemical reactions occurring side by side. Additionally, the interfaces have complex and sundry chemical components in them, with their thickness usually at the nanoscale, and they are always sensitive to air and electron beam irradiation. So, it is difficult to characterize and track the dynamic evolution of its electrochemical charging and discharging process. Cheng et al.[24] obtained the Li/LiPON interface by cryo-FIB and then transferred the sample under vacuum or inert atmosphere for cryo-TEM imaging. Through STEM combined with EDS mapping, gradient distribution of N, P elements is present at the interface between Li metal and LiPON (Figure 2.25a), thickness of which is about 80 nm. High-resolution images showed the transition layers from Li metal to Li_2O, lithium nitride, lithium phosphate, and finally to amorphous LiPON (Figure 2.25b–f). The thickness of each layer can be clearly measured. Besides, XPS depth-profile analysis was used to further prove that the Li/LiPON interface in solid-state batteries presents a multilayered mosaic structure, with inorganic nanocrystals embedded in the amorphous LiPON in each layer. Compared with the porous SEI formed in liquid electrolyte batteries, SEI in ASSBs is more compact, and does not contain any organic alkyl component or LiF. This discovery provides new insights on the composition, structure, and stability of SEI in all-solid-state lithium batteries.

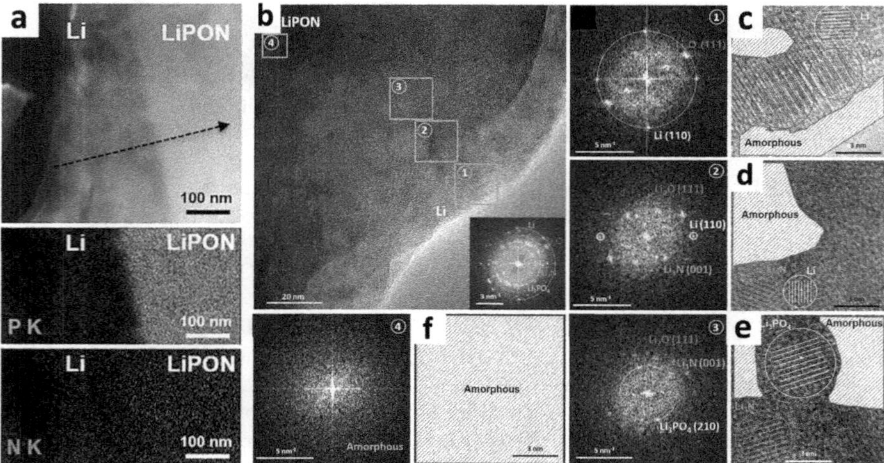

FIGURE 2.25 (a) Cryo-STEM, DF image of Li/LiPON interface and EDS mapping results of P and N signals in the region shown in the STEM image. (b) HRTEM image of the interphase where four regions (regions 1–4) are highlighted by orange squares to indicate different stages of the multilayered structure across the interphase. (c–f) FFT patterns corresponding to regions 1–4 as highlighted in (b), respectively, with the nanostructure schematic overlaying the HRTEM images that correspond to regions 1–4 as highlighted in (b), respectively.

2.5 SUMMARY

2.5.1 Challenges and Issues Associated with the Higher Energy and Safer Batteries

In recent years, with the rapid development of EM and various advanced analytical techniques, people have greatly enriched the fundamental understanding of the micro/nano-structure of key battery materials and interfaces, and deepened the understanding of mechanism of ion migration and structural evolution during cycling. However, there are still many unsolved mysteries about the electro-chemo-mechanical behavior of different cathode, anode, electrolyte and their interfaces. For example, in terms of cathode materials, it is necessary to explore the cathode structure with deeper delithiation at a higher potential, and analyze their structural stability such as surface reconstruction,[82] anionic oxidation, and then seek for solutions. As for the electrolytes, the microscopic analysis of the solvation structure, the mechanism of ion migration in the electrolyte, the decomposition pathway forming SEI, and its interaction with active materials, are needed to be further explored. For the anode materials,[83] researchers are interested in the structural changes of graphite during fast lithiation, conversion reactions, and competitive reaction with Li deposition at a low temperature or during rapid charging. In addition, the growth mode and inhibition scheme of Li dendrites on the anode surface, and the avoidance of dead Li, are worth further discussing and analyzing. Otherwise, it is particularly important to deeply understand the structure and composition of ideal SEI and CEI. Considering that Li_2O, Li_2CO_3, LiF, and other inorganic components have low ionic conductivity, it is still difficult to answer the question whether SEI with high ionic conductivity is achieved through bulk phase or grain boundary. More importantly, the benefits of organic components and LiF in SEI need to be further demonstrated.

In the future, the development of Li-ion battery system toward the goal of higher energy density will result in more in-depth research on Li-S and Li-O_2 batteries. In a Li-S battery,[84] S and its discharging products Li_2S/Li_2S_2 have poor electron conductivity while polysulfide is highly soluble in electrolyte to create a shuttle effect, resulting in sulfide deposition on Li metal anode. Finally, it ends up with complex solid–liquid interfaces. For Li-O_2 battery, oxygen is used as the active substance. It is extremely difficult to characterize the electrochemical behavior at the solid–liquid–gas three-phase interface. Last but not the least, in ASSBs, the research on the mode of ionic conductive channels in solid electrolytes, the characterization of the space charge layer at the cathode/solid electrolyte interface, and the exploration of solid-phase interface reactions under electric-thermal-mechanical field coupling are crucial and challenging tasks in the future.

2.5.2 Future Development of EMs

From ordinary SEM and TEM to advanced STM, AFM, HRTEM, ACTEM, etc., EM has been widely applied in the field of metal-ion batteries nowadays. However, with the deepening of the research on the electrochemical reaction mechanism of batteries, it requires to observe materials at smaller and smaller scales, and characterize solid–liquid interfaces, solid–gas interfaces, and even gas–liquid interfaces. More importantly, we need to ensure the reliability of the results when observing beam-sensitive samples, especially at this period when in-situ characterization of battery reactions is extremely prevalent. As a result, cryo-EM has been urgently introduced into the observation of battery materials, because it has the capability to minimize the irradiation damage caused by electrons. Furthermore, it could observe the solid–liquid interface at the temperature of liquid nitrogen. In the near future, the development of fast cameras and low-dose imaging technology will help the EMs achieve higher signal-to-noise ratio and less irradiation damage. And perhaps cryo-TEM using liquid helium could trap the gas at a low temperature and make observation of the gas–liquid/solid interfaces.[85,86] For example, oxygen will become solid at −218°C, so we are more likely to observe the O_2 evolution from the oxide cathodes, reactions of oxygen cathode, decomposition of SEI, etc. While environmental TEM technology has been successfully reported, further advanced application of in-situ cryo-TEM technology observing electrochemical reactions on the electrodes may make a greater contribution to the research of battery materials.

Meanwhile, future EMs need to be more efficient, more integrated, and smarter.[87] EM alone is no longer sufficient for increasingly sophisticated analysis of electrochemical and interfacial reactions. Then we must combine EM with other advanced methods of phase characterization, such as cryo-microelectron diffraction, cryo-electron tomography, and cryo-electron diffraction pair distribution function. Eventually, statistical analysis method, 3D reconstruction technique, and even artificial intelligence learning technique can be introduced into EM analysis to form powerful characterization systems.

REFERENCES

[1] Chen, Q., C. Dwyer, G. Sheng, et al. 2020. Imaging beam-sensitive materials by electron microscopy. *Advanced Material* 32 (16):e1907619.

[2] Pinna, H., K. Liang, M. Denizart, and B. Jouffrey. 1983. Electronic environment for a field-emission gun in electron-microscopy. *Revue De Physique Appliquee* 18 (10):659–665.

[3] Xu, J., W. Zhang, H. Fan, F. Cheng, D. Su, and G. Wang. 2018. Promoting lithium polysulfide/sulfide redox kinetics by the catalyzing of zinc sulfide for high performance lithium-sulfur battery. *Nano Energy* 51:73–82.

[4] Chen, T., X. Li, H. Wang, et al. 2018. The effect of gradient boracic polyanion-doping on structure, morphology, and cycling performance of Ni-rich $LiNi_{0.8}Co_{0.15}Al_{0.05}O_2$ cathode material. *Journal of Power Sources* 374:1–11.

[5] Gao, Y., D. Wang, Y. C. Li, Z. Yu, T. E. Mallouk, and D. Wang. 2018. Salt-based organic-inorganic nanocomposites: towards a stable lithium metal/$Li_{10}GeP_2S_{12}$ solid electrolyte interface. *Angewandte Chemie International Edition in English* 57 (41):13608–13612.

[6] Wu, Y., and N. Liu. 2018. Visualizing battery reactions and processes by using in situ and in operando microscopies. *Chem* 4 (3):438–465.

[7] Orsini, F., A. Du Pasquier, B. Beaudoin, et al. 1998. In situ Scanning Electron Microscopy (SEM) observation of interfaces within plastic lithium batteries. *Journal of Power Sources* 76 (1):19–29.

[8] Golozar, M., R. Gauvin, and K. Zaghib. 2021. In situ and in operando techniques to study Li-ion and solid-state batteries: micro to atomic level. *Inorganics* 9 (11):85–96.

[9] Cheng, X., Y. Li, T. Cao, et al. 2021. Real-time observation of chemomechanical breakdown in a layered nickel-rich oxide cathode realized by in situ scanning electron microscopy. *ACS Energy Letters* 6 (5):1703–1710.
[10] Chen, D., S. Indris, M. Schulz, B. Gamer, and R. Mönig. 2011. In situ scanning electron microscopy on lithium-ion battery electrodes using an ionic liquid. *Journal of Power Sources* 196 (15):6382–6387.
[11] Shi, H., X. Liu, R. Wu, et al. 2019. In situ SEM observation of structured Si/C anodes reactions in an ionic-liquid-based lithium-ion battery. *Applied Sciences* 9 (5):956–964.
[12] Chen, C. Y., T. Sano, T. Tsuda, et al. 2016. In situ scanning electron microscopy of silicon anode reactions in lithium-ion batteries during charge/discharge processes. *Scientific Reports* 6:36153.
[13] Rong, G., X. Zhang, W. Zhao, et al. 2017. Liquid-phase electrochemical scanning electron microscopy for in situ investigation of lithium dendrite growth and dissolution. *Advanced Materials* 29 (13):1606187.
[14] Nagao, M., A. Hayashi, M. Tatsumisago, T. Kanetsuku, T. Tsuda, and S. Kuwabata. 2013. In situ SEM study of a lithium deposition and dissolution mechanism in a bulk-type solid-state cell with a Li_2S-P_2S_5 solid electrolyte. *Physical Chemistry Chemical Physics* 15 (42):18600–18606.
[15] Krauskopf, T., B. Mogwitz, H. Hartmann, D. K. Singh, W. G. Zeier, and J. Janek. 2020. The fast charge transfer kinetics of the lithium metal anode on the garnet-type solid electrolyte $Li_{6.25}Al_{0.25}La_3Zr_2O_{12}$. *Advanced Energy Materials* 10 (27):2000945.
[16] Wang, H., H. Gao, X. Chen, et al. 2021. Linking the defects to the formation and growth of Li dendrite in all-solid-state batteries. *Advanced Energy Materials* 11 (42):2102148.
[17] Zheng, H., D. Xiao, X. Li, et al. 2014. New insight in understanding oxygen reduction and evolution in solid-state lithium-oxygen batteries using an in situ environmental scanning electron microscope. *Nano Letters* 14 (8):4245–4249.
[18] Hintennach, A., and P. Novák. 2011. A novel combinative Raman and SEM mapping method for the detection of exfoliation of graphite in electrodes at very positive potentials. *Journal of Raman Spectroscopy* 42 (9):1754–1760.
[19] Bessette, S., A. Paolella, C. Kim, et al. 2018. Nanoscale lithium quantification in $Li_XNi_yCo_wMn_ZO_2$ as cathode for rechargeable batteries. *Scientific Reports* 8 (1):17575.
[20] Golozar, M., P. Hovington, A. Paolella, et al. 2018. In situ scanning electron microscopy detection of carbide nature of dendrites in Li-polymer batteries. *Nano Letters* 18 (12):7583–7589.
[21] Golozar, M., A. Paolella, H. Demers, et al. 2019. In situ observation of solid electrolyte interphase evolution in a lithium metal battery. *Communications Chemistry* 2 (1):131.
[22] Liu, Z., Y. K. C. Wiegart, J. Wang, S. A. Barnett, and K. T. Faber. 2016. Three-phase 3D reconstruction of a $LiCoO_2$ cathode via FIB-SEM tomography. *Microscopy and Microanalysis* 22 (1):140–148.
[23] Chen, J. 2021. Advanced electron microscopy of nanophased synthetic polymers and soft complexes for energy and medicine applications. *Nanomaterials (Basel)* 11 (9):2405.
[24] Cheng, D., T. A. Wynn, X. Wang, et al. 2020. Unveiling the stable nature of the solid electrolyte interphase between lithium metal and LiPON via cryogenic electron microscopy. *Joule* 4 (11):2484–2500.
[25] Lee, J. Z., T. A. Wynn, M. A. Schroeder, et al. 2019. Cryogenic focused ion beam characterization of lithium metal anodes. *ACS Energy Letters* 4 (2):489–493.
[26] Zachman, M. J., Z. Tu, S. Choudhury, L. A. Archer, and L. F. Kourkoutis. 2018. Cryo-STEM mapping of solid-liquid interfaces and dendrites in lithium-metal batteries. *Nature* 560 (7718):345–349.

[27] Jaiser, S., J. Kumberg, J. Klaver, et al. 2017. Microstructure formation of lithium-ion battery electrodes during drying – an ex-situ study using cryogenic broad ion beam slope-cutting and scanning electron microscopy (Cryo-BIB-SEM). *Journal of Power Sources* 345:97–107.
[28] Amalraj, F., M. Talianker, B. Markovsky, et al. 2013. Studies of Li and Mn-rich $Li_x[MnNiCo]O_2$ electrodes: electrochemical performance, structure, and the effect of the aluminum fluoride coating. *Journal of the Electrochemical Society* 160 (11):A2220–A2233.
[29] Harris, P. 2018. Transmission electron microscopy of carbon: a brief history. *Carbon* 4 (1):4.
[30] Maiyalagan, T., K. A. Jarvis, S. Therese, P. J. Ferreira, and A. Manthiram. 2014. Spinel-type lithium cobalt oxide as a bifunctional electrocatalyst for the oxygen evolution and oxygen reduction reactions. *Nature Communications* 5:3949.
[31] Gong, Y., J. Zhang, L. Jiang, et al. 2017. In situ atomic-scale observation of electrochemical delithiation induced structure evolution of $LiCoO_2$ cathode in a working all-solid-state battery. *Journal of the American Chemical Society* 139 (12):4274–4277.
[32] Fisher, C. A. J., S. Zheng, A. Kuwabara, et al. 2014. Atomic-level characterization of interfaces in $LiCoO_2$. *ECS Transactions* 58 (13):1–11.
[33] Gu, L., C. Zhu, H. Li, et al. 2011. Direct observation of lithium staging in partially delithiated $LiFePO_4$ at atomic resolution. *Journal of the American Chemical Society* 133 (13):4661–4663.
[34] He, X., L. Gu, C. Zhu, et al. 2011. Direct imaging of lithium ions using aberration-corrected annular-bright-field scanning transmission electron microscopy and associated contrast mechanisms. *Materials Express* 1 (1):43–50.
[35] Wang, X., X. Shen, Z. Wang, R. Yu, and L. Chen. 2014. Atomic-scale clarification of structural transition of MoS_2 upon sodium intercalation. *ACS Nano* 8 (11):11394–11400.
[36] Wang, F., R. Robert, N. A. Chernova, et al. 2011. Conversion reaction mechanisms in lithium ion batteries: study of the binary metal fluoride electrodes. *Journal of the American Chemical Society* 133 (46):18828–18836.
[37] Lin, F., D. Nordlund, T. C. Weng, et al. 2014. Phase evolution for conversion reaction electrodes in lithium-ion batteries. *Nature Communications* 5:3358–3366.
[38] He, Y., M. Gu, H. Xiao, et al. 2016. Atomistic conversion reaction mechanism of WO_3 in secondary ion batteries of Li, Na, and Ca. *Angewandte Chemie International Edition* 55 (21):6244-7.
[39] Wang, R., X. He, L. He, et al. 2013. Atomic Structure of Li_2MnO_3 after Partial Delithiation and Re-Lithiation. *Advanced Energy Materials* 3 (10):1358–1367.
[40] Lu, P., P. Yan, E. Romero, E. D. Spoerke, J. G. Zhang, and C. M. Wang. 2015. Observation of electron-beam-induced phase evolution mimicking the effect of the charge–discharge cycle in Li-rich layered cathode materials used for Li ion batteries. *Chemistry of Materials* 27 (4):1375–1380.
[41] Yan, P., J. Zheng, J. G. Zhang, and C. Wang. 2017. Atomic resolution structural and chemical imaging revealing the sequential migration of Ni, Co, and Mn upon the battery cycling of layered cathode. *Nano Letters* 17 (6):3946–3951.
[42] Song, B., H. Liu, Z. Liu, P. Xiao, M. O. Lai, and L. Lu. 2013. High rate capability caused by surface cubic spinels in Li-rich layer-structured cathodes for Li-ion batteries. *Scientific Reports* 3:3094.
[43] Guo, X. W., X. P. Fang, Y. Sun, L. Y. Shen, Z. X. Wang, and L. Q. Chen. 2013. Lithium storage in carbon-coated SnO_2 by conversion reaction. *Journal of Power Sources* 226:75–81.
[44] Cheng, T., Z. Ma, R. Qian, et al. 2021. Achieving stable cycling of $LiCoO_2$ at 4.6 V by multilayer surface modification. *Advanced Functional Materials* 31 (2):2001974.

[45] Liu, S., Z. Liu, X. Shen, et al. 2018. Surface doping to enhance structural integrity and performance of Li-rich layered oxide. *Advanced Energy Materials* 8 (31):1802105.
[46] Zhang, J. N., Q. Li, C. Ouyang, et al. 2019. Trace doping of multiple elements enables stable battery cycling of $LiCoO_2$ at 4.6 V. *Nature Energy* 4 (7):594–603.
[47] Yuan, Y., K. Amine, J. Lu, and R. S. Yassar. 2017. Understanding materials challenges for rechargeable ion batteries with in situ transmission electron microscopy. *Nature Communications* 8 (1):1–4.
[48] Chen, W., Y. Hong, Z. Zhao, et al. 2021. Directing the deposition of lithium metal to the inner concave surface of graphitic carbon tubes to enable lithium-metal batteries. *Journal of Materials Chemistry A* 9 (31):16936–16942.
[49] Wang, J., F. Fan, Y. Liu, et al. 2014. Structural evolution and pulverization of tin nanoparticles during lithiation-delithiation cycling. *Journal of the Electrochemical Society* 161 (11):F3019–F3024.
[50] Fan, Z., L. Zhang, D. Baumann, et al. 2019. In situ transmission electron microscopy for energy materials and devices. *Advanced Materials* 31 (33):e1900608.
[51] Ma, C., Y. Cheng, K. Yin, et al. 2016. Interfacial stability of Li metal-solid electrolyte elucidated via in situ electron microscopy. *Nano Letters* 16 (11):7030–7036.
[52] Wang, Z., D. Santhanagopalan, W. Zhang, et al. 2016. In situ STEM-EELS observation of nanoscale interfacial phenomena in all-solid-state batteries. *Nano Letters* 16 (6):3760–3767.
[53] Leenheer, A. J., K. L. Jungjohann, K. R. Zavadil, and C. T. Harris. 2016. Phase boundary propagation in Li-alloying battery electrodes revealed by liquid-cell transmission electron microscopy. *ACS Nano* 10 (6):5670–5678.
[54] Yuk, J. M., J. Park, P. Ercius, et al. 2012. High-resolution EM of colloidal nanocrystal growth using graphene liquid cells. *Science* 336 (6077):61–64.
[55] Zhong, L., R. R. Mitchell, Y. Liu, et al. 2013. In situ transmission electron microscopy observations of electrochemical oxidation of Li_2O_2. *Nano Lett* 13 (5):2209–2214.
[56] Deng, L., Y. Hong, Y. Yang, et al. 2021. Sulfurized polyacrylonitrile as a high-performance and low-volume change anode for robust potassium storage. *ACS Nano* 15 (11):18419–18428.
[57] Zhao, X., Y. Hong, M. Cheng, et al. 2020. High performance potassium–sulfur batteries and their reaction mechanism. *Journal of Materials Chemistry A* 8 (21):10875–10884.
[58] Wang, J. W., X. H. Liu, S. X. Mao, and J. Y. Huang. 2012. Microstructural evolution of tin nanoparticles during in situ sodium insertion and extraction. *Nano Letters* 12 (11):5897–5902.
[59] Wang, Q., X. Zhao, C. Ni, et al. 2017. Reaction and capacity-fading mechanisms of tin nanoparticles in potassium-ion batteries. *The Journal of Physical Chemistry C* 121 (23):12652–12657.
[60] Liu, Y., F. Fan, J. Wang, et al. 2014. In situ transmission electron microscopy study of electrochemical sodiation and potassiation of carbon nanofibers. *Nano Letters* 14 (6):3445–3452.
[61] Huang, J. Y., L. Zhong, C. M. Wang, et al. 2010. In situ observation of the electrochemical lithiation of a single SnO(2) nanowire electrode. *Science* 330 (6010):1515–1520.
[62] Liu, X. H., L. Q. Zhang, L. Zhong, et al. 2011. Ultrafast electrochemical lithiation of individual Si nanowire anodes. *Nano Letters* 11 (6):2251–2258.
[63] Liu, X. H., J. W. Wang, S. Huang, et al. 2012. In situ atomic-scale imaging of electrochemical lithiation in silicon. *Nature Nanotechnology* 7 (11):749–756.
[64] He, K., S. Zhang, J. Li, et al. 2016. Visualizing non-equilibrium lithiation of spinel oxide via in situ transmission electron microscopy. *Nature Communications* 7:11441.
[65] Kuhne, M., F. Borrnert, S. Fecher, et al. 2018. Reversible superdense ordering of lithium between two graphene sheets. *Nature* 564 (7735):234–239.

[66] Zhang, L., T. Yang, C. Du, et al. 2020. Lithium whisker growth and stress generation in an in situ atomic force microscope-environmental transmission electron microscope set-up. *Nature Nanotechnology* 15 (2):94–98.

[67] Hwang, S., S. M. Kim, S. M. Bak, et al. 2015. Using real-time electron microscopy to explore the effects of transition-metal composition on the local thermal stability in charged $Li_xNi_yMn_zCo_{1-y-Z}O_2$ cathode materials. *Chemistry of Materials* 27 (11):3927–3935.

[68] Gao, A., X. Li, F. Meng, et al. 2021. In operando visualization of cation disorder unravels voltage decay in Ni-rich cathodes. *Small Methods* 5 (2):e2000730.

[69] Shim, J. H., H. Kang, Y. M. Kim, and S. Lee. 2019. In situ observation of the effect of accelerating voltage on electron beam damage of layered cathode materials for lithium-ion batteries. *ACS Applied Materials & Interfaces* 11 (47):44293–44299.

[70] Wang, X., M. Zhang, J. Alvarado, et al. 2017. New insights on the structure of electrochemically deposited lithium metal and its solid electrolyte interphases via cryogenic TEM. *Nano Letters* 17 (12):7606–7612.

[71] Ilett, M., M. S'Ari, H. Freeman, et al. 2020. Analysis of complex, beam-sensitive materials by transmission electron microscopy and associated techniques. *Philosophical Transactions of the Royal Society A: Mathematical, Physical and Engineering Sciences* 378 (2186):20190601.

[72] Wang, X., Y. Li, and Y. S. Meng. 2018. Cryogenic electron microscopy for characterizing and diagnosing batteries. *Joule* 2 (11):2225–2234.

[73] Xu, W., J. Wang, F. Ding, Fei Ding, et al. 2014. Lithium metal anodes for rechargeable batteries. *Energy & Environmental Science* 7 (2):513–537.

[74] Li, Y., Y. Li, A. Pei, et al. 2017. Atomic structure of sensitive battery materials and interfaces revealed by cryo-electron microscopy. *Science* 358 (6362):506–510.

[75] Wang, X., G. Pawar, Y. Li, et al. 2020. Glassy Li metal anode for high-performance rechargeable Li batteries. *Nature Materials* 19 (12):1339–1345.

[76] He, Y., L. Jiang, T. Chen, et al. 2021. Progressive growth of the solid-electrolyte interphase towards the Si anode interior causes capacity fading. *Nature Nanotechnology* 16 (10):1113–1120.

[77] Huang, W., D. T. Boyle, Y. Li, et al. 2019. Nanostructural and electrochemical evolution of the solid-electrolyte interphase on CuO nanowires revealed by cryogenic-electron microscopy and impedance spectroscopy. *ACS Nano* 13 (1):737–744.

[78] Zhang, B., H. Shi, Z. Ju, et al. 2020. Arrayed silk fibroin for high-performance Li metal batteries and atomic interface structure revealed by cryo-TEM. *Journal of Materials Chemistry A* 8 (48):26045–26054.

[79] Yuan, S., S. Weng, F. Wang, et al. 2021. Revisiting the designing criteria of advanced solid electrolyte interphase on lithium metal anode under practical condition. *Nano Energy* 83:105847.

[80] Zhang, X., S. Weng, G. Yang, et al. 2021. Interplay between solid-electrolyte interphase and (in)active Li_xSi in silicon anode. *Cell Reports Physical Science* 2 (12):100668.

[81] Banerjee, A., X. Wang, C. Fang, E. A. Wu, and Y. S. Meng. 2020. Interfaces and interphases in all-solid-state batteries with inorganic solid electrolytes. *Chemical Reviews* 120 (14):6878–6933.

[82] Zhang, M. J., X. Hu, M. Li, et al. 2019. Cooling induced surface reconstruction during synthesis of high-Ni layered oxides. *Advanced Energy Materials* 9 (43):1901915.

[83] Xie, J., and Y. C. Lu. 2020. A retrospective on lithium-ion batteries. *Nature Communications* 11 (1):2499.

[84] Yang, X., J. Luo, and X. Sun. 2020. Towards high-performance solid-state Li-S batteries: from fundamental understanding to engineering design. *Chemical Society Reviews* 49 (7):2140–2195.

[85] Mitsuoka, K. 2011. Obtaining high-resolution images of biological macromolecules by using a cryo-electron microscope with a liquid-helium cooled stage. *Micron* 42 (2):100–106.

[86] Fujiyoshi, Y. 2011. Electron crystallography for structural and functional studies of membrane proteins. *Journal of Electron Microscopy (Tokyo)* 60(Suppl 1):S149–S159.

[87] Weng, S., Y. Li, and X. Wang. 2021. Cryo-EM for battery materials and interfaces: workflow, achievements, and perspectives. *iScience* 24 (12):103402.

3 Characterizing the Localized Electrochemical Phenomena in Li-Ion Batteries by Using SPM-Based Techniques

Kaiyang Zeng
National University of Singapore

CONTENTS

3.1 INTRODUCTION

Rechargeable Li-ion batteries are the most common and practical energy storage systems for many electronic products at various scales since the first commercialized product by Sony in 1991.[1–7] The Li-ion batteries have been extensively used

DOI: 10.1201/9781003299295-3

in many portable and mobile devices such as mobile phones, laptop computers and cameras, accounting for more than 60% of worldwide market share in portable batteries. Presently, Li-ion batteries are also applied in larger and more durable products such as hybrid electrical vehicles, grid, space satellite and temporary buffering system for renewable energy sources such as solar and wind energy.[3,6,8–12] For those types of applications, prolonged lifetime and higher energy density are very essential and challenging.[3,6,7,13] In addition, due to the increased requirements of today's information-rich society and emerging ecological issues, it is necessary to develop new, low-cost and environmental-friendly materials for Li-ion batteries as well as new-generation energy storage systems beyond Li-ion batteries, such as metal-ion batteries, Na-ion batteries and all-solid-state batteries, those materials including electrode materials and new electrolyte materials, especially solid electrolyte materials.[14–25] In order to develop new energy storage materials and systems, as well as to optimize the performance and safety of the current battery materials, it is necessary to have the comprehensive understanding of the electrochemical phenomena and associated mechanisms ranging from macro- to nanoscales.[26–33] To this extent, Scanning Probe Microscopy (SPM)-based techniques are a unique yet powerful, and sometimes the only one, characterization method to achieve more comprehensive understanding of the properties and performances of the battery materials and devices and the associated aging and failure mechanisms at nanoscales.[34–44] Generally speaking, the most commonly used SPM-based techniques for characterizing battery materials include Atomic Force Microscopy (AFM), Electrochemical Atomic Force Microscopy (EC-AFM), Kelvin Probe Force Microscopy (KPFM), Electrostatic Force Microscopy, Conductive AFM (c-AFM), Piezoresponse Force Microscopy (PFM) or Electrochemical Strain Microscopy (ESM) and Amplitude Modulation-Frequency Modulation (AM-FM) mode. These techniques are widely used for characterizing various functional and mechanical properties of the battery materials and even the systems. There are also many reviews and books/chapters devoted to these topics.[34,35,45–48] To date, these SPM-based techniques have made significant contributions to enhance current understanding of the local electrochemical functionalities and aging mechanisms of the rechargeable Li-ion batteries at meso- to nanoscales. Hence, following the previous reviews/chapters written by the present author,[34,35] this chapter will further summarize some recent advances in this field.

3.2 BRIEFING INTRODUCTION OF RELEVANT SCANNING PROBE MICROSCOPY (SPM)-BASED TECHNIQUES

In this section, the most commonly used SPM-based techniques for characterizing Li-ion battery materials will be summarized first, then the applications of those SPM-based techniques to battery research will be discussed in Sections 3.3 and 3.4.

3.2.1 Atomic Force Microscopy (AFM)

AFM, first invented by Binning et al. in 1986, is the most widely used technique in all of the available SPM techniques.[49] A typical AFM system consists of AFM head with a probe, a piezoelectric actuator, a photodetector and a laser beam as well as the control system. The probe is generally a sharp tip attached at the end of a cantilever. The tip is engaged to the sample surface during the measurements. The principle of AFM operation is that, as the tip scans through the contour of the sample surface, the laser beam is deflected by the cantilever due to the movement of the cantilever caused by surface roughness, with the deflection signal being detected by the photodetector in the AFM system.[35,50] The detected signal is then converted to the surface topography based on the interactive forces between the tip and sample surface. The AFM is now widely used to characterize the surface topography of the samples. Furthermore, AFM can be combined with several other SPM modes, such as EC-AFM to characterize the electrochemical characteristics of the materials.[51,52]

Generally speaking, AFM has primarily two modes of operation, namely contact and dynamic modes.[53] In the contact-mode AFM, the AFM tip is kept in contact with the sample surface during the scanning.[54] Several other SPM modes, such as c-AFM, PFM and ESM, are developed based on contact-mode AFM. For dynamic AFM, there are two main modes: AM and FM.[55,56] AM-AFM, also referred to as the tapping-mode AFM, is one of the most widely used modes for characterizing various materials, especially soft materials or materials with sensitive surface.[57] In this mode, the probe is driven by the piezoelectric actuator at a fixed frequency, usually near or at its free resonant frequency.[58] The oscillation amplitude and the phase shift of the tip motion are measured. During scanning, the tip touches the surface and lifts off alternately. When the tip passes over a bump at the surface, the oscillation amplitude decreases due to less availability of oscillating space. When the tip passes over a depression, the oscillation amplitude increases (approaching to the free air amplitude).[59] This change in the oscillation amplitude is fed back to the controller, which compares the measured amplitude with the pre-set value (set-point) and generates an error signal. The error signal actuates the piezoelectric actuator to adjust the cantilever's vertical position and thereby to keep it at the set-point value. The other mode, FM-AFM, measures the frequency shift due to the tip–sample interaction, which serves as the control signal for the cantilever–sample distance. FM-AFM is usually operated at Ultra-High Vacuum condition to achieve high-resolution images.

During the AFM measurement, height, deflection and phase images can be obtained after each scan. Height image shows the height changes of the sample surface. The deflection image (error signal), which is a measure of the surface irregularities, is used to plot the surface topography of the sample. Phase image is a reflection of the combination of variations of the surface properties such as compositions, mechanical property, adhesion force and dissipation energy.

3.2.2 Surface-Strain-Based SPM Techniques

It is well known that many dielectric materials can show deformation under external electrical field. Obviously, the mechanisms of such deformation are very different for different materials. For example, piezoelectric response is one of the typical examples of the electrically induced deformation behavior; the mechanism is the bias-induced domain orientation and hence the deformation. On the other hand, ionic materials under electrical field can also show large deformation with totally different mechanisms. SPM-based techniques are highly suitable to study the deformation behavior under the external fields, including not only the electrical field, but also the thermal, mechanical, magnetic, optical and many others. The SPM-based techniques used for detecting such types of deformation can be generalized as surface-strain-based SPM techniques (SS-SPM).[60] Piezoresponse Force Microscopy or PFM is the typical example of the SS-SPM technique. PFM was first developed in 1990s and is now widely used to characterize the electromechanical responses of piezo-/ferro-electric materials.[61–63] It was also used to characterize the electrically induced surface deformation in many non-typical piezoelectric materials, such as ZnO.[64–67] A variation in this technique, ESM, which was first named by the researchers in Oak Ridge National Laboratory (USA), is similar to that of the commercial PFM with the primary difference in the mechanisms of bias–strain coupling phenomena,[68–71] and it was now widely used to characterize the various properties of the materials for Li-ion batteries and other energy materials.[68–71]

Figure 3.1 illustrates the basic principle of the ESM; an AC voltage is applied to the sample surface, in this case it is layered $LiCoO_2$, through a conductive tip that induces local fluctuation in lithium ion concentration. This fluctuation of the lithium ion concentration is converted into localized surface oscillation due to the changes in molar volume of the material near the surface.[35,72] Similarly, insertion and extraction of Li-ions in the electrodes of the Li-ion battery generate changes in molar volume of the electrode.[38] For example, in Li_xCoO_2, one of the commonly used electrode materials for Li-ion battery, the lattice parameter changes with the degree of lithiation, x, by approximately 40 pm in the operation region of the Li_xCoO_2 with

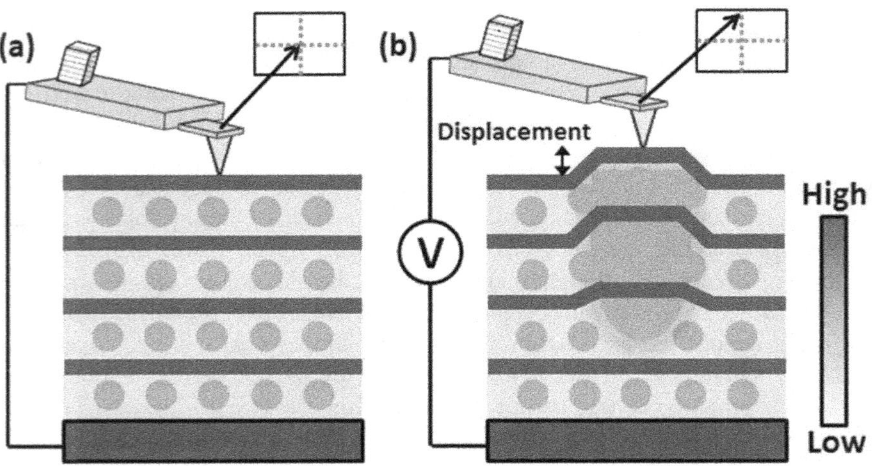

FIGURE 3.1 Schematically diagram showing the principle of Electrochemical Strain Microscopy: (a) The SPM tip is in contact with the sample surface (shown in layered $LiCoO_2$ as an example); and (b) locally applied electrical fields result in Li-ion redistribution in the probed volume, which leads to surface displacements that can be detected by AFM technique. (Reprinted with permission from J. Zhu and K.Y. Zeng, in *Nanotechnology for Sustainable Energy*, ACS Symposium Series, edited by Hu et al., American Chemical Society, 2013, pp. 23–53. Copyright 2013 American Chemical Society.[35])

x varying from 1 to 0.5, as shown by Amatucci et al.[73] The surface oscillation is then recorded by a highly sensitive photodetector, making it possible to map the local ionic motilities with the microstructure by using ESM with resolution unattainable by other traditional methods. Therefore, ESM has been emerging as a powerful technique to probe the localized ionic movements and to investigate the local electrochemical phenomenon associated with ionic dynamics in the battery materials and systems, including anode Si, and various cathodes materials, such as the materials with layered ($LiCoO_2$, $LiNi_{1/3}Co_{1/3}Mn_{1/3}O_2$), spinel ($LiMn_2O_4$) or olivine ($LiFePO_4$) structures.[36,68,74–79] In addition, solid electrolyte materials have also been characterized by using the ESM technique.[80–83]

During the AFM measurement, if a DC bias is applied to the sample surface through a conductive tip that is in direct contact with the sample, the bias at the AFM tip concentrates electric field in a nanometer-scale volume of material, inducing interfacial electrochemical processes at the tip–surface junction and ionic current flow in the material. This technique has been previously used to simulate the effects of the cycling processes in all-solid-state thin-film battery.[84,85] The difference between this technique and the previously discussed ESM is that this technique applies DC voltage to the sample; hence it can be used to study the effects of positive and negative biases to the battery materials, and to simulate the charging/discharging processes for the electrode materials or full-cell battery. If positive bias is applied through the tip, the battery system is similar to the charging process, and therefore, the Li-ions will be forced to move from cathode to anode, whereas if negative bias is applied to the material, this process is similar to the discharge process, and Li-ions are forced to move from anode to cathode. Therefore, with the DC bias, if the measurements are performed with the positive and then negative bias applied in a sequence at a single point within a period of time, and if this time is long enough to cause the Li-ions diffusion in the materials, the biased point and surrounding area are then scanned using the tapping-mode AFM to observe the variation of the surface topography induced by the DC bias. Therefore, the influences of Li-ion intercalation and de-intercalation processes can be studied separately, and these processes cannot be unambiguously distinguished by applying the AC voltage to the specimens during the normal ESM measurements. Furthermore, this method can also be used as an in situ method to investigate the Li-ion dynamics at different locations, such as grain interiors and boundaries; therefore the preferred Li-ion diffusion pathway can be examined.[84,85]

Generally speaking, in the PFM/ESM measurements, the external field-induced electromechanical or electrochemical deformation is usually very weak, only on the order of a few to tens of picometers. These small signals can be amplified by increasing the drive voltage. However, increasing the driving voltage can also potentially damage the samples. Another effective way to boost the weak PFM/ESM signals is to make use of the contact resonant frequency of the SPM cantilever. Generally speaking, the resonant frequency can enhance the signal by roughly the factor Q of the cantilever. However, the resonant frequency often shifts with respect to the location due to its sensitivity to surface morphology and structural heterogeneity. Therefore, driving the tip near the contact resonance at a fixed frequency can sometimes lead to enormous topographic crosstalk. Thus, to maintain the advantages of the contact resonant frequency, adjusting the drive frequency to keep it continually at the contact resonant frequency is required and important to obtain the good ESM images. In the following sections, two methods of tracking contact resonant frequency are discussed; these two methods are found to be effective to track the contact resonant frequency during the scanning. They are not only used for ESM, but can also be used for PFM and other SPM techniques.

3.2.2.1 Dual AC Resonance Tracking

Dual AC resonance tracking (DART) is a dual-frequency technique that can track the contact resonant frequency and then correspondingly adjust the drive frequency of the cantilever.[86,87] DART technique can dramatically reduce the topography cross-talk caused by the resonant frequency shift, which is inevitable in single frequency mode. DART technique drives the tip at two different frequencies: one is below the contact resonant frequency (f_1) and another is above (f_2) as shown in Figure 3.2, which shows a schematic of the two driving frequencies (f_1 and f_2), and two resulting amplitudes (A_1 and A_2), where the resonant frequency is located in the middle of the two driving frequencies.[86] When the resonant frequency shifts downward, A_1 shifts

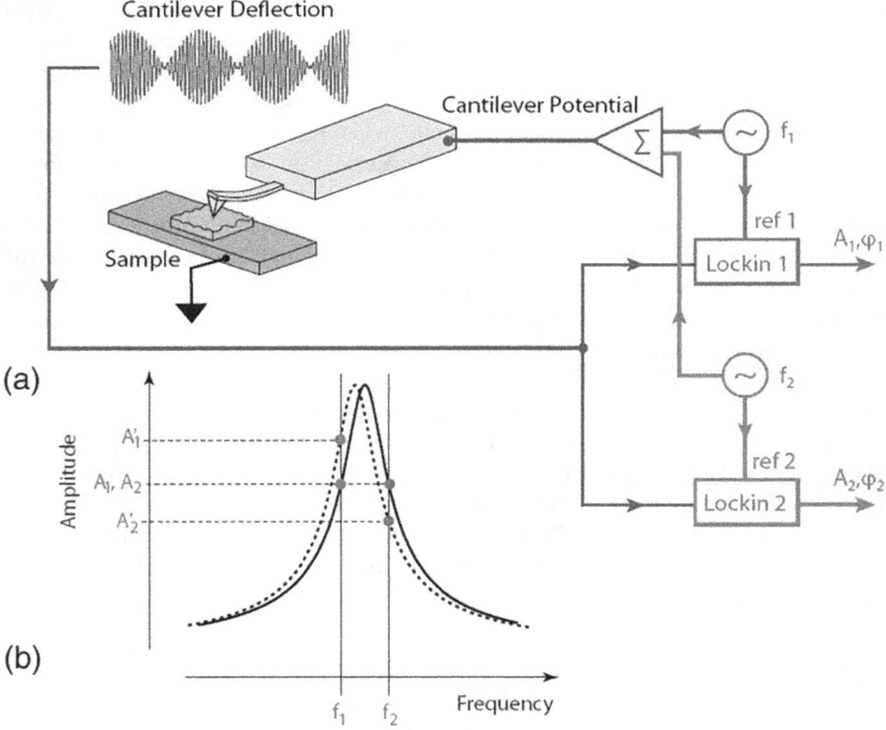

FIGURE 3.2 (a) DART operation principle and feedback control. (b) The feedback control is adjusted to keep the difference between the two driving frequencies zero. (Reprinted from B.J. Rodriguez, C. Callahan, S.V. Kalinin, and R. Proksch, *Nanotechnology*, **18**, 475504 (2007), with permission from Institute of Physics.[86])

up to A_1', and A_2 shifts down to A_2'. The difference between the A_2' and A_1' signal causes the feedback loop to adjust the drive frequency to match the resonance by shifting the drive frequency until those two signals are equal again (i.e., $A_2' - A_1' = 0$ after the adjustment). The recorded responses are directly from the cantilever and usually fitted with the damped simple harmonic oscillator (DSHO) model.[87–89] After the DSHO fitting, images of resonance amplitude, phase, frequency and Q factor can be quantified simultaneously. The resonance amplitude is defined as electrochemical strain, i.e., surface displacement, providing the information of local ionic concentration and diffusivity with resolution unattainable by traditional methods.[69] The resonant frequency is associated with the contact stiffness of the sample surface, whereas the resonance Q factor is a measure of dissipation at the tip—surface junction interaction.[40] The phase image, however, is less well understood so far. The advantages of DART technique include easy to operate and in most of time, can track the resonant frequency reasonably well.

3.2.2.2 Band Excitation Technique

Band Excitation (BE) is another multi-frequency image and analysis technique for PFM/ESM as well as other SPM-based measurements.[90–97] BE excites the system by a digitally synthesized signal having a finite spectral density in a band centered on a resonance peak, as shown in Figure 3.3.[90] The response is detected and Fourier transformed. The ratio of the fast Fourier transforms of the response and excitation signals yields the transfer function of the system. Thus, the contact resonant frequency can

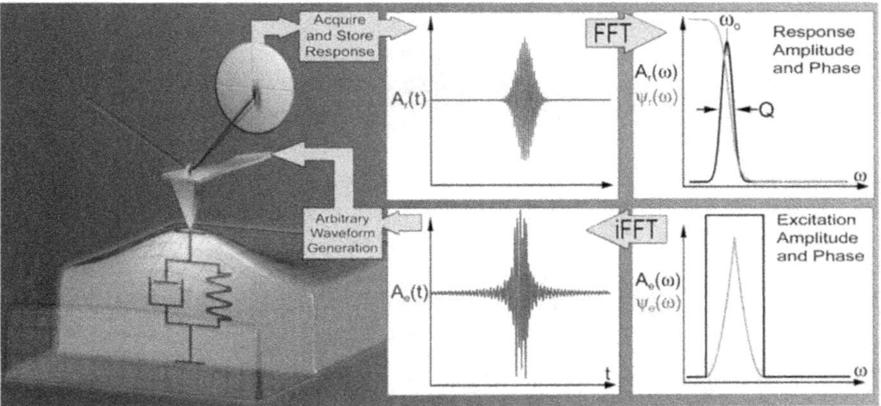

FIGURE 3.3 Operational principle of the BE method in SPM: (a) The excitation signal is digitally synthesized to have a predefined amplitude and phase in the given frequency window; and (b) The cantilever response is detected and Fourier-transformed at each pixel in an image. The ratio of the fast Fourier transform of response and excitation signals yields the cantilever response (transfer function). Fitting the response to the simple harmonic oscillator yields amplitude, resonant frequency and Q factor, which are plotted to yield 2D images, or used as feedback signals. (Reprinted from S. Jesse, S.V. Kalinin, R. Proksch, A.P. Baddorf, and B.J. Rodriguez, *Nanotechnology*, **18**, 435503 (2007), with permission from Institute of Physics.[90])

be accurately tracked. It can therefore eliminate the crosstalk of the surface morphology and enhance signal detection sensitivity. The raw data at each pixel point can also be fitted by post-data analysis processes, i.e., DSHO.[98] Images of resonance amplitude, phase, frequency and Q factor can be obtained as well afterward. Similar to the case of DART technique, the phase image and the energy dissipation image are less well understood so far. However, the BE technique usually takes long scanning time, and many information, especially off contact resonance frequencies are not used in the final analysis. BE technique has been successfully used to characterize electrode materials used for Li-ion batteries.[36,99] Furthermore, the amplitude of surface deformation during the BE-ESM measurements can be related to the diffusion coefficient of the lithium ions in the materials.[99,100]

3.2.3 Conductive AFM

c-AFM is a current-based SPM technique for characterizing the electronic conductance or resistance variations by detecting the current passing through the tip and the sample.[101–105] In the c-AFM module, the specially designed cantilever holder consists of a transimpedance amplifier. A constant voltage is applied between a conductive tip and an electrode at the back of the sample. The current at each tip position is amplified by the amplifiers, enabling the detection of small current. Generally, the c-AFM technique is capable of measuring currents from ~0.5 pA to 10 nA, which allows the investigation of the materials with relatively high resistivity. In addition, c-AFM can simultaneously obtain the topography and current images; therefore the relationship between the topography and conductance can be observed. The concomitantly obtained current and morphology image can provide more detailed information of phase separation, charge generation and transport phenomena. This technique has been applied to characterize various battery materials such as $LiCoO_2$, $LiMn_2O_4$, $LiNi_{0.8}Co_{0.2}O_2$, $Li_{1.2}Co_{0.13}Ni_{0.13}Mn_{0.54}O_2$, Li-ion conducting glass ceramics and others.[37,106–113] Furthermore, by positioning the AFM tip at a point of interest and applying a voltage ramp, the local current–voltage (I–V) curve at that point can be obtained.[106,114-115] In addition, during the I–V curve measurements, if the applied voltages are changed, the I–V curve measurements becomes the first-order reversal curve method (the FORC-IV measurements), which can probe the ionic/electrochemical processes and local ionic dynamics under the external field.[116,117]

3.2.4 SPM-Based Techniques for Characterizing Mechanical Properties

It is well known that the stress and strain can be generated due to the volume changes of the battery materials during the charging/discharging cycles.[117–120] Therefore, the stability of the mechanical properties during the cycle, such as elastic modules, is one of the important properties of the battery materials. Furthermore, for thin-film battery or solid-state battery, the interfacial properties such as adhesion or fracture toughness, are also important to the battery materials and systems. Traditionally, nanoindentation was used to characterize those mechanical properties for battery materials and systems.[121–124] However, nanoindentation can only determine the properties over an area, typically over several tens to hundreds of μm^2; hence, nanoindentation cannot really identify the localized mechanical properties, such as grains and/or grain boundaries, or interfaces. Therefore, SPM-based techniques are ideal for characterizing the localized mechanical properties of battery materials and/or devices.

One of the SPM-based techniques for measuring mechanical properties is Contact Resonance Atomic Force Microscopy technique.[95,125–128] Recently, an AM-FM technique has been developed to measure the mechanical properties of the various functional materials, including battery materials such as electrodes.[129] AM-FM technique operates by exciting the cantilever at two resonances frequencies simultaneously: the fundamental frequency and usually the second but also possibly the third or fourth resonant frequency. The lower cantilever resonant frequency is used for standard tapping-mode imaging, also known as amplitude modulation (AM), and this provides non-invasive data for high-quality topography image. The higher cantilever resonant frequency operates in FM mode, providing information on the elastic tip–sample interaction sensitively.[130] The measured frequency shift can be related to the tip–sample stiffness k_{ts} by using the equation[131]:

$$k_{ts} = \frac{2k_2\Delta f_2}{f_{0,2}}, \tag{3.1}$$

where $f_{0.2}$ is the second resonant frequency measured at free vibration; Δf_2 is the frequency shift of the second resonant mode as the tip interacts with the surface; k_2 is the force constant of the second resonance. The tip–sample stiffness can also be calculated using a mechanical model such as a Hertz contact model, which approximates the shape of an AFM tip as punch shape, and therefore, the contact stiffness can be written as[131]:

$$k_{ts} = 2E_{eff}r_c, \tag{3.2}$$

where r_c is the contact radius of the tip; E_{eff} is the effective elastic modulus. Hence, the frequency shift of the second mode can be converted into tip–sample contact stiffness (N/m) with suitable calibration of the cantilever, or into elastic modulus (GPa) through the calibration of the system with a reference material of known elastic modulus. Basically, higher frequency means greater stiffness or modulus. This technique can be used in characterizing the variations of the mechanical properties of the materials for Li-ion battery during the electrochemical charging/discharging processes.

3.3 CHARACTERIZATION OF ELECTRODES MATERIALS FOR Li-ION BATTERY

In this section, we will summarize the application of the SPM-based techniques to investigate the functionalities of the lithium-ion battery materials. The discussion will be mainly focused on the topographic imaging, current-based probing, electrochemical strain-based detection and the mechanical properties of the materials during the charging/discharging cycles of the battery systems.

3.3.1 In Situ and Ex Situ SPM Characterization

In situ and ex situ SPM characterization works are mainly using AFM and Electrochemical SPM as the primary tools, many earlier work has focused in this area.[132–134] Generally speaking, the battery materials are placed in the electrochemical cell, which is controlled by an external potentiostat, and SPM is mainly used to measure the topographic changes during the electrochemical processes. The operation of a Li-ion battery is typically associated with the changes in the volume, structure and topography of the electrode materials. One of the challenges is the link between the degree of electrochemical conversion (amount of Li-ions intercalated in the lattice) during the charging/discharging processes and the characteristic lattice parameters and molar volume of the material. For example, Clemencon et al. have performed dynamic measurement to show that the step height on Li_xCoO_2 crystals (pressed on a gold foil) was a function of Li-ion concentration during lithium de-intercalation by utilizing an in situ EC-AFM technique.[135] The dimensional change of individual Li_xCoO_2 crystals (pressed on a gold foil) along the C_{hex} axis was in good agreement with previous X-ray powder diffraction results. Evidence of surface instability or structural instability was not found in Li_xCoO_2 single crystals upon de-intercalation to 4.2V versus Li. This work demonstrated the potential of SPM techniques in the study of surface and dimensional change of lithium transition metal oxides. Furthermore, Ramdon et al., using the in situ EC-AFM, have shown that the particle sizes of the Li_xFePO_4 cathode increase/decrease due to lithiation/delithiation during the charging/discharging processes of the cell.[43] After charging, the particle size reduced from an initial size of 0.30 to 0.26 mm^2, whereas after discharging, the monitored particle size increased from 0.26 to 0.33 mm^2. In addition, Wu et al. examined the morphological changes of $LiFePO_4$ thin-film cathode (prepared by radio frequency magnetron sputtering deposition) in aqueous environment during charging/discharging processes by using in situ EC-AFM technique.[136] They observed that the grain size decreased during the charge and increased during discharge processes, respectively.

During the charging/discharging processes of the LIBs, the accumulation of irreversible insertion/extraction of Li-ions contributes to the impedance increase, capacity decay and power fading (aging) of the batteries. Understanding these has resulted in a strong interest for observations of electrode topography under in situ or ex situ conditions. Using in situ EC-AFM, Vidu et al. have monitored the surface dynamics of the composite $LiMn_2O_4$ cathode (75 wt% active material mixed with 25 wt% acetylene black and 5 wt% Teflon) in the organic electrolyte under potentiostatic conditions.[133] They have observed dissolution/precipitation of surface particles during the charging/discharging processes. In addition, Quinlan et al. have monitored the surface of the composite $Li_xMn_2O_4$ cathode (75 wt% active material mixed with 25 wt% acetylene black and 5 wt% Teflon) before and after storage at elevated temperature by using ex situ AFM.[137] They found fine and clearly round-shaped grains covering the larger grains on the surface of the cathode after storage. This surface deterioration is responsible for the corresponding capacity fading. Similarly, Doi et al. have studied surface morphology of $LiMn_2O_4$ thin film (prepared by PLD) by using in situ EC-AFM in the electrochemical environment during cycling.[133] The formation of small round-shaped particles has been identified as the possible origin for capacity fading at elevated temperatures. The understanding of these surface degradation mechanisms allowed the researchers to develop new strategies to increase the thermal stability, such as surface protection, as reported in the subsequent in situ AFM studies.

On the other hand, the surface roughness changes are accompanied with the topography change during the electrochemical cycling. The ex situ SPM measurements are usually performed during the charging/discharging processes in a real battery system, then using AFM to measure the topography changes of the electrode materials after the electrochemical processes. By using this technique, Tang et al. observed some large agglomerates formed on the surface of $LiMn_2O_4$ thin film (prepared by PLD method) after 550 cycles; thus the roughness of the thin film increased.[138] They ascribed this observation to the electrochemical migration and agglomeration of small nano-grains during the cycling processes. In addition, Matsui et al. have employed ex situ AFM to explore the changes in surface morphology during electrochemical operation of $LiCoO_2$ thin film (prepared by RF-sputtering method).[139] They have observed the reversible changes of surface roughness evolution during charging/discharging processes of $LiCoO_2$ electrode films. Moreover, Ramdon et al. have conducted high-resolution morphology mapping on aged and unaged $LiFePO_4$ cathode (extracted from a cylindrical cell) by ex situ AFM.[140] They have found that the aged sample becomes rougher than the unaged sample, which suggested that the aged surface was comprised of smaller particles.

AFM technique has also been exploited in all-solid-state thin-film battery.[36] For example, Kushida et al. have used in situ AFM to observe the dynamic topography of the Al layer capping the Li anode during charging/discharging operations in an all-solid-state Li-ion secondary battery, which consisted of multiple-layers structure ($Al/Li/SiO_2$_15at.%$P_2O_5/LiMn_2O_4$/polycrystalline silicon).[141] The surface of Al electrode showed a cyclic change from scaly to wrinkled structures during the charging/discharging operations, which suggested that the cyclic movement of Li-ions contributed to those changes in surface morphology.

If one applies DC bias to the sample surface or AFM tip by using a conductive tip that is directly contacted with the sample surface, the contact-mode AFM measurement becomes biased AFM measurement. The biased AFM tip concentrates an electric field in a nanometer-scale volume of materials, which can introduce electrochemical processes at the tip–surface junction and ionic current flow in the materials, if the measurements are conducted in the ambient environments. Zhu et al. have applied this technique to simulate the effects of the cycling processes on surface deformation, phase formation and surface potential in an all-solid-state thin-film battery.[84,85] This technique can simulate the charging/discharging processes for the battery materials. When positive bias is applied through the AFM tip, which is similar to the charge process, therefore, the Li-ions are forced to move from cathode to anode, whereas to simulate the discharge process, a negative bias is applied to the material, and then Li-ions are forced to move from anode to cathode. Therefore, when positive and negative biases are applied in sequence at the same area for time long enough to induce Li-ions diffusion, followed by tapping-mode AFM to scan the biased and surrounding region, the Li-ion intercalation and de-intercalation processes can be studied separately (Figure 3.4a–c), and these processes cannot be distinguished if

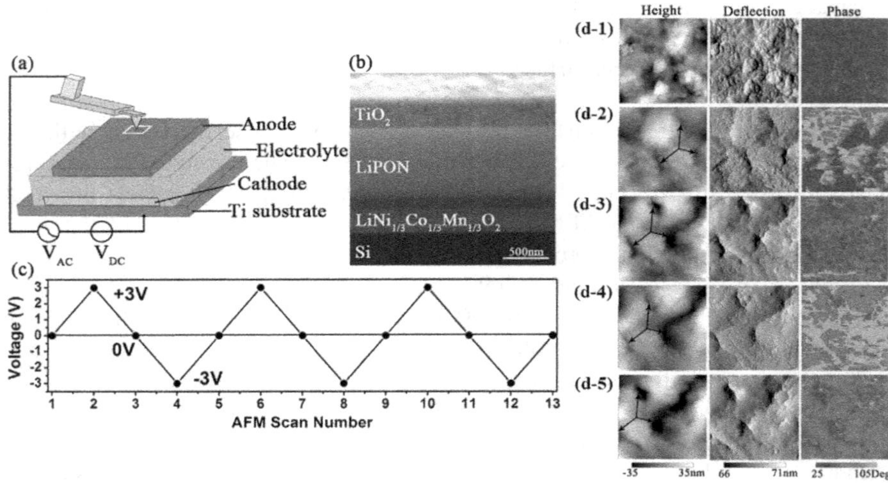

FIGURE 3.4 (a) Schematic of the in situ biased AFM measurement. The bias is applied through the conductive tip to the anode surface. A 1×1 μm^2 area (indicated by the white square) was scanned by the biased AFM tip. (b) The cross-section image of the thin film battery by using SEM. (c) Cyclic electrical field applied to the battery through the AFM tip versus scanning numbers by AFM. (d) The AFM images obtained on the anode surface which is polarized to the cyclic biases; from left to the right column: height, deflection and phase images: (d-1) scan 1, (d-2) scan 2, (d-3) scan 4, (d-4) scan 6 and (d-5) scan 8. (Reprinted from J. Zhu, J.K. Feng, L. Lu, and K.Y. Zeng, *Journal of Power Sources*, **197**, 224 (2012), with permission from Elsevier.[84])

AC voltage is applied. Previous studies have shown that new phases (green color) have been formed and disappeared when applied positive and negative in sequence, respectively, indicating the Li-ion diffusion processes (Figure 3.4d).[84,85] Yang and colleagues have also applied this biased AFM technique to characterize the Li-ion diffusion in a $Li_{1.2}Co_{0.13}Ni_{0.13}Mn_{0.54}O_2$ thin-film cathode under the ambient condition.[113] This work showed significant effects of grain boundaries; also, the image showed that the Li-ions are extracted to the surface when applied the bias but cannot re-enter the materials as this was single layer of cathode (Figure 3.5). The results have provided a new perspective to understand the underlying mechanisms of the cycling performance of the cathode in the battery systems.

Overall, the capability of the in situ and ex situ AFMs to monitor the changes in the surface structure is invaluable in studying the electrochemical processes of the battery systems, both in materials and device levels. However, the chemical identification is generally unavailable and information is limited to surface only. In fact, the topographic measurements, whether in situ or ex situ, are largely focused on using EC-AFM.[51,52,133] The biased AFM can be used to simulate the charging/discharging processes of the battery; it can also study the microstructure evolution during the processes. With the development of various functionality measurement modules in SPM systems, the topographic images are always obtained when measuring various properties of materials, including c-AFM, ESM, KPFM and more; therefore, the

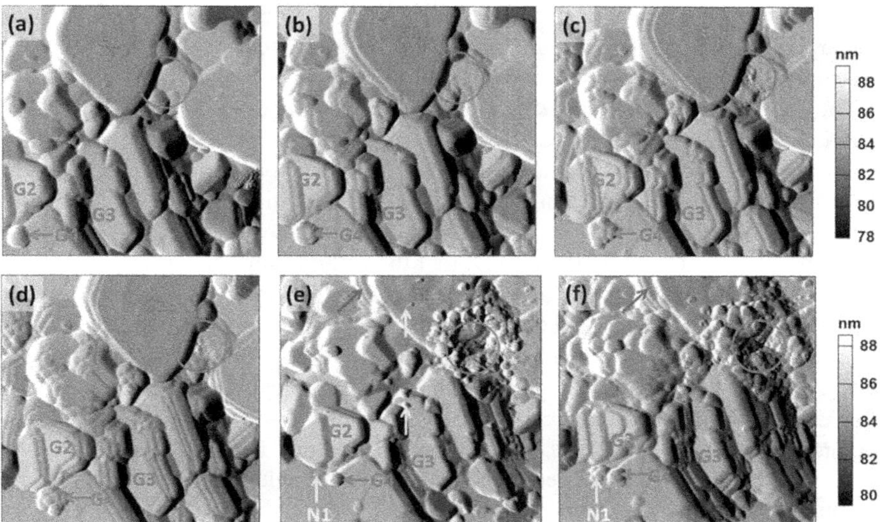

FIGURE 3.5 Biased AFM images of the $Li_{1.2}Co_{0.13}Ni_{0.13}Mn_{0.54}O_2$ thin-film cathode when bias is applied at grain boundary. The biased position is located at the center of the red circle (grain boundary), and the scanning size is 1×1 μm^2. Deflection images of the same area of: (a) before application of the bias; (b) 1st time under bias of +7 V; (c) 2nd time under bias of +7 V; (d) 3rd time under bias of +7 V; (e) under the bias of −7 V; and (f) under the bias of +7V again. (Reproduced from S. Yang, B.G. Yan, L. Lu, and K.Y. Zeng, *RSC Advances*, **6,** 94000 (2016), with permission from Royal Society of Chemistry.[113])

topographic measurement is generally combined with the measurement of properties in recent years; some of those measurements will be reviewed and summarized in more detail below.

3.3.2 Current-Based SPM

Current images are much more sensitive to the formation of conductive layers or paths than the changes in the topography, allowing the detection of new phases. The measurements utilize the current-based SPM technique, or c-AFM.[142] c-AFM is developed based on the contact-mode AFM to simultaneously measure the topography and current by maintaining a constant tip–sample interaction force using a feedback loop. During the c-AFM measurements, a voltage is applied between the sample and AFM tip to form the topography and current image. Kostecki et al. have reported the variation of the conductivity between the top and sides of the grains in $Li_2Mn_4O_9$ thin film (prepared by spin coating).[143] They found that, after long exposure of the $Li_2Mn_4O_9$ film to the electrolyte, a thin insulating passive layer was formed, which could be detected by the ex situ c-AFM technique. They, therefore, proposed a reaction scheme of $Li_2Mn_4O_9$ conversion to MnO_2 accompanied by the Li_2O extraction. A similar study was conducted by Matsuo et al.,[144] who have explored the surface conductivity in $LiMn_2O_4$ thin film (fabricated by spin coating) exposed to electrolyte at elevated temperature by using ex situ c-AFM. The electronically insulating surface layers were detected and found responsible for the electrode decomposition and consequent capacity loss.

The degradation of electrochemical properties of cathodes upon cycling has also been investigated in terms of deterioration of electronic contact resistances within the composite cathodes. Using the c-AFM, Kostecki et al. observed the current variations at grain boundaries and along surface irregularities in composite $LiNi_{0.8}Co_{0.2}O_2$ cathode (consisted of highly conductive graphite and acetylene black, insulating PVDF binders).[108] The authors revealed that the cathode surface electronic conductance diminished significantly after the cell cycling at elevated temperatures, and the rate of the electronic conductance change increased with the cycling temperature. A similar study was conducted on composite $LiNi_{0.8}Co_{0.15}Al_{0.05}O_2$ cathode as well. The current images of the cathode showed a significant surface resistance rise after Li-ion cells were cycled at elevated temperatures. c-AFM imaging of individual agglomerates of $LiNi_{0.8}Co_{0.15}Al_{0.05}O_2$ powder (embedded into an Au foil) revealed poor electronic contact between small crystallites within $LiNi_{0.8}Co_{0.15}Al_{0.05}O_2$ agglomerates. Carbon rearrangement that occurs at the cathode surface of tested cell could lead to poor electronic contact within the agglomerates and, eventually, isolation of active materials. These factors were accounted for the observed cell capacity loss. In addition, Kerlau et al. reported that conductance images of the composite $LiNi_{1/3}Co_{1/3}Mn_{1/3}O_2$ cathode from the cycled cells showed a substantial increase of surface resistance.[145] Moreover, c-AFM was conducted on unaged and aged $LiFePO_4$ electrode (extracted from a commercial cell) by Ramdon et al.[140] The aged sample has shown poor electrical conductivity and hence increased resistance. The loss of the contact with the active material from the substrate, and the rearrangement of the carbon additive were the possible reasons for the lower conduction of the aged cathode. Kuriyama et al. measured locally current and topography changes by applying a negative bias to attract Li-ions toward the surface of $LiMn_2O_4$ cathode film (spin coating) in air.[107] Under the negative bias of 5.5V, electric current abruptly increased, indicating Li ionic conduction. At the same time, part of the scale-shaped grains expanded and flattened. The authors proposed Jahn–Teller instability, which was induced by the repulsion between Mn-e_g and O-2_p electrons, as a possible reason for those observations.

Yang and colleagues have applied the c-AFM technique to study the variation of the conductivity and I–V (current–voltage) characteristics as the function of the microstructure in the $Li_{1.2}Co_{0.13}Ni_{013}Mn_{0.54}O_2$ cathode thin film (Figure 3.6).[113] It was found that the grain boundaries are more conductive than the grain interiors, indicating the enhanced Li-ions diffusivity at the grain boundaries. The measured I–V curves also showed decreased current and increased voltage for current initiation when the bias-applied position moved away from the boundaries. Those results suggested that the larger amount of Li-ions can be induced near the grain boundary and the energy barrier for Li-ions diffusion was lower as the distance from the grain boundary decreases. In another work,[37] the ex situ c-AFM was applied to measure the conductivity variation of the charged/discharged $Li_{1.2}Co_{0.13}Ni_{013}Mn_{0.54}O_2$ cathode thin film. It was found that the conductivity of the thin film cathode was reduced even

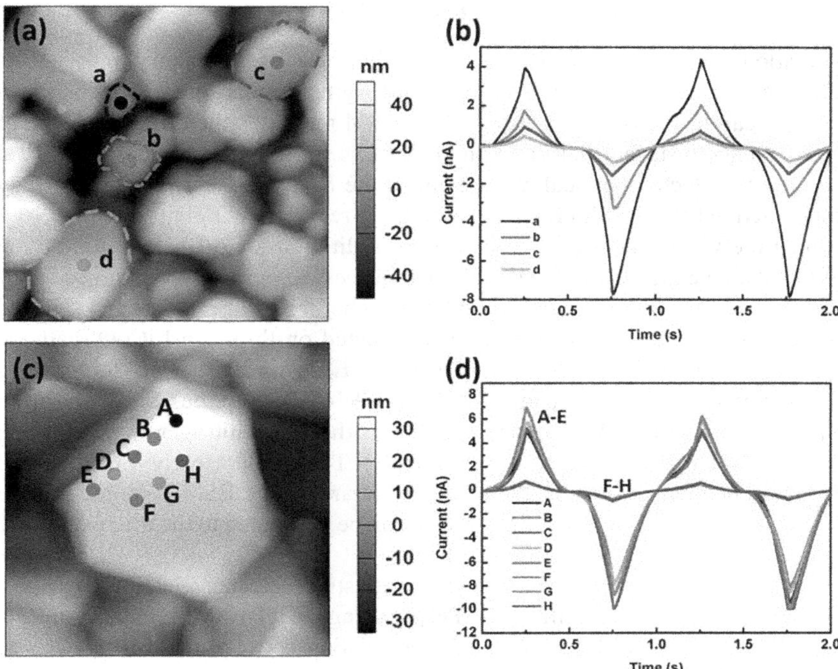

FIGURE 3.6 AFM images and I–V curves measurements at different locations: (a) AFM height image (1.1 × 1.1 μm²) where the center of four selected grains are marked (a–d) for tip locations; (b) current as a function of time at the four locations marked in (a); (c) AFM height image (0.5 × 0.5 μm²) where eight selected points are marked (A–F) for tip locations; and (d) current as a function of time at the eight locations marked in (c). (Reproduced from S. Yang, B.G. Yan, L. Lu, and K.Y. Zeng, *RSC Advances*, **6,** 94000 (2016), with permission from Royal Society of Chemistry.[113])

after first cycle, and the discharge could only recover certain amount of conductivity, indicating the effects of charging/discharging cycles.

Furthermore, c-AFM was conducted on solid Li-ion electrolyte with a general composition of Li_2O-Al_2O_3-SiO_2-P_2O_5-TiO_2-GeO_2.[109–112] Particle formation was observed after the application of larger triangular bias waves. By measuring the total charge transfer during the particle formation and the particle volume, those particles were found to be the Li metal. These studies open pathways for probing nanoscale, electrochemical, bias-induced phenomena in practical Li-ion electrolyte systems.

Overall, current-based SPM or c-AFM is very powerful for exploring the electric properties of the battery materials. The mapping between topography and current images can reveal the detail of the conductivity variation as function of the microstructure; hence the mechanisms of decay of the materials can be understood. The primary limitation of this technique is that the current methods cannot be directly applied to probe the ionic conductivity of the material. The search for feasible method for ionic current mapping is one of the future directions for SPM.

3.3.3 Electrochemical Strain Microscopy Techniques

The operation of the battery is affected by deformation of the battery materials as a result of molar volume changes during the charging/discharging processes. Such deformation can be measured by using ESM. ESM measures the direct coupling of ionic movements to surface deformation (defined as strain), and provides a new tool for mapping the electrochemical phenomena at the nanoscale. The first ESM study was conducted in the thin-film $LiCoO_2$ cathode, as reported by Balke et al. in 2010.[43] By probing the volume changes corresponding to lithium ion diffusion, lithium ion diffusivity at single grains and fixed grain boundaries was spatially visualized. They have shown that the lithium ion diffusivity was higher for certain grains and selected grain boundaries. The subsequent studies conducted on thin-film $LiCoO_2$ cathode have suggested the enormous potential of using ESM measurements for probing ionic transport phenomena in battery materials.[71,75,146] Amorphous Si anode in an all-solid-state thin-film LIB (with the $LiCoO_2$ as cathode and nitrogen-doped lithium phosphate as electrolyte) was also investigated.[41,74] The results clearly illustrated the accumulation of more Li-ions at the sharp boundaries. The ESM response can be used as a direct measure of Li-ion flow; thus it can be correlated to the microstructure changes to identify the high diffusion paths.

In another work, Chen et al. have revealed drastic differences in ESM response between micro- and nanocrystalline $LiFePO_4$ by using DART-ESM.[79] Nanocrystalline area has much higher ESM response. This correlation between ESM response and crystalline morphologies explained the superior capacity observed in LIBs with nanocrystalline $LiFePO_4$ electrode. This study has shown that ESM correlates well with macroscopic battery performances and help understand the mechanisms responsible for performance enhancement.

In addition, Zhu and colleagues investigated the local variations of the Li-ions diffusion and electrochemical activity in the $LiNi_{1/3}Co_{1/3}Mn_{1/3}O_2$ thin-film cathode at different cycling stages by using BE-ESM.[36] The averaged ESM response was significantly decreased after 100 cycles, corresponding to the degradation in electrochemical activity after cycling processes in the thin film cathode (Figure 3.7). This study has shown the capability of ESM to provide further understanding of the origins of battery failure at the nanoscale.

If the Li-ion transport processes are diffusion-controlled and the contribution of the ion migration is minimal, the ESM amplitude can be related to Li diffusion coefficient as[39]:

$$A = 2(1+\nu)\beta \frac{V_{ac}}{\eta} \frac{\sqrt{D}}{\sqrt{\omega}} \tag{3.3}$$

where ν is the Poisson's ratio, V_{ac} is the AC voltage amplitude, ω is the frequency of the applied electrical field, D is the Li ion diffusion coefficient, η describes the linear relationship between the applied field and chemical potential,[39] and $\beta = 0.02349$ is the Vegard coefficient, which can be expressed as an empirical linear relationship between the lattice size and Li ion concentration.[43] Using this relationship, Yang and colleagues have shown that, for $Li_{1.2}Co_{0.13}Ni_{0.13}Mn_{0.54}O_2$ thin-film cathode, the diffusion

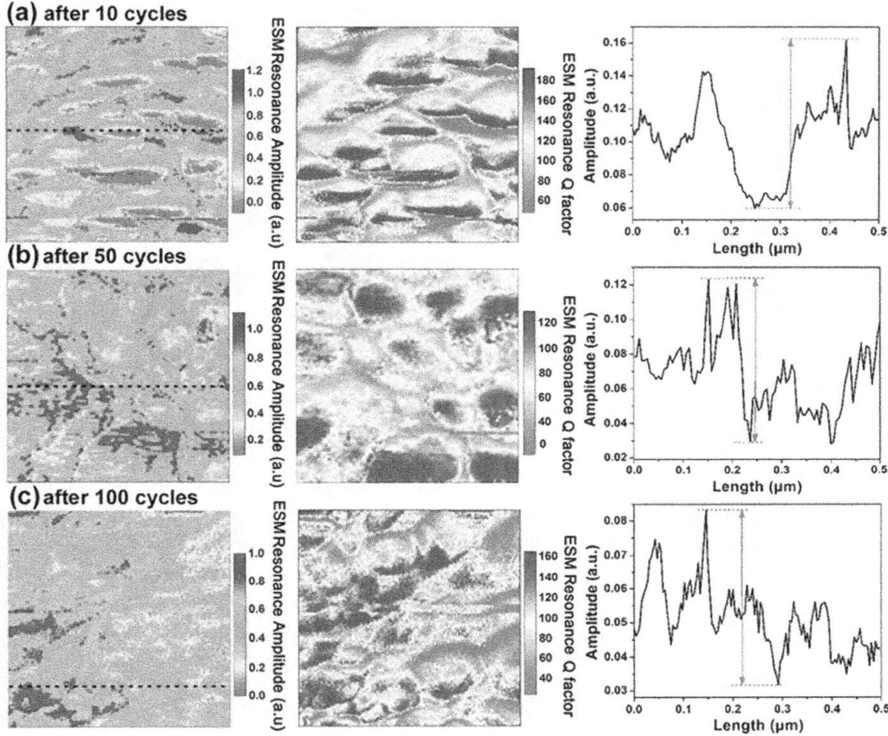

FIGURE 3.7 BE-ESM maps of resonance amplitude and Q factor of $LiNi_{1/3}Co_{1/3}Mn_{1/3}O_2$ thin-film cathode at different cycling stages: (a) 10 cycles, (b) 50 cycles and (c) 100 cycles, as well as the line section profiles of resonance amplitude. The left column is the amplitude map, the middle column is the Q factor map and the right column is the line section profile of the amplitude maps, corresponding to the lines in the left column. (Reprinted with permission from J. Zhu, L. Lu, and K.Y. Zeng, *ACS Nano*, **7,** 1666 (2013), Copyright 2013, American Chemical Society.[36])

coefficient map, derived from the ESM amplitude image, agreed well with that from the PITT measurements (Figure 3.8a–e).[99] The result confirmed that the ESM images can be used to determine the local diffusivity as well as to obtain the corresponding local diffusion coefficients of the cathode materials with certain reasonable assumptions; the diffusion map also suggested that higher deformation regions have higher electrochemical activities. In addition, the diffusion maps can be obtained with the biased cycles to simulate the effects of charging/discharging cycles.[99] It was shown that the local diffusion coefficients decreased continuously with the increased number of ESM scanning, with about 60% of loss of the local diffusivity after the first image, which suggested that the local irreversible topography change and diffusivity loss at the nanoscale can be correlated with each other. Furthermore, the ESM measurements were conducted at different temperatures (Figure 3.8f–j),[100] by deriving the diffusion coefficient maps at various temperatures from the ESM images, the

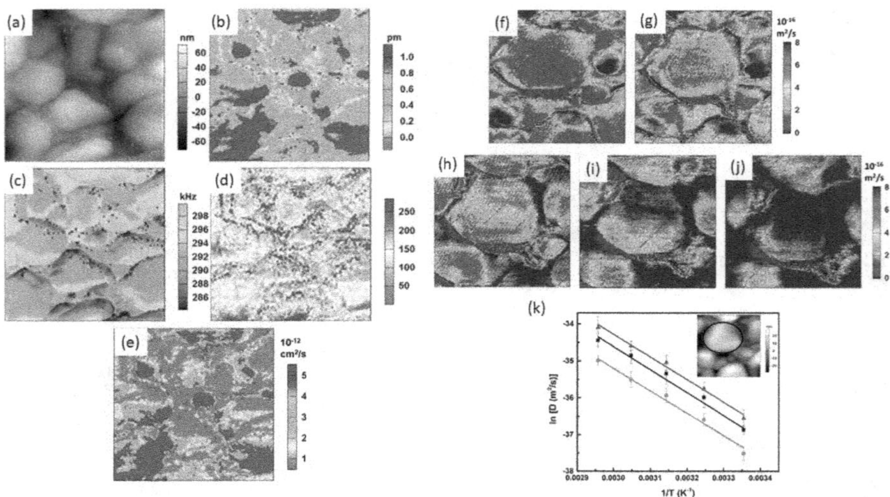

FIGURE 3.8 (a–d) are the typical BE-ESM (with 3 V_{ac} bias) height, resonance amplitude, resonant frequency and Q-factor images, respectively; (e) the calculated diffusion coefficient map. All of the BE-ESM images are obtained with a pixel density of 100×100 over 0.8×0.8 μm^2. (Reprinted from S. Yang, B.G. Yan, T. Li, J. Zhu, L. Lu, and K.Y. Zeng, *Physical Chemistry Chemical Physics*, **17,** 22235 (2015), with permission from Royal Society of Chemistry).[99] The calculated diffusion coefficient maps (600×600 nm^2) at selected temperatures of (f) 25, (g) 35, (h) 45, (i) 55 and (j) 65°C; (k) Arrhenius plot of the diffusion coefficients as a function of inverse of temperature for selected grains. (Reprinted with permission from S. Yang, B. Yan, J. Wu, L. Lu, and K.Y. Zeng, *ACS Applied Materials & Interfaces*, **9,** 13999 (2017), Copyright 2017, American Chemical Society.[100])

activation energy for lithium ions diffusion of local regions (such as grains and grain boundaries), as well as the activation energy map can be developed (Figure 3.8k). A similar measurement was also conducted on the electrolyte material by Chen et al.[147]

ESM in the single frequency mode was implemented to measure the local strain in bulk $LiMn_2O_4$ cathodes of commercial battery cells. ESM response was observed on the $LiMn_2O_4$ active particles. Noise signal was observed on the epoxy resin and the Al current collector. Meanwhile, ESM was conducted on Li-ion conductive glass ceramics (LICGC) with general composition of Li_2O-Al_2O_3-SiO_2-P_2O_5-TiO_2-GeO_2. The ESM response of the LICGC was significantly low. The localized spots with high ESM response were identified as impurity $AlPO_4$ phase.

As discussed above, ESM images, either using DART or BE techniques, are capable of probing Li-ion flows and electrochemical activity in the battery materials at the nanometer level. To date, BE-ESM and DART-ESM have demonstrated its potential for characterizing a variety of LIB materials, including cathodes, anodes and electrolytes, as well as full battery systems.

3.3.4 Characterization of Local Mechanical Properties for Li-Ion Battery Materials

As discussed earlier, it is well known that battery materials undergo large volume changes during the electromechanical processes through the charging/discharging cycles. Therefore, mechanical property of the battery materials is always one of the important properties which determines the lifetime of the battery.[117,119,120] Generally speaking, the degradation of the mechanical properties may reduce the interfacial adhesion and the contact between the electrode and electrolyte (solid or electrolyte) and hence may cause cracking in the electrode which reduces the battery life. Furthermore, the individual particles in the electrode materials may be broken and fall off from the electrode during the charging/discharging processes, resulting in the graduate failure of the battery electrode materials.[148] In addition, nanoindentation experiments have shown that, for both $LiMn_2O_4$, and $LiNi_{1/3}Co_{1/3}Mn_{1/3}O_2$ thin-film cathodes, the elastic modulus and hardness have significantly reduced (by 30%–50%) after a number of charging/discharging cycles.[149] However, nanoindentation experiments can only determine the mechanical properties over an area, but cannot really identify the localized mechanical properties of individual components or microstructure, such as grains and/or grain boundaries, or interfaces.

Therefore, in recent years, SPM-based techniques, such as AM-FM techniques, are used to characterize the changes in the localized mechanical properties of battery materials and/or devices. For example, Yang and colleagues utilized the AM-FM technique,[37] and showed the detail of the changes of the elastic modulus of the $LiNi_{1/3}Co_{1/3}Mn_{1/3}O_2$ thin-film cathodes at different stages of galvonostatic charging/discharging processes; the elastic modulus has reduced to less than half of that of the as-deposited film even after one cycle of charge cycle and recovered somewhat, but was still lower than that of the as-deposited sample by about 30% (Figure 3.9j–l), it indicated the degradation of the mechanical properties during the first cycle was very significant. The results also agreed well with the observation and nanoindentation experiments on $LiMn_2O_4$ and $LiNi_{1/3}Co_{1/3}Mn_{1/3}O_2$ thin-film cathodes as well as RuO_2 and TiO_2 thin-film anode materials.[36,121–124] In fact, the changes in the mechanical properties can be associated with the changes in the topography and current, as well as the electrochemical deformation of the cathode materials (Figure 3.9a–i).[37]

3.4 CHARACTERIZATION OF SOLID ELECTROLYTE FOR Li-ION BATTERY

One of the biggest concerns on Li-ion battery is the widely used liquid electrolyte, which causes safety issues for the Li-ion battery. Hence, solid electrolytes are developed to overcome the safety issue; however, the ionic conductivities of most solid electrolytes are at least one or two orders smaller than those of the liquid electrolyte, which leads to the significant issue of battery performance when using solid electrolyte. Extensive studies show that probing the bias-induced reactivity of the solid electrolytes is still largely confined to the macroscopic length scales. For example,

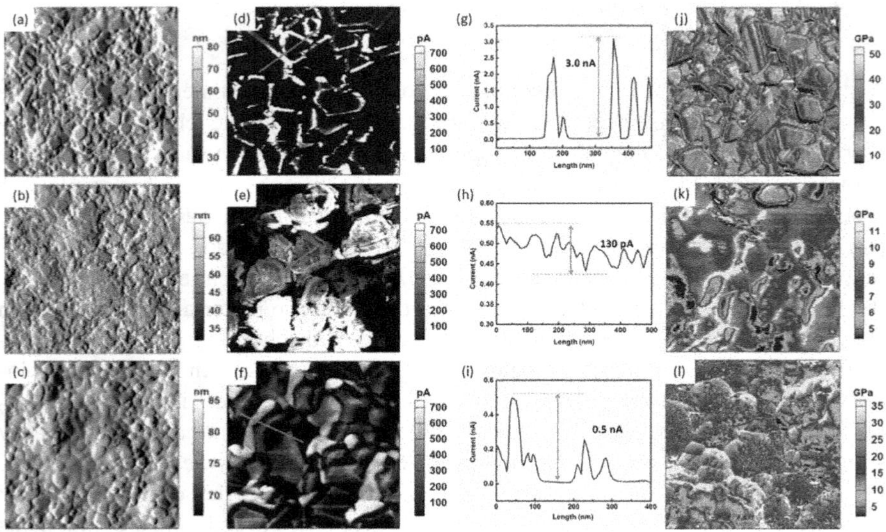

FIGURE 3.9 AFM deflection images of $Li_{1.2}Co_{0.13}Ni_{0.13}Mn_{0.54}O_2$ cathode films taken at different stages of galvanostatic charging/discharging processes: (a) as-deposited, (b) charged and (c) discharged; the current maps of $Li_{1.2}Co_{0.13}Ni_{0.13}Mn_{0.54}O_2$ thin-film cathode with 3V DC bias over the scanning area of 1×1 μm^2 taken at different stages of galvonostatic charging/discharging processes: (d) as-deposited, (e) charged and (f) discharged. (g), (h) and (i) the line section profiles in current images, corresponding to the lines in (d), (e) and (f), respectively; the AM-FM measured and calculated elastic modulus maps taken at different stages of galvanostatic charging/discharging processes in $Li_{1.2}Co_{0.13}Ni_{0.13}Mn_{0.54}O_2$ thin-film cathode: (j) as-deposited, (k) charged and (l) discharged. (Reprinted from S. Yang, J.X. Wu, B.G. Yan, L. Li, Y. Sun, L. Lu, and K.Y. Zeng, *Journal of Power Sources*, **352,** 9 (2017) with permission from Elsevier.[37])

the electrochemical performance of the electrolytes is usually studied by the EIS methods, where the electrolytes are coated with two plane electrodes and the ionic transportation is driven under an external electric field on the two plane electrodes. Therefore, the ionic conductivity of the solid electrolytes is derived by the impedance response, and a statistical and average macro-level conduction of the mobile ions is reflected in this method. However, the nonuniform structure of the solid electrolytes means that different components can play different roles on the overall performance. It is worth pointing out that none of research objectives in this thesis is a uniform matter. For example, the polymer-based solid electrolytes consist of polymer matrix and lithium salt and the solid composite electrolytes contain additional inorganic fillers. Even the inorganic ceramic solid electrolytes, such as $Li_{1.5}Al_{0.5}Ge_{1.5}(PO_4)_3$ or LAGP and $Na_{1+x}Zr_2Si_xP_{3-x}O_{12}$ ($0 \leq x \leq 3$) or NZSP, have complex grain and gain boundaries features. In this case, the SPM-based techniques employed in this thesis have shown their unique advantages in studying the microscopic and local electrochemical properties of these solid electrolytes with multiple structural features.

The microscopic features of the solid-state electrolytes are usually investigated by the electron microscopies such as SEM and TEM. However, these methods provide only static snapshot of the material structure or in selected cases beam-induced dynamics,[88] rather than the real electrochemical responses. In addition, they always require complicated fabrication processing, and the limited electron permeability of the solid-state electrolytes puts them in risk of destruction by the high-energy electron beam. On the other hand, the SPM-based techniques are capable of focusing an electrical stimulus on the nanometers region and resolving associated structural and functional responses. Moreover, they are non-destructive and can be operated on both inert and ambient conditions.

The chemo-mechanical properties would strongly affect the interfacial stability in practical batteries. On one hand, the mechanical stress strongly influences the interfacial contact and causes some side reactions. On the other hand, electrochemical reactions such as lithium deposition would cause mechanical deformation and stress at electrode and solid electrolytes interfaces.[88] Based on the unique advantages of SPM techniques in probing the local multifunctional response of the solid matters, some interesting findings have been obtained by studying the local electrochemical and nano-mechanical properties of these typical solid-state electrolytes. These findings of the local performance of the solid-state electrolytes are instructive for the future design of novel solid-state electrolytes as well as their application in the battery cells.

Generally speaking, ionic conductivity is one of the major parameters for electrolyte materials, as the electrolyte should be good ionic conductor whereas should be electric insulator. Hence, some of the SPM-based techniques, such as c-AFM, are not suitable for characterizing the conductivity of the electrolyte materials. For characterizing solid electrolyte, the main SPM techniques include AFM, ESM, bi-modal AFM, AM-FM, etc., which are mainly used to characterize surface topography, deformation and mechanical properties of materials.

Wang and colleagues have recently applied various SPM-based techniques, including AFM and ESM, to characterize the nanoscale electrochemical deformation within grains and grain boundary in the NASICON-structured (sodium (Na) Super Ionic Conductor) $Li_{1.5}Al_{0.5}Ge_{1.5}(PO_4)_3$ (LAGP) solid electrolyte.[80] They have observed the different electrochemical responses at grains and grain boundaries, which can be correlated with different lithium ion conductivities inside the grains and along the grain boundaries; the diffusional coefficient within the glassy phase at the grain boundary was estimated to be approximately 6.7 times higher than that at the grains based on Eq. (3.5), indicating that the diffusion of the lithium ions in the glassy phase is much faster than that within the grains (Figure 3.10). This work also found that the electrochemical strain induced by external bias might cause the strain mismatch between the solid electrolyte and electrode, which could result in significant influence on the performance of the all-solid-state battery in long-term services.

The SPM-based characterization on solid electrolyte can also be extended to the solid electrolyte used for Na^+ battery. Wang and colleagues recently applied bi-modal AFM and ESM to characterize NASICON/Poly(ethylene oxide) composite solid electrolytes (NASICON/PEO CEs).[83] The bi-modal AFM can clearly show the different phases in the composite electrolyte, and again ESM showed the Na ion diffusion faster along the interface between the NASICON-PEO interfaces than that within pure PEO or pure NASICON ceramics (Figure 3.11).

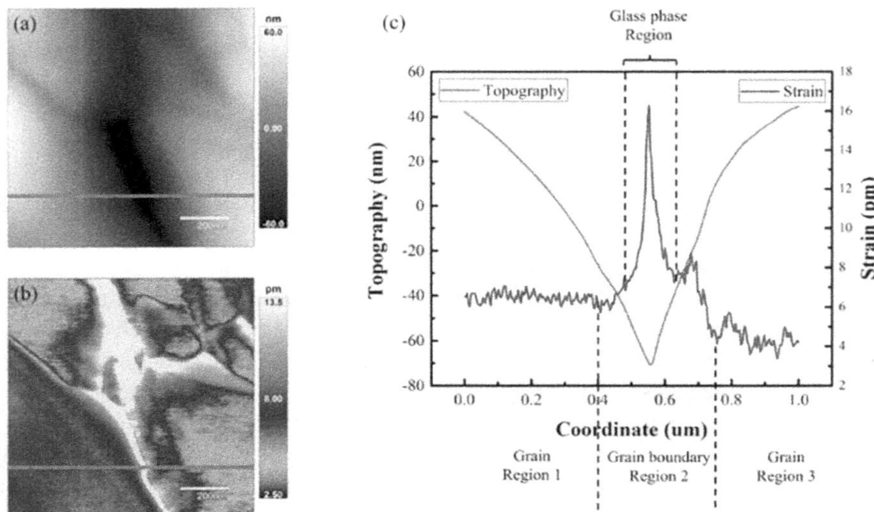

FIGURE 3.10 The ESM image of LAGP solid electrolyte: (a) topography, (b) electrochemical strain and (c) a line analysis of topography and electrochemical strain. (Reprinted from Z.T. Wang, M. Kotobuki, L. Lu, and K.Y. Zeng, *Electrochimica Acta*, **334,** 135553 (2020), with permission from Elsevier.[80])

In a recent work, Sun and colleagues have applied SPM-based techniques, including ESM and c-AFM, to characterize another NASICON structured solid electrolyte NZSP ($Na_{1+x}Zr_2Si_xP_{3-x}O_{12}$, $0 \leq x \leq 3$).[81] Although solid electrolyte is an electric insulator, but c-AFM can be used to examine the metal reduction process in ionic conductors. The results showed that the over-potential applied on the surface of the electrolyte can lead to the extraction of Na^+ ions toward the surface of the material, and the Na^+ ions could quickly spread over the sample surface and get oxidized in air to form the final products which were attached on the sample surface. Hence, this work can be used to study the ionic wettability in the materials, which is also suitable for Li-ions-based solid electrolyte (Figure 3.12).

The mechanical properties of solid electrolyte materials can also be characterized by using AM-FM technique. To obtain the elastic modulus maps from the AM-FM images, one would need a reference material which has the modulus similar to the measured materials.[95,148,150,151] If the materials can be considered as homogeneous structure, this measurement is usually not a problem. However, for composite materials, such as those ceramic-polymer hybrid electrolyte materials, the elastic modulus mapping from AM-FM images is very challenging, as the ceramic and polymer materials usually have large differences in elastic moduli values; one reference material usually can only be used to extract the elastic modulus of one component in the composite. Therefore, one would need two different reference materials for modulus mapping of the composite, one for ceramic and one for polymers, and each time can

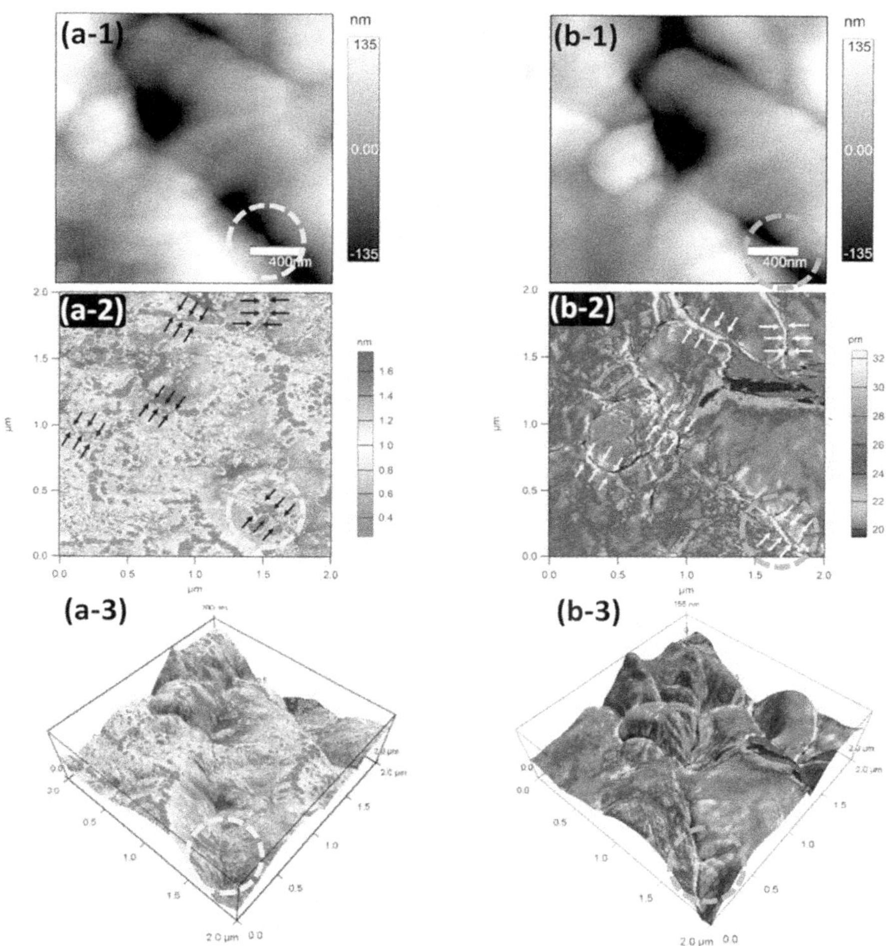

FIGURE 3.11 Bimodal AFM: (a-1) topography, (a-2) amplitude and (a-3) 3D topography profile overlapped with amplitude scale; ESM at the same position: (b-1) topography, (b-2) amplitude and (b-3) 3D topography profile overlapped with amplitude scale. (Reprinted from Y.M. Wang, Z.T. Wang, J. Sun, F. Zheng, M. Kotobuki, T. Wu, K.Y.T. Zeng, and L. Lu, *Journal of Power Sources*, **454,** 227949 (2020), with permission from Elsevier.[83])

only extract the elastic modulus of one component; hence the processes need to be repeated twice, and then re-synthesize the image from two processes. This will bring difficulty and errors in the measurements. The method which can avoid this problem is to analyze the force–distance curve obtained from AFM measurements.[53,152] The analysis of the force–distance curve can derive the elastic modulus and adhesion force of the materials based on contact mechanics theory and can be applied to the composite materials with certain limitation of reduced resolutions.

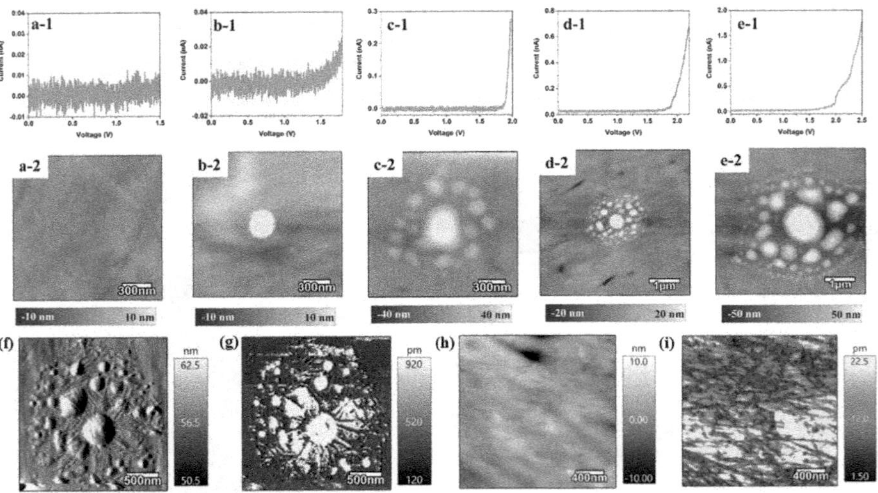

FIGURE 3.12 Current and topography evolution of NZSP as a function of the voltage. I–V curves with the gradual increase of voltage from 0V to: (a-1) 1.5V, (b-1) 1.8V, (c-1) 2.0V, (d-1) 2.2V and (e-1) 2.5V; the topography after each linear voltage sweep is captured as (a-2) to (e-2), correspondingly. Bimodal images of the products formed on NZSP sample at a bias of 2.2V. (f) First eigenmode amplitude and (g) second eigenmode amplitude. The first eigenmode shows the cantilever feedback signal related to topography, while the second eigenmode amplitude shows contrast in the material composition. Local ionic dynamics of the surface represented by: (h) Topography and (i) corresponding ESM amplitude image of the sample. (Reprinted with permission from Q. Sun, J.A.S. Oh, L. Lu, and K.Y. Zeng, *ACS Applied Materials & Interfaces*, **13,** 46588 (2021), Copyright 2021, American Chemical Society).

3.5 CONCLUSION REMARKS

This chapter reviews and summarizes a number of principles, functions and applications of various SPM-based techniques in the research of Li-ion batteries, including the in situ and ex situ characterization of electrodes, half and full battery cells, solid electrolyte and all-solid-state-batteries. However, this field is developing fast and many new phenomena need to be studied. In addition, new characterization techniques are also needed. The in situ and ex situ SPM-based characterization of the battery materials, electrode and/or electrolyte, provides the local properties which can be associated with the microstructure and functionality of the materials and devices; the results obtained can be used to understand the various underlying mechanisms associated with the functionality of the batteries, such as Li-ion diffusion coefficients, grain boundary effects and temperature effects of the electrode materials. It can reveal the changes in conductivity, diffusion activities and the mechanical properties of electrode materials during the charging/discharging cycles. In addition, it can also reveal various properties of electrolyte materials as well as interfaces between electrode and electrolyte, especially for all-solid-state batteries. It is believed that the SPM-based techniques will be used more and more widely to characterize the battery-related, or in more broad sense, energy storage materials in the future.

ACKNOWLEDGMENT

The author would like to thank my former post-doctoral researchers and graduate students, in particular, Drs. Jing Zhu, Tao Li, Shan Yang, Zongting Wang and Qiaomei Sun, for their hard work and results presented in this chapter. The author also wants to thank the collaborators, Drs. Yumei Wang, Sam Oh, Jianguo Sun, Prof. Masashi Kotobuki (Ming Chi University of Technology) and Prof. Li Lu (National University of Singapore), for their contribution, help and discussion during the courses of the projects related to this chapter. This work was supported by the Ministry of Education, Singapore through National University of Singapore under Academic Research Funds (R-265-000-305-112, R-265-000-406-112, R-265-000-596-112 and R-265-000-A27-112).

REFERENCES

1. J. B. Goodenough and Y. Kim, *Chemistry of Materials*, **22**, 587 (2010).
2. J. B. Goodenough and K. S. Park, *Journal of the American Chemical Society*, **135**, 1167 (2013).
3. J. Ma, Y. Li, N. S. Grundish, J. B. Goodenough, Y. Chen, L. Guo, Z. Peng, X. Qi, F. Yang, and L. Qie, *Journal of Physics D: Applied Physics*, **54**, 183001 (2021).
4. N. Nitta, F. Wu, J. T. Lee, and G. Yushin, *Materials Today*, **18**, 252 (2015).
5. A. Patil, V. Patil, D. W. Shin, J. W. Choi, D. S. Paik, and S. J. Yoon, *Materials Research Bulletin*, **43**, 1913 (2008).
6. J. B. Goodenough and H. C. Gao, *Science China-Chemistry*, **62**, 1555 (2019).
7. J. B. Goodenough, *Nature Electronics*, **1**, 204 (2018).
8. D. Choi, N. Shamim, A. Crawford, Q. Huang, C. K. Vartanian, V. V. Viswanathan, M. D. Paiss, M. J. E. Alam, D. M. Reed, and V. L. Sprenkle, *Journal of Power Sources*, **511**, 230419 (2021).
9. G. J. la O', *Chemical Engineering Progress*, **116**, 36 (2020).
10. Z. M. Zhang, W. F. Fang, and R. J. Ma, *Etransportation*, **2**, 100032 (2019).
11. Y. Miao, P. Hynan, A. von Jouanne, and A. Yokochi, *Energies* **12**, 1074 (2019).
12. R. Morello, R. Di Rienzo, R. Roncella, R. Saletti, R. Schwarz, V. R. H. Lorentz, E. R. G. Hoedemaekers, B. Rosca, and F. Baronti, *in 44th Annual Conference of the IEEE Industrial Electronics Society (IECON 2018)*, Washington, DC, 2018, p. 4949.
13. M. Shahjalal, P. K. Roy, T. Shams, A. Fly, J. I. Chowdhury, M. R. Ahmed, and K. L. Liu, *Energy*, **241**, 122881 (2022).
14. Z. Bakenov, K. Zhu, J. Liu, and H. Yadegari, *Frontiers in Energy Research*, **9**, 658875 (2021).
15. Q. Zhao, S. Stalin, C.-Z. Zhao, and L. A. Archer, *Nature Reviews Materials*, **5**, 229 (2020).
16. M. Li, C. Wang, Z. Chen, K. Xu, and J. Lu, *Chemical Reviews*, **120**, 6783 (2020).
17. N. Zhao, W. Khokhar, Z. Bi, C. Shi, X. Guo, L.-Z. Fan, and C.-W. Nan, *Joule*, **3**, 1190 (2019).
18. D. Saritha, in *1st International Conference on Manufacturing, Materials Science and Engineering*, CMR Inst Technol, Kandlakoya, India, 2019, **19**, p. 726.
19. F. Zheng, M. Kotobuki, S. F. Song, M. O. Lai, and L. Lu, *Journal of Power Sources*, **389**, 198 (2018).
20. C. Zhao, L. Liu, X. Qi, Y. Lu, F. Wu, J. Zhao, Y. Yu, Y.-S. Hu, and L. Chen, *Advanced Energy Materials*, **8**, 1703012 (2018).
21. W. Zhou, Y. Li, S. Xin, and J. B. Goodenough, *ACS Central Science*, **3**, 52 (2017).

22. J.-J. Kim, K. Yoon, I. Park, and K. Kang, *Small Methods*, **1**, 1700219 (2017).
23. L. S. Roselin, R. S. Juang, C. T. Hsieh, S. Sagadevan, A. Umar, R. Selvin, and H. H. Hegazy, *Materials*, **12**, 1229 (2019).
24. Q. Li, J. E. Chen, L. Fan, X. Q. Kong, and Y. Y. Lu, *Green Energy & Environment*, **1**, 18 (2016).
25. H. Kim, H. Kim, Z. Ding, M. H. Lee, K. Lim, G. Yoon, and K. Kang, *Advanced Energy Materials*, **6**, 1600943 (2016).
26. S. Hemavathi, in *International Conference on Computing, Power and Communication Technologies (GUCON 2019)*, New Delhi, India, 2019, **609**, p. 157.
27. L. Wang, R. Xie, B. Chen, X. Yu, J. Ma, C. Li, Z. Hu, X. Sun, C. Xu, and S. Dong, *Nature Communications*, **11**, 1 (2020).
28. Y. Li, Z. Gao, F. Hu, X. Lin, Y. Wei, J. Peng, J. Yang, Z. Li, Y. Huang, and H. Ding, *Small Methods*, **4**, 2000111 (2020).
29. L. He, Q. Sun, C. Chen, J. A. S. Oh, J. Sun, M. Li, W. Tu, H. Zhou, K. Y. Zeng, and L. Lu, *ACS Applied Materials & Interfaces,* **11**, 20895 (2019).
30. G. Hou, X. Ma, Q. Sun, Q. Ai, X. Xu, L. Chen, D. Li, J. Chen, H. Zhong, Y. Li, Z. Xu, P. Si, J. Feng, L. Zhang, F. Ding, and L. Ci, *ACS Applied Materials & Interfaces,* **10**, 18610 (2018).
31. Y. Yuan, K. Amine, J. Lu, and R. Shahbazian-Yassar, *Nature Communications*, **8**, 15806 (2017).
32. D. E. Demirocak, S. S. Srinivasan, and E. K. Stefanakos, *Applied Sciences-Basel*, **7**, 731 (2017).
33. M. M. Kabir and D. E. Demirocak, *International Journal of Energy Research*, **41**, 1963 (2017).
34. K. Y. Zeng, T. Li, and T. Tian, *Journal of Physics D: Applied Physics*, **50**, 313001 (2017).
35. J. Zhu and K. Y. Zeng, in *Nanotechnology for Sustainable Energy*, edited by Hu et al. American Chemical Society, 2013, p. 23.
36. J. Zhu, L. Lu, and K. Y. Zeng, *ACS Nano*, **7**, 1666 (2013).
37. S. Yang, J. X. Wu, B. G. Yan, L. Li, Y. Sun, L. Lu, and K. Y. Zeng, *Journal of Power Sources*, **352**, 9 (2017).
38. E. Strelcov, S. M. Yang, S. Jesse, N. Balke, R. K. Vasudevan, and S. V. Kalinin, *Nanoscale*, **8**, 13838 (2016).
39. N. Balke, S. Kalnaus, N. J. Dudney, C. Daniel, S. Jesse, and S. V. Kalinin, *Nano Letters*, **12**, 3399 (2012).
40. S. Guo, S. Jesse, S. Kalnaus, N. Balke, C. Daniel, and S. V. Kalinin, *Journal of the Electrochemical Society*, **158**, A982 (2011).
41. S. Jesse, N. Balke, E. Eliseev, A. Tselev, N. J. Dudney, A. N. Morozovska, and S. V. Kalinin, *ACS Nano*, **5**, 9682 (2011).
42. S. V. Kalinin, S. Jesse, A. T. Tselev, A. Kumar, T. M. Arruda, S. Guo, R. Proksch, *Materials Today*, **14**, 548 (2011).
43. N. Balke, S. Jesse, A. N. Morozovska, E. Eliseev, D. W. Chung, Y. Kim, L. Adamczyk, R. E. García, N. Dudney, and S. V. Kalinin, *Nature Nanotechnology*, **5**, 749 (2010).
44. T. Li and K. Y. Zeng, *Advanced Materials*, **30**, 1803064 (2018).
45. D. A. Bonnell and S. V. Kalinin, *Scanning Probe Microscopy for Energy Research,* World Scientific Publishing Company, 2013.
46. C. Shen, *Atomic Force Microscopy for Energy Research,* CRC Press, 2022.
47. S.-Y. Lang, Z.-Z. Shen, J. Wan, and R. Wen, in *Atomic Force Microscopy for Energy Research*, edited by C. Shen, CRC Press, 2022, p. 241.
48. Q. Zeng and K. Y. Zeng, in *Atomic Force Microscopy for Energy Research*, edited by C. Shen, CRC Press, 2022, p. 105.
49. G. Binnig and C. F. Quate, *Physical Review Letters*, **56**, 930 (1986).

50. A. F. Raigoza, J. W. Dugger, and L. J. Webb, *ACS Applied Materials & Interfaces*, **5**, 9249 (2013).
51. M. A. O'Connell and A. J. Wain, *Analytical Methods*, **7**, 6983 (2015).
52. M. Y. Kang, D. Momotenko, A. Page, D. Perry, and P. R. Unwin, *Langmuir*, **32**, 7993 (2016).
53. R. Reifenberger, *Fundamentals of Atomic Force Microscopy*, World Scientific, 2016.
54. Y. Seo and W. Jhe, *Reports on Progress in Physics*, **71**, 016101 (2008).
55. R. P. Garcia, *Surface Science Reports*, **47**, 197 (2002).
56. F. J. Giessibl, *Reviews of Modern Physics*, **75**, 949 (2003).
57. M. Jaafar, D. Martínez-Martín, M. Cuenca, J. Melcher, A. Raman, and J. Gómez-Herrero, *Beilstein Journal of Nanotechnology*, **3**, 336 (2012).
58. M. Lee and W. Jhe, *Physical Review Letters*, **97**, 036104 (2006).
59. N. Jalili and K. Laxminarayana, *Mechatronics*, **14**, 907 (2004).
60. J. Y. Li, J. F. Li, Q. Yu, Q. N. Chen, and S. H. Xie, *Journal of Materiomics*, **1**, 3 (2015).
61. A. Gruverman and S. V. Kalinin, *Journal of Materials Science*, **41**, 107 (2006).
62. E. Soergel, *Journal of Physics D: Applied Physics*, **44**, 464003 (2011).
63. A. Gruverman, M. Alexe, and D. Meier, *Nature Communications*, **10**, 1661 (2019).
64. T. S. Herng, M. F. Wong, D. Qi, J. Yi, A. Kumar, A. Huang, F. C. Kartawidjaja, S. Smadici, P. Abbamonte, C. Sanchez-Hanke, S. Shannigrahi, J. M. Xue, J. Wang, Y. P. Feng, A. Rusydi, K. Y. Zeng, and J. Ding, *Advanced Materials*, **23**, 1635 (2011).
65. T. S. Herng, A. Kumar, C. S. Ong, Y. P. Feng, Y. H. Lu, K. Y. Zeng, and J. Ding, *Scientific Reports*, **2**, 587 (2012).
66. J. X. Xiao, T. S. Herng, J. Ding, and K. Y. Zeng, *Acta Materialia*, **123**, 394 (2017).
67. J. Xiao, T. S. Herng, Y. Guo, J. Ding, N. Wang, and K. Y. Zeng, *Journal of Materiomics*, **5**, 574 (2019).
68. A. N. Morozovska, E. A. Eliseev, N. Balke, and S. V. Kalinin, *Journal of Applied Physics*, **108**, 053712 (2010).
69. S. Jesse, A. Kumar, T. M. Arruda, Y. Kim, S. V. Kalinin, and F. Ciucci, *MRS Bulletin*, **37**, 651 (2012).
70. S. V. Kalinin and A. N. Morozovska, *Journal of Electroceramics*, **32**, 51 (2014).
71. N. Balke, E. A. Eliseev, S. Jesse, S. Kalnaus, C. Daniel, N. J. Dudney, A. N. Morozovska, and S. V. Kalinin, *Journal of Applied Physics*, **112**, 052020 (2012).
72. S. V. Kalinin and N. Balke, *Advanced Materials*, **22**, E193 (2010).
73. G. G. Amatucci, J. M. Tarascon, and L. C. Klein, *Journal of the Electrochemical Society*, **143**, 1114 (1996).
74. N. Balke, S. Jesse, Y. Kim, L. Adamczyk, A. Tselev, I. N. Ivanov, N. J. Dudney, and S. V. Kalinin, *Nano Letters*, **10**, 3420 (2010).
75. N. Balke, S. Jesse, Y. Kim, L. Adamczyk, I. N. Ivanov, N. J. Dudney, and S. V. Kalinin, *ACS Nano*, **4**, 7349 (2010).
76. Q. N. Chen, Y. Ou, F. Ma, and J. Li, *Applied Physics Letters*, **104**, 242907 (2014).
77. S. Y. Luchkin, K. Romanyuk, M. Ivanov, and A. L. Kholkin, *Journal of Applied Physics*, **118**, 072016 (2015).
78. D. O. Alikin, A. V. Ievlev, S. Y. Luchkin, A. P. Turygin, V. Y. Shur, S. V. Kalinin, and A. L. Kholkin, *Applied Physics Letters*, **108**, 113106 (2016).
79. Q. N. Chen, Y. Y. Liu, Y. M. Liu, S. H. Xie, G. Z. Cao, and J. Y. Li, *Applied Physics Letters*, **101**, 063901 (2012).
80. Z. T. Wang, M. Kotobuki, L. Lu, and K. Y. Zeng, *Electrochimica Acta*, **334**, 135553 (2020).
81. Q. Sun, J. A. S. Oh, L. Lu, and K. Y. Zeng, *ACS Applied Materials & Interfaces*, **13**, 46588 (2021).
82. Q. Sun, L. He, F. Zheng, Z. Wang, S. J. An Oh, J. Sun, K. Zhu, L. Lu, and K. Y. Zeng, *Journal of Power Sources*, **471**, 228468 (2020).

83. Y. Wang, Z. Wang, J. Sun, F. Zheng, M. Kotobuki, T. Wu, K. Y. Zeng, and L. Lu, *Journal of Power Sources*, **454**, 227949 (2020).
84. J. Zhu, J. K. Feng, L. Lu, and K. Y. Zeng, *Journal of Power Sources*, **197**, 224 (2012).
85. J. Zhu, K. Y. Zeng, and L. Lu, *Journal of Applied Physics* **111**, 063723 (2012).
86. B. J. Rodriguez, C. Callahan, S. V. Kalinin, and R. Proksch, *Nanotechnology*, **18**, 475504 (2007).
87. A. Gannepalli, D. G. Yablon, A. H. Tsou, and R. Proksch, *Nanotechnology*, **22**, 355705 (2011).
88. N. Liu, R. Dittmer, R. W. Stark, and C. Dietz, *Nanoscale*, **7**, 11787 (2015).
89. S. Xie, A. Gannepalli, Q. N. Chen, Y. Liu, Y. Zhou, R. Proksch, and J. Li, *Nanoscale*, **4**, 408 (2012).
90. S. Jesse, S. V. Kalinin, R. Proksch, A. P. Baddorf, and B. J. Rodriguez, *Nanotechnology*, **18**, 435503 (2007).
91. S. Jesse and S. V. Kalinin, *Journal of Physics D: Applied Physics*, **44**, 464006 (2011).
92. M. P. Nikiforov, S. Jesse, A. N. Morozovska, E. A. Eliseev, L. T. Germinario, and S. V. Kalinin, *Nanotechnology*, **20**, 395709 (2009).
93. A. U. Kareem and S. D. Solares, *Nanotechnology*, **23**, 015706 (2012).
94. S. Jesse, R. K. Vasudevan, L. Collins, E. Strelcov, M. B. Okatan, A. Belianinov, A. P. Baddorf, R. Proksch, and S. V. Kalinin, in *Annual Review of Physical Chemistry, Vol. 65*, edited by M. A. Johnson and T. J. Martinez, Published by Annual Reviews, San Mateo, CA, 2014, p. 519.
95. T. Li and K. Y. Zeng, *Nanoscale*, **6**, 2177 (2014).
96. Y. T. Liu, R. K. Vasudevan, K. K. Kelley, D. Kim, Y. Sharma, M. Ahmadi, S. V. Kalinin, and M. Ziatdinov, *Machine Learning-Science and Technology*, **2**, 045028 (2021).
97. L. Collins, S. Jesse, N. Balke, B. J. Rodriguez, S. V. Kalinin, and Q. Li, *Applied Physics Letters*, **106**, 104102 (2015).
98. K. Romanyuk, S. Y. Luchkin, M. Ivanov, A. Kalinin, and A. L. Kholkin, *Microscopy and Microanalysis*, **21**, 154 (2015).
99. S. Yang, B. G. Yan, T. Li, J. Zhu, L. Lu, and K. Y. Zeng, *Physical Chemistry Chemical Physics*, **17**, 22235 (2015).
100. S. Yang, B. Yan, J. Wu, L. Lu, and K. Y. Zeng, *ACS Applied Materials & Interfaces*, **9**, 13999 (2017).
101. R. Berger, H. J. Butt, M. B. Retschke, and S. A. L. Weber, *Macromolecular Rapid Communications*, **30**, 1167 (2009).
102. D. C. Coffey, O. G. Reid, D. B. Rodovsky, G. P. Bartholomew, and D. S. Ginger, *Nano Letters*, **7**, 738 (2007).
103. J. Y. Park, S. Maier, B. Hendriksen, and M. Salmeron, *Materials Today*, **13**, 38 (2010).
104. G. Benstetter, R. Biberger, and D. Liu, *Thin Solid Films*, **517**, 5100 (2009).
105. W. Lu and K. Y. Zeng, *Functional Materials Letters*, **11**, 1830002 (2018).
106. X. J. Zhu, C. S. Ong, X. X. Xu, B. L. Hu, J. Shang, H. L. Yang, S. Katlakunta, Y. W. Liu, X. X. Chen, L. Pan, J. Ding, and R. W. Li, *Scientific Reports*, **3**, 1084 (2013).
107. K. Kuriyama, A. Onoue, Y. Yuasa, and K. Kushida, *Surface Science*, **601**, 2256 (2007).
108. R. Kostecki and F. McLarnon, *Electrochemical and Solid State Letters*, **5**, A164 (2002).
109. T. M. Arruda, A. Kumar, S. V. Kalinin, and S. Jesse, *Nanotechnology*, **23**, 325402 (2012).
110. A. Kumar, T. M. Arruda, A. Tselev, I. N. Ivanov, J. S. Lawton, T. A. Zawodzinski, O. Butyaev, S. Zayats, S. Jesse, and S. V. Kalinin, *Scientific Reports*, **3**, 1621 (2013).
111. J. Kruempelmann, D. Dietzel, A. Schirmeisen, C. Yada, F. Rosciano, and B. Roling, *Electrochemistry Communications*, **18**, 74 (2012).
112. T. M. Arruda, A. Kumar, S. Jesse, G. M. Veith, A. Tselev, A. P. Baddorf, N. Balke, and S. V. Kalinin, *ACS Nano*, **7**, 8175 (2013).
113. S. Yang, B. G. Yan, L. Lu, and K. Y. Zeng, *RSC Advances* **6**, 94000 (2016).

114. A. Nevosad, M. Hofstaetter, M. Wiessner, P. Supancic, and C. Teichert, in *Conference on Oxide-based Materials and Devices IV (Proceedings of SPIE)*, San Francisco, CA, 2013, **8626**, p. 862618.
115. E. Strelcov, Y. Kim, S. Jesse, Y. Cao, I. N. Ivanov, I. I. Kravchenko, C. H. Wang, Y. C. Teng, L. Q. Chen, Y. H. Chu, and S. V. Kalinin, *Nano Letters*, **13**, 3455 (2013).
116. W. H. Lu, J. X. Xiao, L. M. Wong, S. J. Wang, and K. Y. Zeng, *ACS Applied Materials & Interfaces,* **10**, 8092 (2018).
117. J. Liu, G. Z. Cao, Z. G. Yang, D. H. Wang, D. Dubois, X. D. Zhou, G. L. Graff, L. R. Pederson, and J. G. Zhang, *Chemsuschem*, **1**, 676 (2008).
118. J. Liu, B. Reeja-Jayan, and A. Manthiram, *The Journal of Physical Chemistry C*, **114**, 9528 (2010).
119. J. Liu, J. G. Zhang, Z. G. Yang, J. P. Lemmon, C. Imhoff, G. L. Graff, L. Y. Li, J. Z. Hu, C. M. Wang, J. Xiao, G. Xia, V. V. Viswanathan, S. Baskaran, V. Sprenkle, X. L. Li, Y. Y. Shao, and B. Schwenzer, *Advanced Functional Materials*, **23**, 929 (2013).
120. Y. Qi, L. G. Hector, C. James, and K. J. Kim, *Journal of the Electrochemical Society*, **161**, F3010 (2014).
121. J. Zhu, K. B. Yeap, K. Y. Zeng, and L. Lu, *Thin Solid Films*, **519**, 1914 (2011).
122. J. Zhu, K. Y. Zeng, and L. Lu, *Electrochimica Acta*, **68**, 52 (2012).
123. J. Zhu, K. Y. Zeng, and L. Lu, *Journal of Solid State Electrochemistry*, **16**, 1877 (2012).
124. J. Zhu, K. Y. Zeng, and L. Lu, *Metallurgical and Materials Transactions A - Physical Metallurgy and Materials Science*, **44A**, 26 (2013).
125. Q. Li, S. Jesse, A. Tselev, L. Collins, P. Yu, I. Kravchenko, S. V. Kalinin, N. Balke, *ACS Nano*, **9**, 1848 (2015).
126. G. Stan, S. Krylyuk, A. V. Davydov, M. D. Vaudin, L. A. Bendersky, and R. F. Cook, *Ultramicroscopy*, **109**, 929 (2009).
127. E. Dillon, K. Kjoller, and C. Prater, *Microscopy Today*, **21**, 18 (2013).
128. S. Bradler, A. Schirmeisen, and B. Roling, *Journal of Applied Physics*, **122**, 065106 (2017).
129. R. García and R. Pérez, *Surface Science Reports*, **47**, 197 (2002).
130. Y. Naitoh, Z. Ma, Y. J. Li, M. Kageshima, and Y. Sugawara, *Journal of Vacuum Science & Technology*, B **28**, 1210 (2010).
131. R. Garcia and R. Proksch, *European Polymer Journal*, **49**, 1897 (2013).
132. R. Vidu, F. T. Quinlan, and P. Stroeve, *Industrial & Engineering Chemistry Research*, **41**, 6546 (2002).
133. T. Doi, M. Inaba, H. Tsuchiya, S. K. Jeong, Y. Iriyama, T. Abe, and Z. Ogumi, *Journal of Power Sources*, **180**, 539 (2008).
134. Y. Tian, A. Timmons, and J. R. Dahn, *Journal of the Electrochemical Society*, **156**, A187 (2009).
135. A. Clémençon, A. T. Appapillai, S. Kumar, and Y. Shao-Horn, *Electrochimica Acta*, **52**, 4572 (2007).
136. J. Wu, W. Cai, and G. Shang, *Nanoscale Research Letters*, **11**, 1 (2016).
137. F. T. Quinlan, K. Sano, T. Willey, R. Vidu, K. Tasaki, and P. Stroeve, *Chemistry of Materials*, **13**, 4207 (2001).
138. S. B. Tang, M. O. Lai, and L. Lu, *Journal of Power Sources*, **164**, 372 (2007).
139. M. Matsui, K. Dokko, and K. Kanamura, *Journal of Power Sources*, **177**, 184 (2008).
140. S. Ramdon and B. Bhushan, *Journal of Colloid and Interface Science*, **380**, 187 (2012).
141. K. Kushida and K. Kuriyama, *Applied Physics Letters*, **84**, 3456 (2004).
142. A. Avila and B. Bhushan, *Critical Reviews in Solid State and Materials Sciences*, **35**, 38 (2010).
143. R. Kostecki, F. Kong, Y. Matsuo, and F. McLarnon, *Electrochimica Acta*, **45**, 225 (1999).

144. Y. Matsuo, R. Kostecki, and F. McLarnon, *Journal of the Electrochemical Society*, **148**, A687 (2001).
145. M. Kerlau, M. Marcinek, V. Srinivasan, and R. M. Kostecki, *Electrochimica Acta*, **52**, 5422 (2007).
146. D. W. Chung, N. Balke, S. V. Kalinin, and R. E. Garcia, *Journal of the Electrochemical Society*, **158**, A1083 (2011).
147. Q. N. Chen, S. B. Adler, and J. Y. Li, *Applied Physics Letters*, **105, 201602** (2014).
148. T. Li, B. H. Song, L. Lu, and K. Y. Zeng, *Physical Chemistry Chemical Physics*, **17**, 10257 (2015).
149. K. Y. Zeng and J. Zhu, *Mechanics of Materials*, **91**, 323 (2015).
150. Y. Sun, L. H. Vu, N. Chew, Z. Puthucheary, M. E. Cove, and K. Y. Zeng, *ACS Biomaterials Science & Engineering,* **5**, 478 (2019).
151. Y. Sun, Z. G. Hu, D. Zhao, and K. Y. Zeng, *ACS Applied Materials & Interfaces,* **9**, 32202 (2017).
152. B. D. Cappella, *Surface Science Reports*, **34**, 1 (1999).

4 Atom Probe Tomography

Yimeng Chen
CAMECA Instrument Inc.

CONTENTS

DOI: 10.1201/9781003299295-4

4.1 APT ANALYSIS OF Li-ION BATTERIES

Li-ion batteries for energy storage have received a great deal of attention.[1–6] Microstructure and chemistry analysis are keys to understanding the factors controlling the electrical capacity, degradation mechanisms, and safety of Li-ion batteries. Being the lightest metal, Li poses challenges in quantification for many techniques due to its low atomic weight and small atomic/ionic radius. Atom probe employs a time-of-flight (TOF) mass spectrometer, in principle, providing equal sensitivity to all elements in the periodic table.[7] Depending on the volume analyzed, the detection limit for impurities or dopants can reach the few parts-per-million (ppm) level. Along with the three-dimensional compositional mapping at subnanometer resolution, atom probe tomography (APT) is well suited for many of the applications in Li-ion batteries where the distribution of Li and the composition of surfaces or interfaces are critical for the electrochemical performance.

For battery materials, the application of APT is still relatively new. One of the first relevant applications of APT was $LiCoO_2$,[8] where APT was used as a complementary technique with transmission electron microscopy and electron energy loss spectroscopy (TEM-EELS) for chemical analysis. The APT composition profiles across a thin film deposited by ion beam sputtering on needle-shaped tungsten tips revealed the tendency of Li and O to segregate to the film surface. Other Li-containing materials like Li-borate glass have been analyzed by APT in the above-mentioned work. Inspired by the initial work on the thin film Li oxides, Diercks et al. reported the first APT analysis of commercial-grade $LiCoO_2$ cathode materials.[9] APT was carried out to investigate the capacity loss after 1,000 electrochemical cycles. Lithium depletion from the grains and Li accumulation toward the internal grain boundaries were reported for the cycled samples. Later, $LiMn_2O_4$, $LiNi_{0.5}Mn_{1.5}O_4$ in spinel structure,[4,10] and $Li_{1.2}Ni_{0.5}Mn_{0.6}O_2$, $Li_{1.2}Ni_{0.15}Mn_{0.55}Co_{0.1}O_2$[2,11] in layered structure were explored, providing important insights into chemistry, structure, and phase transportation for cathode materials. Around the same time, the effects of laser wavelength and pulse energy on compositional accuracy were systematically investigated using $LiFePO_4$ cathode materials.[12] The work indicated the need to explore the optimal acquisition conditions for reliable results.

In addition to cathode materials, Si-SiO powders before and after Li insertion/extraction were analyzed with APT showing Li atoms trapped in the amorphous SiO phase which provides explanations to the irreversible capacity fade after the first cycle.[13] Furthermore, the grain boundaries in the $Li(AlTi)_2(PO_4)_3$ solid-state ceramic electrolyte were investigated using APT, scanning transmission electron microscopy (STEM), and simulations based on the first-principles density-functional theory.[14] APT was able to provide remarkable information on grain boundary structure and composition showing an asymmetric distribution of Li across the grain boundaries that were several nanometers wide. The Li enrichment can be well correlated to the crystalline-to-amorphous phase transition predicted by simulations, which was supported by correlative analysis with high-resolution scanning transmission electron microscopy (HR-STEM).

In summary, many studies have now shown that Li-ion batteries can be analyzed with APT successfully and produce excellent results. Developments in sample preparation methods using the focused ion beam (FIB) for particle samples with porosity[15–17] allows the analysis of concentration gradients within cathode particles, which is challenging by using other bulk compositional analysis techniques.[3] For instance, inductively coupled plasma-mass spectrometry requires at least hundred milligrams of materials to measure the average composition from a large population of particles. Beyond using APT for local compositions, there are numerous studies investigating structural evolution and capacity degradation after successive charge/discharge cycles.[1,3,5] Of note, electron microscopy used in conjunction with APT brings in a synergistic effect to resolve the crystallographic structure and the chemical information for nano-sized features that would have been challenging to analyze.[18–20] Correlative analysis could greatly advance our understanding of the Li distribution and atomic structure on the relevant length scale.[3,5]

4.2 INTRODUCTION TO APT

4.2.1 Technology Roadmap

APT inherently refers to three-dimensional analysis, but that was not the case when the concept of probing atoms was introduced for the first time. The first one-dimensional (1D) atom probe was invented in 1967 at Pennsylvania State University by Erwin Müller, John Panitz, and Brooks McLane.[21] At that time, the atom probe was a 1D analytical instrument that collected hundreds of atoms per day with approximately a 1-nm field-of-view (FOV), where the entire probe hole is a single pixel. Later, the three-dimensional capability was pursued and made possible in 1972 by John Panitz in his Imaging Atom Probe (IAP).[22] The instrument collected three-dimensional data by combining a series of two-dimensional maps of a single preselected atom type (or isotope) recorded on a microchannel plate (MCP). Since the field desorption process is destructive, the sequence of two-dimensional images being collected provides depth information. A revolution in atom probe technology came when the concept of position-sensitive detector was adapted to the previous IAP-like instruments. The Position-Sensitive Atom Probe was the first fully operational atom probe with true three-dimensional capability developed in 1988 by Cerezo, Godfrey, and Smith at the University of Oxford.[23] The improvement in data collection speed to hundreds of atoms per minute and the enlarged wide FOV to ~15 nm were spectacular for that time period. Inspired by the idea of Scanning Atom Probe proposed by Nishikawa in 1993,[24] Kelly and Larson et al. carried out the development further, aiming for higher data collection rates and larger FOVs without sacrificing mass resolving power (MRP). The first commercial voltage pulsing Local Electrode Atom Probe (LEAP®) system adopted by Oak Ridge National Laboratory (ORNL, USA) was operational in 2003. The local-electrode geometry combined with a 100kHz voltage pulser allowed millions of atoms per hour to be collected with a 150nm FOV.

4.2.2 Laser Pulsing

Historically, the application field of APT had been limited to specimens with high electrical conductivity due to the necessity of generating field evaporation with voltage pulsing. This requirement limited the technique to metals and highly doped semiconductors.[25] The pioneering work performed by Kellogg and Tsong in the early 1980s demonstrated the use of laser pulsing to assist field evaporation for metals and semiconductors.[26,27] The first laser-pulsed three-dimensional atom probe was built in 1988 by Liddle et al.[28] and used for studies of compound semiconductors. The concept was not widely adopted until the first suitable small-spot laser pulser became commercially available in the early 2000s. Eventually, laser-assisted atom probes grew in popularity given the compelling reasons for analyzing insulating materials or fragile metal specimens that were low-yielding with voltage pulsing. The first commercial laser atom probes were offered in 2006.[29] The advent of faster small spot-size laser systems at ultraviolet (UV) wavelength with picosecond or femtosecond pulse durations greatly expanded the applications field of APT. Examples cover a wide range of materials including silicon-based compounds, semiconductors,[30,31] ceramics,[32] and oxides that include Li-ion battery materials.[1,2,33,34]

Laser-pulsing APT can be further classified by the laser wavelength used. Over the course of instrumentational development, commercially available atom probes have seen a steady progression toward shorter wavelengths. The LEAP 3000™ initially was equipped with a 532nm green laser and progressed to a UV wavelength of 355nm[35] for the CAMECA LEAP 4000™ and 5000™. Recently, CAMECA introduced a deep-UV 266nm laser systems with the LEAP and Invizo 6000® products.[36] In addition to these commercial systems, there are other home-built and prototype APT systems operated in some universities around the word, for example, the 355nm UV laser in-house system at the University of Münster,[37] and the wavelength-tunable (the second and the third harmonics of 1,030nm IR lasers) systems at the University of Rouen[38] and the National Institute for Materials Science in Japan.[39]

4.2.3 Spatial Resolution and Chemical Sensitivity

APT is often compared with (scanning) transmission electron microscopy ((S)TEM) and secondary ion mass spectrometry (SIMS) regarding the capability in compositional mapping. TEM gives two-dimensional atomic resolution averaged over the thickness of specimens of 100nm or less.[40] Analytical TEM using electron dispersive spectroscopy (EDS) is particularly less sensitive to light elements, because of the low fluorescence yield and the high absorption of low-energy X-rays by the matrix. By contrast, SIMS provides high chemical sensitivity on the order of parts-per-billion (ppb) as a function of depth but loses spatial resolution in the lateral directions.[41]

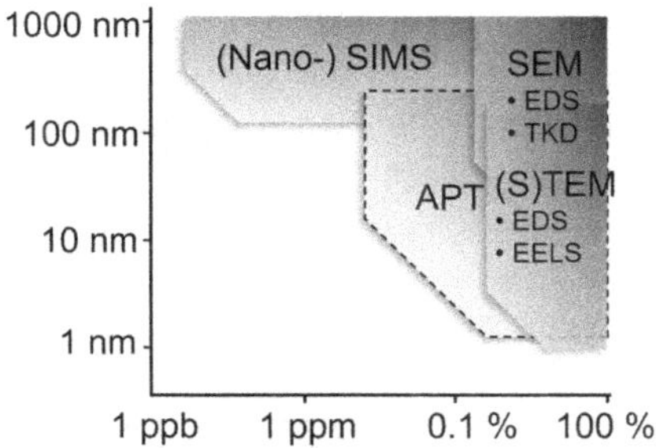

FIGURE 4.1 Map of the analytical instrumentation space modified from [42] and [43]. Atom probe tomography lies in the regime that bridges SIMS and TEM for combined spatial resolution and chemical sensitivity toward the physical detection limit on the bottom left corner.

Among the different techniques that can provide either spatial or chemical information with nanometer resolution, the strength of APT is evident as shown Figure 4.1. APT occupies a unique space in terms of bridging the gap between (S)TEM, with energy-dispersive X-ray spectroscopy and EELS, and SIMS. Several key points worth mentioning are emphasized in this figure. The APT data are inherently three-dimensional, where the relative locations are determined for individual atoms. The highly divergent field in the vicinity of the specimen apex provides an extremely high magnification that enables the resolution of individual atomic positions. This high spatial resolution is coupled with a very high analytical sensitivity that can reach the parts-per-million scale. Moreover, all atoms are detected with equal efficiency, including light elements such as Li, and thus composition determinations are typically straightforward with no applied corrections.

4.2.4 Working Principles

4.2.4.1 Field Evaporation

APT is inherently three-dimensional with the lateral information derived from the impact position on a two-dimensional position-sensitive detector, and the depth information calculated from the volume increment from each ion detected. Figure 4.2 illustrates the working mechanism of an APT instrument in general. The system operates under ultra-high vacuum (UHV) condition with typical base pressures in the analysis chamber of less than 1×10^{-10}Torr. The setup consists of a needle-shaped specimen, a local electrode placed in close proximity to the specimen, and a two-dimensional position-sensitive detector for ion detection. For analysis, the specimen is cryogenically cooled (typically between 20K and 80K) to minimize atomic thermal vibration and suppress surface migration.

Electric fields sufficient for field ionization are achieved by a combination of a base DC voltage up to 15kV and specimens prepared in nanoscale needle-shaped geometry. The apex of radius R causes a high electric field F in V/nm following the phenomenological formula[45]:

$$F = \frac{V}{kR}, \tag{4.1}$$

where R is the specimen apex radius, and k is a dimensionless shape factor that depends on geometric considerations.[46] For spheres, k equals 1 giving the surface field of V/R. For a conventional needle-shaped emitter well away from its surroundings, k is often assumed to be between 5 and 8.[47] For the LEAP systems with local electrodes, a smaller k of ~3–4 is achieved that allows higher field values to be generated at the apex of the specimen.

To control the rate of field evaporation, the applied DC voltage can be adjusted such that the generated electric field is just below the threshold for significant field evaporation. Thermal or voltage pulses are applied to trigger the ionization of surface atoms and record the time of departure. The ionized atoms are then accelerated through the local electrode aperture toward a two-dimensional position-sensitive ion detector. The ion trajectories follow a projection law that easily achieves a magnification of ~10^6. In APT, the specimen itself is the primary optic that affects ion trajectories and spatial resolution.

4.2.4.2 Ion Detection

The detector consists of two MCPs and a delay-line anode assembly (Figure 4.2). Ion detection efficiency is typically 37%–80%, primarily governed by the open area on the MCP and the transmission ratio of ions through the metal mesh on the flight path that modifies the electrostatics for energy compensation.[48] When an ion enters the open area on the front MCP, the impact event on the sidewalls of the channels generates a cascade of 100-fold amplified secondary electrons. The amplitude of

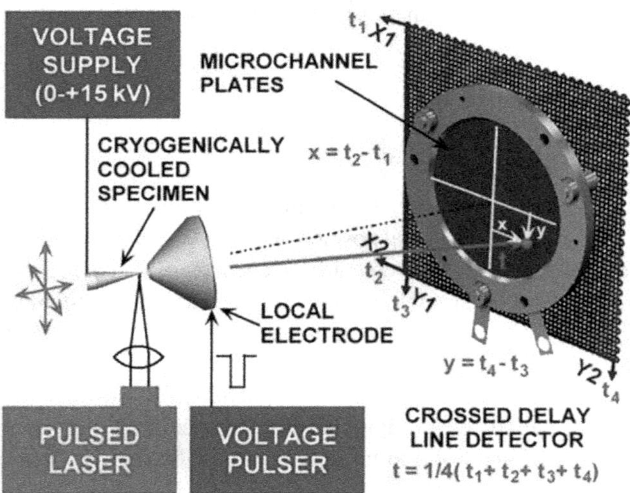

FIGURE 4.2 Schematic principle of the local electron atom probe. The APT needle-shaped specimen is facing a position-sensitive detector through the aperture on a conical-shaped local electrode. The detector combines an assembly of microchannel plates and delay-line anodes. The modern LEAP systems use three anodes to calculate the ion impact position. The evaporation of the specimen can be initiated by voltage or laser pulses. (Reprinted from Miller et al.[44])

the electron cloud is then further amplified through the back MCP creating secondary electrons on the order of 10^4–10^6 per ion. The resulting electron cloud spreads over a few millimeters and impacts the delay-line anode wires behind. On the anode wires, electrical pulses are detected at both ends of each wire as discrete timing events. Therefore, the position of the ion's impact is determined from the arrival times measured at each end. These times are converted into distances in the *X* and *Y* coordinates on the detector. The time resolution on a delay-line detector is equivalent to the spatial resolution across the detector. The resulting timing precision provides approximately 800 × 800 pixels across the detector face which corresponds to a spatial resolution of around 2 Å for a typical 100 nm FOV.

4.2.4.3 Time-of-Flight Mass Spectrometry

For ion identification, APT relies on TOF mass spectrometry. To measure the TOF, the field evaporation event must be constrained to a time window on the order of nanoseconds. The total flight time of ions depends on several key operating parameters including DC voltage, ion charge state, and the physical distance between the specimen and the detector. Since the electric field gradient is very high near the specimen apex, a field-evaporated ion gains most of its kinetic energy when departing from the specimen's surface, then travels at a constant velocity along the trajectory line as illustrated in Figure 4.2. To first order, the functional dependence of the TOF can be derived by setting the final kinetic energy equal to the total potential gain,

$$neV = \frac{1}{2}mv^2, \tag{4.2}$$

where n is the charge state, e is the electron charge, V is the total voltage applied to the specimen at the point of evaporation, m is the mass of the ion, and v is the speed of the ion after the full potential gain. With traveling over a distance L (flight path) at a constant speed, by rearranging, the mass-to-charge state ratio (m/n) can be determined as:

$$\frac{m}{n} = \frac{2eV}{v^2} = \frac{2eV}{L^2}t^2, \tag{4.3}$$

where t is TOF and $v=L/t$. A full electrostatic solution is required for a more accurate relationship. To measure this flight time with sufficient mass spectrometry resolution, a timing uncertainty for the ions must be on the order of nanoseconds or better. For either voltage or laser, the field evaporation event must be pulsed on this time scale, and the timing electronics must be able to resolve sub-nanosecond flight time differences.

4.2.5 Tomographic Reconstruction

Reconstruction is the process of converting the positions and arrival sequence of ions on a two-dimensional detector into their relative positions in three dimensions that represent the original atomic arrangement in the specimen. The standard methodology is based on a point-projection model.[49] The reconstruction process assumes that the specimen shape has a spherical end form on a circular cone. The field-evaporated ions are geometrically projected onto the two-dimensional detector. This model has been improved to accommodate the larger fields of view of modern instruments,[50] but this basic algorithm remains the foundation of atom probe reconstructions today.

Note that several models have been explored to describe the projection of ions.[50,51] For the sake of simplicity, small FOV expressions are used to illustrate the main points, as shown in Figure 4.3. Consequently, assuming that ions are projected along straight trajectories, a direct relationship can be established between the specimen surface position and the detector space impact position using the magnification.

Hence, the standard methodology involves calculating the magnification to convert the detector hit position (X and Y) into real-space coordinates (x and y). In a point-projection model, where the projection center is located at the center of the apex sphere with radius R, the magnification (M) is simply the distance between the apex center and the detector ($L+R$) divided by the radius of the specimen (R) as illustrated in Figure 4.3. For needle-shaped specimens with a hemispherical end form, due to the presence of the shank, the ion trajectories originating from the specimen surface are compressed, thus reducing the magnification by a factor of ξ (image compression factor, $\xi > 1$). This results in the projection center shifting from the center of the apex circle to ξR, as highlighted by the black dot along the specimen's axis in Figure 4.3. The projection magnification can now be converted to

$$M = \frac{L+\xi R}{\xi R} \approx \frac{L}{\xi R}. \tag{4.4}$$

It is generally accepted that the nominal value of ξ is between 1 and 2 for the projection center which falls on a point between radial projection and stereographic projection along the specimen's axis.

Before the depth information for each ion is calculated, an estimate of the imaged specimen surface area is required. This is accomplished by converting the detected surface area (S_D) to the specimen surface area (S_a) using a geometrical model of the apex and the magnification derived from Equation 4.4. Hence, the analyzed surface area can be expressed as follows:

$$S_a = \frac{S_D}{M^2} \tag{4.5}$$

For each reconstructed ion, the depth coordinate (z) is adjusted by a small increment equal to the atomic volume (Ω) divided by the surface area assuming that the volume of the reconstructed ion is spread uniformly over the detected area. Combining the above equations with the instantaneous value of the radius derived from Equation 4.1 (for radius evolution using the voltage applied), the depth increment can be written as:

$$dz = \frac{\Omega}{\varepsilon S_a} = \frac{\Omega L^2}{\varepsilon S_D \xi^2 R^2} \tag{4.6}$$

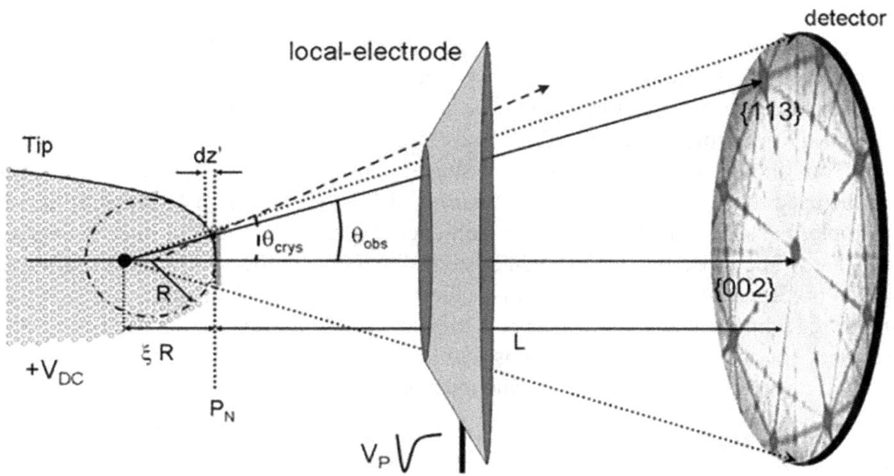

FIGURE 4.3 Schematic view of the point projection where the projection point is located along the axis of the specimen, at a distance of ξR from the plane of tangency (P_N), causing the compressed ion trajectory. (Reprinted from Gault et al.[52])

Correction to the analyzed volume is necessary by considering the detection efficiency of ε. The backward projection generates atomic positions within the plane tangent to the specimen apex. A subsequent correction (dz' denoted in Figure 4.3) must be applied to account for the hemispherical curvature from where the ion is evaporated. The depth z of a given ion is calculated by sequentially cumulating the depth increment dz and adding the corrective term dz'. Given these points, a brief introduction to the APT reconstruction regarding the Bas et al. protocol[49] has been provided. A detailed reconstruction procedure can be found elsewhere, for the interested readers.[7,53]

4.3 SPECIMEN PREPARATION

Specimen preparation is a key aspect of many analytical techniques, and APT is no exception. There has been explosive growth in the adoption of APT in the past two decades owing to the development of specimen preparation techniques. The concept of site-specific sample preparation methods is adopted from preparing TEM specimens. This method utilizes a dual-beam system of scanning electron microscope and FIB (SEM-FIB). Instead of thinning electron transparent lamella specimens for TEM, APT requires needled-shaped specimens with the apex radius smaller than 100 nm which have been traditionally achieved by electrochemical etching.[54] The advantages of using ion beam milling instead of the traditional electropolishing technique include the reduction of chemical residues from electrolytes, reduction of preferential and asymmetric etching that produce bladed specimens, and most importantly for Li-ion batteries, the applicability for less conductive specimens and site-specific preparation.[55]

The attempts made by Waugh et al.[56] and Alexander et al.[57] to prepare specimens for atom probe using direct ion beams were the precursor to the dual-beam FIB work that came in the late 1990s. FIB-based specimen preparation is now regularly used to prepare multiple site-specific atom probe specimens in a single lift-out. SEM is used to identify a region of interest (ROI) and the ion beam is used to cut the specimen free from the surface and mill it into the shape required for APT. Correlative analysis can be used to target microstructural features of interest identified by employing, for example, backscattered imaging, electron backscatter diffraction, or energy-dispersive X-ray spectroscopy in the SEM system.[58,59] Site-specific specimen preparation with a precision of less than 100 nm is achieved with the aid of the above-mentioned imaging techniques and controlled FIB milling.[60] A similar approach has been used to target specific features or locations within Li-ion batteries which are often in the form of aggregated micro or nanoscale particles.[3,14]

4.3.1 FIB-Based Lift-Out Method for Large Particles

The use of modern SEM-FIB instruments to perform lift-out specimen preparation has revolutionized the atom probe.[61] There were several early methods for specimen extraction, but eventually a modification of the lift-out method developed for preparing fast and reliable TEM specimens became a standard approach.[62,63] Wedges can be lifted out from the surface region of samples and mounted onto flat-top posts for APT. A wide variety of substrates are suitable to receive APT specimens. Electropolished tungsten wires and TEM half-grids are among those that are commonly used for correlative analysis with TEM. Additionally, the microtip coupons introduced by CAMECA greatly increase the productivity of APT analysis. The presence of many microtips per coupon enables quick manufacture using FIB and batch transfer into LEAP systems equipped with a local electrode. As each APT dataset typically analyzes a ~$100 \times 100 \times 500\,nm^3$ section of data from microscopic-sized specimens, it is particularly important to gain representative observations that would characterize the structure as whole via collecting a statistically significant number of datasets.

Atom probe specimens from large particles can be prepared using the lift-out method with little or no modifications to the process. For particles larger than 10 μm in radius, trench cutting can be applied directly to extract wedge specimens from the surface. Figure 4.4 illustrates the steps adopted from the standard lift-out method that is applied to prepare APT specimens from $Li(NiMnCo)O_2$ (NMC) cathode particles. Note that, rather than protecting the ROI with FIB-deposited platinum (FIB-Pt) as shown in Figure 4.4b, an optional protective capping layer can be applied over the entire surface that reduces the damage caused by ion milling, mitigates the charging effect, and, most importantly, enables the analysis of the top surface. For example, a passivation layer of Ag or Ni deposited by sputter coating was used by Chae et al.[3] and Vissers et al.,[10] respectively, on cathode particles. The wedge extraction region is typically 2–3 μm wide with a length that can be limited by the geometry of the particle in some cases. After identifying the ROI, the surrounding material is removed to create a wedge-shaped cantilever. The wedge in Figure 4.4b has been cut free from the top, down, left, and bottom sides, leaving only the right side attached to the bulk particle. To extract the wedge, a micromanipulator is lowered into contact with the

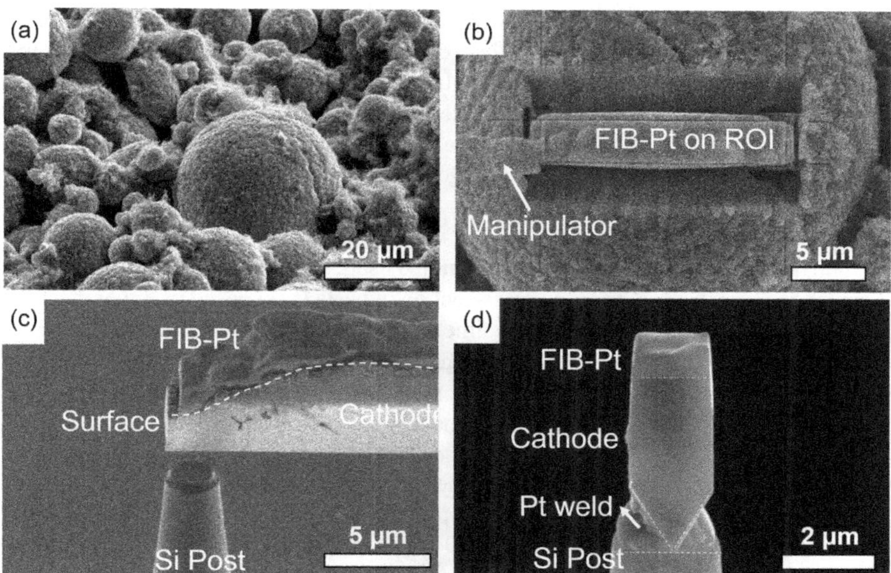

FIGURE 4.4 Illustration of steps involved in a stand lift-out method procedure. (a) Li-ion battery cathode particles of ~20 μm in size surrounded by polyvinylidene fluoride (PVDF) binder and conductive carbon. (b) A FIB-deposited protective strip is placed over the extraction region and the material around is then removed by ion milling to produce a ~20 μm long cantilevered wedge. The wedge is attached to an in situ micromanipulator for lift-out extraction. (c) The micromanipulator is used to position the wedge above the carrier post with a 2 μm wide flat surface on the top (side view). (d) The wedge is welded to the carrier post using FIB-deposited Pt and then cut free from the remaining wedge for propagation to additional posts.

free end of the cantilevered wedge for attachment with FIB-deposited Pt or other gas injection systems. The wedge is cut free from the particle by ion milling then 'lifted-out'. Then, the wedge is repositioned above a carrier tip. Figure 4.4c shows a flat-topped Si post on a microtip array. The design of the microtip array, which carries multiple posts with identical geometry, allows rapid specimen preparation. The wedge centered on the post is then lowered to contact the flat surface on the post. FIB-deposited Pt is deposited at the wedge-to-post interface for welding. After mounting, the remaining wedge is cut with the ion beam from the mounted region and then moved to the next post. The mounting procedure is repeated until the wedge is gone. From a single lift-out, a 20 μm wedge as shown in Figure 4.4b can produce 10–12 APT specimens.

4.3.2 Edge to Center Specimen Preparation for Large Particles

APT is a technique for destructively analyzing the elemental identities and position of atoms with an extremely small volume of material, typically $\sim 100 \times 100 \times 500\,\text{nm}^3$. Therefore, the chemical gradient happening over a length scale beyond micrometers cannot be captured within a single APT dataset. Given the small volumes of materials examined by APT, individual datasets can be obtained from multiple grains or particles to better represent the overall composition and microstructure. Using the wedge lift-out method described in the previous section, multi-point analysis on the surface region can greatly improve the confidence level of the particle composition. Furthermore, compositional heterogeneity may also be studied with multiple samples, as some studies optimize the capacity and thermal stability by introducing the so-called core–shell structure.[64,65]

With no modification to the lift-out technique itself, sequential specimens from edge to center can be extracted from the cross-section of particles exposed by either mechanical polishing or ion milling. Such specimen preparation techniques to investigate the uniformity of composition within a cathode particle are illustrated in Figure 4.5. In this image, ~15 μm cathode particles on conductive carbon tape are attached to a 45° pre-tilted holder to enable tilting to larger negative angles. That allows cross-sectional milling parallel to the pre-tilted stage, as how the particle cross-section is exposed in Figure 4.5a. The SEM contrast reveals the agglomeration of ~500 nm primary particles within the secondary particle of ~17 μm in size. According to the particle sizes, a rough estimation indicates around 1,000 secondary particles being exposed to the cross-sectional surface viewed from the top. The standard lift-out method was used to extract a 15 μm long wedge from the particle cross-section. The process successfully yielded nine mounted specimens, at the interval distance of 1.5 μm, from the wedge. The APT analysis on these specimens serves as a survey for the interior composition in the direction of edge-center-edge.

4.3.3 Methodologies for Nanoparticles

Nanofabrication design strategies have shown improvements in the electrochemical properties of Li-ion batteries, where the increase in the surface area-to-volume ratio leads to a greatly improved power density and a reduced charging time.[66] However, the preparation of APT specimens from nanoparticles is an ongoing challenge,

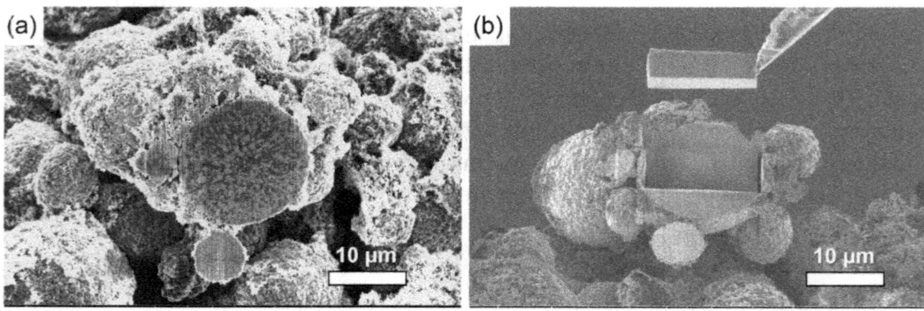

FIGURE 4.5 (a) Cross-sectional image of the interior of a secondary cathode particle exposed by ion milling. (b) Illustration of extracting a ~15 μm long wedge specimen from cross-section using the standard lift-out method. In the subsequent mounting procedure, nine APT specimens were mounted to individual posts.

because it requires the preparation of a dense, solid, needle-shaped specimen that can withstand the high electric field applied. Although the specimen preparation is not trivial, a few papers have reported data from different particles.[67–71]

By modifying the conventional lift-out method, Devaraj et al.[33] developed an innovative direct lift-out and transfer method to prepare APT specimens from $LiMnO_2$ cathode nanoparticles of ~500 nm in size. Qu et al.[72] adopted the same approach to study the $Li_{1.2}Mn_{0.57}Ni_{0.17}Co_{0.06}O_2$ nanoparticles. The steps involved in the direct lift-out process are partially illustrated in Figure 4.6. To prepare the particles on an adequate substrate, a suspension of nanoparticles is dispersed on a clean Si substrate (Figure 4.6a). By carefully controlling the micro-manipulator, a ~500 nm nanoparticle is picked up and transferred to a flat-top post (Figure 4.6b and c). Platinum deposition using electron beam or ion beam is applied for encapsulation (Figure 4.6e). The final APT specimen after sharpening is shown in Figure 4.6f, where the darker contrast at the apex is FIB-Pt that protects the brighter nanoparticle beneath. Unfilled regions between the particle and the post may be present as the gaps with small openings could be difficult for FIB-Pt to access. Additional Pt deposition from the side with the electron beam can be applied to reinforce the unfilled regions.

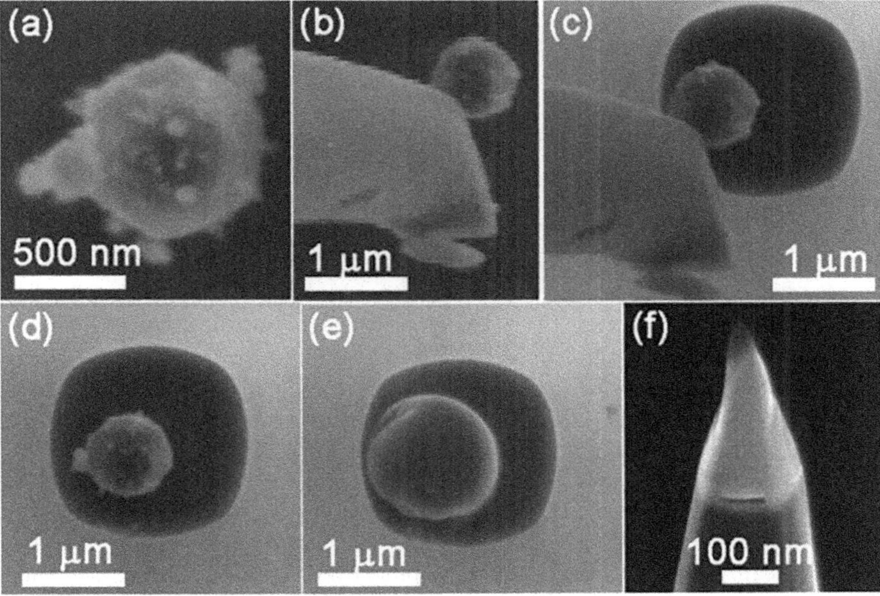

FIGURE 4.6 Illustration of the direct pick-up and transfer method for APT specimen preparation. (a) SEM image of the ~500 nm diameter $LiMnO_2$ nanoparticle. (b) A selected particle is attached to the manipulator. (c and d) The nanoparticle is transferred and dropped onto a flat-top post carrier. (e) Image of nanoparticle coated with electron-beam-deposited Pt. (f) The final needle specimen after annular milling. A gap at the particle-to-post interface is evident. (Adopted from Devaraj et al.[33])

The pick-up method mentioned above is claimed to be effective for particles of size between 100 and 500 nm, where particles larger than 500 nm will be difficult to pick up with the Van der Waals forces, and particles smaller than 100 nm pose difficulties in the transfer, encapsulation, and annular milling process.[72] An alternative for particles under 100 nm is to embed spatially separated nanoparticles in a metal or oxide matrix. Examples include electrophoresis method,[72] sputter deposition and cross-section lift-out method,[73] atomic layer deposition encapsulation method,[74] and metallic electrodeposition method which was applied to ~50 nm $Li_4Ti_5O_{12}$ powders as an anode material.[75] These methods enable the analysis of complete nanoparticles. However, agglomerated particles that are weakly bonded should be avoided as these cause difficulty with the APT analysis yield.[43] APT analysis of nanoparticles is an area of ongoing development. As the application space grows, the APT community is working toward establishing clear guidelines on selecting methods and encapsulation materials for nanoparticles for a variety of different nanoparticle and encapsulation matrix materials.

4.3.4 Sharpening and Cleaning

4.3.4.1 Sharpening

In 1998, Larson et al. proposed two methods to sharpen bulk materials into a needle shape for APT.[55] One method involves an ion beam scanning in an annular pattern, with successively smaller patterns applied until a sharp needle is formed. This method, termed annular milling, has become the standard method used by the community. A second method involves a sort of carving away at the sample from the side until a needle is formed. The aforementioned methods are schematically shown in Figure 4.7 for sharpening spinel $LiNi_{0.5}Mn_{1.5}O_4$ particles.[4] For the annular milling method, annular milling patterns are directly applied along the vertical axis. The annular milling pattern should start with an outer diameter exceeding the width of the mounted specimen, hence, to prevent forming parasitic side peaks that may become the source of noise if they emit ions under high electric field. Gradually, the inner and outer diameters of the donut-shaped patterns are reduced to produce pointed needle-shaped specimens for APT. The annular milling method has the benefit of producing small-radius specimens. Semiautomated annular milling is feasible using pre-defined milling recipes. Chen et al. demonstrated several benefits of recipe-based milling including reduced milling time, producing APT specimens with a robust geometry for field evaporation, and producing APT specimens with similar geometry which is an important factor for specimen-to-specimen comparison [submitted, Y. Chen 2022 M&M].

For the carving method, ion beam was applied from the side to shape the mounted particle into a triangle shape as shown in Figure 4.7b. Followed by further side cuts from different directions around the vertical axis, the desired cone shape of the APT specimen is achieved. This method ensures a well-controlled shank angle, which can be difficult with annular milling if the milling rate is significantly different between the mount and the post materials. For both methods, it is recommended to successively reduce the Ga^+ ions beam current as the pattern size becomes smaller for more control over the final tip shape.

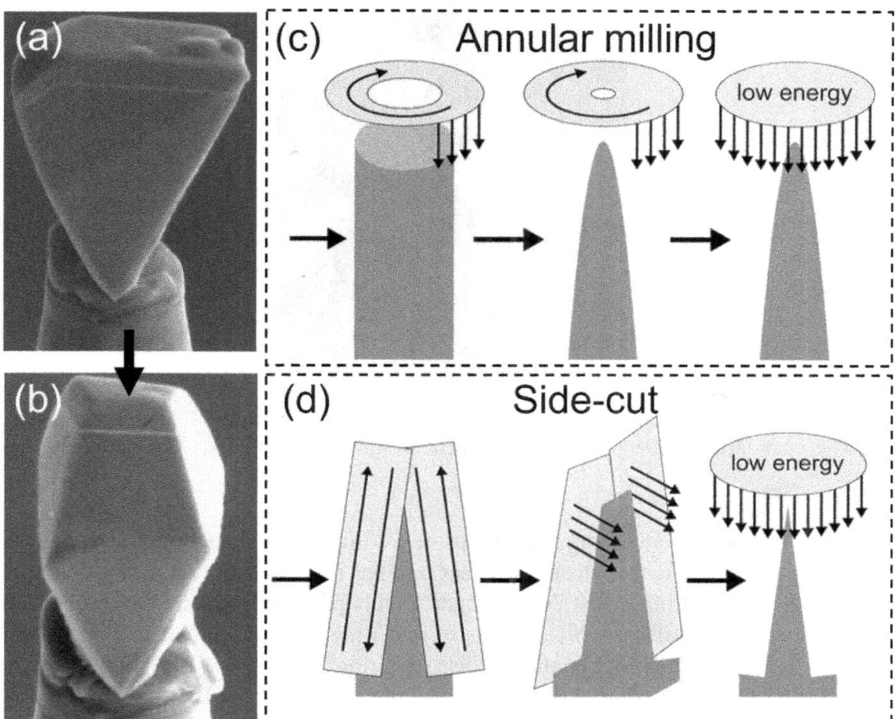

FIGURE 4.7 Secondary electron images of $LiNi_{0.5}Mn_{1.5}O_4$, (a) without and (b) with side-cuts, mounted to a post using a modified lift-out method for particles. Illustration of (c) annular milling and (d) side-cuts, the two sharpening methods using a focused Ga^+ ion beam. (SEM images of the $LiNi_{0.5}Mn_{1.5}O_4$ cathode particle modified from Vissers et al.[10]; sharpening procedures modified from Maier et al.[4])

4.3.4.2 Low-Energy Ion Beam Cleaning

Finishing with a low-energy ion milling step has been found to be critical to the preparation of quality specimens. The implant and damage region created by a 30 keV Ga^+ ion beam have been shown to extend significantly into the surface of the specimen, while <5 keV ions limit damage to less than ~5 nm. The goals of the low-energy ion cleaning step are to (i) remove the damage introduced by the annular milling or cutting steps where the 30 keV beam was used, (ii) position the apex of the specimen at or slightly above the ROI, and (iii) narrow the width at the ROI so that the voltage to sustain field evaporation will not exceed what the specimen can tolerate.

Figure 4.8 shows a typical needle-shaped APT specimen produced by the aforementioned sharpening process. The material mounted on the Si post was prepared from the secondary particles of $Li(NiMnCo)O_2$ (NMC) cathode that are ~10 μm in size. FIB-based deposition was used to weld the mount to the post. The grains exposed to the shank are primary particles, where the particle boundaries are clearly visible in the SEM image. In the image, the protective layer has been milled away during the low-energy cleaning step. If the outermost surface is the subject of APT analysis, a

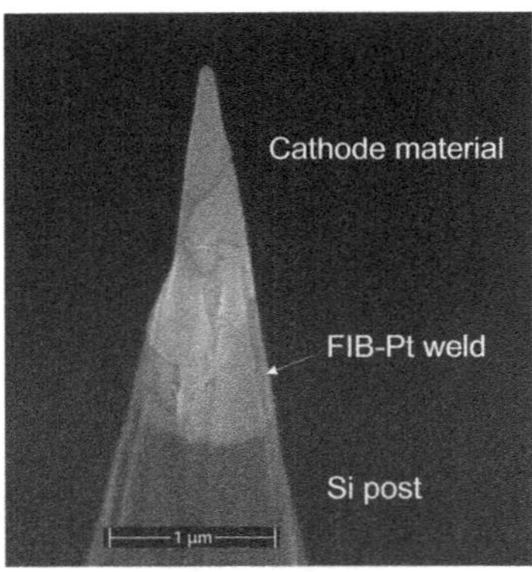

FIGURE 4.8 Ready-to-analyze APT specimen after annular milling and low-energy cleaning. The boundaries between the primary particles are visible in darker contrast along the shank.

thin layer of protective coating must be preserved in the final specimen. In general, protective layers present a strong contrast against the cathode materials, which is convenient for targeting. Most current instruments provide live viewing of electron images during ion milling which assists in real-time end-point control. FIB instruments with STEM capability enable precise targeting as good as 10 nm. For example, Vissers et al.[10] performed an APT analysis on the surface of the $Li_xNi_{1.5}Mn_{0.5}O_4$ spinel cathode with a core–shell structure (Co coating). A 25–30 nm Ni layer deposited by ion beam sputtering was found to be effective in protecting the original surface. In the subsequent APT analysis, the Ni layer emerged first before the cathode material, verifying a successful analysis of the Co-coated surface.

4.3.5 Cryogenic Vacuum Transfer to Atom Probe

As shown in the specimen preparation section, APT requires specialized specimen preparation resulting in needle-shaped specimens with very high surface-to-volume ratios. For some applications, changes related to exposed surfaces, or the bulk temperature history of the specimen has an effect on the analysis results. Lithium-containing materials are typically both air- and temperature-sensitive such that they may experience rapid oxidation and surface contamination during transportation between different microscopy instruments. Furthermore, for correlative analysis, a carefully controlled environment is critical.

Due to growing interest in studying air- and temperature-sensitive materials, a UHV/cryogenic transfer system[76,77] (vacuum cryo-transfer module, VCTM) has been developed based on a collaboration between the Max Planck Institute for Steel Research (Germany), CAMECA (the United States), and Ferrovac (Switzerland). Figure 4.9 shows a VCTM system integrated to a specially modified LEAP 5000 system. The transfer design is based on the Ferrovac mobile UHV Suitcase, customized to transfer samples held in a standard LEAP specimen carrier. The module features a portable, battery-operated ion pump and a non-evaporable getter to maintain UHV

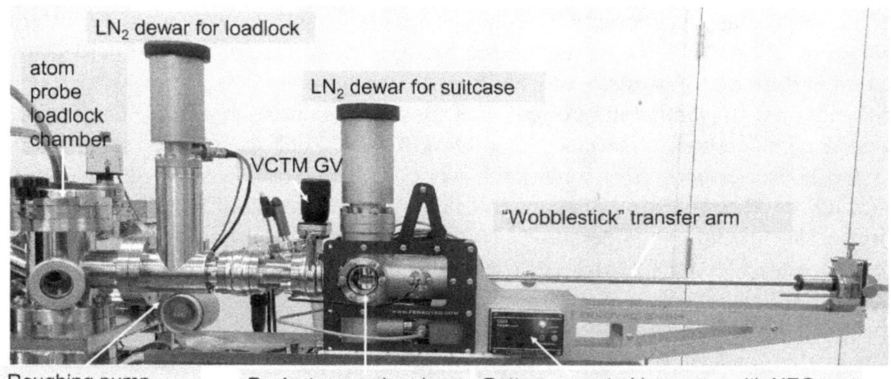

FIGURE 4.9 Image of the CAMECA® vacuum cryo-transfer module (VCTM) attached to a modified LEAP 5000 system. The VCTM is a portable, cryogenically cooled, UHV unit that enables fast transfer or short-term storage from a treatment or preparation system which is compatible with atom probe systems equipped with a VCTM docking station.

conditions, and a cold stage to maintain low-temperature conditions. Fast transfer to other treatment, preparation, and microscopy instruments can be enabled through the compatible connection port. Temperature and vacuum control prevent the formation of crystalline ice on the prepared samples, which opens the possibility of APT analysis for electrolytes that have been frozen.

However, the implementation of VCTM on Li-ion batteries is still in its preliminary stage for Li-containing materials. Kim et al.[75] reported a major delithiation during the APT analysis for specimens that underwent sharpening at cryogenic temperatures and FIB-APT transport under UHV. On the contrary, the specimens sharpened at room temperature and transferred through air under ambient laboratory conditions did not show any issues with delithiation. The prevention of delithiation was attributed to modifications to the surface of the specimens that introduce a shielding effect to prevent field penetration, as discussed by Kim et al. However, the nature or the possible cause of this surface modification is unclear calling for a deeper investigation to clarify the mechanism behind delithiation.

4.3.6 A Word on the Data Acquisition Conditions

Because each material has different field ionization characteristics, the data acquisition conditions are often chosen by balancing data quality, analysis success rates, and data collection speed. Luckily in the literature, the acquisition conditions of APT for Li-ion batteries are not too diverse. In this section, we will provide a summary on the common acquisition conditions that the users can refer to as a starting point.

4.3.6.1 Base Temperature Considerations

The operation of rechargeable Li-ion batteries relies on mobile Li-ions that rapidly move between the cathode and the anode. However, an adequate compositional analysis shall not induce mobilized Li-ions that influence the validity of the Li distribution in the specimens. In general, APT analyses are performed at cryogenic temperatures between 20K and 80K. However, the exact temperature at the instance of field evaporation is difficult to estimate, especially considering laser illuminations which temporarily raise the temperature of the apex. Base temperatures between 30K and 40K were used for cathode materials ($LiMn_2O_4$ @30K,[4] $LiCoO_2$ @35–40K,[9] Li(MnNiCo) O_2 @40K[72]). Furthermore, a remarkable work conducted by Pfeiffer et al.[78] with APT showed that the mobility of Li-ions in $LiMn_2O_4$ are suppressed below the detection limit at the specimen temperature of 30K. In general, lower base temperatures are recommended for APT analysis of specimens that contain mobile Li atoms.

4.3.6.2 Detection Rate

Practically, an APT experiment controls data acquisition parameters such as DC voltage, laser pulse energy, and temperature to maintain a steady rate of detected ion events called detection rate. It is worth noting that the detection rate is different from the evaporation rate, where the latter represents the number of ions that are evaporated across the entire surface, but partially interact with other portions of the vacuum system and do not hit the active area of the detector. Although more precisely controlled evaporation algorithms such as constant charge-state ratio (CSR) or constant field exist, the constant detection rate mode will only be considered here, since it has been used on Li-ion battery materials in the literature.

Although higher detection rates generally promote higher data quality by collecting more signal over the same amount of background noise[79] (increased signal-to-noise ratio), at the same time, higher detection rates result in more stress on the specimen that leads to mechanical failures.[80–82] Materials that contain a large number of weakly bonded interfaces, or voids acting as stress concentrators, pose a risk to undesired specimen ruptures. In the literature, conservative detection rates between 0.2% and 0.5% are often used for Li-ion battery materials in layered structures.[5,6,72,83] However, higher detection rates between 0.8% and 1.0% can be used for cathode materials with spinel structures,[4,78] which appear to be more robust.

4.3.6.3 Laser Pulsing

The thermally based evaporation enabled by laser pulsing allows for APT analyses of electrical insulators including Li-ion battery materials. Significant improvements in yield and data size can be anticipated by reducing the activation energy barrier for field evaporation through raising specimen temperatures. However, there is always a trade-off. Using higher than necessary laser pulse energy can introduce degraded MRP and add complexity to interpretation of the mass spectrum. The abundance of hydride-containing peaks and complex clusters of ions increases as the laser pulse energy is increased. Most importantly, compositional bias can be introduced by using higher than needed laser pulse energies.

The influence of laser pulse energy on the chemical composition of $LiCoO_2$ cathode particles was investigated by Diercks et al.[9] (supplementary material) on a LEAP 4000XSi system. Between 1 and 100 pJ, $LiCoO_2$ exhibited a laser energy dependence that showed increased lithium composition at high laser energies. The apparent composition of Li is balanced by a joint decrease in transition metals and oxygen. Therefore, Diercks et al. conducted APT analyses using laser pulse energy between 1 and 2 pJ to obtain the correct compositional measurements. However, laser–material interaction in the presence of a high electric field is, indeed, a complicated phenomenon. An opposite trend was reported for $LiFePO_4$,[12] where a stochiometric composition was achieved by increasing the laser pulse energy, and a mixed trend of how the Li concentration fluctuates with the laser pulse energy was reported for $Li(Ni_{0.8}Co_{0.15}Al_{0.05})O_2$ and $Li(Ni_{0.8}Mn_{0.1}Co_{0.1})O_2$.[84] Although the selection of laser pulse energy was either not discussed or justified, most of the work, conducted on the LEAP 4000 & 5000 systems, used the energy range between 0.5 and 30 pJ, where Mohanty et al. 0.5 pJ,[11] Diercks et al. 1–2 pJ,[9] Kim et al. 5–25 pJ,[75] Qu et al. 20 pJ,[72] Chae et al. 25 pJ,[3] Lee et al. 30 pJ,[5] and Choi et al. 30 pJ.[6] Note that data points from home-built systems[4,13,78] are not added here for comparison due to the variation in the spot size of lasers. Furthermore, Santhanagopalan et al. investigated the effects of laser wavelength between green and UV lasers,[12] which is provided for interested readers.

4.3.6.4 Summary

In terms of field evaporation, each specimen may be unique, and different material types do not perform similarly under equivalent APT analysis conditions. The task is to achieve an acceptable analysis yield while achieving the best composition accuracy and spatial resolution possible. As these may be opposing goals in some cases, the choice of experimental parameters is often an exercise in choosing various trade-offs. For new or unknown materials, it is highly recommended to first conduct a systematic experiment to evaluate analysis parameters. For Li-ion battery materials in the oxide form, the laser pulse energy is the most influential parameter to investigate. At constant detection rates, trade-offs in data quality summarized in Figure 4.10 represent a general trend of how APT specimens respond to varying the laser pulse energy. Datasets acquired with lower laser pulse energies generally provide better compositional and spatial accuracy by reducing the strength of the thermal component in the process of field evaporation. However, there are two main concerns associated with applying higher voltages to sustain field evaporation. First, the likelihood of specimen fractures will increase with the higher field stresses applied to the specimen apex. Second, the abundance of unsynchronized evaporations between laser pulses will cause the background noise to increase. Decreases in the number of complex ions can ease peak identification and ranging. Comparison of CSRs between various datasets can provide indirect indications of apex temperature and field strength.[85]

In AP Suite 6, the APT data analysis software, programming experimental metrics is possible through the new 'chain acquisition' and 'scripted acquisition' features. Users can specify a list of specimens on a microtip array to be analyzed using predefined acquisition conditions.[86] 'Scripted acquisition' allows varying acquisition conditions in live, during data collection, after fulfilling the predefined conditions in the scripts. For example, lowering the detection rate when an interface approaches in the analysis FOV can be scripted to help increase the analysis yield. In addition to collecting data at constant detection rate, advanced data acquisition modes like constant field and constant CSR have been made available in the atom probe control software. This allows further constrained experimental variables, such as variability in FOV and effectiveness of laser pulse energy, to be investigated.[87]

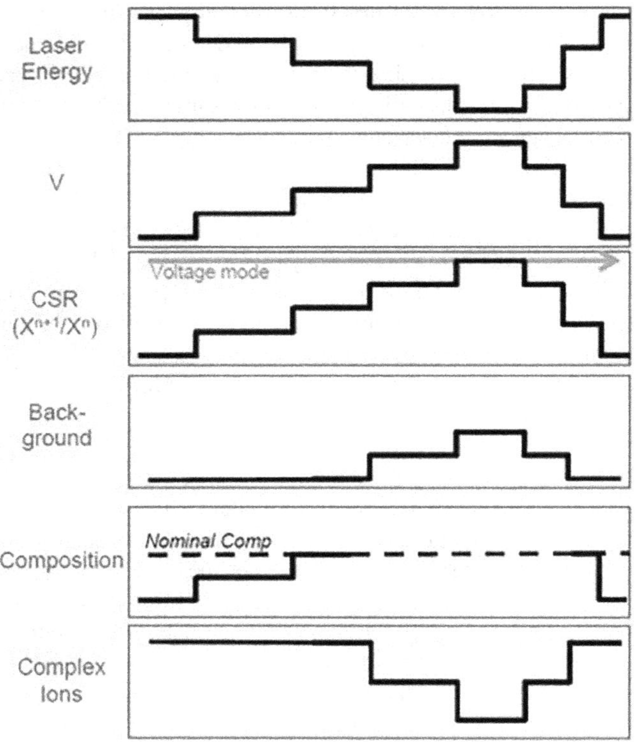

FIGURE 4.10 From top to bottom, the Y-axes of the blue boxes represent commonly observed responses of voltage, charge-state ratio (higher charge over lower charge), background level, composition, and prevalence of complex ions in response to changing the laser pulse energy. (Courtesy of Dr. Stephan Gerstl, ETH Zurich.)

4.4 CASE STUDIES

The global lithium-ion battery market has been growing dramatically and is expected to continue. This is largely driven by the increased use of electric vehicles, grid storage, and portable consumer electronics, where the Li-ion battery offers the highest energy density and output voltage among commercial rechargeable battery systems.[88] Performance improvements require the development of new battery components, as well as understanding and addressing the mechanisms that result in performance degradation with repeated charging and discharging cycles.

The ability to obtain three-dimensional compositional information at the nanoscale has led APT to a host of different applications in lithium-ion battery research. This section is not intended to be a complete review of the literature. Included here are some interesting examples from the literature that illustrate how APT has been used for the analysis of materials used in Li-ion batteries. Among those, the oxide cathode materials mostly studied by APT are the focus of this section.

4.4.1 Layered NMC Cathode Materials

The development of cathode materials has seen a shift in the chemistries of different materials. $LiCoO_2$ was introduced by Goodenough[89] and became the first commercialized cathode material used in Li-ion batteries by Sony in 1990. Driven by the high cost of Co, low thermal stability, and low capacity, $LiNiO_2$ was introduced as an alternative.[90] This was due to the high capacity and low cost attributed to Ni, compared with Co. However, a high Ni content exhibits low thermal stability and a reduction in the Li pathway caused by Li/Ni intermixing. Thus, Co and Mn were introduced for structural stability, cost reduction, and safety.[91] This gave rise to the $Li(NiMnCo)O_2$ (NMC) layered oxides that are currently being used in both portable electronic and electric transport applications.

4.4.1.1 Experimental

The composition and fine-scale microstructure of two types of NMC cathode materials were investigated. Cathode tapes with 90% active particles, 5% binder, and 5% conductive carbon were obtained from NEI Corporation (product names EB-50E-622 and EB-50E-811). The nominal compositions are $Li(Ni_xMn_yCo_z)O_2$ X:Y:Z=6:2:2 and X:Y:Z=8:1:1 (hereafter referred to as NMC622 and NMC811). SEM confirms that the particles are ~10 µm in diameter (Sample preparation Section 4.3.1, Figure 4.4a), and EDS analysis on the surface of these materials verifies compositionally homogeneous at the micrometer scale. The standard APT specimen preparation method (Section 4.3.1) with FIB lift-out was used to analyze the subsurface region on various particles. On average, each lift-out region extracts five atom probe specimens from a single particle. Multiple particles from each NMC material were examined to better understand the variation from particle to particle. The analysis conditions for data collection were systematically studied and optimized to balance data quality and yield. A combination of a laser pulse energy of 1.0 pJ and a detection rate of 0.2% ions per pulse at a base temperature of 30K was used for APT analysis.

4.4.1.2 Mass-to-Charge Spectrum

Figure 4.11 shows a typical mass spectrum of the NMC622 material. Peaks for molecules of Co, Ni, and Mo with different charge states were found, as are typically

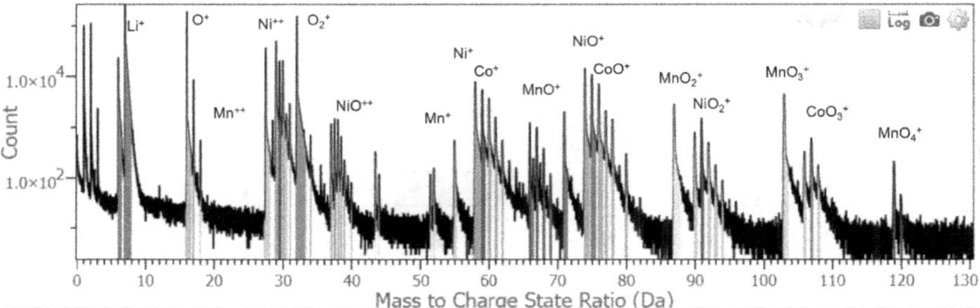

FIGURE 4.11 Mass spectrum example of NMC622 cathode material acquired with LEAP 5000 XR at 30K base temperature, 0.002 atom/pulse detection rate, using 1.0 pJ laser pulse energy.

observed for NMC oxides.[5,33] Peaks at 17, 18, and 19 Da are most likely to be hydroxide-related species from residual gas in the chamber, which are field-ionized at the specimen apex and do not originate from the analyzed material itself. Low concentration (<1 at.%) FIB-Ga at 69 Da from milling and low-energy cleaning are present but only on the surface of the reconstructed volumes. If desirable, Ga-containing regions can be removed from the analysis to avoid analyzing ion-beam damage regions.

4.4.1.3 Compositional Analysis of NMC 622

On average, the 43 measurements studied in this work yielded 6 million ions per dataset. Figure 4.11 shows a typical mass spectrum of NMC611 showing the constituent elements. All peaks could be assigned to an element or a molecule. Li was only detected in the single charge state, and O was detected as single oxygen ions and molecular ions. Additionally, Ni, Co, and Mo are favorable for the formation of molecular ions with O. The observed ratio between the Li isotopes $^{6}Li^{+}$ and $^{7}Li^{+}$ after subtracting the exponentially decaying noise level is $^{7}Li/^{6}Li$=12.0–12.2, which is in good agreement with the natural abundance ratio of 12.16. Overlapping peaks between NiH and Co exist for single and molecular ions with O. Decomposition of peaks algorithm in the AP Suite 6™ data analysis software was used to obtain the best compositional accuracy by decomposing the overlapping species using the Ni isotope abundance. Peak decomposition is not necessary for Li, O, and Mn related peaks, as they do not overlap. For ranging, the leading sides were relatively easy to determine, owing to the sharp rise in ion counts typical for the data acquired in laser pulsing mode. The trailing sides were determined using 3× the background count or to the point of interfering with the next peak in mass spectra. APT specimens from multiple particles provide a robust compositional measurement that takes compositional fluctuations into account. The overall compositions of 12 specimens from 3 particles (4 from particle A, 3 from particle B, and 5 from particle C) from NMC622 are reported in Figure 4.12. The average composition is 27.9 ± 4.2%Li, 17.0 ± 1.5%Ni, 4.6 ± 0.7%Mn, 4.6 ± 0.7%Co, and 45.4 ± 3.4%O, where the errors are ±2 sigma

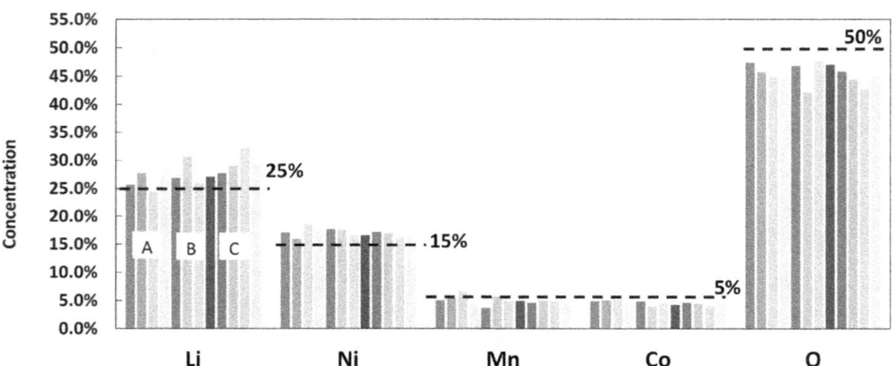

FIGURE 4.12 Compositions of NMC 622 particles A, B, and C. The dashed lines indicate the nominal composition of the materials. Ni, Mn, and Co are in good agreement with the nominal composition. The rise in Li above the expected composition is compensated by the opposite trend in oxygen.

standard deviation with a 95% confidence level. This translates to a measured ratio of $Li_{1.1}(Ni_{0.7}Mn_{0.2}Co_{0.2})O_{1.8}$, compared to the nominal ratio of $Li_{1.0}(Ni_{0.6}Mn_{0.2}Co_{0.2})O_{2.0}$ for NMC622. Transition metals are in excellent agreement, and Li and O are in good agreement with the expected composition. In fact, there are reports on cathode materials synthesized with extra Li that helps improve the performance in electrochemical cycles.[92] Regarding the lower ratio of O, the deficit of O is commonly observed in APT analysis. Commonly accepted explanations include the asynchronous DC evaporation, formation of neutral molecules, multi-hit events (co-evaporation), and dissociation of molecular species during the TOF that could explain the loss of atoms.[93–97]

4.4.1.4 NMC622 & 811 Comparison

During the manufacturing process, changes in raw materials or process conditions could affect the composition, and thus affect battery performance. Therefore, accurately measuring the main elements and impurities that may be present in the samples is an important step in battery manufacturing. After confirming a successful and reliable compositional analysis of the NMC particles, the compositional ratios of NMC622 and NMC811 are compared in Figure 4.13. The nominal composition would be one part of Li, one part of Ni+Mn+Co, and two parts of O. The numbers (622 and 811) represent the ratios between transition metals in the sequence of Ni, Mn, and Co. Composition analysis of APT is conducted by counting the number of ions that fall within specified ranges in the mass spectrum. Using the ion counts, the atomic ratio between Ni, Mn, and Co would be

$$\frac{N_{\mathrm{Ni}}}{N_{\mathrm{Total}}} : \frac{N_{\mathrm{Mn}}}{N_{\mathrm{Total}}} : \frac{N_{\mathrm{Co}}}{N_{\mathrm{Total}}}. \tag{4.7}$$

Here, N_i (i = Ni, Mn, or Co) is the ion count. N_{Total} is the cumulated count of Ni, Mn, and Co.

The ratios in Figure 4.13 were normalized to 10 to match the name notation. From each population, the compositional variation is well below the calculated ±2 sigma

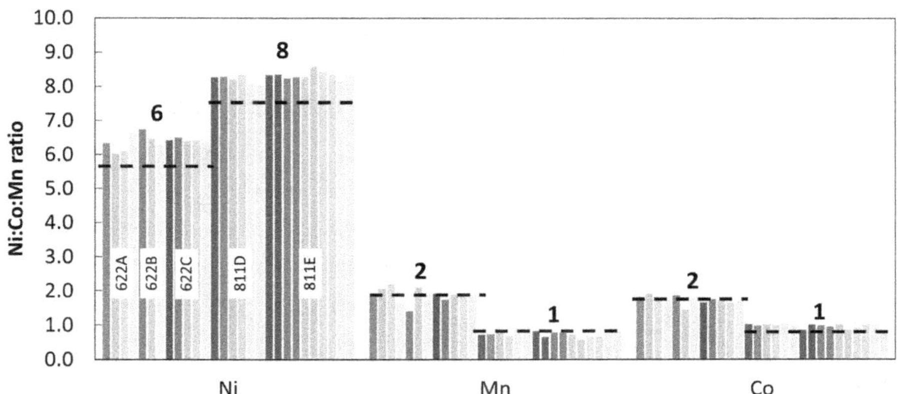

FIGURE 4.13 Composition of NMC 622 particles A, B, and C along with NMC811 particles D and E. The dashed lines indicate materials' nominal compositions.[83]

statistic errors, indicating good homogeneity from location to location, and particle to particle. According to the results, the compositions of NMC622 and NMC811 are easily differentiated using APT.

4.4.1.5 Li Distribution and Concentration Profiles

APT analysis is ideal for studying fine-scale features. Although most of the specimens present a homogeneous distribution of the constituent element as shown in Figure 4.14a, a small portion of the data shows Li-enriched features. Figure 4.14b shows an example of a planar feature with a significantly higher Li concentration highlighted using 30% Li isosurfaces in purple. A 1D concentration profile (Figure 4.14c) taken across the Li-enriched feature shows a peak value of 38 at.% Li compared to the matrix value of 27 at.%. The feature is slightly depleted with the transition metals, and a significant decrease of 5 at.% O is observed. Chae et al.[3] reported similar types of compositional fluctuation, as plates or line-shaped features. Defects observed by HR-STEM, such as dislocations and anti-phase domain boundaries, are in morphologies similar to those reported by APT. These defects serving as Li accommodation sites could possibly explain the Li-enriched features observed by APT.

4.4.1.6 Interface Analysis of Primary Particles

The NMC particles may contain impurities from the synthetic process.[10] The presence and concentration of impurities is a key factor to the cycle life, storage capacity, and even safety of the Li-ion batteries. Trace amounts of Al which exclusively segregated to the boundaries between the primary particles were observed in NMC622. There are two connected interfaces in Figure 4.15a where only one showed Al segregation. The resulting 1D concentration profile is presented in Figure 4.14b in log scale to emphasize the Al content. The Al concentration at the interface is measured to be 1.0 at.% with a concentration below the detection limit inside the primary particles on each side. In addition to Al, the boundary region is also enriched in Li and Mn with reduced Ni, Co, and O. The distinct segregation behaviors of Ni and Co versus those of Mn and Al are remarkable to observe. The presence of Al in NMC622 is unintended but is likely to influence Li transport along and across interfaces between primary particles, which would affect battery performance, as Al doping is reported

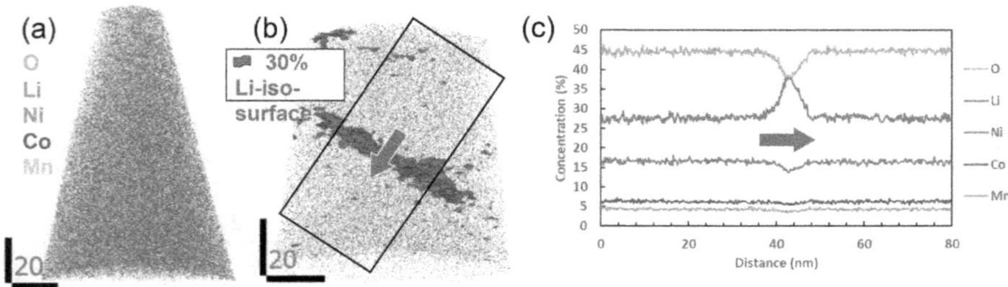

FIGURE 4.14 (a) A typical APT reconstruction of the NMC material shows a homogeneous distribution of the constituent elements. (b) A Li-enriched plate-like feature revealed with 30% Li-isosurfaces. (c) A 1D concentration profile taken across the Li-enriched feature in (b) along the black box. Scale bars are in nanometers.

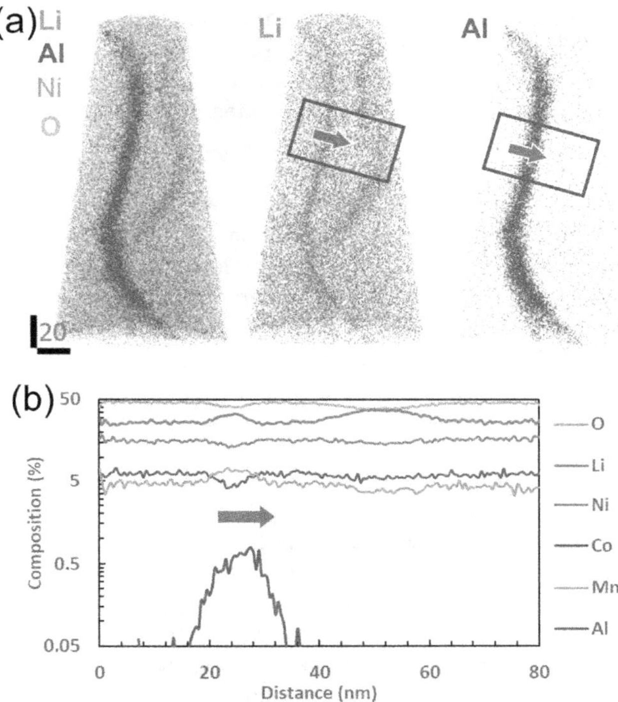

FIGURE 4.15 (a) The APT reconstruction of NMC 622 that contains boundaries between the primary particles. Atom maps of Li and Al reveal distinct segregation behaviors. (b) 1D concentration across the boundaries quantitatively shows the Li and Al enrichment on the boundaries.

to be beneficial to NMCs, showing improved cycling and capacity stability, as well as air stability for $Li(Ni_{0.85}Co_{0.15-x}Al_{x)}O_2$.[98]

4.4.1.7 Summary

In conclusion, APT has been shown to provide valid and useful data from NMC cathode materials for Li-ion batteries. It can identify the matrix composition and reveal fine-scale compositional variations. The two types of NMCs can be quantitatively and precisely differentiated by using their metal-to-metal ratios. Li-enriched features at low number densities may correlate with defects in NMCs reported in the literature. Unexpected segregation of Al is observed at the interfaces between the NMC622 primary particles.

4.4.2 Charge/Discharge Cycles

Increases in battery performance require the development of new battery components, as well as understanding and addressing the mechanisms that result in performance degradation with repeated charging and discharging cycles. Evaluation of batteries and battery components requires a variety of analytical methods that study materials and components on various length scales. Correlating the Li distribution and atomic structure at the relevant length scales over the operation of a battery life-cycle is essential for understanding the degradation process and providing the tools for battery development.

4.4.2.1 Li Migration and Transition Metal Loss

Choi et al.[6] investigated the lithiation/delithiation behavior of a composite cathode in all-solid-state batteries (ASSBs). Significant losses in discharge capacity and Coulombic efficiency (the ratio of the total charge extracted from the battery over the total charge put into the battery in a full cycle) in the initial cycling were confirmed by electrochemical performance testing. Atom probe analysis was carried out on the near-surface region of the cathode with the aim of better understanding the structural and compositional stability during the charge/discharge events. Using the standard lift-out method described in Section 4.3.1, APT specimens of the uncycled sample were prepared from the pristine powder sample, with particle size of ~5 μm in diameter. Atom probe specimens of the cycled sample were prepared directly from particles embedded in the $Li_7P_3S_{11}$ (LPS) solid-state electrolyte. The atomic ratios between Li, Ni, Co, and Mn (normalized to Li+Ni+Co+Mn=1) in the pristine (P), charged (C), and discharged (D) conditions are compared with the stoichiometric ratios (S) in Figure 4.16. The reason for excluding O from the comparison was not discussed; however, *a* ~7 at.% O deficit was reported elsewhere (the supplementary material of Ref. [6]) for the discharged condition. The pristine sample measured

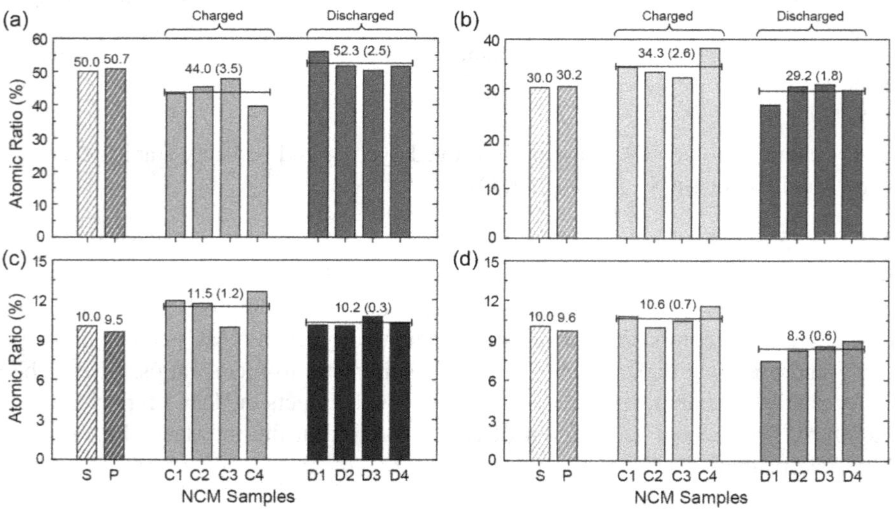

FIGURE 4.16 Concentrations of the constituent elements of (a) Li, (b) Ni, (c) Co, and (d) Mn measured with APT. 'S', 'P', 'C#', and 'D#' on the horizontal axis, present 'stoichiometric', 'pristine', 'charged', and 'discharged' conditions, respectively. Multiple datapoints were acquired from each sample with their mean values and standard deviations overlayed on the bar graph. (Adopted from Choi et al.[6])

Li:Ni:Co:Mn=0.5:0.3:0.1:0.1 (without oxygen) within ±1% deviation from the stoichiometric ratio. The Li ratio decreased from 0.51 to 0.44 in the charged condition, then increased to 0.52 in the discharged condition after the initial charge/discharge cycle. The trend of Li fluctuation is the expected behavior for cathodes resulting from delithiation/lithiation. As illustrated in Figure 4.17, during operation, Li atoms stored in the anode create a transport through the electrolyte to the cathode. In the cathode, Li anions are oxidized, creating a compound that can store the arriving Li-ions. When the cell is recharged after output, the flow of ions is in the opposite direction and Li-ions are reduced back to the Li metals for storage in the anode.

By comparing the atomic ratios, the authors suggested a mechanism of transport of Mn from the cathode to the solid-state electrolyte during cycling. The dissolution of transition metal species is known as an irreversible electrochemical reaction that occurs across the electrode surface.[99] This could lead to the loss of active material and an increase of internal resistance, and consequently, the fade of the overall energy and power of the battery cells. Such observations of the transition metal dissolutions are important for understanding the chemical and structural stability during charge/discharge cycles.

4.4.2.2 Evolution of Li Concentration Gradient

Chae et al.[3] performed APT analyses on the NMC ($Li_{1.04}Ni_{0.80}Mn_{0.15}Co_{0.05}O_2$) cathode materials in various states to assess their structural and compositional evolution after

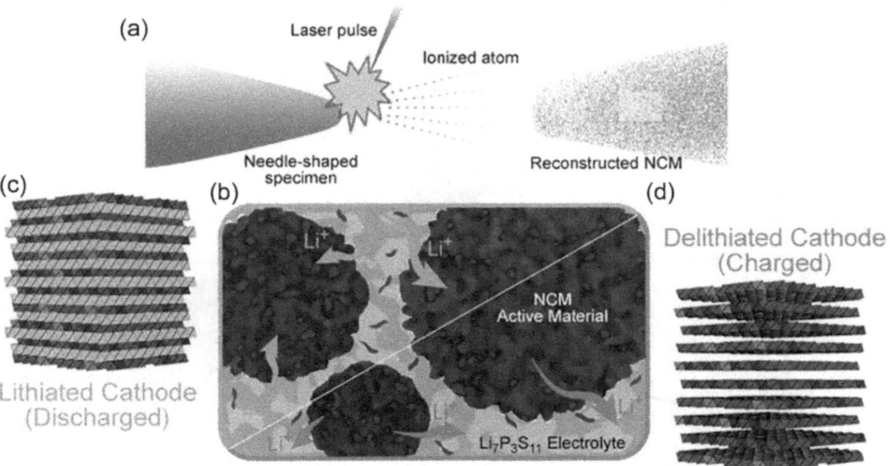

FIGURE 4.17 (a) Schematic illustration of APT analysis of a NMC cathode material in laser pulsing mode. (b) Transport of Li between the active cathode material and the LiPS solid-state electrolyte during operation where Li-ions diffuse from the anode into the cathode in the events of charging and take the opposite route in the events of discharging. (c) Li intercalation and (d) deintercalation in layered structure cathodes associated with charge/discharge cycles. (Adopted from Choi et al.[6])

going through 300 charge/discharge cycles. Similar to what was reported in the previous section by Choi et al., the concentration of Li exhibited a significant decrease after cycling confirmed by both APT and inductively coupled plasma-atomic emission spectroscopy (ICP-AES). As shown in Figure 4.18a, the Li concentration as a function of depth was investigated, where multiple APT specimens were analyzed to improve the statistics. As shown in Figure 4.18b, a radial gradient was observed in the Li content where the surface of particles had the lowest Li content in the cycled state. Furthermore, the gradient in Li loss was found to increase with increasing the number of cycles. When APT and STEM were combined, structural degradation associated with Li accumulation was confirmed similar to the nano-sized features in Figure 4.14b.

Based on the results, the degradation mechanism can be described as the migration of transition metals to the Li accommodation sites[3,100] There is a direct link between the amount of Li depletion and the degree of structural disorder. As the cathode delithiates during charging, transition metal ions moving into the Li sites trigger a phase change to rock-salt or spinel-like structure. This phase transformation inhibits the subsequent Li intercalation upon discharging causing the reduction in capacity and preventing a full charge. Thus, both the depth of the gradient and the extent of Li depletion increase with the increasing number of cycles owing to the lack of Li accommodation sites.

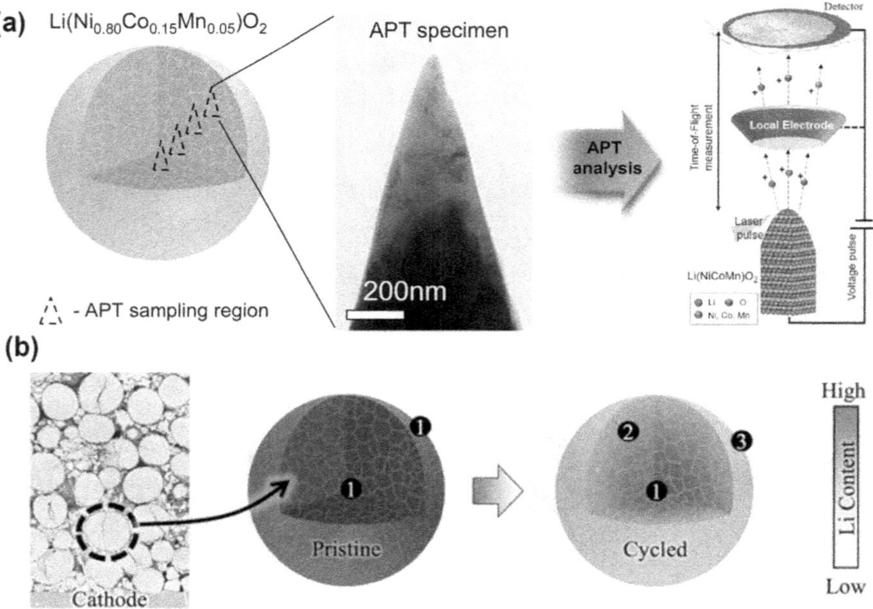

FIGURE 4.18 (a) Schematic illustrations of site-specific analysis along the depth of secondary particles to correlate structural degradation using STEM with Li concentration quantification using APT. (b) Schematic illustrations of the evolution of Li distribution before and after cycling. The Li concentration gradient along the radial direction, which is lower at the surface and higher in the central area, evolves during the cycling events. (Adapted from Refs. [3, 100].)

4.4.2.3 Summary

The aforementioned studies provided insight into the mechanisms of Li-ion battery capacity degradation that originates from structural defects created by structural instability in high-voltage NMC cathodes. Li removal from metal oxides with layered structures causes an irreversible structural change associated with the formation of secondary phases and defects. This information provides possible routes to improve the current technology and contribute to the development of new battery components.

4.4.3 Spinel $LiMn_2O_4$ Cathode Materials

The development of new cathodic materials for Li-ion batteries is an active research field. $LiCoO_2$ is a classic cathode material for Li-ion batteries, but there are concerns about the cost, safety, and toxicity due to the presence of Co. Manganese spinel, $LiMn_2O_4$, is a low-cost alternative that is safer and more environmentally friendly.[101] This cathode material has been used in some commercial Li-ion battery cells, but the loss of capacity over time, especially at high temperatures, has been an issue.[102] The main contributing factors of the capacity fade appears to be resulting from the dissolution of Mn[103] and structural distortions.[104] Improved capacity has been achieved by doping the material with other transition metals to form mixed metal spinels (e.g. $LiNi_{0.5}Mn_{1.5}O_4$).[105]

4.4.3.1 Atomic Resolution

Maier et al.[4] performed APT analyses on non-doped $LiMn_2O_4$ (LMO) and doped $LiNi_{0.5}Mn_{1.5}O_4$ (LNMO) of several micrometers grain size. A custom-built system at the University of Göttingen, equipped with a UV (355 nm) laser, was used that typically yielded datasets of ~10 million ions. The specimens for APT analysis were prepared using the lift-out method described in Section 4.3.1. Prior to APT analysis, specimens were investigated with TEM to determine the crystallographic orientation of the tip axis in regard to the evaporation direction. The volume reconstructed from a specimen with its $\left[11\bar{1}\right]$ crystallographic orientation aligned with the field evaporation direction shows lattice planes in the atom maps of Li, Mn, and O (evaporation direction is from left to right in Figure 4.19). Although low-index crystallographic lattices are commonly observed for metallic samples, they are typically not resolved for oxide materials. There are only a few studies of oxide materials in which lattice plane reconstructions were obtained.[106–108] In this work, Maier et al. employed a fast Fourier transform (FFT)[109] that is capable of detecting weak periodicities in atomic arrangement in all directions. Li, Mn, and O all yielded periodic arrangements in the FFT analysis at similar angles normal to the evaporation direction, which can be correlated with the $\left(11\bar{1}\right)$ planes.

4.4.3.2 In Situ Li Deintercalation

During APT analyses, the specimens are kept at cryogenic temperatures typically between 20K and 80K to suppress surface migration and bulk diffusion of ions in the presence of a high electric field. In laser-assisted evaporation, experimental evidence has shown that thermal pulsing dominates the process where specimens are heated

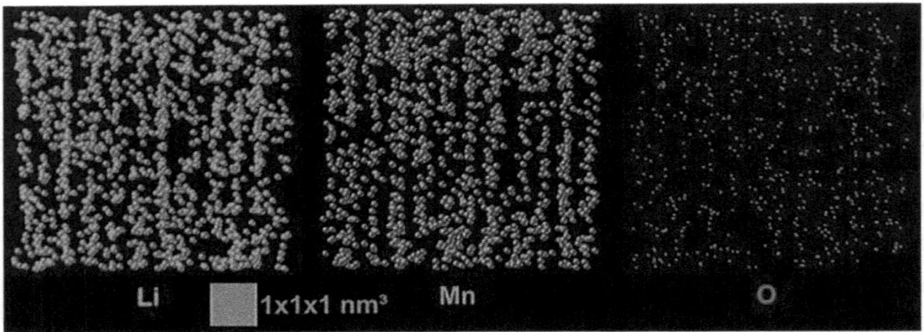

FIGURE 4.19 APT reconstruction detail showing lattice plane resolution for Li, Mn, and O along the horizontal direction. The maps were rotated by a small angle (4°) to synchronize with the local maximum observed in the FFT analysis corresponding to the $\left(11\bar{1}\right)$ planes. (Adapted from Maier et al. [4])

for a very short duration of time.[110,111] In the case of Li-containing materials, it is discussed that the intense electric field necessary for field evaporation may also drive atomic migration attributed to field penetration on insulating materials.[8,112]

To better understand the field-driven Li migration, Pfeiffer et al.[78] modified the APT analysis to compare the Li mobility at a conventional temperature of 30K versus that at a much higher temperature of 298K. For the in situ Li deintercalation (a reversible inclusion of ions into materials with a layered structure) experiment at 30K, specimens after a conventional APT analysis in laser pulsing mode were held at ~80% direct current (DC) voltage to sustain data collection at a reduced evaporation rate for a very long period of time. The concept is that while the specimen still evaporates and keeps getting blunter, if Li-ions are mobile at the base temperature of 30K, the relative ratio of Li/Mn should increase over time, caused by the Li migration either driven by the electric field penetration into the specimen or by a gradient in the chemical potential. The experiment kept at a constant DC voltage and a constant laser pulse energy continued for 180 hours, while ions evaporating from the specimen were identifiable based on their mass-to-charge ratios. After applying a correction to the O supplied from the residual gas in the analysis chamber, no changes in the Li/Mn ratios were confirmed during the course of analysis, indicating no Li transport under the given experiment conditions at the specimen temperature of 30K.

Li mobility at room temperature was investigated using the same concept mentioned above. The specimens were warmed to 298K, and a higher laser pulse energy was used. In the mass spectrum, Li-ions were exclusively detected for the first several hours followed by a significant decrease in the overall detection rate until the deintercalation experiment was terminated at 120 hours. After the deintercalation experiment, the microstructure of the LMO specimen was characterized with the conventional APT analysis at 30K. As indicated by the Li density contour map in Figure 4.20a, where Li exhibits an inhomogeneous distribution showing depletion in the first ~120 nm followed by an alternating structure with Li-rich and Li-depleted regions. The Li/Mn ratio along the center of the reconstruction is shown in Figure 4.20b. A continuously increasing Li/Mn ratio from 0.02 to 0.12 was observed in the

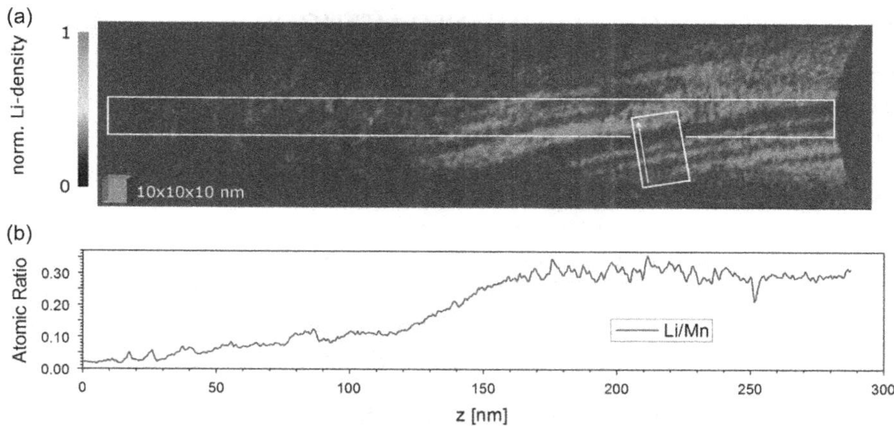

FIGURE 4.20 (a) Li density in a 1 nm thick slice through the reconstruction of a lithium-manganese-oxide (L_xMO) specimen after in situ deintercalation for 120 hours at room temperature. (b) Ratio of Li to Mn (atomic) calculated from the cylindrical volume in the center of the (a). (Adapted from Pfeiffer et al. [78])

first half of the reconstructed volume. After the transition region around 120–160 nm, in addition to some fluctuation, the Li/Mn overall ratio exhibits no dependence on the depth in the dataset. In comparison, the pristine LMO exhibited a homogeneous distribution of Li with Li/Mn ratio between 0.45 and 0.50 in contrast to the stochiometric ratio of 0.5 for spinel $LiMn_2O_4$.[4] The decreased Li concentration was attributed to the in situ deintercalation experiment carried out at room temperature. The Li/Mn ratio taken from the small box along the arrow indicated a sharp transition across various values between 0.1, 0.3, and 0.5. The mechanism behind the formation of parallel Li-enriched and Li-depleted regions between 150 and 300 nm in Figure 4.20a is not clear, but this likely indicates a non-equilibrium stage of the deintercalated L_xMO produced by the in situ delithiation. The possibility of preferential Li transport along a particular crystalline orientation in the spinel structure can be considered to interpret the Li-enriched and Li-depleted features.

4.4.3.3 Summary

APT analysis demonstrated high spatial resolution for spinel LMO where the $\left(11\bar{1}\right)$ crystallographic planes were resolved. The influence of Li-ion mobility during APT analysis was also investigated. For conventional APT analysis at 30K, the mobility of Li-ions is suppressed below the detection limit, so that the microstructure of NMC can be reliably characterized. Acquiring data at room temperature is uncommon but was conducted intentionally to introduce the delithiation. At 298K, a substantial increase in mobility can be achieved, allowing in situ deintercalation of the material in the atom probe without affecting the Mn–O host structure. Such methodological approaches are promising for in situ deintercalation and subsequent characterization of non-equilibrium microstructures in the spinel LMO. Insights into the microstructural evolution accompanying deintercalation processes can be obtained.

4.5 CORRELATIVE AND COMBINED METHODS

The development in correlating atom probe with other analytical techniques in material science has received a tremendous amount of attention. Many new techniques have been introduced that have provided new insights to understand the structure–function relationship. The research of Li-ion batteries requires an approach to address both the structural characterization and the local chemical information. Although microscopy technologies exist to observe each of these characteristics, none of them can perform the task simultaneously. A fine synergy is to combine APT with (S)TEM techniques, with the size of APT specimens readily fulfilling the requirement of electron transparency. At the same time, such correlative approaches pose new challenges regarding hardware modification, specimen preparation, ROI targeting, and data interpretation.

APT remains challenged to deliver accurate length scales, which is one of the strengths of electron microscopy. As described in Section 4.2.5, the selection of APT reconstruction parameters, such as initial radius, shank angle, and evaporation field, directly impacts the reconstruction outcome. Knowing the exact geometry of specimens can be very useful. As demonstrated by Maier et al., TEM images of a LMO specimen were taken prior and after APT analysis to constrain the APT reconstruction for improved reconstruction accuracy.[113]

In the same work by Maier et al., the occurrence of a two-phase region consisting of LiNiMnO and Li-free NiMnO was analyzed with TEM and APT on the same APT specimen.[4] This two-phase structure appears to be commonplace for synthesized cathode NMC with the layered structure. Lee et al. and Devaraj et al., respectively, reported a similar structure as that produced by phase transformation in cycled NMC materials using nano-beam diffraction and EDS analysis in APT specimens.[2,5] The APT results confirmed the composition of the rhombohedral-phase Li(NiCoMn)O_2, and the two sub-products of the phase transformation as the Li$(NiCoMn)_2O_4$ spinel phase and the (NiCoMn)O rock-salt-like phase. Correlative analysis offers a solution to combine the structural information by (S)TEM and the chemical information by APT.

As described in Section 4.4.2, APT and HR-STEM have been used correlatively to explore the evolution of the local Li composition and the corresponding structural changes at the atomic scale. The analysis provided direct evidence of the behavior of Li-ions during cycling, suggesting that the capacity to accommodate Li-ions is determined by the degree of structural disordering.

The field of correlative microscopy for APT is still relatively young but holds promise for future efforts to investigate nanoscale materials. Seamlessly transferring specimens between different types of microscopy instruments is still challenging for needle-shaped atom probe specimens. The development of home-made and commercialized fixtures has been made available to ease the difficulty of specimen handling.[114–116] The capability of the various approaches has yet to be explored and justified considering the best choice of strategy and workflow.

4.6 FUTURE DEVELOPMENT

APT is rapidly developing new applications in Li-ion battery research. The ability to characterize three-dimensional nanoscale chemistry with high sensitivity has revealed many properties controlling the performance of Li-ion batteries. Here, we discuss future directions in the development of the technique.

Experimental evidence shows benefits of the 355 nm UV laser in comparison to the 532 nm green laser in characterizing $LiFePO_4$ cathode materials using APT.[12] Further improvements in data quality using shorter wavelength lasers can be anticipated. In late 2021, CAMECA introduced LEAP 6000 XR and Invizo 6000 systems equipped with lasers at the fourth-harmonic wavelength of 257 nm. Preliminary results on a multilayer structure of Si/SiO_2 have demonstrated significant improvements in specimen yield compared to the LEAP 5000 systems equipped with 355 nm lasers. For Li-ion battery materials that contain interfaces and boundaries, the benefits of the deep UV on the analysis yield, evaporation uniformity, and compositional accuracy are subject to further investigation.

The development of new transfer devices and analysis protocols is driven by the requirement to analyze air- and temperature-sensitive materials that are directly relevant to the research of Li-ion batteries. There has been a recent surge in the use of cryogenic and/or vacuum specimen preparation and transfer systems to broaden the scope of research enabled by APT.[117] Such ability has the potential to allow many exciting new experiments such as compositional mapping across electrolyte–solid interfaces, surface composition in chemical reactions, and nanoparticle capsulation in vitrified liquids. Successful specimen preparation will require effective protocols for vitrification and cryogenic lift-out. Substantial developments in optimization of acquisition conditions, new reconstruction algorithms and data interpretations will also be critical to handle data acquired from vitrified liquids or electrolytes.

Correlative analyses of APT with other analytical techniques have recently been adopted by an increasing number of researchers to obtain specimen geometries, structures, and crystallographic orientations prior to APT analysis. The combined results provide information on microstructure and composition of a material that a single technique alone cannot provide. Information on the geometry and morphology of features contained in an APT specimen can greatly improve the accuracy of the three-dimensional reconstructions.[18,118] Although limited in resolution, electron beam imaging of FIB-SEM systems, especially in the high-resolution immersion mode, allows easy access to the specimen geometry. Transmission electron backscatter diffraction (tEBSD) or transmission Kikuchi diffraction (TKD) to provide crystal orientation during or at the end of the sharpening procedure has been regularly implemented for targeting grain or phase boundaries.[58] Higher spatial resolution information can be gathered by using TEM. APT analysis provides great synergies with TEM, since the radius of the specimens complies with the thickness requirement of electron transparency. Analytical techniques such as EDS and EELS have also been used to obtain preliminary composition. To seamlessly correlate APT and

TEM, efforts have been made to either combine or integrate the two techniques to provide intermittent imaging or diffraction analysis to assess specimen shape and crystallography.[119]

From the control software point of view, advanced automation features such as specimen alignment and chain acquisition have been added to enhance the productivity of APT. Taking advantage of the microtip arrays developed for the LEAP system, a significant improvement in data productivity can be achieved when combined with the advanced automation features. Conditions can be set to automatically start the next specimen in the chain, which would allow the instruments to operate without user interactions. Scripted experimental control has been introduced to allow the instrument to change acquisition conditions in response to the real-time data acquisition.[86] This also can facilitate the process of parameter optimization by screening a wide range of acquisition conditions to evaluate the quality of the data matric. Such features also allow data to be collected at optimal quality and speed when a different phase or feature emerges in the analyzed volume.

In addition, there is an urgent need for the development of APT standards for materials that contain high-mobility Li-ions. Many studies have raised their concerns about the quantification of Li, as there is evidence of in situ delithiation caused by the electrostatic field applied during APT analysis.[8,75,78] Indeed, many studies reported Li concentrations 5–10 at.% above stochiometric compositions (see Table B.1 in Parikh's thesis[84]). As discussed in Section 4.3.6, collecting data at low base temperatures with low laser pulse energies can close the compositional gap for some material systems. This calls for further investigation on the influence of acquisition conditions on the compositional accuracy with APT.

4.7 CONCLUDING REMARKS

Material characterization of various cell components can include many different analytical techniques, but one technique that is rapidly growing in use for the analysis of materials is APT. The capabilities that stand out are its unique combination of very high spatial resolution for analysis (~0.2 nm locally in many specimens) with high analytical sensitivity (single atom detection and parts-per-million fractions). Furthermore, the aspect that makes APT important for battery applications is its ability to detect Li. The accumulation of Li-ions on nano-sized features or particle boundaries has been successfully analyzed by APT. Reliable compositional accuracy can be achieved using the optimized data acquisition conditions even for nano-sized particles. Furthermore, the distribution of impurities and dopants at low concentrations can be quantitatively studied using APT.

Through site-specific specimen preparation methods, the compositional gradients inside individual particles can be studied, which brings important insights into the mechanism of Li migration and capacity fading. The concentration gradient of Li after cycling observed by APT has been attributed to the formation of a disordered structure owing to the occupancy of transition metals on the Li accommodation sites. Using chemical information, APT informs phase transformations from the layered structure to rock-salt or spinel-like disordered phases.

APT also presents great potential in performing in situ deintercalation experiments by means of inducing ion mobility at room temperature with the help of high electric fields. Valuable information on the composition, structure integrity, and ultimately the mechanism of Li deintercalation/intercalation can be investigated using APT.

In summary, there are many positive factors that are driving a wider adoption of APT toward greater achievements, and I look forward to continued contributions that APT will bring to the Li-ion battery research.

ACKNOWLEDGMENT

I sincerely thank Dr. Shen for the opportunity to contribute to the Li-ion battery community. I also thank my colleagues Katherine Rice, Robert Ulfig, and Samantha Hestad at CAMECA Instruments, Inc. for comments that greatly improved the strength and expertise of this chapter on Atom Probe Tomography.

REFERENCES

[1] Song B, Sui T, Ying S, Li L, Lu L, Korsunsky AM. 2015. Nano-structural changes in Li-ion battery cathodes during cycling revealed by FIB-SEM serial sectioning tomography. *J. Mater. Chem. A*. 3:18171–18179.

[2] Devaraj A, Gu M, Colby R, et al. 2015. Visualizing nanoscale 3D compositional fluctuation of lithium in advanced lithium-ion battery cathodes. *Nat. Commun*. 6:8014.

[3] Chae B-G, Park SY, Song JH, Lee E, Jeon WS. 2021. Evolution and expansion of Li concentration gradient during charge–discharge cycling. *Nat. Commun*. 12:3814.

[4] Maier J, Pfeiffer B, Volkert CA, Nowak C. 2016. Three-dimensional microstructural characterization of lithium manganese oxide with atom probe tomography. *Energy Technol*. 4:1565–1574.

[5] Lee JY, Kim JY, Cho HI, et al. 2018. Three-dimensional evaluation of compositional and structural changes in cycled LiNi1/3Co1/3Mn1/3O2 by atom probe tomography. *J Power Sources*. 379:160–166.

[6] Choi S, Yun B-N, Jung WD, et al. 2019. Tomographical analysis of electrochemical lithiation and delithiation of LiNi0.6Co0.2Mn0.2O2 cathodes in all-solid-state batteries. *Scr Mater*. 165:10–14.

[7] Larson DJ, Prosa TJ, Ulfig RM, Geiser BP, Kelly TF. 2013. *Local Electrode Atom Probe Tomography: A User's Guide*. New York: Springer.

[8] Schmitz G, Abouzari R, Berkemeier F, et al. 2010. Nanoanalysis and ion conductivity of thin film battery materials. *Zeitschrift für Physikalische Chemie*. 224:1795–1829.

[9] Diercks DR, Musselman M, Morgenstern A, et al. 2014. Evidence for anisotropic mechanical behavior and nanoscale chemical heterogeneity in cycled LiCoO2. *J. Electrochem. Soc*. 161:F3039–F3045.

[10] Vissers DR, Isheim D, Zhan C, Chen Z, Lu J, Amine K. 2016. Understanding atomic scale phenomena within the surface layer of a long-term cycled 5V spinel electrode. *Nano Energy*. 19:297–306.

[11] Mohanty D, Mazumder B, Devaraj A, et al. 2017. Resolving the degradation pathways in high-voltage oxides for high-energy-density lithium-ion batteries; Alternation in chemistry, composition and crystal structures. *Nano Energy*. 36:76–84.

[12] Santhanagopalan D, Schreiber DK, Perea DE, et al. 2015. Effects of laser energy and wavelength on the analysis of LiFePO4 using laser assisted atom probe tomography. *Ultramicroscopy*. 148:57–66.

[13] Sepehri-Amin H, Ohkubo T, Kodzuka M, et al. 2013. Evidence for nano-Si clusters in amorphous SiO anode materials for rechargeable Li-ion batteries. *Scr. Mater.* 69:92–95.

[14] Stegmaier S, Schierholz R, Povstugar I, et al. 2021. Nano-scale complexions facilitate Li dendrite-free operation in LATP solid-state electrolyte. *Adv. Energy Mater.* 2100707. doi:10.1002/aenm.202100707

[15] Heck PR, Pellin MJ, Davis AM, et al. 2010. Atom-Probe Tomographic Analyses of Presolar Silicon Carbide Grains and Meteoric Nanodiamonds – First Results on Silicon Carbide. In: *41st Lunar and Planetary Science Conference.* The Woodlands, Texas.

[16] Gordon LM, Cohen MJ, Joester D. 2013. Towards atom probe tomography of hybrid organic-inorganic nanoparticles. *Microsc. Microanal.* 19:952–953.

[17] Kim S-H, El-Zoka AA, Gault B. 2021. A liquid metal encapsulation for analyzing porous nanomaterials by atom probe tomography. *Microsc. Microanal.* 1–9. doi:10.1017/S1431927621012964

[18] Herbig M, Choi P-P, Raabe D. 2015. Combining structural and chemical information at the nanometer scale by correlative transmission electron microscopy and atom probe tomography. *Ultramicroscopy.* 153:32–39.

[19] Inoue K, Yoshida K, Nagai Y, Kishida K, Inui H. 2021. Correlative atom probe tomography and scanning transmission electron microscopy reveal growth sequence of LPSO phase in Mg alloy containing Al and Gd. *Sci. Rep.* 11:3073.

[20] Stoffers A, Cojocaru-Mirédin O, Seifert W, Zaefferer S, Riepe S, Raabe D. 2015. Grain boundary segregation in multicrystalline silicon: correlative characterization by EBSD, EBIC, and atom probe tomography. *Prog. Photovolt. Res. Appl.* 23:1742–1753.

[21] Müller EW, Panitz JA, McLane SB. 1968. The atom-probe field ion microscope. *Rev. Sci. Instrum.* 39:83–86.

[22] Panitz JA. 1973. The 10 cm atom probe. *Rev. Sci. Instrum.* 44:1034–1038.

[23] Cerezo A, Godfrey TJ, Smith GDW. 1988. Application of a position-sensitive detector to atom probe microanalysis. *Rev. Sci. Instrum.* 59:862–866.

[24] Nishikawa O, Kimoto M. 1994. Toward a scanning atom probe - computer simulation of electric field. *Appl. Surf. Sci.* 76/77:424–430.

[25] Sijbrandij S, Larson DJ, Miller M. 1997. Atom probe field ion microscopy of high resistivity materials. *Ultramicroscopy* 4(S2):90–91.

[26] Kellogg GL, Tsong TT. 1980. Pulsed-laser atom-probe field-ion microscopy. *J. Appl. Phys.* 51:1184–1194.

[27] Kellogg GL. 1981. Determining the field emitter temperature during laser irradiation in the pulsed laser atom probe. *J. Appl. Phys.* 52:5320–5326.

[28] Liddle JA, Norman A, Cerezo A, Grovenor CRM. 1988. Pulsed laser atom probe analysis of ternary and quaternary III-V epitaxial layers. *J. Phys. (Paris) Colloq.* 49:509–514.

[29] Kelly TF, Thompson K, Marquis EA, Larson DJ. 2006. Atom probe tomography defines mainstream microscopy at the atomic scale. *Microsc. Today.* 14(4):34–40.

[30] Larson DJ, Prosa TJ, Lawrence D, Geiser BP, Jones CM, Kelly TF. 2011. Atom Probe Tomography for Microelectronics. In: Haight R, Ross F, Hannon J, eds. *Handbook of Instrumentation and Techniques for Semiconductor Nanostructure Characterization,* 2:407–477, London: World Scientific Publishing.

[31] Lauhon LJ, Adusumilli P, Ronsheim P, Flaitz PL, Lawrence D. 2009. Atom-probe tomography of semiconductor materials and device structures. *MRS Bulletin.* 34:738–743.

[32] Chen YM, Ohkubo T, Kodzuka M, Morita K, Hono K. 2009. Laser-assisted atom probe analysis of zirconia/spinel nanocomposite ceramics. *Scr. Mater.* 61:693–696.

[33] Devaraj A, Szymanski C, Yan P, et al. 2015. Nanoscale characterization of Li-ion battery cathode nanoparticles by atom probe tomography correlated with transmission electron microscopy and scanning transmission X-ray microscopy. *Microsc. Microanal.* 21:685–686.

[34] Sharifi-Asl S, Lu J, Amine K, Shahbazian-Yassar R. 2019. Oxygen release degradation in Li-ion battery cathode materials: mechanisms and mitigating approaches. *Adv. Energy Mater.* 9:1900551.
[35] Bunton JH, Olson JD, Lenz DR, Larson DJ, Kelly TF. 2010. Optimized laser thermal pulsing of atom probe tomography: LEAP 4000X. *Microsc. Microanal.* 16(S2):10–11.
[36] Prosa T, Lenz D, Martin I, Reinhard D, Larson D, Bunton J. 2021. Evaporation-field differences with deep-UV atom probe tomography. *Microsc. Microanal.* 27(S1):1262–1264.
[37] Stender P, Oberdorfer C, Artmeier M, Pelka P, Spaleck F, Schmitz G. 2007. New tomographic atom probe at University of Muenster, Germany. *Ultramicroscopy.* 107:726–733.
[38] Gault B, Vurpillot F, Vella A, et al. 2006. Design of a femtosecond laser assisted tomographic atom probe. *Rev. Sci. Instrum.* 77:043705/1-8.
[39] Hono K, Ohkubo T, Chen YM, et al. 2011. Broadening the applications of the atom probe technique by ultraviolet femtosecond laser. *Ultramicroscopy.* 111:576–583.
[40] Wang ZL. 2000. Transmission electron microscopy of shape-controlled nanocrystals and their assemblies. *J. Phys. Chem. B.* 104:1153–1175.
[41] McPhail DS. 2006. Applications of secondary ion mass spectrometry (SIMS) in materials science. *J. Mater. Sci.* 41:873–903.
[42] Kelly TF. 2019. Atom-Probe Tomography. In: Hawkes PW, Spence JCH, eds. *Springer Handbook of Microscopy*, 2. Cham: Springer International Publishing. doi:10.1007/978-3-030-00069-1_15
[43] Gault B, Chiaramonti A, Cojocaru-Mirédin O, et al. 2021. Atom probe tomography. *Nat. Rev. Methods Primers.* 1:51.
[44] Miller MK, Forbes RG. 2009. Tutorial review: atom probe tomography. *Mater. Charact.* 60:461–469.
[45] Gomer R. 1961. *Field Emission and Field Ionization.* Cambridge, MA: Harvard University Press.
[46] Loi ST, Gault B, Ringer SP, Larson DJ, Geiser BP. 2013. Electrostatic simulations of a local electrode atom probe: the dependence of tomographic reconstruction parameters on specimen and microscope geometry. *Ultramicroscopy.* 132:107–113.
[47] Sakurai T, Müller EW. 1977. Field calibration using the energy distribution of a free-space field ionization. *J. Appl. Phys.* 48:2618–2625.
[48] Prosa TJ, Geiser BP, Lawrence D, Olson D, Larson DJ. 2014. Developing detection efficiency standards for atom probe tomography. *SPIE Proc.* 9173:917307.
[49] Bas P, Bostel A, Deconihout B, Blavette D. 1995. A general protocol for the reconstruction of 3D atom probe data. *Appl. Surf. Sci.* 87/88:298–304.
[50] Geiser BP, Larson DJ, Oltman E, et al. 2009. Wide-field-of-view atom probe reconstruction. *Microsc. Microanal.* 15(S2):292–293.
[51] Gault B, Haley D, de Geuser F, et al. 2011. Advances in the reconstruction of atom probe tomography data. *Ultramicroscopy.* 111:448–457.
[52] Gault B, Moody MP, de Geuser F, et al. 2009. Advances in the calibration of atom probe tomographic reconstruction. *J. Appl. Phys.* 105:034913/1-9.
[53] Gault B, Moody MP, Cairney JM, Ringer SP. 2012. *Atom Probe Microscopy*, 160. New York: Springer.
[54] Melmed AJ. 1991. The art and science and other aspects of making sharp tips. *J. Vac. Sci. Technol.* B9:601–609.
[55] Larson DJ, Foord DT, Petford-Long AK, et al. 1999. Field-ion specimen preparation using focused ion-beam milling. *Ultramicroscopy.* 79:287–293.
[56] Waugh AR, Payne S, Worrall GM, Smith GDW. 1984. In situ ion milling of field ion specimens using a liquid metal ion source. *J. Phys.* 45:207–209.
[57] Alexander KB, Angelini P, Miller MK. 1989. Precision ion milling of field-ion specimens. *J. Phys. (Paris), Colloq.* 50:549–554.

[58] Chen Y, Rice KR, Prosa TJ. 2016. Site-specific sample preparation using correlative microscopy: APT and tEBSD. *Micros. Analysis Suppl.* May/June:S4.
[59] Guo W, Sneed BT, Zhou L, et al. 2016. Correlative energy-dispersive X-ray spectroscopic tomography and atom probe tomography of the phase separation in an Alnico 8 alloy. *Microsc. Microanal.* 22:1251–1260.
[60] Larson DJ, Lawrence D, Olson D, et al. 2011. From the Store Shelf to Device-Level Atom Probe Analysis: An Exercise in Feasibility. In: *36th International Symposium for Testing and Failure Analysis*, 189–197. ASM International.
[61] Kelly TF, Larson DJ. 2012. The second revolution in atom probe tomography. *MRS Bulletin.* 37:150–158.
[62] Miller MK, Russell KF, Thompson K, Alvis R, Larson DJ. 2007. Review of atom probe FIB-based specimen preparation methods. *Microsc. Microanal.* 13:428–436.
[63] Thompson K, Lawrence DJ, Larson DJ, Olson JD, Kelly TF, Gorman B. 2007. In-situ site-specific specimen preparation for atom probe tomography. *Ultramicroscopy.* 107:131–139.
[64] Sun Y-K, Kim D-H, Yoon CS, Myung S-T, Prakash J, Amine K. 2010. A novel cathode material with a concentration-gradient for high-energy and safe lithium-ion batteries. *Adv. Funct. Mater.* 20:485–491.
[65] Hou P, Zhang H, Zi Z, Zhang L, Xu X. 2017. Core–shell and concentration-gradient cathodes prepared via co-precipitation reaction for advanced lithium-ion batteries. *J. Mater. Chem. A.* 5:4254–4279.
[66] Heo K, Im J, Kim S, et al. 2020. Effect of nanoparticles in cathode materials for flexible Li-ion batteries. *J. Ind. Eng. Chem.* 81:278–286.
[67] Jiang K, Back S, Akey AJ, et al. 2019. Highly selective oxygen reduction to hydrogen peroxide on transition metal single atom coordination. *Nat. Commun.* 10:1–11.
[68] Yang Q, Danaie M, Young N, et al. 2019. Atom probe tomography of Au–Cu bimetallic nanoparticles synthesized by inert gas condensation. *J. Phys. Chem. C.* 123:26481–26489.
[69] Felfer P, Li T, Eder K, et al. 2015. New approaches to nanoparticle sample fabrication for atom probe tomography. *Ultramicroscopy.* 159(Part 2):413–419.
[70] Kim S-H, Kang PW, Park OO, et al. 2018. A new method for mapping the three-dimensional atomic distribution within nanoparticles by atom probe tomography (APT). *Ultramicroscopy.* 190:30–38.
[71] Sundell G, Hulander M, Pihl A, Andersson M. 2019. Atom probe tomography for 3D structural and chemical analysis of individual proteins. *Small.* 15:1900316.
[72] Qu J, Yang W, Wu T, et al. 2022. Atom probe specimen preparation methods for nanoparticles. *Ultramicroscopy.* 233:113420.
[73] Heck PR, Stadermann FJ, Isheim D, et al. 2014. Atom-probe analyses of nanodiamonds from Allende. *Meteorit. Planet Sci.* 49:453–467.
[74] Larson DJ, Giddings AD, Wu Y, et al. 2015. Encapsulation method for atom probe tomography analysis of nanoparticles. *Ultramicroscopy.* 159:420–426.
[75] Kim S-H, Antonov S, Zhou X, et al. 2022. Atom probe analysis of electrode materials for Li-ion batteries: challenges and ways forward. *J. Mater. Chem. A.* 10:4926–4935.
[76] Rice KP, Ulfig RM, Maier U, Passey RG. 2019. Cryogenic UHV specimen preparation for APT: a transfer solution. *Microsc. Microanal.* 25:528–529.
[77] Cairney JM, McCarroll I, Chen Y-S, et al. 2019. Correlative UHV-cryo transfer suite: connecting atom probe, SEM-FIB, transmission electron microscopy via an environmentally-controlled glovebox. *Microsc. Microanal.* 25:2494–2495.
[78] Pfeiffer B, Maier J, Arlt J, Nowak C. 2017. *In situ* atom probe deintercalation of lithium-manganese-oxide. *Microsc Microanal.* 23:314–320.
[79] Miller MK, Forbes RG. 2014. *Atom-Probe Tomography: The Local Electrode Atom Probe.* Boston, MA: Springer US.

[80] Smith PJ, Smith DA. 1970. Preliminary calculations of the electric field and the stress on a field-ion specimen. *Philos. Mag.* 21:907–912.

[81] Birdseye PJ, Smith DA. 1970. The electric field and the stress on a field-ion specimen. *Surf. Sci.* 23:198–210.

[82] Mikhailovskij IM, Wanderka N, Storizhko VE, Ksenofontov VA, Mazilova TI. 2009. A new approach for explanation of specimen rupture under high electric field. *Ultramicroscopy.* 109:480–485.

[83] Chen Y, Clifton P, Prosa T. 2021. Matrix composition and fine-scale structure analysis of NMC Li-ion battery using atom probe tomography. *Microsc. Microanal.* 27:2476–2478.

[84] Parikh P. 2019. *Exploring Nanoscale Chemical Distributions for Energy Applications Using Atom Probe Tomography.* San Diego: University of California.

[85] Kingham DR. 1982. The post-ionization of field evaporated ions: a theoretical explanation of multiple charge states. *Surf. Sci.* 116:273–301.

[86] Reinhard DA, Payne TR, Strennen EM, et al. 2019. Atom probe tomography productivity enhancements. *Microsc. Microanal.* 25:522–523.

[87] Prosa TJ, Oltman E. 2021. Study of LEAP® 5000 Deadtime and Precision via Silicon Pre-Sharpened-Microtip™ standard specimens. *Microsc. Microanal.* 1–19. doi:10.1017/S143192762101206X

[88] Placke T, Kloepsch R, Dühnen S, Winter M. 2017. Lithium ion, lithium metal, and alternative rechargeable battery technologies: the odyssey for high energy density. *J. Solid State Electrochem.* 21:1939–1964.

[89] Goodenough JB. 2007. Cathode materials: a personal perspective. *J. Power Sources.* 174:996–1000.

[90] Broussely M, Perton F, Biensan P, et al. 1995. LixNiO2, a promising cathode for rechargeable lithium batteries. *J. Power Sources.* 54:109–114.

[91] Schipper F, Erickson EM, Erk C, Shin J-Y, Chesneau FF, Aurbach D. 2016. Recent advances and remaining challenges for lithium ion battery cathodes. *J. Electrochem. Soc.* 164:A6220.

[92] Wei X. 2017. Effects of lithium content on structure and electrochemical properties of Li-rich cathode material Li1.2+xMn0.54Ni0.13Co0.13O2. *Int. J. Electrochem. Sci.* 12:5636–5645.

[93] Devaraj A, Colby R, Hess W, Perea DE, Thevuthasan S. 2013. Role of photoexcitation and field ionization in the measurement of accurate oxide stoichiometry by laser-assisted atom probe tomography. *J. Phys. Chem. Lett.* 4:998–1003.

[94] Bachhav M, Danoix R, Danoix F, Hannoyer B, Ogale S, Vurpillot F. 2011. Investigation of wustite (Fe1-xO) by femtosecond laser assisted atom probe tomography. *Ultramicroscopy.* 111:584–588.

[95] Kirchhofer R, Teague MC, Gorman BP. 2013. Thermal effects on mass and spatial resolution during laser pulse atom probe tomography of cerium oxide. *J. Nucl. Mater.* 436:23–28.

[96] Cappelli C, Smart S, Stowell H, Pérez-Huerta A. 2021. Exploring biases in atom probe tomography compositional analysis of minerals. *Geostand. Geoanal. Res.* (n/a).

[97] Bachhav M, Danoix F, Hannoyer B, Bassat JM, Danoix R. 2013. Investigation of O-18 enriched hematite (α-Fe2O3) by laser assisted atom probe tomography. *Int. J. Mass Spectrom.* 335:57–60.

[98] Zhou K, Xie Q, Li B, Manthiram A. 2021. An in-depth understanding of the effect of aluminum doping in high-nickel cathodes for lithium-ion batteries. *Energy Storage Mater.* 34:229–240.

[99] Vetter J, Novák P, Wagner MR, et al. 2005. Ageing mechanisms in lithium-ion batteries. *J. Power Sources.* 147:269–281.

[100] Gault B, Poplawsky JD. 2021. Correlating advanced microscopies reveals atomic-scale mechanisms limiting lithium-ion battery lifetime. *Nat Commun.* 12:3740.

[101] Kim DK, Muralidharan P, Lee H-W, et al. 2008. Spinel LiMn2O4 Nanorods as Lithium Ion Battery Cathodes. *Nano Lett.* 8:3948–3952.
[102] Xia Y, Sakai T, Fujieda T, et al. 2001. Correlating capacity fading and structural changes in Li1+ y Mn2- y O 4- δ spinel cathode materials: a systematic study on the effects of Li/Mn ratio and oxygen deficiency. *J. Electrochem. Soc.* 148:A723.
[103] Börner M, Klamor S, Hoffmann B, et al. 2016. Investigations on the C-rate and temperature dependence of manganese dissolution/deposition in LiMn2O4/Li4Ti5O12 lithium ion batteries. *J. Electrochem. Soc.* 163:A831.
[104] Gummow RJ, de Kock A, Thackeray MM. 1994. Improved capacity retention in rechargeable 4 V lithium/lithium-manganese oxide (spinel) cells. *Solid State Ionics.* 69:59–67.
[105] Oh SH, Chung KY, Jeon SH, Kim CS, Cho WI, Cho BW. 2009. Structural and electrochemical investigations on the LiNi0. 5- xMn1. 5- yMx+ yO4 (M= Cr, Al, Zr) compound for 5 V cathode material. *J. Alloys Compd.* 469:244–250.
[106] Gault B, Moody MP, Cairney JM, Ringer SP. 2012. Atom probe crystallography. *Mater. Today.* 15:378–386.
[107] Yeoh WK, Gault B, Cui XY, et al. 2011. Direct observation of local potassium variation and its correlation to electronic inhomogeneity in (Ba_{1-x}K_{x})Fe_{2}As_{2} Pnictide. *Phys. Rev. Lett.* 106:247002.
[108] Ceguerra AV, Breen AJ, Cairney JM, Ringer SP, Gorman BP. 2021. Integrative atom probe tomography using scanning transmission electron microscopy-centric atom placement as a step toward atomic-scale tomography. *Microsc. Microanal.* 27:140–148.
[109] Araullo-Peters VJ, Breen A, Ceguerra AV, Gault B, Ringer SP, Cairney JM. 2015. A new systematic framework for crystallographic analysis of atom probe data. *Ultramicroscopy.* 154:7–14.
[110] Vurpillot F, Houard J, Vella A, Deconihout B. 2009. Thermal response of a field emitter subjected to ultra-fast laser illumination. *J. Phys. D: Appl. Phys.* 42:125502.
[111] Vella A, Vurpillot F, Gault B, Menand A, Deconihout B. 2006. Evidence of field evaporation assisted by nonlinear optical rectification induced by ultrafast laser. *Phys. Rev. B.* 73:165416/1-6.
[112] Greiwe G-H, Balogh Z, Schmitz G. 2014. Atom probe tomography of lithium-doped network glasses. *Ultramicroscopy.* 141:51–55.
[113] Maier J, Pfeiffer B, Volkert CA, Nowak C. 2016. Three-dimensional microstructural characterization of lithium manganese oxide with atom probe tomography. *Energy Technol.* 4:1565–1574.
[114] Felfer P, Alam T, Ringer SP, Cairney JM. 2012. A reproducible method for damage-free site-specific preparation of atom probe tips from interfaces. *Microsc. Res. Tech.* 75:484–491.
[115] HBSAdmin. 2014. Atom Probe Tomography Workflow. Hummingbird Scientific. Available at: https://hummingbirdscientific.com/atom-probe-tomography/. Accessed August 2, 2022.
[116] Model 2050 | Fischione. Available at: https://www.fischione.com/products/holders/model-2050-axis-rotation-tomography-holder. Accessed August 2, 2022.
[117] McCarroll IE, Bagot PAJ, Devaraj A, Perea DE, Cairney JM. 2020. New frontiers in atom probe tomography: a review of research enabled by cryo and/or vacuum transfer systems. *Mater. Today Adv.* 7:100090.
[118] Haley D, Petersen T, Ringer SP, Smith GDW. 2011. Atom probe trajectory mapping using experimental tip shape measurements. *J. Microsc.* 244:170–180.
[119] Kelly TF, Miller MK, Rajan K, et al. 2011. Toward atomic-scale tomography: the ATOM project. *Microsc. Microanal.* 17(Suppl 2):708–709.

5 In Situ X-Ray Diffraction Studies on Lithium-Ion and Beyond Lithium-Ion Batteries

N. Angulakshmi
Gyeongsang National University

Murugavel Kathiresan and A. Manuel Stephan
CSIR-Central Electrochemical Research Institute

CONTENTS

5.1 INTRODUCTION

X-ray diffraction (XRD) has been widely employed as a powerful technique to investigate the crystallographic information of electrode materials, including lattice parameters, phase transition, and strain.[1] These parameters are closely related to the properties of battery electrodes, e.g., energy density, cycling life, and rate capability. Furthermore, in situ/operando synchrotron-based XRD has been extensively employed to analyze the electrochemical reaction mechanism of different types of electrode materials for lithium-ion, lithium-sulfur and lithium-air/O_2 with a focus on the structural and phase evolution under the operating conditions.

Depending upon the working principles, three typical XRD techniques are classified as X-ray powder, single-crystal, and X-ray Laue diffraction. The first two techniques use monochromatic X-ray, while third applies polychromatic X-ray to test materials combined with 2D detectors.[2] Compared with the laboratory XRD, SXRD shows higher brilliance and flux and tunable energies of X-ray beam due to the synchrotron resource, which can strengthen the intensity of signals and shorten testing

DOI: 10.1201/9781003299295-5

time. The high intensity of spatially resolved (SR X-ray) beam flux makes SXRD suitable for microprobe characterization (X-ray microdiffraction) to track microstructure of electrode or electrode materials evolution during cycling.[3] Interestingly, owing to the large range of tunable energies of SR X-ray beams, energy-dispersive X-ray diffraction has been intensively used to probe the different sections of electrode materials with the ability to penetrating the testing batteries with metal cases.[4] SXRD can provide a high level of structural detail and time resolution for electrode and electrolyte materials. In situ and operando SXRD has been intensively carried out to study the electrochemical reaction mechanism of insertion, conversion, and alloy-type electrode materials with a focus on tracking structural and phase evolution and the formation of defects and strain in lithium-ion (LIBs),[5] Li-S/Se,[6] and Li-O_2[7] batteries with cycling and operation at various temperatures.

The operando XRD technique has been deliberated as a potential analytic tool amongst other operando techniques as it sheds light on the phase transitions, changes in the lattice constant, and crystal evolutions that have a close association with the redox reactions occurring on the electrodes during galvanostatic charge/discharge. Compared with the laboratory-based XRD, the synchrotron-based XRD has the advantages of high flux and tunable X-ray energy, which can greatly improve the signal intensity and shorten the testing time. The high flux of focused X-ray beam also makes synchrotron-based XRD suitable for microprobe to investigate the structural evolution of electrode materials during cycling with spatial evolution. Operando XRD characterization of battery materials such as cathode and anode throws light not only on the fundamental understanding of these materials' structural changes and degradation but also on the development of novel electrode designs. In operando XRD, both reflection and transmission modes are used to monitor crystal structure and phase changes of electrodes during battery cycling. Different in situ cell configurations were designed with an X-ray transparent window, such as beryllium disk, polymer films (Kapton, Mylar), and aluminum foil. A beryllium window has the advantages of maintaining contact between cell components and is nearly 100% X-ray transparent due to disk rigidity and low X-ray absorption. Unfortunately, corrosion of beryllium was observed, especially at high voltage >4.2V; moreover, beryllium is highly toxic and expensive compared to any other materials. Kapton, Mylar, and aluminum are inexpensive and less toxic. They also allow X-rays to pass through with certain absorption; however, it is difficult to keep cell pressurized with a large area window owing to their flexibility. Aluminum foil is used as a current collector in LIBs and also used as an X-ray window to reduce undesired diffraction peaks in the spectra. When the synchrotron-based high-energy X-rays are used, the transmission mode can be applied simultaneously to follow the evolution of both cathode and anode materials.

5.2 OPERANDO STUDIES ON CATHODE MATERIALS

Lithium iron phosphate ($LiFePO_4$) is one of the most investigated cathode materials for LIBs owing to its high theoretical capacity (170mAh/g), long cycle life, good thermal stability, high reversibility, low cost, non-toxicity, and high safety characteristics. Both $LiFePO_4$ (LFP) and $FePO_4$ (FP) have similar crystal structure described by space group Pnma, but different lattice parameters which undergo a two-phase reaction process between $LiFePO_4$ and $FePO_4$. In the early stages, the nature of these materials toward Li^+ insertion and performance was unknown. However, certain strategies were adopted on the material characteristics to improve the rate performance of $LiFePO_4$. Modification of electrochemical performance however needs a proper understanding of ion-diffusion kinetics and phase-transformation mechanisms of materials. In this aspect, the operando method such as XRD would be of great help in understanding the materialistic as well as electrochemical aspects of battery materials.

Table 5.1 lists the application and development of operando XRD characterization for rechargeable batteries (cathode and anode).

Kim et al.[27] used the temperature-controlled operando XRD to assess the phase stability of binary Fe and Mn olivine materials, and the authors found the thermal behavior of partly charged olivine material is affected by the Fe/Mn ratio. Lee et al.[28] used operando XRD to investigate the structural progression of Li-excess nickel-titanium molybdenum oxides upon charge/discharge. Operando XRD results along with XANES spectroscopy revealed that the first charge of $Li_{1.2}Ni_{1/3}Ti_{1/3}Mo_{2/15}O_2$ to 4.8V involves Ni^{1+}/Ni^{3+} oxidation, oxygen loss, and oxygen oxidation, while discharging, Mo^{6+} and/or Ti^{4+} undergo reduction. In a similar study, Wu et al. used operando XRD to study the structural progression of Na_xTiO_2 during electrochemical deintercalation.[29] The results showed a reversible O3–O′3 phase transition with an

TABLE 5.1
Classification of the Electrode Materials by Operando XRD Characterization in Rechargeable Batteries

Classification		Materials	Energy-Storage Systems	Analytical Viewpoint	Ref.
Cathode materials	Polyanionic compounds	$LiFePO_4$	LIBs	Phase transition	9,10
		$LiMn_{0.25}Fe_{0.75}PO_4$	LIBs	Phase transition	11
		$Na_3V_2(PO_4)_3$	NIBs	Structure evolution	12
		$K_3V_2(PO_4)_3$	NIBs	Structure evolution	13
	Layered oxides	$NaNiO_2$	NIBs	Phase transition	14
		V_2O_5/NaV_6O_{15}	LIBs	Phase transition	15
		P2-Na_xCoO_2	NIBs	Phase transition	16
		$Na_{0.7}Fe_{0.5}Mn_{0.5}O_2$	NIBs	Structure evolution	17
		$K_{0.7}Fe_{0.5}Mn_{0.5}O_2$	KIBs	Structure evolution	18
Anode materials	Insertion type	$NaTiO_2$	NIBs	Phase transition	19
		$Li_4Ti_5O_{12}$	NIBs	Reaction mechanism	20
		$Na_3Ti_2(PO_4)_3$	NIBs	Structure evolution	21
	Conversion type	Mn_3O_4	LIBs	Phase transition	22
		$Fe_3O_4/VO_x/G$	LIBs	Structure evolution	23
		$Co_2V_2O_7$	LIBs	Reaction mechanism	24
	Alloying type	Si	LIBs	Reaction mechanism	25
		Sb	NIBs	Reaction mechanism	26

Source: Reproduced with permission from ref. [8].

uncommon lattice parameter variation coupled with complicated Na vacancy orderings. The unusual lattice parameter variation renders a constant interslab distance and slightly changing in-plane Ti–Ti distance in the O′3 phase in a series of second-order phase transitions. Wang et al. studied high-energy XRD and synchrotron-based operando transmission X-ray microscopy to understand the dissymmetric phase alteration and structure-evolution mechanism of layered $NaNiO_2$ material.[30] The authors found a phase transformation as well as the deformation of $NaNiO_2$ in the range of 3.0V–4.0V during the first cycling, which was responsible for the irreversible capacity loss.

Orikasa et al. used time-resolved XRD to investigate the phase transition of $LiFePO_4$ at room temperature.[31] It was found that a metastable crystal phase transitorily arises between $LiFePO_4$ and $FePO_4$, and this metastable phase accounts for the high rate capability of $LiFePO_4$. Similarly, Liu et al. probed the phase transformation of $LiFePO_4$ using time-resolved operando XRD at room temperature.[9] Results indicate that the high rate performance of $LiFePO_4$ is due to the existence of a facile non-equilibrium single-phase-transient pathway. Yan et al. used operando two-dimensional XRD to study the phase-transformation routes of $LiFePO_4/FePO_4$ at various cyclic voltammetry scan rates as well as temperatures.[10] To perform such an experiment, a special electrochemical cell with an X-ray-transparent beryllium window was used. The cell was kept in an argon-filled glove box. During the galvanostatic charge-discharge process, XRD signals were directly collected on a planar detector. The results exposed the existence of intermediate phases during Li insertion/extraction processes at low temperatures.

The operando XRD patterns of $LiFePO_4$ at various scan rates such as 1.4, 2.8, and 4.2 mV/s at 293K are shown in Figure 5.1. It is quite apparent that between the 2θ range of 41.7° and 42.8°, continuous positive intensities were observed in the selected diffraction patterns at a scan rate of 4.2 mV/s (Figure 5.2c). This observation indicates the possible existence of intermediate phases between $LiFePO_4$ and $FePO_4$. Further, the phase transition of $LiFePO_4$ and $FePO_4$ was also probed by operando XRD[2] under similar conditions as shown in Figure 5.2d and e. It is evident that the formation of intermediate phases is specified by broad asymmetric reflections of (211), (311), and (121). Also, the selected diffraction peaks shift to a higher angle after cycling, signifying a decrease in the unit-cell volume.

The ion-diffusion coefficients, as well as the structure evolution of $LiFePO_4$ at different temperatures, are shown in Figure 5.2a–c. The ion-diffusion coefficient decreases to some extent with respect to the temperature decrease from 293K to 273K indicating the formation of intermediate phases. The corresponding lattice parameter variations are shown in Figures 5.2b and c. It is found that the parameter b decreases and increases with time in the cathodic and anodic processes when cycled at 273K which is indicative of a solid-solution reaction between the intermediate phases of $LiFePO_4$ and $FePO_4$. These studies clearly indicate that the formation of intermediate phases can prevent the deprivation of the ion-diffusion coefficient at low temperatures.

Chen et al. employed the operando XRD technique with synchrotron-based XRD to understand the phase transformation of $LiMn_{0.25}Fe_{0.75}PO_4$.[11] The operando XRD patterns of $LiMn_{0.25}Fe_{0.75}PO_4$ as a function of lithium content x during the first

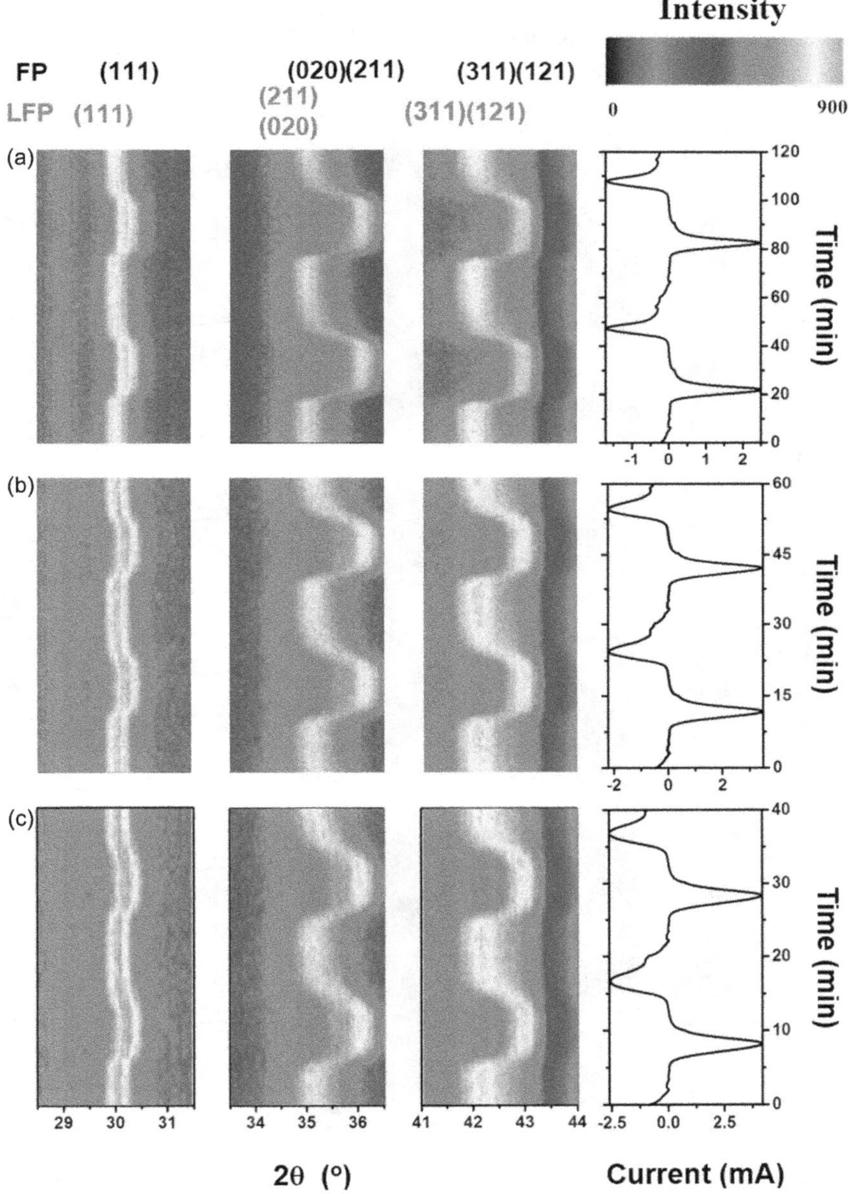

FIGURE 5.1 The image plot of diffraction patterns for (111), (211), (020), (311), and (121) reflections during the two CV cycles under different scan rates of 1.4 (a and d), 2.8 (b and e), and 4.2 mV/s (c and f) at temperatures of 293K (a–c) and 273K (d–f). The corresponding current curves are plotted on the right. LFP represents $LiFePO_4$; FP represents $FePO_4$. ((a–f) Reproduced with permission.[10] Copyright 2016, Elsevier.)

(Continued)

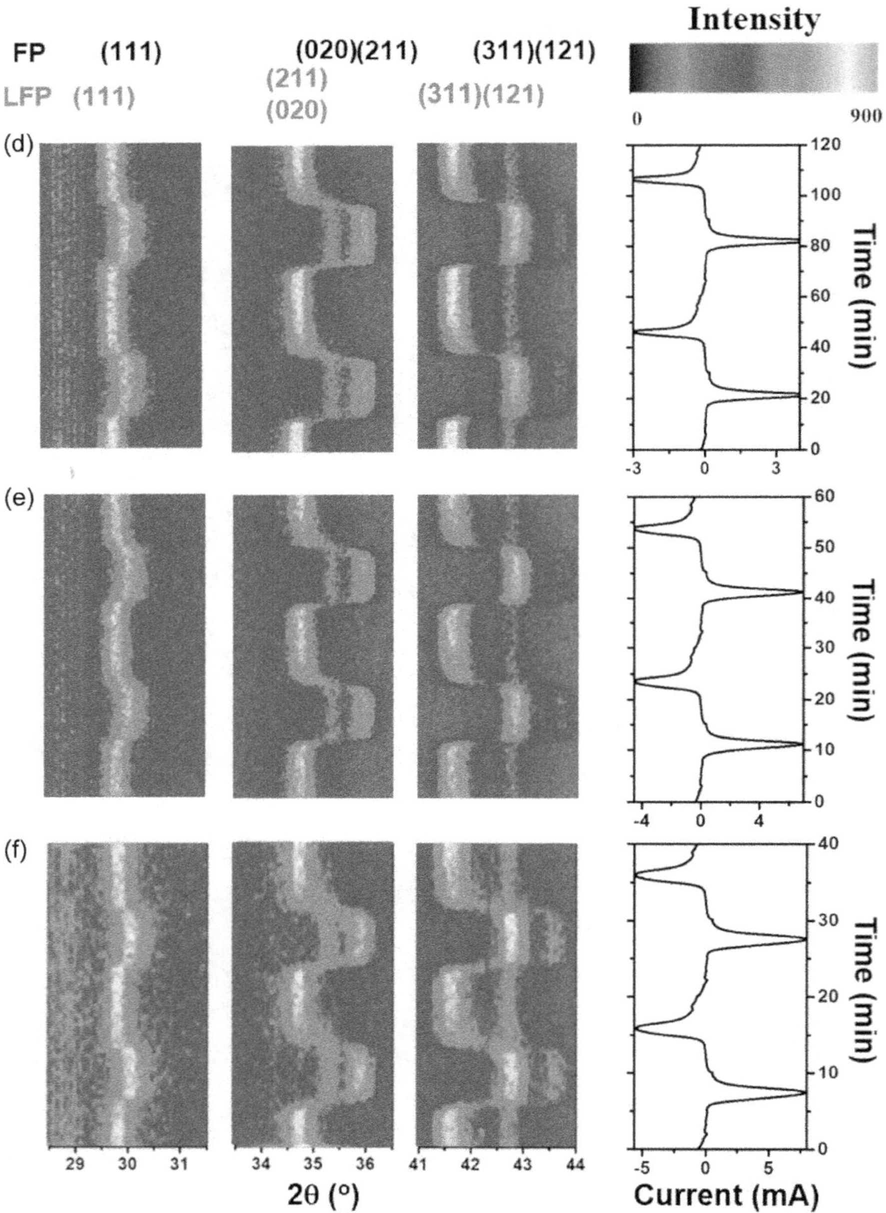

FIGURE 5.1 (*CONTINUED*) The image plot of diffraction patterns for (111), (211), (020), (311), and (121) reflections during the two CV cycles under different scan rates of 1.4 (a and d), 2.8 (b and e), and 4.2 mV/s (c and f) at temperatures of 293K (a–c) and 273K (d–f). The corresponding current curves are plotted on the right. LFP represents $LiFePO_4$; FP represents $FePO_4$. ((a–f) Reproduced with permission.[10] Copyright 2016, Elsevier.)

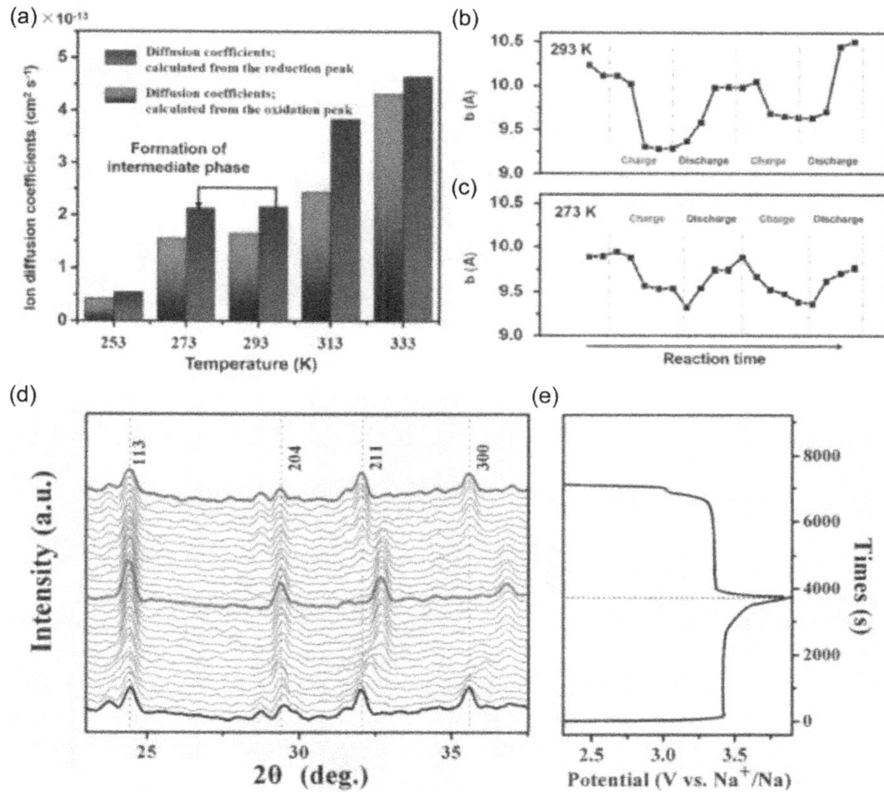

FIGURE 5.2 (a) Ion-diffusion coefficients of $LiFePO_4$ at different temperatures. (b and c) The unit cell parameter *b* as a function of reaction time, obtained from Rietveld refinement at temperatures of 293K (b) and 273K (c) with a scan rate of 1.4 mV/s. (d) The operando XRD patterns of three-dimensional $Na_3V_2(PO_4)_3$ nanofiber network for a full charge/discharge cycle in a voltage range of 2.3V–3.9V. (e) The corresponding time–potential curve. ((a–c) Reproduced with permission.[10] Copyright 2016, Elsevier. (d, e) Reproduced with permission.[12] Copyright 2016, Elsevier.)

and second cycles are shown in Figure 5.3a, b (charge/discharge at 0.05 C), and c, d (charge/discharge at 0.5 C), respectively. Apparently, a two-phase transformation occurs during delithiation caused by the Fe^{2+}/Fe^{3+} redox reaction. These phase transformations were indicated by operando XRD. It is evident that a peak starts to appear at $x=0.62$ corresponding to $Li_xMn_{0.25}{}^{2+}Fe_{0.75}{}^{3+}PO_4$ (020), representing the co-occurrence of $Li_xMn_{0.25}{}^{2+}Fe_{0.75}{}^{2+}PO_4$ and $Li_xMn_{0.25}{}^{2+}Fe_{0.75}{}^{3+}PO_4$ phases, whereas when $x=0.30$, the phase exclusively contains $Li_xMn_{0.25}{}^{2+}Fe_{0.75}{}^{3+}PO_4$. Similarly, another two-phase transformation occurs during lithiation caused by Mn^{2+}/Mn^{3+} redox reactions. The $Li_xMn_{0.25}{}^{3+}Fe_{0.75}{}^{3+}PO_4$ (020) peak starts to appear when $x=0.14$, whereas at $x=0.02$ the phase exclusively contains $Li_xMn_{0.25}{}^{3+}Fe_{0.75}{}^{3+}PO_4$.

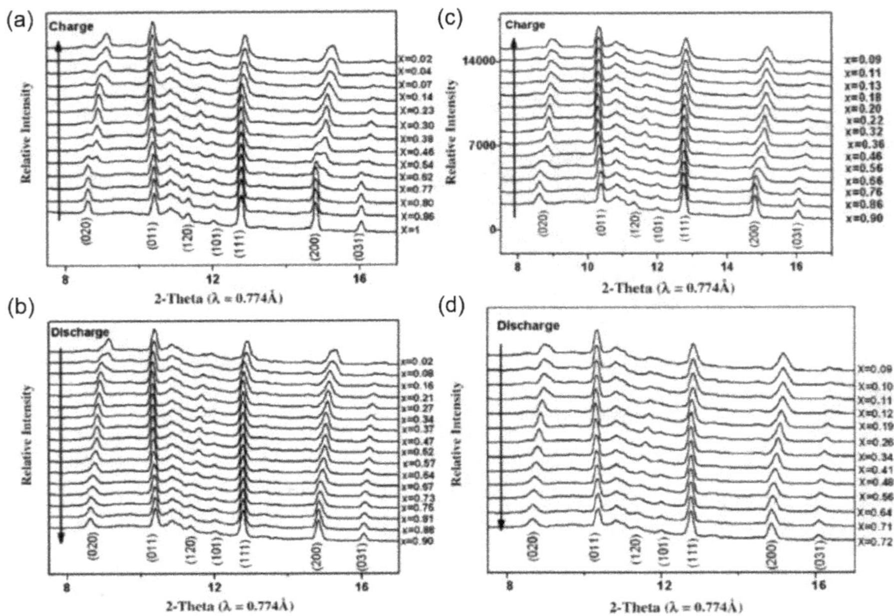

FIGURE 5.3 The operando XRD patterns of $Li_xMn_{0.25}Fe_{0.75}PO_4$ as a function of the lithium content x during the first cycle (charge/discharge at 0.05 C (a and b)) and second cycle (charge/discharge at 0.5 C (c and d)). ((a–d) Reproduced with permission.[11] Copyright 2009, Elsevier.)

Layered oxide compounds are the most widely used cathodes for LIBs. In order to achieve high capacity and long life, numerous efforts have been made to understand the crystal structure change at different stages of charge-discharge process, with operando XRD studies making significant contributions in revealing various phase transformations in different cathode materials, especially in the $LiCoO_2$ family. Generally, operando XRD investigations revealed a complicated phase transitions involved in the delithiation of $LixCoO_2$ ($0 \leq x \leq 1$). Figure 5.4 plots the XRD spectra collected during the first charge at rate of C/4.5 in which phases H1, H2, O1a, and O1 are observed as a function of cell voltage, where H1 is the fully lithiated phase at $x = 1$ and O1 is the completely delithiated phase at $x = 0$, and the others are partially delithiated with $0 < x < 1$. The phase-transformation steps are: H1→H2→O1a→O1.[32,33] The H1 and H2 have hexagonal lattice in space group $R\bar{3}m$, with similar lattice parameter a, but different in c, $c_{H1} = 1.089$A and $c_{H2} = 14.370$A, respectively.[34,35] The peak (003) has been shifted to the lower 2θ direction as lattice parameter c is increased during the H1 to H2 phase transition. The positions of (101) and (102) are almost unchanged. As the charge voltage reached ~4.6V. (100) and (101) peaks of O1a phase (hexagonal lattice, $R\bar{3}m$, $a = 2.823$A. $c = 27.07$A) appeared and their intensities increased thereafter at the expense of H2 phase. H2 phase completely disappeared to O1 phase started to form, finally, all the O1a phase-transformed to O1 phase (hexagonal lattice, $a = 2.83$A, $c = 4.24$A, $P\bar{3}m1$).[36] A monoclinic phase formed near $x = 0.5$ and other

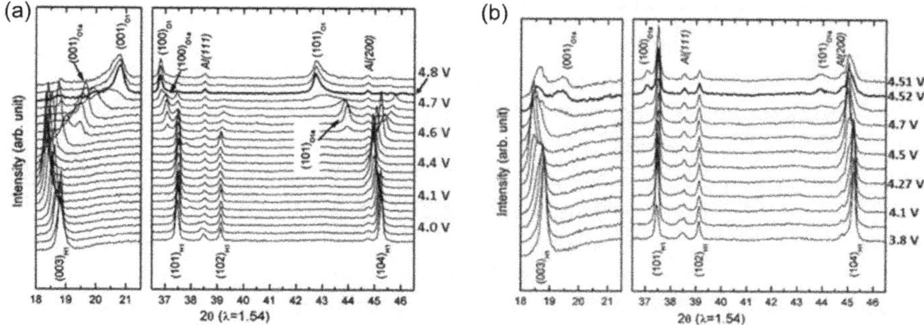

FIGURE 5.4 (a) XRD spectra during the first charge to 4.8V at C/4.5, demonstrating a complete phase transformation from H1 to O1 and (b) XRD spectra from 5th charge to 4.8V at C/4.5, showing an incomplete phase transition between H1 and O1. (Adapted from ref. [33])

transitional phase, such as H2a, were also detected via operando XRD in different studies under various experimental conditions.[37]

Figure 5.4b is the spectra recorded during the 5th charge to 4.8V at C/4.5.[38] The spectra show no O1 phase formation at the end of charge; instead, the existing phases are Ola and H2 suggesting a delayed phase transformation, which indicates the loss of capacity in the previous cycles. The electrode was further cycled to 8.35V at C/4.5 during the 12th cycle (not shown here).[36] At 4.8V, the major phases are H1 and H2, Ola was first observed at 5.04V, and Ola to O1 phase transition completed ~6.05V, which demonstrates the further delayed phase transformation compared to the 5th cycle. After, the transformation to O1 is completed at 6.05V, no new phase was found even up to 8.35V, indicating all Li ions or almost all Li ions are removed from the structure. The charging curves show that the first voltage plateaus are ~4.0, 4.1, and >4.7V for the 1st, 5th, and 12th cycles, respectively, indicating an increase in the polarization of cell caused by the impedance augmentation due to the electrolyte decomposition at high voltage.

5.3 OPERANDO STUDIES ON ANODE MATERIALS

Transition metal oxides were extensively studied as conversion-type materials in LIBs owing to their high theoretical capacity, low cost, and eco-friendliness. Nevertheless, these transition metal oxides undergo large volume change, and as a result, low conductivity during insertion/extraction of ions hampers their practical applications. To understand the electrochemical reaction mechanism so as to improve and evade these issues, operando analysis techniques were utilized. Lowe et al. used the operando synchrotron-based XRD and X-ray absorption spectra (XAS) to study the electrochemical reaction mechanism of the Mn_3O_4 anode.[22] It was observed that several intermediate phases were formed during lithiation. Mn_3O_4 surface reacts first, with simultaneous reduction of both Mn_3O_4 and $LiMn_3O_4$ all through the high-voltage regime.

An et al. prepared amorphous vanadium oxides supported hierarchical porous Fe_3O_4/graphene nanowires (Fe_3O_4/VO_x/G-P NWs).[23] This new system provided fast effective electron transport, lithium-ion diffusion, and exceptional stress relaxation, displaying larger lithium-storage performance. To probe the reaction mechanisms, in situ XRD technique was utilized (Figure 5.5). The in situ cell was first discharged to 0V and then charged to 3.0V at a constant current of 100 mA/g at room temperature. The as-prepared composites in situ cell was discharged to 0V and then charged to 3.0V using a constant current of 100 mA/g at 25°C and was never removed from the diffractometer. During the first discharge the diffraction peaks 30.1° and 35.5° from the Fe_3O_4/VO_x/G-P NWs were shifted to a lower angle owing to the expansion of the electrode host matrix arises due to the insertion of Li atoms. Additionally, two diffraction peaks become more weaker and finally disappeared, and no further diffraction peak could be seen during the following charge and discharge process. This is attributed to the amorphous nature of the electrode formed upon Li+insertion. Further, the phase transformations were studied by TEM analysis.

Luo et al. reported the preparation of binary metal oxide ($Co_2V_2O_7$) nanosheets.[24] Operando XRD was studied to understand the reaction mechanism during cycling (Figure 5.5). It is apparent that the discharge curve shows three plateaus. The peak

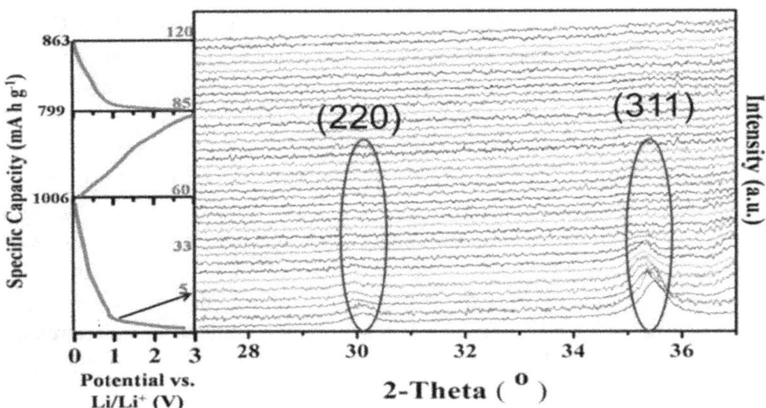

FIGURE 5.5 In situ XRD patterns of Fe_3O_4/VO_x/G-P NWs. (Reproduced with permission.[23] Copyright 2014, American Chemical Society.)

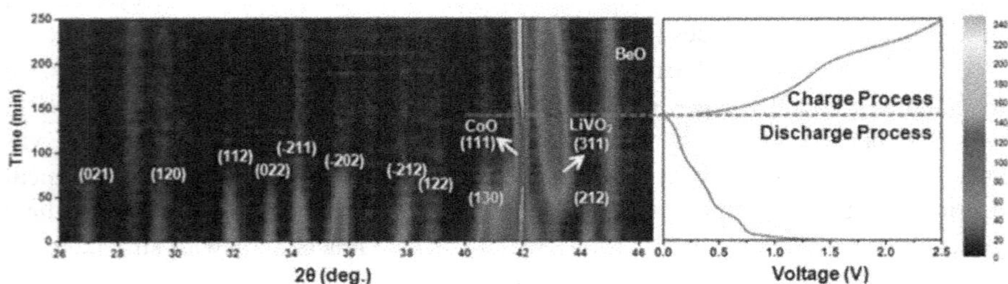

FIGURE 5.6 In situ X-ray diffraction patterns of CoVO-1 and (right) corresponding discharge/charge curve. (Reproduced with permission.[24] Copyright 2016, American Chemical Society.)

intensity of $Co_2V_2O_7$ diminished and no further changes were observed in the first plateau, related to the formation of a solid electrolyte interphase (SEI). In the second plateau, the diffraction peaks of $Co_2V_2O_7$ disappeared and a new peak formation was observed. The new peak at 43° was indexed to $LiVO_2$ (311). Based on the operando XRD, the reaction mechanism was proposed. In the initial discharge process, metallic Co and $Li_{1+x}VO_2$ were formed. In the subsequent charging process, CoO from metallic Co and delithiation of vanadium oxides were observed.

Silicon (Si) is a promising anode material for LIB owing to its abundance and high theoretical capacity; however, it undergoes large volume changes during the lithiation/delithiation process which hampers its practical applications. Such large volume expansion leads to the cracking of Si surface, pulverization altogether causing damage to the electrode leading to poor performance. Misra et al. studied the lithium-ion insertion/extraction mechanism of Si nanowire anode using operando synchrotron-based XRD.[25] At low potentials, a metastable $Li_{15}Si_4$ phase was detected and it was confirmed that the presence of the crystalline phase reduces the cycling stability of Si. The operando XRD results of the first and second lithiation/delithiation cycle for Si on a stainless-steel mesh cell is shown in Figure 5.7a–g. Upon lithiation of Si nanowires, the diffraction peak intensity of Si(111) decreases signifying the formation of amorphous Li_xSi and the disappearance of crystalline Si. The diffraction peaks of $Li_{15}Si_4$ were observed at low potentials (<20 mV). The formation of $Li_{15}Si_4$ and LiAu phases are due to the reactions of Li, Au, and Si particles.

In general, it is very difficult to get a fundamental understanding of battery materials without systematic and sustainable research efforts. Particularly for the electrodes, the charge and discharge processes involve both structural and electron state evolutions, which often interact with each other. In addition to the most extensively studied electrochemical properties, more attention is being paid to study the structural evolution rather than analyzing their electronic states. For example, the occupied and unoccupied electron states of electrolyte determine the upper limit of the open-circuit voltage of a thermodynamically stable battery cell; the electron state configuration of lithium transition-metal oxide-based cathodes fundamentally determines the safety and intrinsic voltage limit and the phase stability and transformation. Yang and co-workers[38] have reviewed the recent studies on the LIB cathodes, anodes, and solid electrolyte interfaces by soft X-ray absorption and emission spectroscopy.

5.4 BEYOND LITHIUM-ION BATTERIES

5.4.1 Lithium-Sulfur Batteries

Major advances have been achieved in lithium-battery technologies over the past three decades by the invention of new materials and designs through novel approaches, experimental and predictive reasoning, and structural and chemical reactions. Nevertheless, even those being proposed for the next generation of battery products are unable to meet the long-term performance for hybrid electric vehicles. The materials, their limitations, and the uncertainties in the insufficiently validated electrochemical couples and materials may take up the LIB technology to an incremental rather than exponential. Therefore, beyond lithium-ion, systems such as lithium-air/O_2, lithium-sulfur, and sodium-ion batteries (SIB) are considered as potential

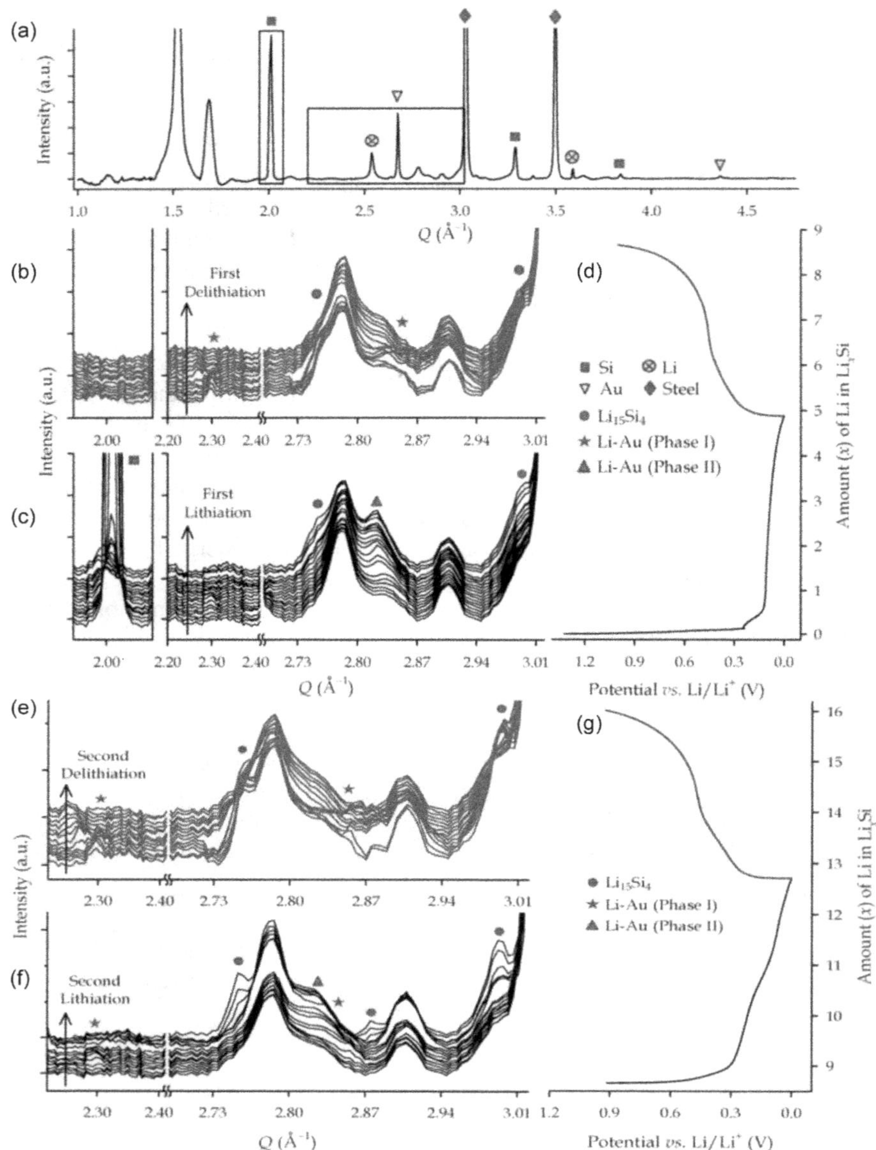

FIGURE 5.7 (a–g) Operando XRD pattern for a stainless-steel (SS) mesh cell: (a) at the start of the lithiation cycle; (b–f) zoomed-in sections of Q-space regions shown in (a) for the first and second delithiation/lithiation cycles, respectively. (d and g) Voltage profile showing the first and second cycles, respectively. (Reproduced with permission.[25])

alternatives to the state-of-the-lithium-ion batteries. The working principles of lithium-sulfur,[39–41] lithium-air[42] and SIBs[42] their types, advantages, and limitations are described in numerous review articles which is beyond the scope of this chapter.

Unlike lithium-ion systems which functions by lithium intercalation/deintercalation mechanisms, lithium-sulfur reactions occur by dissimilation process at the electrode surface to form lithium polysulfide and lithium oxide products, respectively.

During the discharge process of lithium-sulfur batteries, sulfur is reduced to lower order polysulfide, Li_2S, by accepting the Li ions through the electrolyte and electrons via external circuit at the cathode by undergoing dissimilation process, and while charging, a reverse reaction takes place. The overall reaction is shown in Equation 5.1

$$S8 + 16Li + 16e \Leftrightarrow 8Li_2S \tag{5.1}$$

The working principles, advantages, and limitations of lithium-sulfur batteries can be understood from review articles published elsewhere.[40,41]

In the case of lithium-sulfur batteries, the redox reaction is accompanied by phase transformation of active material (solid–liquid phases), structural and morphological changes that occur at the positive electrode. As a consequence, researchers have investigated the Li-S system by means of XRD, via an ex situ methodology.[43] Although several reports appear on the XRD studies of ex situ studies on Li-S batteries, only a couple of research groups have reported the complete studies of Li-S batteries with in situ XRD analysis.[44,45]

Apparently both studies recorded formation of crystalline S_8 at fully charged electrodes. Canas et al.[45] observed the appearance of crystalline Li_2S at 60% depth of discharge; on the contrary, no Li_2S was identified by Nelson and co-workers[46] upon complete discharge. In order to get a clearer insight into structural changes occurring inside the Li-S battery while cycling, Walus and et al.[47] have successfully analyzed the structural changes inside Li-S batteries using the XRD technique. The in situ synchrotron-based results clearly indicated the formation of crystalline Li_2S on the positive electrode at lower discharge plateau and its complete consumption during successive cycle. The authors also confirmed that soluble polysulfides are oxidized into solid S_8 at the end of each charge and recrystallized appeared in the form of another allotrope namely monoclinic β-sulfur.

5.4.2 Sodium-Ion Batteries

Contrary to lithium, sodium is one of the most abundant elements on the Earth's crust and the resources are unlimited. Additionally, sodium is the second-lightest and -smallest alkali metal next to lithium. On the basis of material abundance and standard electrode potential, rechargeable sodium batteries (NIB) are identified as a potential alternative to LIBs. NIBs are operable at ambient temperature, and metallic sodium is not used as the negative electrode, which is different from commercialized high-temperature sodium-based technology. Most commonly, NIBs are composed of sodium insertion materials as cathode with aprotic solvent as electrolyte and, therefore, are free from metallic sodium. The basic architecture, components, and charge storage mechanisms of NIBs are basically the same except that lithium-ions are replaced with sodium ions.

By employing in situ XRD, the structural evolution of $Na_{1.2}Ni_{0.2}Mn_{0.2}Ru_{0.4}O_2$ during sodium deintercalation/intercalation was investigated by Su and co-workers.[48] The selected portions of the resulting XRD patterns along with the first charge/discharge curve are illustrated in Figure 5.3a. The appearance of invariant peaks at 38.4° and 44.7° represent aluminum foil, which was used as current collector and X-ray window. The peaks at 30° and 38° are attributed to the impurities present in the sample which are electrochemically inactive, and did not influence the electrochemical properties of $Na_{1.2}Ni_{0.2}Mn_{0.2}Ru_{0.4}O_2$. It is apparent from Figure 5.8a when sodium ions are deintercalated from the lattice, a new hexagonal O_3 phase, labeled as O3′, appears, associated with new (00l) (l = 3, 6, 9) peaks at lower angles, corresponding to an increase of the interplanar distance. The intensity of the diffraction peaks of O_3 phase are significantly reduced and disappear until 0.14 mol sodium ions deintercalated. Upon deintercalation the (00l) (l = 3, 6, 9) peaks of the O3′ phase shift to lower angles and the (104) peak shifts to a higher angle, corresponding to the typical enlargement of the interplanar distance and decrease of the intraplanar distance in O3 type structure. The evolution of lattice parameters is illustrated in Figure 5.8b. While charging to 0.26 mol, another new phase of hexagonal P3 appears. Almost all of the O3′ phase transforms to P3 phase with the gliding mechanism of MO_2 slabs at the plateau close to 2.6V. After that, the sample undergoes a solid solution reaction of P3 phase. The phase transition of O3–O3′–P3 during charge and P3–O3′ during discharge has been observed in the first cycle, which is in accordance with the voltage plateaus and slops in the galvanostatic charge/discharge curve of Figure 5.9a.

Shen et al.[49] designed a novel electrochemical cell for powder XRD studies of LIBs and sodium-ion batteries in operando with high time resolution using a conventional powder X-ray diffractometer. The studies can be made for both anode and cathode electrode materials in reflection mode. The cell design closely mimics that of standard battery testing coin cells and allows obtaining powder XRD patterns under representative electrochemical conditions. Interestingly, the cell uses graphite as the X-ray window instead of beryllium, and can be easily operated and maintained. The lithium insertion/extraction in two spinel-type LIB electrode materials ($Li_4Ti_5O_{12}$ anode and $LiMn_2O_4$ cathode) and sodium extraction from a layered NIB cathode material, $Na_{0.84}Fe_{0.56}Mn_{0.44}O_2$ have been extensively studied.

Reduced graphene oxide (rGO) homogenously wrapped nickel diselenide ($NiSe_2$/rGO) hybrid was prepared by a facile one-spot hydrothermal method as potential anode material for SIBs.[50] The $NiSe_2$/rGO hybrid delivered a high reversible capacity (433 mAh/g at 100 mA/g). Further, in situ XRD analysis and ex situ SEM/TEM measurement revealed that the high capacity of $NiSe_2$/rGO was originated from the combined Na intercalation and conversion reactions indicating that $NiSe_2$/rGO hybrid is a promising anode candidate for NIBs.

In situ XRD techniques, such as real-time X-ray analytical micro-furnace (RT-XAMF) and time-resolved X-ray diffraction (TRXRD), are widely employed to identify the synthesis process and thermal stability of cathode materials.[51] Using the RT-XAMF technique, Kim and co-workers reported the synthesis process of $NaNiO_2$ cathode material for Na-ion batteries. The authors observed a phase transition from rhombohedral $NaNiO_2$ at temperatures above 243°C to monoclinic $NaNiO_2$ at temperatures below 243°C during heat treatment. In addition, the structural instability

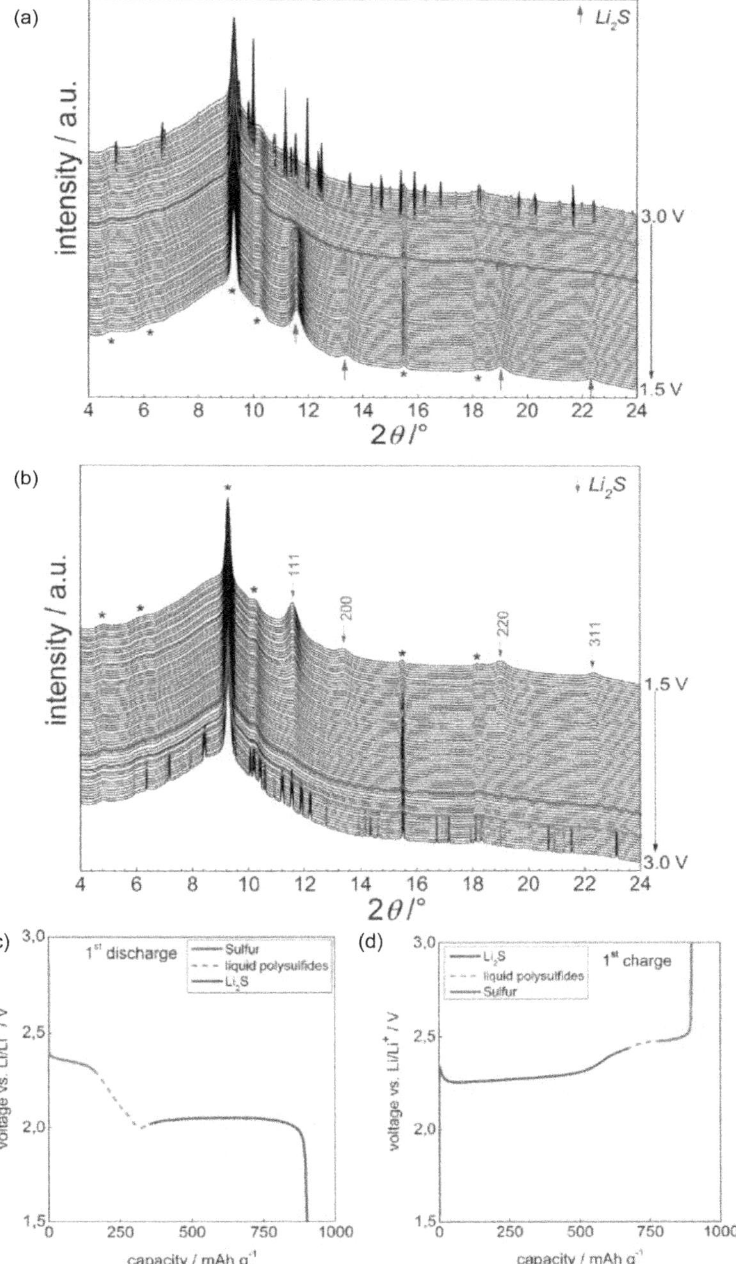

FIGURE 5.8 In situ XRD patterns of the complete cell during 1st discharge (a) and charge (b); corresponding electrochemical plots (c and d). Peaks associated with packaging are marked by *. Bold lines indicate moments of solid-phase appearance–disappearance. (Adapted from ref. [47].)

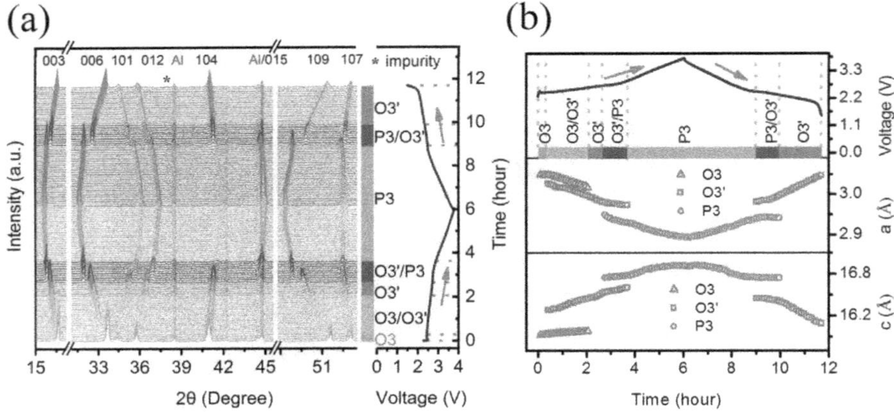

FIGURE 5.9 (a) In situ XRD patterns of the $Na_{1.2}Ni_{0.2}Mn_{0.2}Ru_{0.4}O_2$/Na cell under a current rate of 1/12 C, corresponding to the voltage–time curve given on the right. For comparison, the intensities from 15° to 34° are weakened to its fifth; (b) the corresponding lattice parameters evolutions during the in situ experiments. (Adapted from ref.[48].)

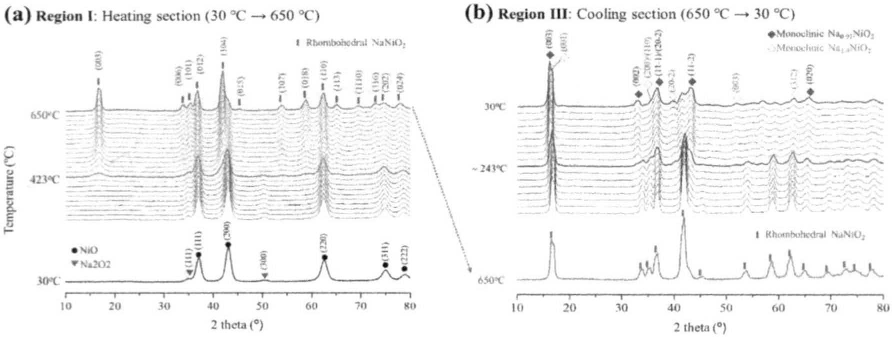

FIGURE 5.10 X-ray diffraction patterns of Na2O2 and NiO mixture during (a) heating from 30°C to 650°C and (b) cooling from 650°C to 30°C at a rate of 1°C/minute by real-time X-ray analytical micro-furnace technique. (Adapted from ref. [50].)

of the monoclinic $NaNiO_2$ phase changes to the Na-deficient $Na_{0.91}NiO_2$ phase. The TRXRD technique was used to investigate the thermal stability of the desodiated $Na_{1\_}xNiO_2$ (x=0.09, 0.5) cathodes in the presence of an electrolyte. It was confirmed that the structural changes of desodiated $Na_{1\_}xNiO_2$ were relatively simple compared to those of the Ni-based cathode material in Li-ion batteries. First, the layered structure of $Na_{1\_}xNiO_2$ at room temperature turns into an MO-type rock salt phase (NiO) and subsequently into a metallic phase (Ni) without the appearance of spinel-type ($Li_{1\_}xM_2O_4$ and M_3O_4) intermediates, which are typically observed in lithium-nickel-based oxides.

Shadike et al.[52] reviewed both static (ex situ) and real-time (in situ or in operando) techniques for analyzing the NIBs. The authors also focused on characterization

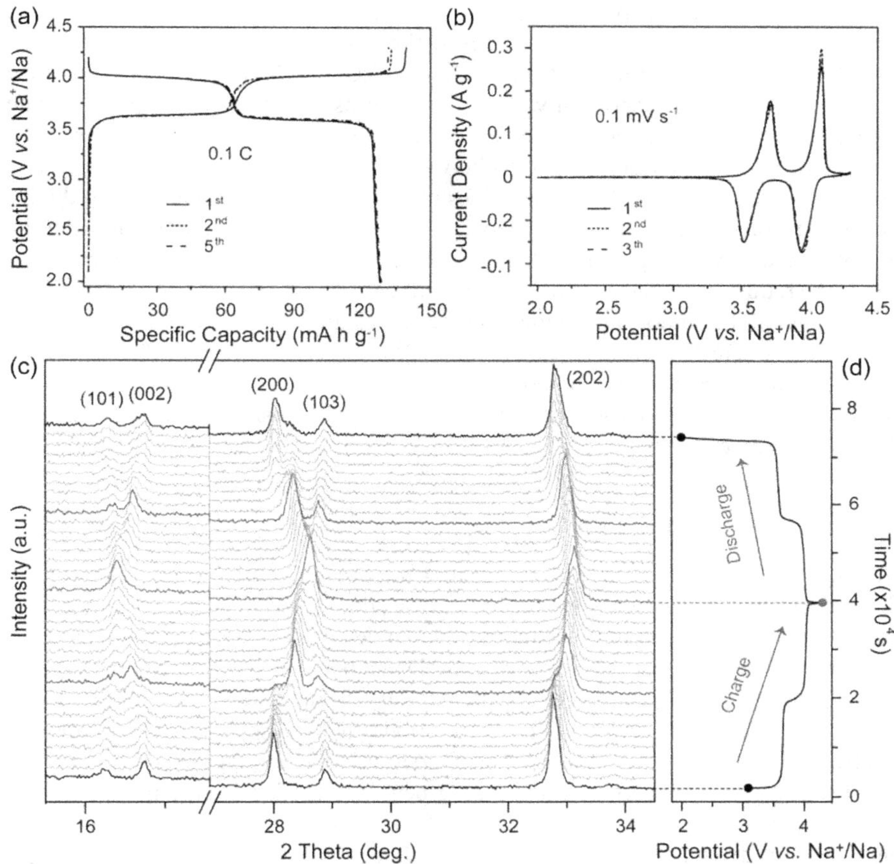

FIGURE 5.11 Structural evolutions during Na intercalation/extraction of NVPF-NTP. (a and b) Representative galvanostatic charge/discharge (a) and CV curves of initial cycles (b) for the prepared NVPF-NTP. The rate for galvanostatic test is 0.1 C, CV scan rate is 0.1 mV/s, and both potential windows are 2.0V–4.3V versus Na^+/Na. (c) The in situ XRD patterns collected during the 1st cycle of NVPF-NTP electrode under a 0.1 C rate with sampling interval of 30 minutes. (d) Galvanostatic charge/discharge curve of the 1st cycle. (Adapted from ref.[53].)

techniques applied to the fundamental investigation of NIB systems with an emphasis on the results so far obtained.

To elucidate the evolutions of a high-voltage cathode composed of $Na_3V_2(PO_4)_2O_2F$ nano-tetraprisms (NVPF-NTP) crystal structure during Na intercalation/extraction, in situ XRD test with a sampling interval of 30 minutes was carried out by Guo et al.[53] on the 1st cycle at 0.1 C. The structural evolutions while cycling was monitored via the shifts/changes on the selected peaks of (101), (002), (200), (103), and (202). For the diffraction of (200) planes, the peak at 27.997° weakened gradually and then split with the formation of a new peak at the higher 2θ angle of ≈28.219° in the first charge plateau. When fully charged to 4.3V, the XRD pattern could be well indexed to pure-phase $NaV_2(PO_4)_2O_2F$, a two-Na-extracted product of NVPF-NTP. As a result, the charging process achieved the two-Na (two-electron) full extraction through the transformation from $Na_3V_2(PO_4)_2O_2F$ to $Na_2V_2(PO_4)_2O_2F$ and then $NaV_2(PO_4)_2O_2F$. More importantly, the XRD patterns can return to the pristine state before charging during the subsequent discharging (Na re-intercalation) process, demonstrating that the transformations between the three compounds are completely reversible. Results revealed that the values of a decreased continuously with the Na extraction from crystal lattice, originating from the decreasing cationic radius of the transition metal V owing to its increased valence; whereas the values of c increased due to the reduction of Na ions, and thereby increased the repulsive force between the pseudolayers. The authors also found that the change in the NVPF-NTP unit-cell volume was less 2.56%, indicating the extremely low lattice strain of NVPF-NTP while Na intercalation/extraction. The ultrastable cycling performance of NVPF-NTP as cathode material for NIBs was attributed to the high reversibility and low strain of the crystal lattice.

5.5 SUMMARY AND OUTLOOK

Neutron scattering is another powerful tool to investigate the structural information of electrodes. It is worthwhile to point out that neutron scattering techniques are more sensitive to the light elements and can detect the distribution of Na in the electrode materials effectively. However, XRD and neutron diffractions are not good in studying amorphous electrode. On the contrary, X-ray and neutron pair distribution function analyses are potential tools in providing local structural information of electrode materials, even for amorphous or aperiodic disordered systems.

Although the in situ XRD studies bring new information for the more complex understanding of the working mechanism of LIB, Li-S and SIBs, more information on the electrode changes upon cycling such as morphology modification and presence of amorphous Li_2S could not be detected by XRD. Therefore, other complementary techniques such as in situ SEM, transmission microscopy, and electrochemical impedance spectroscopy can be used.

REFERENCES

[1] Cheetham, A.K.; Wilkinson, A.P. 1992. Synchrotron X-ray and neutron diffraction studies in solid-state chemistry. *Angew. Chem. Int. Ed.*, 31(12), 1557–1570.

[2] Kunz, M.; Tamura, N.; Chen, K.; MacDowell, A.A.; Celestre, R.S.; Church, M.M.; Fakra, S.; Domning, E.E.; Glossinger, J.M.; Kirschman, J.L. 2009. A dedicated superbend x-ray microdiffraction beamline for materials, geo-, and environmental sciences at the advanced light source. *Rev. Sci. Instrum.*, 80(3), 035108 (1–10).

[3] Robert, R.; Zeng, D.; Lanzirotti, A.; Adamson, P.; Clarke, S.J.; Grey, C.P. 2012. Scanning X-ray fluorescence imaging study of lithium insertion into copper based oxysulfides for Li-ion batteries. *Chem. Mater.*, 24(14), 2684–2691.

[4] Krishenbaum, K.C.; Bock, D.C.; Zhong, Z.; Marchilok, A.C.; Takeuchi, K.J.; Takeuchi, E.S. 2015. Electrochemical reduction of $Ag_2VP_2O_8$ composite electrodes visualized via in situ energy dispersive X-ray diffraction (EDXRD): unexpected conductive additive effects. *J. Mater. Chem. A*, 3, 18027–18035.

[5] Arachi, Y.; Kobayashi, H.; Emura, S.; Nakata, Y.; Tanaka, M.; Asai, T.; Sakaebe, H.; Tatsumi, K.; Kageyama, H. 2005. Li de-intercalation mechanism in $LiNi_{0.5}Mn_{0.5}O_2$ cathode material for Li-ion batteries. *Solid State Ionics*, 176(9–10), 895–903.

[6] Cui, Y.; Abouimrane, A.; Sun, C.-J.; Ren, Y.; Amine, K. 2014. Li-Se battery: absence of lithium polyselenides in carbonate-based electrolyte. *Chem. Commun.*, 50, 5576–5579.

[7] Ruan, K.R.; Trahey, L.; Okasinski, J.S.; Burrell, A.K.; Ingram, B.J. 2013. In situ synchrotron X-ray diffraction studies of lithium oxygen batteries. *J. Mater. Chem. A*, 1, 6915–6919.

[8] Wei, X.; Wang, X.; An, Q.; Han, C.; Mai, L. 2017. Operando X-ray diffraction characterization for understanding the intrinsic electrochemical mechanism in rechargeable battery materials. *Small Methods*, 1(5), 1700083.

[9] Liu, H.; Strobridge, F.C.; Borkiewicz, O.J.; Wiaderek, K.M.; Chapman, K.W.; Chupas, P.J.; Grey, C.P. 2014. Capturing metastable structures during high rate cycling of $LiFePO_4$ nanoparticle electrodes. *Science*, 344(6191), 1252817.

[10] Yan, M.; Zhang, G.; Wei, Q.; Tian, X.; Zhao, K.; An, Q.; Zhou, L.; Zhao, Y.; Niu, C.; Ren, W.; He, L.; Mai, L. 2016. *In operando* observation of temperature-dependent phase evolution in lithium-incorporation olivine cathode. *Nano Energy*, 22, 406–413.

[11] Chen, Y.-C.; Chen, J.-M.; Hsu, C.-H.; Yeh, J.-W.; Shih, H.C.; Chang, Y.-S.; Sheu, H.-S. 2009. Structure studies on $LiMn_{0.25}Fe_{0.75}PO_4$ by in-situ synchrotron X-ray diffraction analysis. *J. Power Sources*, 189(1), 790–793.

[12] Ren, W.; Zheng, Z.; Xu, C.; Niu, C.; Wei, Q.; An, Q.; Zhao, K.; Yan, M.; Qin, M.; Mai, L. 2016. Self-sacrificed synthesis of three-dimensional $Na_3V_2(PO_4)_3$ nanofiber network for high-rate sodium-ion full batteries. *Nano Energy*, 25, 145–153.

[13] Wang, X.; Niu, C.; Meng, J.; Hu, P.; Xu, X.; Wei, X.; Zhou, L.; Zhao, K.; Luo, W.; Yan, M.; Mai, L. 2015. Novel $K_3V_2(PO_4)_3/C$ bundled nanowires as superior sodium-ion battery electrode with ultrahigh cycling stability. *Adv. Energy Mater.*, 5(17), 1500716.

[14] Wang, L.; Wang, J.; Zhang, X.; Ren, Y.; Zuo, R.; Yin, G.; Wang, J. 2017. Unravelling the origin of irreversible capacity loss in $NaNiO_2$ for high voltage sodium ion batteries. *Nano Energy*, 34, 215–223.

[15] Niu, C.; Liu, X.; Meng, J.; Xu, L.; Yan, M.; Wang, X.; Zhang, G.; Liu, Z.; Xu, X.; Mai, L. 2016. Three dimensional V_2O_5/NaV_6O_{15} hierarchical heterostructures: controlled synthesis and synergistic effect investigated by *in situ* X-ray diffraction. *Nano Energy*, 27, 147–156.

[16] Berthelot, R.; Carlier, D.; Delmas, C. 2011. Electrochemical investigation of the P_2–Na_xCoO_2 phase diagram. *Nat. Mater.*, 10(1), 74–80.

[17] Yabuuchi, N.; Kajiyama, M.; Iwatate, J.; Nishikawa, H.; Hitomi, S.; Okuyama, R.; Usui, R.; Yamada, Y.; Komaba, S. 2012. P_2-type $Na_x[Fe_{1/2}Mn_{1/2}]O_2$ made from earth-abundant elements for rechargeable Na batteries. *Nat. Mater.*, 11(6), 512–517.

[18] Wang, X.; Xu, X.; Niu, C.; Meng, J.; Huang, M.; Liu, X.; Liu, Z.; Mai, L. 2017. Earth abundant Fe/Mn-based layered oxide interconnected nanowires for advanced K-ion full batteries. *Nano Lett.*, 17(1), 544–550.

[19] Wu, D.; Li, X.; Xu, B.; Twu, N.; Liu, L.; Ceder, G. 2015. $NaTiO_2$: a layered anode material for sodium-ion batteries. *Energy Environ. Sci.*, 8, 195–202.

[20] Sun, Y.; Zhao, L.; Pan, H.; Liu, X.; Gu, L.; Hu, Y-S.; Li, H.; Armand, M.; Ikuhara, Y.; Chen, L.; Huang, X. 2013. Direct atomic-scale confirmation of three-phase storage mechanism in $Li_4Ti_5O_{12}$ anodes for room-temperature sodium-ion batteries. *Nat. Commun.*, 4, 1870.

[21] Xu, C.; Xu, Y.; Tang, C.; Wei, Q.; Meng, J.; Huang, L.; Zhou, L.; Zhang, G.; He, L.; Mai, L. 2016. Carbon-coated hierarchical $NaTi_2(PO_4)_3$ mesoporous microflowers with superior sodium storage performance. *Nano Energy*, 28, 224–231.

[22] Lowe, M.A.; Gao, J.; Abruna, H.D. 2013. *In operando* X-ray studies of the conversion reaction in Mn_3O_4 lithium battery anodes. *J. Mater. Chem. A*, 1, 2094–2103.

[23] An, Q.; Lv, F.; Liu, Q.; Han, C.; Zhao, K.; Sheng, J.; Wei, Q.; Yan, M.; Mai, L. 2014. Amorphous vanadium oxide matrixes supporting hierarchical porous Fe_3O_4/graphene nanowires as a high-rate lithium storage anode. *Nano Lett.*, 4(11), 6250–6256.

[24] Luo, Y.; Xu, X.; Zhang, Y.; Chen, C.-Y.; Zhou, L.; Yan, M.; Wei, Q.; Tian, X.; Mai, L. 2016. Graphene oxide templated growth and superior lithium storage performance of novel hierarchical $Co_2V_2O_7$ nanosheets. *ACS Appl. Mater. Interfaces*, 8(4), 2812–2818.

[25] Misra, S.; Liu, N.; Nelson, J.; Hong, S.S.; Cui, Y.; Toney, M.F. 2012. *In situ* X-ray diffraction studies of (De)lithiation mechanism in silicon nanowire anodes. *ACS Nano*, 6(6), 5465–5473.

[26] Darwiche, A.; Marino, C.; Sougrati, M.T.; Fraisse, B.; Stievano, L.; Monconduit, L. 2012. Better cycling performances of bulk Sb in Na-ion batteries compared to Li-ion systems: an unexpected electrochemical mechanism. *J. Am. Chem. Soc.*, 134(51), 20805–20811.

[27] Kim, J.; Park, K.Y.; Park, I.; Yoo, J.K.; Hong, J.; Kang, K.K. 2012. Thermal stability of Fe–Mn binary olivine cathodes for Li rechargeable batteries. *J. Mater. Chem.*, 22, 11964–11970.

[28] Lee, J.; Seo, D.H.; Balasubramanian, M.; Twu, N.; Li, X.; Ceder, G. 2015. A new class of high-capacity cation-disordered oxides for rechargeable lithium batteries: Li–Ni–Ti–Mo oxides. *Energy Environ. Sci.*, 8, 3255–3265.

[29] Wu, D.; Li, X.; Xu, B.; Twu, N.; Liu, L.; Ceder, G. 2015. $NaTiO_2$: a layered anode material for sodium-ion batteries. *Energy Environ. Sci.*, 8, 195–202.

[30] Wang, L.; Wang, J.; Zhang, X.; Ren, Y.; Zuo, R.; Yin, G.; Wang, J. 2017. Unravelling the origin of irreversible capacity loss in $NaNiO_2$ for high voltage sodium ion batteries. *Nano Energy*, 34, 215–223.

[31] Orikasa, Y.; Maeda, T.; Koyama, Y.; Murayama, H.; Fukuda, K.; Tanida, H.; Arai, H.; Matsubara, E.; Uchimoto, Y.; Ogumi, Z. 2013. Direct observation of a metastable crystal phase of Li_xFePO_4 under electrochemical phase transition. *J. Am. Chem. Soc.*, 135(15), 5497–5500.

[32] Sun, X.; Yang, X.Q.; Mcbreen, J.; Gao, Y.; Yakovleva, M.V.; Xing, X.K.; Daroux, M.L. 2001. New phases and phase transitions observed in over-charged states of $LiCoO_2$-based cathode materials. *J. Power Sources*, 97–98, 274–276.

[33] Chung, K.Y.; Yoon, W.-S.; Mcbreen, J.; Yang, X.Q.; Oh, S.Y.; Shin, H.C.; Cho, W.I.; Cho, B.W. 2007. In situ X-ray diffraction studies on the mechanism of capacity retention improvement by coating at the surface of $LiCoO_2$. *J. Power Sources*, 174(2), 619–623.

[34] Yang, X.Q.; Sun, X.; Mcbreen, J. 2000. New phases and phase transitions observed in $Li_{1-x}CoO_2$ during charge: in situ synchrotron X-ray diffraction studies. *Electrochem. Commun.*, 2(2), 100–103.

[35] Reimers, J.N.; Dahn, J.R. 1992. Electrochemical and in situ X-ray diffraction studies of lithium intercalation in $LixCoO_2$. *J. Electrochem. Soc.*, 139(8), 2091–2097.

[36] Ohzuku, T.; Ueda, A. 1994. Solid-state redox reactions of LiCoO (R3m) for 4-volt secondary lithium cells. *J. Electrochem. Soc.*, 141(11), 2972–2977.

[37] Chung, K.Y.; Yoon, W.-S.; Lee, H.S.; McBreen, J.; Yangm, X.-Q.; Oh, S.H.; Ryu, W.H.; Lee, J. L.; Cho, W.I.; Cho, B.W. 2006. In situ XRD studies of the structural changes of ZrO_2-coated $LiCoO_2$ during cycling and their effects on capacity retention in lithium batteries. *J. Power Source*, 163(1), 185–190.

[38] Yang, W.; Liu, X.; Qiao, R.; Velasco, P.O.; Spear, J.D.; Roseguo, L.; Pepper, J.X.; Chuang, Y.-D.; Jonathan, D.D.; Hussain, Z. 2013. Key electronic states in lithium battery materials probed by soft X-ray spectroscopy. *J. Electr. Spectros. Rel. Phen.*, 190, 64–74.

[39] Bruce, P.G.; Freunberger, S.A.; Hardwick, L.J.; Tarascon, J.-M. 2012. Li-O_2 and Li-S batteries with high energy storage. *Nat. Mater.*, 11, 19–29.

[40] Ji, X.; Nazar, L.F. 2010. Advances in Li–S batteries. *J. Mater. Chem.*, 20(44), 9821–9826.
[41] Christensen, J.; Albertus, P.; Sanchez-Carrera, R.S.; Lohmann, T.; Kozinsky, B.; Liedtke, R.; Ahmed, J.; Kojic, A. 2012. A critical review of Li/Air batteries. *J. Electrochem. Soc.*, 159(2), R1–R30.
[42] Girishkumar, G.; McCloskey, B.; Luntz, A.C.; Swanson, S.; Wilckel, W. 2010. Lithium–air battery: promise and challenges. *J. Phys. Chem. Lett.*, 1(14), 2193–2203.
[43] Yabuuchi, N.; Kubota, K.; Dahbi, N.; Komaba, S. 2014. Research development on sodium-ion batteries. *Chem. Rev.*, 114(23), 11636–11682.
[44] Choi, Y.-J.; Chung, Y.-D.; Baek, C.-Y.; Kim, K.-W.; Ahn, H.-J.; Ahn, J.-H. 2008. Effects of carbon coating on the electrochemical properties of sulfur cathode for lithium/sulfur cell. *J. Power Sources*, 184(2), 548–552.
[45] Canas, N.A.; Wolf, S.; Wagner, N.; Friedrich, K.A. 2013. In-situ X-ray diffraction studies of lithium-sulfur batteries. *J. Power Sources*, 226, 313–319.
[46] Nelson, J.; Misra, S.; Yang, Y.; Jackson, A.; Liu, Y.; Wang, H.; Dai, H.; Andrews, J.C.; Cui, Y.; Toney, M.F. 2012. In operando X-ray diffraction and transmission X-ray microscopy of lithium sulfur batteries. *J. Am. Chem. Soc.*, 134(14), 6337–6343.
[47] Walus, S.; Barchas, C.; Colin, J.-F.; Martin, J.-F.; Alloin, F. 2013. New insight into the working mechanism of lithium-sulfur batteries: in situ and operando X-ray diffraction characterization. *Chem. Commun.*, 49, 7899–7901.
[48] Su, N.; Lyu, Y.; Guo, B. 2018. Electrochemical and in-situ X-ray diffraction studies of $Na_{1.2}Ni_{0.2}Mn_{0.2}Ru_{0.4}O_2$ as a cathode material for sodium ion batteries. *Electrochem. Commun.*, 87, 71–75.
[49] Shen, Y.; Pedersen, E.E.; Christensen, M.; Iversena, B.B. 2014. An electrochemical cell for in operando studies of lithium/sodium batteries using a conventional x-ray powder diffractometer. *Rev. Sci. Instrum.*, 85, 104103.
[50] Ou, X.; Li, J.; Zheng, F.; Wu, P.; Pan, Q.; Xiong, X.; Yang, C.; Liu, M. 2017. In situ X-ray diffraction characterization of $NiSe_2$ as a promising anode material for sodium ion batteries. *J. Power Sources*, 343, 483–491.
[51] Kim, D.H.; Kim, J.Y.; Park, J.; Chung, K.Y. 2022. RT-XAMF and TR-XRD studies of solid-state synthesis and thermal stability of $NaNiO_2$ as cathode material for sodium-ion batteries. *Ceramics Intl.*, 48(14), 19675–19680.
[52] Shadike, Z.; Zhao, E.; Zhou, Y.N.; Yu, X.; Yang, Y.; Hu, E.; Bak, S.; Gu, L.; Yang, X.-Q. 2018. Advanced characterization techniques for sodium-ion battery studies advanced characterization techniques for sodium-ion battery studies. *Adv. Energy Mater.*, 8(17), 1702588.
[53] Guo, J.-Z.; Wang, P.-F.; Wu, X.-L.; Zhang, X.-H.; Yan, Q.; Chen, H.; Zhang, J.-P.; Guo, Y.-G. 2017. High-energy/power and low-temperature cathode for sodium-ion batteries: in situ XRD study and superior full-cell performance. *Adv. Mater.*, 29(33), 1701968.

6 ICP-Based Techniques for LIBs Characterization

Sascha Nowak
University of Münster

CONTENTS

6.1 BASIC PRINCIPLES OF ICP-BASED TECHNIQUES

A plasma is a gas whose atoms or molecules are dissociated to a certain degree into the corresponding cations and electrons. In addition to the classical aggregate states of solid, liquid and gas, it is referred to as the "fourth" aggregate state. One can imagine this state as further heating of solid matter to a liquid, to a gas and then beyond the gas state. It should be noted, however, that there are also so-called cold plasmas, with temperatures of only 40°C–50°C. Here, at low pressures, only the electrons are accelerated and excited by the high-frequency electromagnetic radiation. Ninety nine percent of the matter in the universe are in this aggregate state and thus the plasma state is the most frequent mass state of the matter. The kinetic gas temperature in an inductively coupled plasma (ICP) thereby assumes values up to 6,000 K. These temperature conditions cause a complete separation of electrons and atomic nuclei [1–3].

DOI: 10.1201/9781003299295-6

The ICP with argon operation is characterized by a number of advantages. Almost all elements of the periodic table can be excited and ionized, allowing fast, simultaneous multi-element determination. It is also independent of the sample type 6 and has low detection limits, wide dynamic linear working ranges and good precision of the measurement results (relative standard deviation of 0.2%–3%)10, as well as a robustness that makes the method independent of the sample matrix [1,2,4,5].

6.1.1 Sample Preparation and Introduction

Acid digestion is the standard approach for transference of solid samples into liquids. Acid digestion usually involves the dissolution of a sample in a hot acid, or a mixture of hot acids. Heating acids is performed on a hot plate, or by using a microwave digestion system that employs pressurized vessels to produce even higher reaction temperatures. Complete digestion and no loss of analytes during these steps are mandatory for later quantitative measurements. Afterward the sample introduction is the next and equally important step. The analytical performance depends sustainably on the sample introduction system. Many problems and inferences have their origin in the sample introduction step. Therefore, the sample introduction system is often called the "Achilles heel" of the analytical system, and it has to be reconsidered for every sample to be analyzed. The sample introduction system ensures that the sample is introduced into the plasma without affecting its stability. Liquid, solid and also gaseous substances can be introduced into the plasma via a suitable sample introduction system or preparation [6,7].

The introduction of liquid samples plays the central role. The analyte is present in dissolved form and is introduced into the plasma as an aerosol. Therefore, the respective solution must first be converted into small droplets and then transported further to the plasma by means of a gas stream. Ideally, this is done in the form of very small droplets, as this facilitates any drying and all further steps in the plasma.

A wide variety of atomizer types are available for this purpose. In addition to the classic nebulizers, the pneumatic nebulizers, there is ultrasonic nebulization, heated spray chambers with a desolvation system, electrothermal vaporization and the low sample throughput methods (direct injection nebulization, high efficiency nebulization, micro-concentric nebulization). As previously mentioned, the quality of the aerosol is a contributing factor to a stable plasma and thus a stable signal. Droplet size and droplet size distribution are essential in this respect. The intensity of the achievable signals is limited by the aerosol yield. Thus, if the nebulization efficiency is low, the intensity will not reach an optimal value either. The theoretical requirements, a homogeneous droplet distribution with high aerosol yield at the same time, can hardly be realized in practice. Therefore, the best possible compromise between the two factors is sought with the different nebulizer types [8–10]. Acid digestion is the most common step to dissolve the materials for analysis. This can be done either in vessels or with the help of a microwave system. The most common for battery materials are aqua regia or mixtures of HCl and HNO_3. These are applied for most of the cathode materials like NCA, NMC or LFP [11–14]. LTO on the other hand needs special treatment. HF can dissolve the material; however, microwave digestion with a mixture of H_2O_2 and H_2SO_4 is also possible. However, the exact conditions for cell opening, subsequent sample preparation, e.g. the electrode digestion step, and operating parameters for the used instruments are seldom stated thoroughly in literature [15].

6.1.2 Excitation and Ionization

First, most plasmas must be ignited by the supply of free charge carriers. In an ICP a Tesla coil is used. By coupling electromagnetic energy, the free charge carriers are accelerated and, through collision processes, contribute to the ionization of plasma particles, which in turn are accelerated in the electric field and lead to a cascade of collision processes. The continuous supply of energy thus contributes to the maintenance of the plasma. In this process, the number of recombinations of ions and electrons and the constant formation of new charged particles balance each other out, so that one can speak of a stable plasma discharge. However, there is no equilibrium, neither a chemical nor a thermodynamic equilibrium. The plasma should rather be understood as a complex dynamic system in which the distribution of species (atomic, molecular and ionic) on energetic states is very difficult to predict. The processes that are important for the maintenance of the plasma are presented in the following. This is done on the basis of a classical argon discharge that also occurs in an ICP [16,17].

Impact ionization (or three-state ionization) (Equation 9.1):

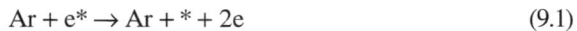

$$\mathrm{Ar} + \mathrm{e}^{*} \rightarrow \mathrm{Ar} + {}^{*} + 2\mathrm{e} \tag{9.1}$$

Radiative recombination (or absorption) (Equation 9.2):

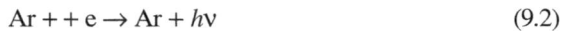

$$\mathrm{Ar} + + \mathrm{e} \rightarrow \mathrm{Ar} + h\nu \tag{9.2}$$

where e: electron

e*: Electron with high kinetic energy

Ar: argon atom

Ar*: Argon atom in excited state

Ar+: Argon cation, single positively charged

h: Planck's quantum of action

v: frequency of the emitted radiation.

In addition to the particles of argon, other species are present. In particular, due to the introduction of an analyte, which is atomized as a result of the drastic conditions. These particles are transformed into excited energetic states by a variety of collision processes.

Collisions between plasma gas (Ar) and analyte (A) (Equation 9.3):

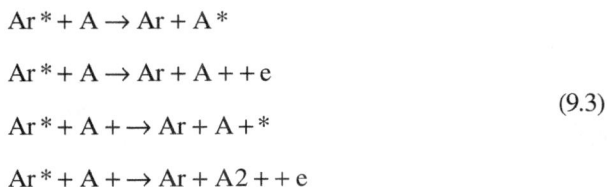

$$\begin{aligned}
&\mathrm{Ar}^{*} + \mathrm{A} \rightarrow \mathrm{Ar} + \mathrm{A}^{*} \\
&\mathrm{Ar}^{*} + \mathrm{A} \rightarrow \mathrm{Ar} + \mathrm{A} + + \mathrm{e} \\
&\mathrm{Ar}^{*} + \mathrm{A} + \rightarrow \mathrm{Ar} + \mathrm{A} + {}^{*} \\
&\mathrm{Ar}^{*} + \mathrm{A} + \rightarrow \mathrm{Ar} + \mathrm{A}2 + + \mathrm{e}
\end{aligned} \tag{9.3}$$

Collisions between electrons and analyte (Equation 9.4):

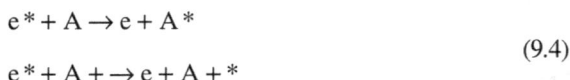

$$\begin{aligned} e^* + A &\rightarrow e + A^* \\ e^* + A+ &\rightarrow e + A+^* \end{aligned} \tag{9.4}$$

Release of excitation energy (Equation 9.5):

$$\begin{aligned} A^* &\rightarrow A + h\nu A(1) \\ A+^* &\rightarrow A+ + h\nu A(2) \end{aligned} \tag{9.5}$$

where A: analyte atom
A*: Analyte atom in excited state
A+: Analyte cation
A+*: excited-state analyte cation
Ar*: energetic, metastable argon atoms
$\nu A(1)$: frequency of the radiation emitted by the analyte atom
$\nu A(2)$: frequency of the radiation emitted by the singly charged analyte cation

6.1.3 Detection

6.1.3.1 ICP-OES

The processes in the ICP elevate atoms and ions to excited states. When the atoms return back to their ground state via relaxation, element selective emission is emitted because of discrete energetic transitions. Due to the multiple excited states that can be populated, a spectral line pattern is emitted. The states are depending on the electronic structure of the element. The emissions are transferred to the spectral analyzer by the optical interface. Two setups are available with different systems of plasma observing, the axial and radial viewing. Compared to the radial plasma viewing, the axial plasma viewing provides limits of detection (LODs) about one order of magnitude better. Additionally, it provides excellent sensitivities, but the plasma tail (or recombination zone) has to be cut off. This can be achieved by a shear or counter gas flow to avoid spectral interferences and matrix effects. In contrast, side-on observation of the plasma discharge is less prone to matrix effects, such as caused by easily ionized elements (EIE). Very robust detection conditions can be achieved by optimizing the observation height. The element specific radiation is then separated by wavelength with a mono- or polychromator and eventually detected [2,5].

6.1.3.2 ICP-MS

The ICP-mass spectrometry (MS) has several advantages compared to ICP-optical emission spectroscopy (OES). The LODs are typically enhanced by up to three orders of magnitude and the working range is expanded up to ten orders of magnitude. The full detection power is limited by blank values and can only be achieved under clear-room conditions. A disadvantage compared to ICP-OES is the decreased matrix tolerance, so that the total dissolved solids should be below 0.1% in the sample solutions [18,19].

Concerning the mass separation capabilities, quadrupole mass spectrometers (Q-ICP-MS), the most commonly used mass analyzer, offer a resolution of nearly 300 m/Δm. In comparison, a high-resolution sector-field ICP-MS (SF-ICP-MS) offers a resolution of up to 10,000 m/Δm and can resolve many of the spectral interferences in ICP-MS. Also coupling to a TOF mass analyzer can increase the resolution. In the high-resolution modus, the ion count rate is decreased by 95% compared to the low-resolution modus (300 m/Δm), but because of the overall ion transmission and the selectivity, better LODs can be achieved compared to the Q-ICP-MS. Thus, the instruments scan over the mass-spectrum, multielement measurements are rather performed sequentially than simultaneous compared to multicollector instruments [20,21].

One approach to minimize polyatomic interferences in ICP-MS was the collision–reaction-cell ICP-MS. Between the ion lenses and the quadrupole, the cell was inserted into the ion beam. Either through inert multiple collision or chemical reactions, the influence of polyatomic interferences should be decreased. The cell consists of an ion guide (hexapole or octopole) or a mass filter (quadrupole) and is filled either with an inert gas like xenon or helium or a reactive gas like H_2, O_2 or NH_3. The decrease of the polyatomic interferences is then achieved by discrimination (collision cell) or by conversion of either the analyte or the interference into other species (reaction cell). However, unwanted side reactions can take place when reactive gases are used [22,23]. A more recent approach to resolve problematic interferences is the triple quadrupole ICP-MS (TQ-ICP-MS). These instruments can operate in MS-MS mode, which offers the potential to improve the removal of spectral interferences compared to conventional quadrupole ICP-MS with a collision–reaction cell, or high-resolution ICP-MS [24,25].

6.2 ICP-MS AND ICP-OES IN LIB RESEARCH

6.2.1 Bulk Analysis

The determination of impurities and the true stoichiometries of the active material is a reoccurring topic in the literature. Also with the rising interest in recycling, determination of valuable metals contents gain more attention [26–35]. However, standalone reports are scarce and ICP-based methods are only mentioned in most publications with regard to short section. Furthermore, it is a frequent phenomenon, where elemental analysis is commonly referred to just ICP without any information about setup, parameters or figures of merit. Here, the more detailed reports are summarized to show the different, possible applications for these methods.

In 2002, Kida et al. reported investigations on lithium secondary batteries using $LiNi_{0.7}Co_{0.3}O_2$ and graphite/coke hybrid carbon. After long-term cycling, the cells were disassembled and the cathode was investigated by ICP spectroscopy and AAS [36]. The ratio of Li, Ni and Co was determined and compared to the reference composition. They observed a decrease in the lithium concentration inside the cathode, which was correlated to the growth of the SEI at the anode and corresponding Li loss during cycling.

In 2004, Wohlfahrt-Mehrens et al. studied the dissolution of lithium-nickel-cobalt mixed oxides for several self-synthesizing oxides [37]. The composition of lithium metal oxides and the dissolution of cobalt, nickel and manganese were analyzed by ICP analysis. In this study, the concentrations of dissolved nickel and cobalt were below the LOD, while manganese was not mentioned at all.

Later in 2012, Zheng at al. used EDX and XRD measurements in combination with ICP to investigate the correlation between dissolution behavior and electrochemical cycling performance for in NCM-based cells [38]. The dissolution of the metals into the electrolyte was analyzed by ICP with a given sample preparation. The operating cell voltage, i.e., state of charge (SOC) was found to be one of the important factors influencing the dissolution rate of metal ions. The determined concentration of Mn was considerably higher than that of Co and Ni for all SOCs, which implies that Mn was easier removed from the NCM structure in comparison to both other metals.

The effect of the manganese contamination on the lithium-ion battery anode SEI properties was intensively studied in 2013 by Delacourt et al. with ICP-OES and soft X-ray absorption spectroscopy (XAS) [39]. An electrochemical cell setup was specifically designed for these experiments. As cathode material, the manganese-free $LiyNi_{0.8}Co_{0.15}Al_{0.05}O_2$ (NCA) was chosen because it is much more resistant according to the authors against dissolution under the used conditions.

Jung et al. described in 2014 the effects of metal impurities in the electrolyte [40]. Within the experimental conditions, the determined Mn/Li mole ratio by ICP-OES in the film was found to be in the range 0.09–0.21, with Mn species (by soft XAS) being nearly exclusively in the +2 oxidation state.

In the same year, Kobayashi et al. reported the "degradation factor" in Mn-based/graphite lithium ion batteries reflecting the decrease in capacity during cycling [41]. To verify the findings, the lithium and manganese concentration in the electrodes was analyzed by ICP-OES. The amount of lithium in the discharged anode was attributed to electrochemically inactive lithium although no quantitative values were presented.

Furthermore in 2013, Takahara et al. further improved a previously developed GD-OES method for the quantification of lithium [42]. The objects of investigation were NCM cathodes and hard-carbon-based anodes. The results were validated by additional ICP-MS measurements. The obtained GD-OES intensity was linearly correlated to those obtained with the ICP-MS setup. For further improvement of the correlation coefficients the matrix element ratios were taken into account (Li/C for the anodes and Li/Co for the cathodes). As samples, the cells were cycled to a state of health (SOH) of 0%, 25%, 50%, 75% and 100%.

Also in 2016, Ghanbari et al. applied GD-OES in combination with ICP-OES for the analysis of anodes from lithium ion batteries, as well [43]. The recommendation from Takahara et al. regarding the addition of H_2 was included in the experiments. The anodes were sampled from pouch cells with NCM cathodes and graphite anodes. Calibration was carried out by lab-coated electrodes with known Li content, which was verified by ICP-OES. The overall error of the calibration was 18%.

In 2017, Vortmann-Westhoven recently published a comprehensive analysis by ICP-OES of all single components of a lithium ion battery to determine the Li distribution [12]. The experiments were carried out with two different cell types, T-cells of the Swagelok®-type and pouch bag cells. Subject to investigation were anode, cathode, separator and the electrolyte, as well as washing solutions from the housing and insulating foils. It could be shown that for the T-cell setup 16% of lithium was lost because of the used geometry and therefore that full lithium ion battery cell measurements are more appropriate in pouch bag cells. Furthermore, it was stated that 2%–4% of lithium was immobilized in the CEI at the cathode and 2%–3% in the SEI and inserted/deposited lithium in/at the anode.

More recently, Fu et al. published a report about the determination of metal impurity elements in lithium hexafluorophosphate [4]. By eliminating the interference with reaction gas mixtures, a new method based on ICP-MS/MS was developed to determine 18 metal impurity elements in $LiPF_6$. In the MS-MS mode, by comparing the interference elimination effects of the reaction gas mixtures systems and single reaction gas systems the reaction gas mixtures were selected to eliminate the spectral interference. The LODs of 18 metal impurity elements were in the range of 0.30–63.8 ng/L.

6.2.2 (Nano)-Particle Analysis

Single-particle ICP-MS and ICP-OES have been developed as emerging techniques to characterize suspended particles in different media and applications reported numerous times. Kröger et al. used single-particle ICP-OES in 2021 to investigate cathode materials for the first time [44]. Since single particle analysis in aqueous media was not feasible, due to potentially occurring Li^+-H^+ exchange reactions, they developed a sample introduction approach by a gravitational counter-flow classification with an adjustable vertical rising argon flow. The size LOD of $Li(Ni_{1/3}Co_{1/3}Mn_{1/3})O_2$ (NCM111) particles was estimated to be 0.5 µm, which corresponds to analyte masses of 30 and 180 fg for Li and Mn, respectively. Furthermore, the results for the matrix-matched external calibration were accompanied by a high coefficient of determination and high reproducibility, which demonstrates the applicability of this approach.

In the following year, Kröger et al. applied the method for the direct investigation of the interparticle-based state-of-charge distribution of polycrystalline NMC532 [45]. For partially delithiated particles, severe micro-cracking in the particle interior was observed, which implied the buildup of kinetic hindrances for the Li and electron transport, potentially explaining the heterogeneous Li extraction. However, the micro-cracks also provide more channels for the percolation of electrolyte in the particle interior, which can lead to new electrochemically active surface areas that promote improved Li transport kinetics.

6.3 COMBINATION WITH LASER ABLATION (SURFACE ANALYSIS)

6.3.1 Basic Principles and Background

Laser ablation (LA) can be applied as a sampling technique, which is based on sputtering solid materials using a laser beam. Coupled to a mass spectrometer it is a powerful tool for elemental analysis of solid samples. A focused discharge of a laser beam, produced by a pulsed laser source, enables a controlled ablation of particulate

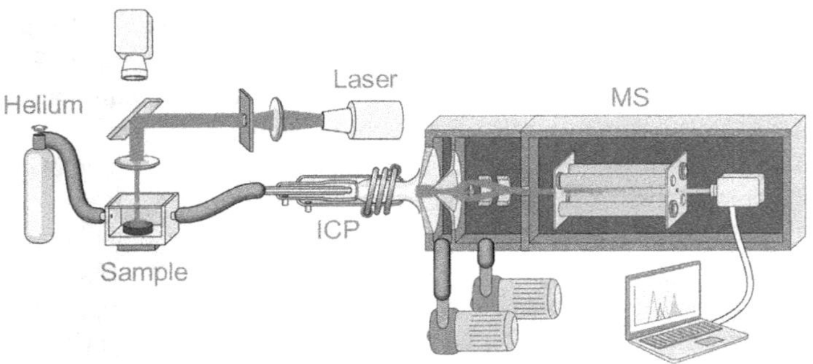

FIGURE 6.1 Typical setup of a LA-ICP-MS system.

matter from the sample. A carrier gas like helium transports the produced aerosol to the ICP-MS or ICP-OES, where an excitation/ionization and subsequent detection takes place. The basic setup for LA-ICP-MS is depicted in Figure 6.1.

With focus on accurate analysis via LA-ICP-MS, three basic conditions have to be met, namely (i) a representative aerosol composition, (ii) high transport efficiencies and (iii) complete decomposition of particles in the ICP. Especially the aerosol composition poses a fundamental problem, as fractionation effects occur during ablation. In an effort to improve the authenticity of analysis, modern development focused on laser wavelengths and pulse duration. Presently, neodymium-doped yttrium aluminum garnet (Nd:YAG) laser systems utilizing wavelengths of 266 nm and especially 213 nm with pulse lengths in the nanosecond range which are most commonly used. However, their tendency to produce micrometer-sized particles, and low evaporation efficiencies for highly transparent materials, limit the quality of analysis. Therefore, a shift to vacuum UV laser, such as the 193 nm argon fluoride (ArF)-type excimer, seemed to be beneficial with regard to particle size distribution and aerosol conditions. The possibility of laser adjustments in three dimensions (*x*, *y* and *z*) allow for lateral as well as depth resolution investigations. Lateral resolutions from the single- to the three-digit micrometer range can be achieved while depth resolutions below 1 μm are possible. Compared to other surface-sensitive techniques, LA offers the ability to scan samples with a size of several cm^2 in a short period of time. Another significant advantage is the minimal effort in terms of sample preparation. Therefore, phenomena like the transition metal dissolution (TMD) and the loss of active lithium are primary examples for possible application of LA-ICP-based techniques in lithium ion battery research.

Despite these major advantages, LA-based techniques are not well established for the analysis of battery materials. This is mainly due to the complex and rough constitution of the samples which makes the elemental quantification of the electrodes difficult. Elemental fractionation and different optical properties of materials can lead to different analyte intensities, despite the concentrations being the same in standard and sample material. In order to minimize the influence of these factors, matrix-matched standards or reference materials are necessary for reliable quantification. However, LA was already implemented in other fields of research, such surface analysis of biological samples, medical research and geology [46–53].

6.3.2 Application

As mentioned earlier, there are not many studies about the direct analysis of LIB materials. However, the processing and structuring of electrolyte by LA processes are steadily mentioned in the last years. However, some studies were reported and will be discussed in the following section.

In 2017, Schwieters et al. reported the investigation of the lithium loss in SEI in lithium ion batteries by means of LA-ICP-MS and ICP-OES [54]. They quantified the lithium content in a reference and aged graphite anodes when different graphite intercalation compounds (GICs) where reached (LiC30, LiC18, LiC12 and LiC6). Furthermore, self-made standards were used due to lack of commercially available reference materials or standards. To evaluate the homogeneity of the calibration standards, 39 circle-shaped spots with a size of 110 μm in diameter each, were applied for $^{7}Li/^{13}C$ depth profiling. Monitoring the ^{63}Cu signal gave good evidence regarding the ablation depth because the current collector consisted of copper. Once the ^{63}Cu signal increased rapidly the ablation was considered as complete. The lithium in the SEI, excluding the residual $LiPF_6$ in the pores of the electrode and after formation of the four GICs and subsequent delithiation, was determined between 0.69 and 0.93 wt%. Despite the slight differences in morphology and structure between the custom-made external standards (MCMB and hard carbon) and the investigated GIC samples (KS6 type), the developed method showed satisfying results regarding the lithium quantification (Figure 6.2).

The same group applied LA-ICP-MS to previously aged carbonaceous anodes [55]. These electrodes are treated by cyclic aging in a lithium ion cell setup against $Li_1[Ni_{1/3}Mn_{1/3}Co_{1/3}]O_2$ = NMC111 to elucidate factors that influence TMD of the cathode and subsequent deposition on the anode. The investigations were carried out by visualizing the ^{7}Li and transition metal patterns (^{60}Ni, ^{55}Mn and ^{59}Co) of whole coin and pouch-bag electrodes. The determined elemental concentration was directly correlated to the applied upper cut-off voltage with higher deposition of Li and TMs at elevated voltages. While ^{7}Li showed a more homogeneous pattern, the transition metal distribution was inhomogeneous but showing a similar pattern for all transition metals of the same sample. An unequal pressure distribution, resulting in a nonparallel electrode alignment, on the electrode stack was identified to be responsible for the inhomogeneous transition metal deposition pattern. This uneven electrode orientation resulted in different diffusion pathways for the transition metal migration with regard to the spatial distances (Figure 6.3).

Based on these previous studies, Harte et al., in 2019, adapted and improved the method and expanded the measurement possibilities to the cathode and separator [56]. By the optimization of different parameters, the authors achieved several improvements. Higher scanning speeds had an observable influence on the resolution of the obtained image and an overall saving of 60% with regard to time and gas consumption could be achieved. In general, the developed method has proven to be effective and significantly faster for mapping of anodes than with previously used settings. The obtained good resolution allowed the investigation of deposition patterns between electrodes and separator. Klein and coworkers applied the method recently for visualization experiments of the transition metal and lithium distribution with regard to separators and additives.

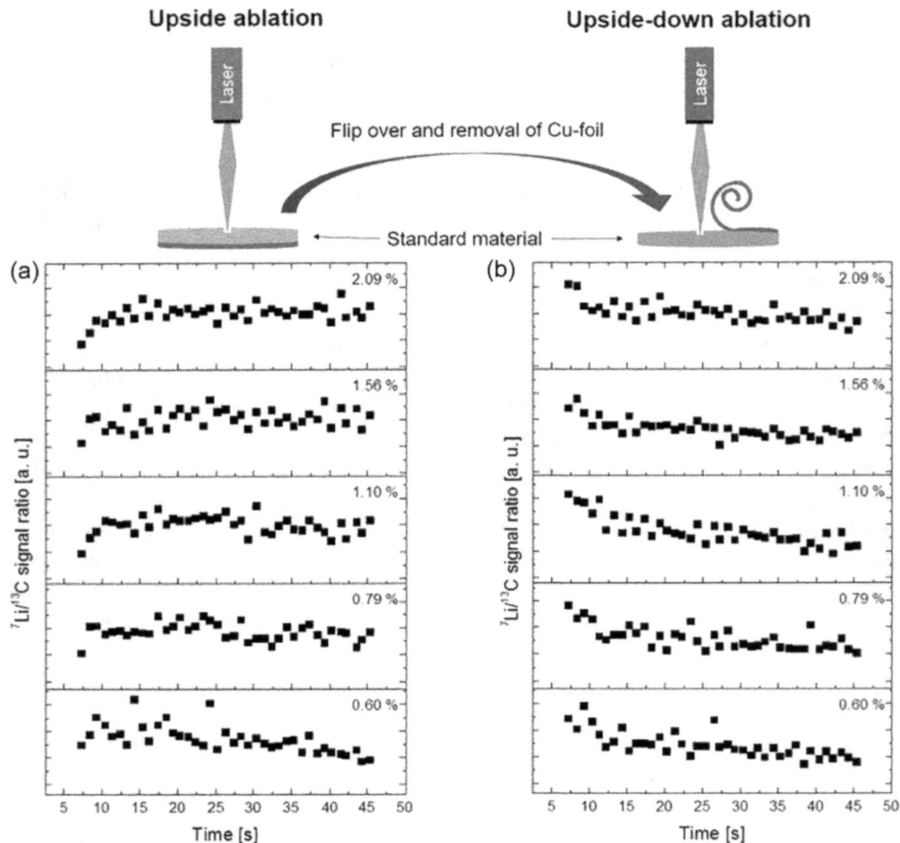

FIGURE 6.2 $^{7}Li/^{13}C$ depth profile sequences of solid custom-made external standards (0.60–2.09 wt% lithium) in (a) upside and (b) in upside-down direction. (Reprinted with permission from ref. [54], Copyright Elsevier.)

Compared to the used LA-ICP-MS methods, Lürenbaum et al. developed a LA-ICP-OES method to investigate the lithium distribution and transition metal depositions on cycled electrodes [57]. The quantification was performed by using self-prepared matrix-matched standards; correlation coefficients higher than $r^2 = 0.99$ could be achieved with reasonably low standard deviations. However, the investigated transition metal contents are significantly higher than previously reported, which could be explained by measurement inaccuracies, as no correction for varying ablation yields or fractionation effects (such as transport efficiencies) was applied. In order to improve the precision of this method, possible internal standards need to be evaluated and applied in future works. Furthermore, the method was capable of visualizing and determining local deposition of lithium (Li plating) in a cell suffering from short circuits (Figure 6.4).

Henschel et al. examined the phytoremediation capabilities of *Alyssum murale* regarding commonly applied metals and constituents in lithium ion battery cells in

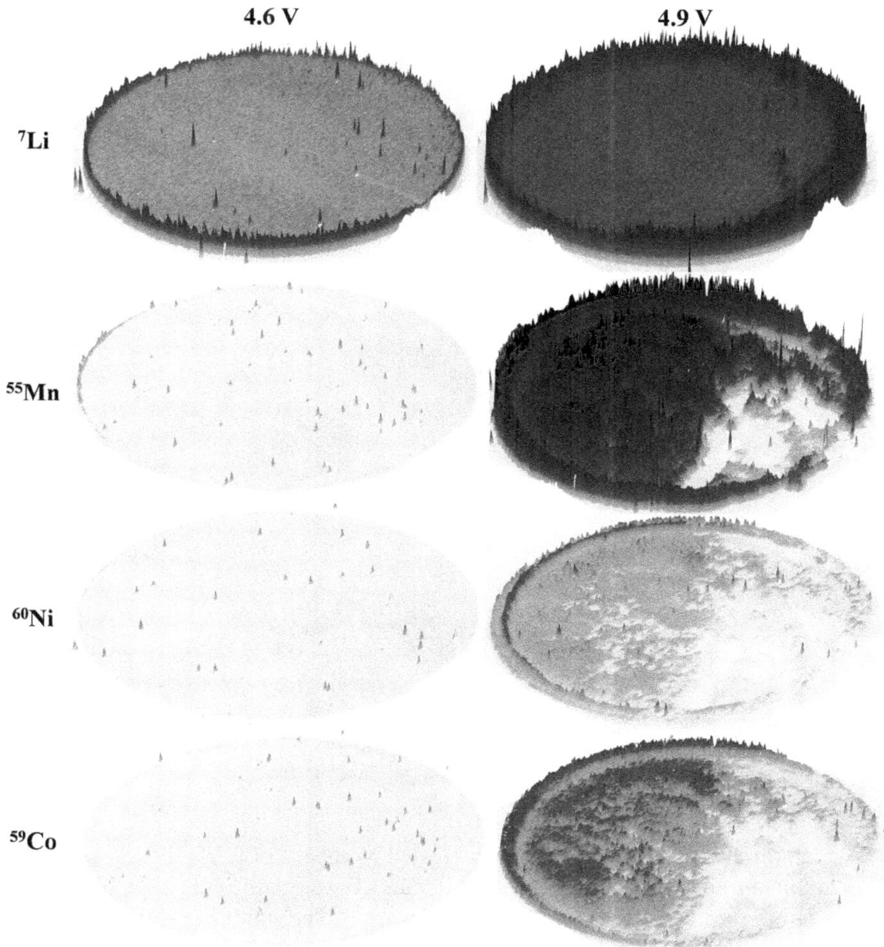

FIGURE 6.3 LA-ICP-MS results of four mapping experiments recording the ^{7}Li signal intensity via 113 line scans. Varying upper cut-off voltages of 4.6V and 4.9V for the cycles 4–53 were applied. (Reprinted with permission from ref. [55], Copyright Elsevier.)

the context of recycling [58]. A contamination approach with the dissolved transition metals Ni, Co and Mn in irrigation water showed translocation and hyperaccumulation for all metals up to 3 wt% in the dry matter of different plant tissues (leaf, stem, root). In the following accumulation phase, metal concentrations in all plant compartments increased significantly (leaves > 20%) and soil concentrations decreased (10%–50%), proving the active uptake of Ni, Co and Mn in *A. murale* and its phytoremediation capabilities. The hyperaccumulation of plants represents a low-cost and green technology to reduce environmental pollution of landfills and disused mining regions with low environmental regulations. Plants were cultivated in a conservatory for 46 days whilst soils were contaminated stepwise with dissolved transition metal species via the irrigation water (Figure 6.5).

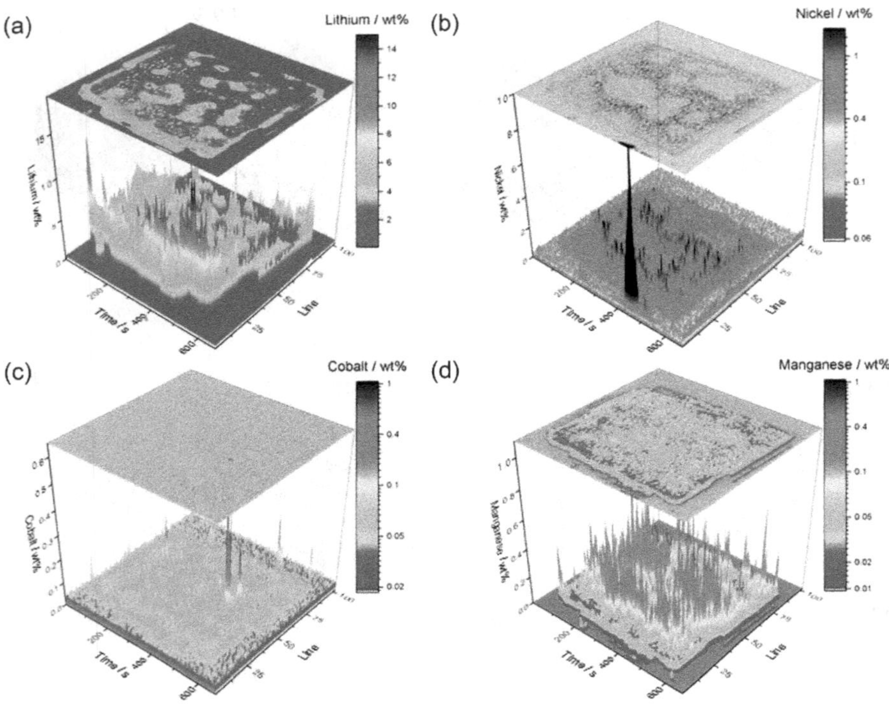

FIGURE 6.4 Quantitative results of the ablated graphitic anode A; concentration is depicted in wt%; (a) lithium, (b) nickel, (c) cobalt and (d) manganese [57].

The lateral distribution of Li was investigated by means of LA-ICP-MS, and a co-localization of transition metals observed with μXRF was confirmed. They performed three ablation runs to obtain more depth information. The small Li cation was apparently distributed in the whole leaf and presumably present in the cytosol and water-bearing tissues (xylem/phloem). This assumption was based on the finding of reduced Li signals in parched leaf areas. Considering the low concentrations of transition metals, the assumption of effective vacuolar sequestration of Ni and Co, and hence, no co-localization with trichome structures is consistent with the literature. Overall, the accumulation of toxic transition metals in the leaf tip is consistent with the interpretation of hyperaccumulation as an elemental-defense strategy of *A. murale*.

6.4 COMBINATION WITH CHROMATOGRAPHIC TECHNIQUES (SPECIATION ANALYSIS)

6.4.1 Basic Principles and Background

"Speciation analysis" is defined by the International Union of Pure and Applied Chemistry (IUPAC) as analytical activities of identifying and/or measuring the quantities of one or more individual chemical species in a sample. The definition of a

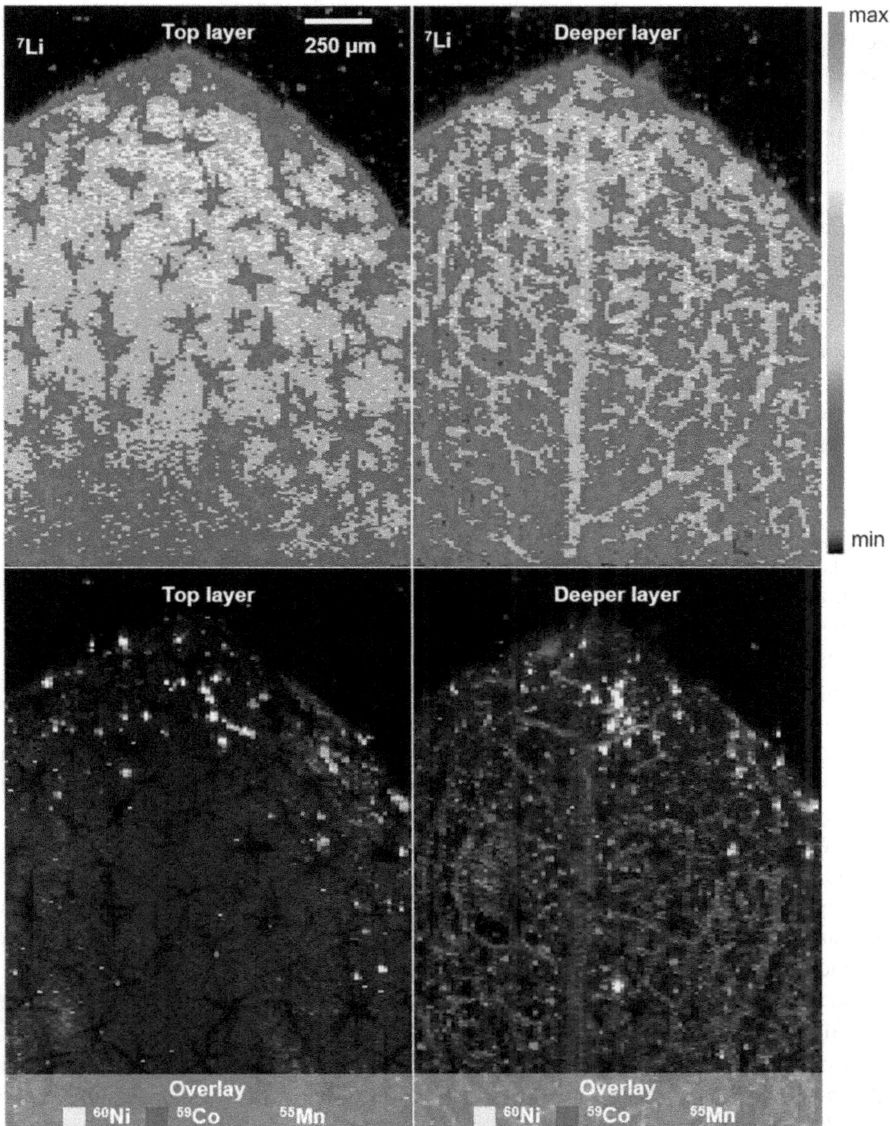

FIGURE 6.5 LA-ICP-MS image showing the Li distribution in the surface layer and deeper mesophyll (top), as well as the respective overlay of transition metal distribution (Ni, Co, Mn; bottom) in a leaf of *Alyssum murale* [58].

chemical species is a "specific form of an element defined as to isotopic composition, electronic or oxidation state, and/or complex or molecular structure." The composition and structure of the specific forms determine their properties, including their toxicity [59]. Therefore, not only the elemental content in a sample is important, but also the identification of the chemical environment of the target element. However, the ICP-based instrumentation can only provide species-independent determination due to efficient atomization in the plasma. Elemental species have to be separated and introduced into the plasma in a time-resolved manner. Therefore, a number of chromatographic (liquid, gas or ion) and electrophoretic devices have been coupled with ICP-based systems.

The literature cites numerous reports about aging products and mechanisms for lithium ion batteries [37,60–64]. As for the electrolyte, these products include HF stemming from hydrolysis of the electrolyte salt $LiPF_6$, subsequent formation of acidic and non-acidic (fluoro)phosphates (OPs). Unfortunately, for analysis of most of the described substances no commercial standards are available which complicates the quantification, and therefore hyphenated techniques, i.e., coupling of analysis techniques, are essential, especially since several identified compounds are potentially toxic and quantification is absolutely necessary to identify the impact of these compounds on the human body [65–69].

6.4.2 Application

Like LA, there are a very few investigations where hyphenated techniques between chromatographic and ICP-based methods were applied. In 2013, Terborg et al. were the first to report hyphenated systems for the analysis of lithium ion battery electrolytes [70]. They investigated the thermal decomposition and hydrolysis of $LiPF_6$. To obtain elemental information about the detected peaks, an ion chromatography system was coupled to an ICP-OES and the phosphorus content was monitored. In combination with the corresponding ion chromatography-electrospray ionization experiments the combined data were evaluated and several hydrolysis products could be assigned.

Kraft et al. further improved the separation of the decomposition products in 2015 by applying a two-dimensional ion chromatography method using a heart-cut mode [71]. Furthermore, after identification of several organophosphates, the authors carried out the hyphenation to an ICP-MS system. The focus of the analysis was three commercially available electrolytes. It was shown that five baseline separated peaks for phosphorous could be detected by comparison of the retention times with the IC-ESI-MS results.

In comparison to the two phosphorus-based reports, Grützke et al. reported in 2015 an IC/ICP-MS method for the investigation of boron-based degradation products [72]. During their investigation of an electrolyte from a field-tested hybrid electric vehicle, a mass to charge ratio (*m/z*) of 87 was observed. Since the m/z could not assigned to the otherwise identified organophosphates, additional IC/ICP-MS measurements were undertaken. With the help of additional ICP-OES measurements, the concentration of the assigned $LiBF_4$ was determined and therefore it was concluded that the $LiBF_4$ was added as a second conducting salt, i.e., in addition to $LiPF_6$.

The first report dealing with quantitative analysis was from 2017 by Menzel et al., based on the work of Kraft et al. from 2015. It employed the simultaneous coupling of

2D-IC with electrospray-ionization mass spectrometry (ESI-MS) and with ICP-MS. Besides thermal aging, a first attempt of quantification of ionic organo(fluoro)phosphates in battery systems, a quantification from cycled Li-metal cells was carried out. The results showed that the addition of water and high temperature caused a more severe aging than cycling does. However, the authors suggested additional measurements by gas chromatography coupled to ICP-MS to get a broader overview.

Stenzel et al. provided these measurements in 2018 [73]. In this study, the organophosphates formed were investigated with a sector field mass spectrometer coupled to a GC. With this setup, it was possible to overcome the need for molecular standards and to perform quantification after separation of species using only one external standard. Different resolution modes ($R > 300$ and $R > 4{,}000$) and different plasma conditions were applied to detect possible interferences originating from aerosol entry, GC solvents, sample matrix or decomposition products.

In the next year, Stenzel et al. presented an approach with a hydrophilic interaction liquid chromatography (HILIC) approach to investigate several decomposition products [74]. A total of 16 different O(F)P compounds could successfully be quantified and the formation of OFPs exceeds the amount of non-fluorine-containing OPs by a factor of up to 15.

Henschel et al. showed a combined approach of both previous methods for the characterization of electrolytes from several electric vehicles [75]. Quantitative data revealed the need for a toxicological relevance estimate of these compounds concerning the hazard for user, environment, accident scenario and recycling, as well as a possible second-life application. However, quantities remain decisively under the amounts generated by thermal aging (Figure 6.6).

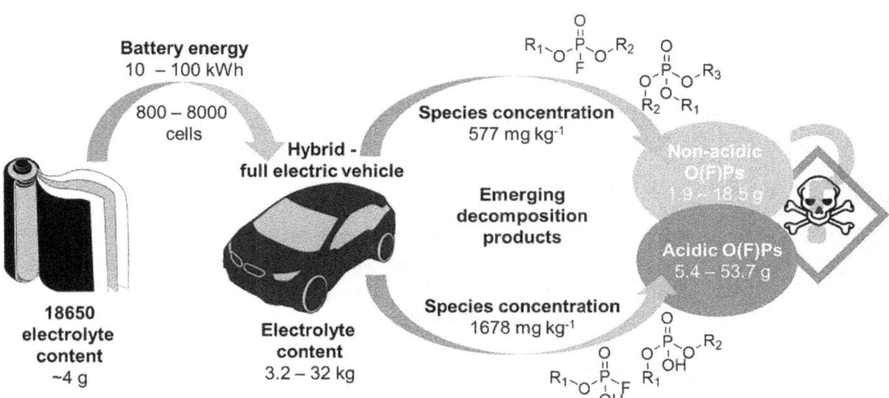

FIGURE 6.6 Electrolyte amount calculation from small hybrid to BEVs battery size with range extender (10–100 kWh). The quantities of potentially hazardous O(F)Ps are extrapolated from HILIC-ICP-SF-MS and GC-ICP-SF-MS results. (Reprinted with permission from ref. [75], Copyright Elsevier.)

6.5 SUMMARY AND OUTLOOK

6.5.1 Next-Generation Batteries

The analysis of next generation of batteries like lithium-sulfur or all-solid-state batteries will have an impact on the ICP-based methods. While all-solid-state batteries will have no liquid electrolyte, the decomposition product analysis will no longer be accessible by hyphenated techniques for example. However, surface techniques like LA with the abilities to perform depth profiles will gain additional importance. In comparison, the application of lithium-sulfur with similar electrolyte systems will offer several challenges for the adaption of reported methods. Calibration standards and reference materials which are needed for the current state of the art will be mandatory for the next generation of materials too. Furthermore, providing information about sample preparation, figures of merit and instrument parameters are needed for a rapid and safe adaptation to this new class of materials.

6.5.2 Outlook on Future Instrumentation

New instrumentation can offer several advantages for the community. Especially for the relatively new and seldom reported analysis of LA or particle-analysis high-resolution instruments like time-of-flight mass spectrometers will provided new measurement possibilities, e.g., better resolution, interference control and lower LODs. However, it seems that most of the newer instruments, which are used in other research areas only, slowly adapted to lithium-ion battery investigations. It is quite surprising that although elemental analysis of cathode concentrations or stoichiometries is frequently used, setups for surface or speciation analysis, e.g., determining the aging status of a cell, are only rarely reported by a handful of groups. With regard to machine learning, only few references are available so far [76–79]. However, this approach can be helpful for the determination of battery materials. Nevertheless, the missing information on the sample preparation and analysis condition need to be updated in order to apply these steps for future analysis.

REFERENCES

[1] P.W.J.M. Boumans, *Inductively coupled plasma emission spectroscopy*, Wiley, 1987.

[2] A. Montaser, D.W. Golightly, *Inductively coupled plasmas in analytical atomic spectrometry*, 2nd Edition, VCH, Weinheim, 1987.

[3] S. Seshadri, *Fundamentals of plasma physics*, American Elsevier Publishing Co., Inc., 1973, 558 p.

[4] L. Fu, H. Xie, J. Huang, X. Chen, L. Chen, Determination of metal impurity elements in lithium hexafluorophosphate using inductively coupled plasma tandem mass spectrometry based on reaction gas mixtures, *Spectrochimica Acta Part B: Atomic Spectroscopy*, 181 (2021) 106217.

[5] P. Boumans, N.W. Barnett, *Inductively coupled plasma emission spectroscopy, part 1: methodology, instrumentation and performance: Horwood, Chichester*, Elsevier, 1987, xi+ 584 pp.

[6] E. Chudinov, I. Ostroukhova, G. Varvanina, Acid effects in ICP-AES, *Fresenius' Zeitschrift für analytische Chemie*, 335 (1989) 25–33.

[7] J. Broekaert, Trends in optical spectrochemical trace analysis with plasma sources, *Analytica Chimica Acta*, 196 (1987) 1–21.
[8] M. Holderbeke, Evaluation of a commercially available microconcentric nebulizer for inductively coupled plasma mass spectrometry, *Journal of Analytical Atomic Spectrometry*, 11 (1996) 543–548.
[9] S.-H. Nam, J.-S. Lim, A. Montaser, High-efficiency nebulizer for argon inductively coupled plasma mass spectrometry, *Journal of Analytical Atomic Spectrometry*, 9 (1994) 1357–1362.
[10] R.H. Scott, V.A. Fassel, R.N. Kniseley, D.E. Nixon, Inductively coupled plasma-optical emission analytical spectrometry, *Analytical Chemistry*, 46 (1974) 75–80.
[11] M. Evertz, C. Lurenbaum, B. Vortmann, M. Winter, S. Nowak, Development of a method for direct elemental analysis of lithium ion battery degradation products by means of total reflection X-ray fluorescence, *Spectrochim Acta B*, 112 (2015) 34–39.
[12] B. Vortmann-Westhoven, M. Winter, S. Nowak, Where is the lithium? Quantitative determination of the lithium distribution in lithium ion battery cells: Investigations on the influence of the temperature, the C-rate and the cell type, *Journal of Power Sources*, 346 (2017) 63–70.
[13] C. Zhan, J. Lu, A. Jeremy Kropf, T. Wu, A.N. Jansen, Y.K. Sun, X. Qiu, K. Amine, Mn(II) deposition on anodes and its effects on capacity fade in spinel lithium manganate-carbon systems, *Nature Communications*, 4 (2013) 2437.
[14] M. Evertz, F. Horsthemke, J. Kasnatscheew, M. Börner, M. Winter, S. Nowak, Unraveling transition metal dissolution of $Li1.04Ni1/3Co1/3Mn1/3O_2$ (NCM 111) in lithium ion full cells by using the total reflection X-ray fluorescence technique, *Journal of Power Sources*, 329 (2016) 364–371.
[15] S. Nowak, M. Winter, Elemental analysis of lithium ion batteries, *Journal of Analytical Atomic Spectrometry*, 32 (2017) 1833.
[16] L. Burton, M. Blades, A simple method for calculating deviations from local thermodynamic equilibrium in the inductively coupled plasma, *Spectrochimica Acta Part B: Atomic Spectroscopy*, 45 (1990) 139–144.
[17] M. Blades, B. Caughlin, Z. Walker, L. Burton, Excitation, ionization, and spectral line emission in the inductively coupled plasma, *Progress in Analytical Spectroscopy*, 10 (1987) 57–109.
[18] R. Thomas, *Practical Guide to ICP-MS*, Scientific Solutions Gaithersburg, Maryland, USA Marcel Dekker INC, New York, USA, 2004.
[19] R.S. Houk, V.A. Fassel, G.D. Flesch, H.J. Svec, A.L. Gray, C.E. Taylor, Inductively coupled argon plasma as an ion source for mass spectrometric determination of trace elements, *Analytical Chemistry*, 52 (1980) 2283–2289.
[20] L. Moens, F. Vanhaecke, J. Riondato, R. Dams, Some figures of merit of a new double focusing inductively coupled plasma mass spectrometer, *Journal of Analytical Atomic Spectrometry*, 10 (1995) 569–574.
[21] D.P. Myers, G. Li, P. Yang, G.M. Hieftje, An inductively coupled plasma—time-of-flight mass spectrometer for elemental analysis. Part I: Optimization and characteristics, *Journal of the American Society for Mass Spectrometry*, 5 (1994) 1008–1016.
[22] S.D. Tanner, V.I. Baranov, A dynamic reaction cell for inductively coupled plasma mass spectrometry (ICP-DRC-MS). II. Reduction of interferences produced within the cell, *Journal of the American Society for Mass Spectrometry*, 10 (1999) 1083–1094.
[23] N. Yamada, J. Takahashi, K.I. Sakata, The effects of cell-gas impurities and kinetic energy discrimination in an octopole collision cell ICP-MS under non-thermalized conditions, *Journal of Analytical Atomic Spectrometry*, 17 (2002) 1213–1222.
[24] V. Balaram, Strategies to overcome interferences in elemental and isotopic geochemical analysis by quadrupole inductively coupled plasma mass spectrometry: A critical evaluation of the recent developments, *Rapid Communications in Mass Spectrometry*, 35 (2021) e9065.

[25] L. Balcaen, E. Bolea-Fernandez, M. Resano, F. Vanhaecke, Inductively coupled plasma–Tandem mass spectrometry (ICP-MS/MS): A powerful and universal tool for the interference-free determination of (ultra) trace elements–A tutorial review, *Analytica Chimica Acta*, 894 (2015) 7–19.
[26] J.-I. Yamaki, Y. Shinjo, T. Doi, S. Okada, The rate equation for oxygen evolution by decomposition of $LixCoO_2$ at elevated temperatures, *Journal of the Electrochemical Society*, 161 (2014) A1648–A1654.
[27] C. Deng, L. Liu, W. Zhou, K. Sun, D. Sun, Characterization of Li [Ni1/3Co1/3Mn1/3] O_2 synthesized via salvolatile coprecipitation for lithium-ion batteries, *Electrochemical and Solid-State Letters*, 10 (2007) A279–A282.
[28] X. Huang, X. He, C. Jiang, G. Tian, Morphology evolution and impurity analysis of $LiFePO_4$ nanoparticles via a solvothermal synthesis process, *RSC Advances*, 4 (2014) 56074–56083.
[29] J.-M. Kim, N. Kumagai, S. Komaba, Improved electrochemical properties of Li 1 + x (Ni 0.3 Co 0.4 Mn 0.3) $O_2 - \delta$ (x = 0, 0.03 and 0.06) with lithium excess composition prepared by a spray drying method, *Electrochimica Acta*, 52 (2006) 1483–1490.
[30] T.-F. Yi, C.-L. Hao, C.-B. Yue, R.-S. Zhu, J. Shu, A literature review and test: Structure and physicochemical properties of spinel $LiMn_2O_4$ synthesized by different temperatures for lithium ion battery, *Synthetic Metals*, 159 (2009) 1255–1260.
[31] L.H. Chi, N.N. Dinh, S. Brutti, B. Scrosati, Synthesis, characterization and electrochemical properties of 4.8 V $LiNi_{0.5}Mn_{1.5}O_4$ cathode material in lithium-ion batteries, *Electrochimica Acta*, 55 (2010) 5110–5116.
[32] J.-W. Liu, X.-H. Li, Z.-X. Wang, H.-J. Guo, W.-J. Peng, Y.-H. Zhang, Q.-Y. Hu, Preparation and characterization of lithium hexafluorophosphate for lithium-ion battery electrolyte, *Transactions of Nonferrous Metals Society of China*, 20 (2010) 344–348.
[33] H. Bae, Y. Kim, Technologies of lithium recycling from waste lithium ion batteries: A review, *Materials Advances*, 2 (2021) 3234–3250.
[34] M.A.J.S. Roldán-Ruiz, M.A.L. Ferrer, M.A.C.N. Gutiérrez, F.D. Monte, Highly efficient p-toluenesulfonic acid-based deep-eutectic solvents for cathode recycling of Li-ion batteries, *ACS Sustainable Chemistry & Engineering*, 8 (2020) 5437–5445.
[35] Y.I. Mesbah, N. Ahmed, B.A. Ali, N.K. Allam, Recycling of Li–Ni–Mn–Co hydroxide from spent batteries to produce high-performance supercapacitors with exceptional stability, *ChemElectroChem*, 7 (2020) 975–982.
[36] Y. Kida, A. Kinoshita, K. Yanagida, A. Funahashi, T. Nohma, I. Yonezu, Study on capacity fade factors of lithium secondary batteries using $LiNi_{0.7}Co_{0.3}O_2$ and graphite–coke hybrid carbon, *Electrochimica Acta*, 47 (2002) 4157–4162.
[37] M. Wohlfahrt-Mehrens, C. Vogler, J. Garche, Aging mechanisms of lithium cathode materials, *Journal of Power Sources*, 127 (2004) 58–64.
[38] H. Zheng, Q. Sun, G. Liu, X. Song, V.S. Battaglia, Correlation between dissolution behavior and electrochemical cycling performance for $LiNi_{1/3}Co_{1/3}Mn_{1/3}O_2$-based cells, *Journal of Power Sources*, 207 (2012) 134–140.
[39] C. Delacourt, A. Kwong, X. Liu, R. Qiao, W. Yang, P. Lu, S. Harris, V. Srinivasan, Effect of manganese contamination on the solid-electrolyte-interphase properties in Li-ion batteries, *Journal of the Electrochemical Society*, 160 (2013) A1099–A1107.
[40] S.-K. Jung, H. Gwon, J. Hong, K.-Y. Park, D.-H. Seo, H. Kim, J. Hyun, W. Yang, K. Kang, Understanding the degradation mechanisms of $LiNi_{0.5}Co_{0.2}Mn_{0.3}O_2$ cathode material in lithium ion batteries, *Advanced Energy Materials*, 4 (2014) 1300787.
[41] Y. Kobayashi, T. Kobayashi, K. Shono, Y. Ohno, Y. Mita, H. Miyashiro, Decrease in capacity in Mn-based/graphite commercial lithium-ion batteries I. Imbalance proof of electrode operation capacities by cell disassembly, *Journal of The Electrochemical Society*, 160 (2013) A1181–A1186.

[42] H. Takahara, M. Shikano, H. Kobayashi, Quantification of lithium in LIB electrodes with glow discharge optical emission spectroscopy (GD-OES), *Journal of Power Sources*, 244 (2013) 252–258.

[43] N. Ghanbari, T. Waldmann, M. Kasper, P. Axmann, M. Wohlfahrt-Mehrens, Inhomogeneous degradation of graphite anodes in Li-ion cells: A postmortem study using glow discharge optical emission spectroscopy (GD-OES), *The Journal of Physical Chemistry C*, 120 (2016) 22225–22234.

[44] T.-N. Kröger, S. Wiemers-Meyer, P. Harte, M. Winter, S. Nowak, Direct multielement analysis of polydisperse microparticles by classification-single-particle ICP-OES in the field of lithium-ion battery electrode materials, *Analytical Chemistry*, 93 (2021) 7532–7539.

[45] T.-N. Kröger, P. Harte, S. Klein, T. Beuse, M. Börner, M. Winter, S. Nowak, S. Wiemers-Meyer, Direct investigation of the interparticle-based state-of-charge distribution of polycrystalline NMC532 in lithium ion batteries by classification-single-particle-ICP-OES, *Journal of Power Sources*, 527 (2022) 231204.

[46] E.R. Schenk, J.R. Almirall, Elemental analysis of glass by laser ablation inductively coupled plasma optical emission spectrometry (LA-ICP-OES), *Forensic Science International*, 217 (2012) 222–228.

[47] J. Koch, D. Günther, Review of the state-of-the-art of laser ablation inductively coupled plasma mass spectrometry, *Applied Spectroscopy*, 65 (2011) 155A–162A.

[48] M. Aramendía, M. Resano, F. Vanhaecke, Isotope ratio determination by laser ablation-single collector-inductively coupled plasma-mass spectrometry. General capabilities and possibilities for improvement, *Journal of Analytical Atomic Spectrometry*, 25 (2010) 390.

[49] B. Rusk, Laser ablation ICP-MS in the earth sciences: Current practices and outstanding issues. Paul Sylvestor, Editor. Pp. 356. Mineralogical Association of Canada. Short Course Volume 40. 2008. ISBN: 9-0-921294-49-8. Price US 55.00(nonmembers), US 44.00 (members), *Economic Geology*, 104 (2009) 601–602.

[50] C.C. Garcia, H. Lindner, K. Niemax, Laser ablation inductively coupled plasma mass spectrometry - Current shortcomings, practical suggestions for improving performance, and experiments to guide future development, *Journal of Analytical Atomic Spectrometry*, 24 (2009) 14.

[51] J. Pisonero, D. Günther, Femtosecond laser ablation inductively coupled plasma mass spectrometry: Fundamentals and capabilities for depth profiling analysis, *Mass Spectrometry Reviews*, 27 (2008) 609.

[52] J.S. Becker, Applications of inductively coupled plasma mass spectrometry and laser ablation inductively coupled plasma mass spectrometry in materials science, *Spectrochimica Acta Part B: Atomic Spectroscopy*, 57 (2002) 1805–1820.

[53] T. Pettke, C.A. Heinrich, A.C. Ciocan, D. Günther, Quadrupole mass spectrometry and optical emission spectroscopy: Detection capabilities and representative sampling of short transient signals from laser-ablation, *Journal of Analytical Atomic Spectrometry*, 15 (2000) 1149.

[54] T. Schwieters, M. Evertz, M. Mense, M. Winter, S. Nowak, Lithium loss in the solid electrolyte interphase: Lithium quantification of aged lithium ion battery graphite electrodes by means of laser ablation inductively coupled plasma mass spectrometry and inductively coupled plasma optical emission spectroscopy, *Journal of Power Sources*, 356 (2017) 47–55.

[55] T. Schwieters, M. Evertz, A. Fengler, M. Borner, T. Dagger, Y. Stenzel, P. Harte, M. Winter, S. Nowak, Visualizing elemental deposition patterns on carbonaceous anodes from lithium ion batteries: A laser ablation-inductively coupled plasma-mass

spectrometry study on factors influencing the deposition of lithium, nickel, manganese and cobalt after dissolution and migration from the Li-1 [Ni1/3Mn1/3Co1/3]O-2 and LiMn1.5 Ni0.5O_4 cathode, *Journal of Power Sources*, 380 (2018) 194–201.
[56] P. Harte, M. Evertz, T. Schwieters, M. Diehl, M. Winter, S. Nowak, Adaptation and improvement of an elemental mapping method for lithium ion battery electrodes and separators by means of laser ablation-inductively coupled plasma-mass spectrometry, *Analytical and Bioanalytical Chemistry*, 411 (2019) 581–589.
[57] C. Lurenbaum, B. Vortmann-Westhoven, M. Evertz, M. Winter, S. Nowak, Quantitative spatially resolved post-mortem analysis of lithium distribution and transition metal depositions on cycled electrodes via a laser ablation-inductively coupled plasma-optical emission spectrometry method, *Rsc Advances*, 10 (2020) 7083–7091.
[58] J. Henschel, M. Mense, P. Harte, M. Diehl, J. Buchmann, F. Kux, L. Schlatt, U. Karst, A. Hensel, M. Winter, S. Nowak, Phytoremediation of soil contaminated with lithium ion battery active materials—A proof-of-concept study, *Recycling*, 5 (2020) 26.
[59] W. Quiroz, Speciation analysis in chemistry, *ChemTexts*, 7 (2021) 7.
[60] S. Nowak, M. Winter, The role of cations on the performance of lithium ion batteries: A quantitative analytical approach, *Accounts of Chemical Research*, 51 (2018) 265–272.
[61] S. Nowak, M. Winter, Elemental analysis of lithium ion batteries, *Journal of Analytical Atomic Spectrometry*, 32 (2017) 1833–1847.
[62] M. Broussely, P. Biensan, F. Bonhomme, P. Blanchard, S. Herreyre, K. Nechev, R. Staniewicz, Main aging mechanisms in Li ion batteries, *Journal of Power Sources*, 146 (2005) 90–96.
[63] J. Vetter, P. Novák, M.R. Wagner, C. Veit, K.C. Möller, J.O. Besenhard, M. Winter, M. Wohlfahrt-Mehrens, C. Vogler, A. Hammouche, Ageing mechanisms in lithium-ion batteries, *Journal of Power Sources*, 147 (2005) 269–281.
[64] Y.P. Stenzel, F. Horsthemke, M. Winter, S. Nowak, Chromatographic techniques in the research area of lithium ion batteries: current state-of-the-art, *Separations*, 6 (2019).
[65] F.M. Raushel, Chemical biology: Catalytic detoxification, *Nature*, 469 (2011) 310–311.
[66] S. Silver, The toxicity of dimethyl-, diethyl-, and diisopropyl fluorophosphate vapors, *Journal of Industrial Hygiene and Toxicology*, 30 (1948) 307–311.
[67] H.L. Tripathi, W.L. Dewey, Comparison of the effects of diisopropylfluorophosphate, sarin, soman, and tabun on toxicity and brain acetylcholinesterase activity in mice, *Journal of Toxicology and Environmental Health, Part A Current Issues*, 26 (1989) 437–446.
[68] R.T. Delfino, T.S. Ribeiro, J.D. Figueroa-Villar, Organophosphorus compounds as chemical warfare agents: A review, *Journal of the Brazilian Chemical Society*, 20 (2009) 407–428.
[69] P. Patnaik, *A comprehensive guide to the hazardous properties of chemical substances*, John Wiley & Sons, 2007.
[70] L. Terborg, S. Weber, F. Blaske, S. Passerini, M. Winter, U. Karst, S. Nowak, Investigation of thermal aging and hydrolysis mechanisms in commercial lithium ion battery electrolyte, *Journal of Power Sources*, 242 (2013) 832–837.
[71] V. Kraft, M. Grützke, W. Weber, J. Menzel, S. Wiemers-Meyer, M. Winter, S. Nowak, Two-dimensional ion chromatography for the separation of ionic organophosphates generated in thermally decomposed lithium hexafluorophosphate-based lithium ion battery electrolytes, *Journal of Chromatography A*, 1409 (2015) 201–209.
[72] M. Grützke, V. Kraft, B. Hoffmann, S. Klamor, J. Diekmann, A. Kwade, M. Winter, S. Nowak, Aging investigations of a lithium-ion battery electrolyte from a field-tested hybrid electric vehicle, *Journal of Power Sources*, 273 (2015) 83–88.
[73] Y.P. Stenzel, M. Winter, S. Nowak, Evaluation of different plasma conditions and resolutions for understanding elemental organophosphorus analysis via GC-ICP-SF-MS, *Journal of Analytical Atomic Spectrometry*, 33 (2018) 1041–1048.

[74] Y.P. Stenzel, J. Henschel, M. Winter, S. Nowak, A new HILIC-ICP-SF-MS method for the quantification of organo(fluoro)phosphates as decomposition products of lithium ion battery electrolytes, *RSC Advances*, 9 (2019) 11413–11419.

[75] J. Henschel, F. Horsthemke, Y.P. Stenzel, M. Evertz, S. Girod, C. Lurenbaum, K. Kosters, S. Wiemers-Meyer, M. Winter, S. Nowak, Lithium ion battery electrolyte degradation of field-tested electric vehicle battery cells - A comprehensive analytical study, *Journal of Power Sources*, 447 (2020).

[76] T.R. Holbrook, D. Gallot-Duval, T. Reemtsma, S. Wagner, Machine learning: our future spotlight into single-particle ICP-ToF-MS analysis, *Journal of Analytical Atomic Spectrometry*, 36 (2021) 2684–2694.

[77] R. Grimmig, S. Lindner, P. Gillemot, M. Winkler, S. Witzleben, Analyses of used engine oils via atomic spectroscopy - Influence of sample pre-treatment and machine learning for engine type classification and lifetime assessment, *Talanta*, 232 (2021) 122431.

[78] J.A. Carter, L.M. O'Brien, T. Harville, B.T. Jones, G.L. Donati, Machine learning tools to estimate the severity of matrix effects and predict analyte recovery in inductively coupled plasma optical emission spectrometry, *Talanta*, 223 (2021) 121665.

[79] N.L. da Costa, J.P.B. Ximenez, J.L. Rodrigues, F. Barbosa, R. Barbosa, Characterization of Cabernet Sauvignon wines from California: Determination of origin based on ICP-MS analysis and machine learning techniques, *European Food Research and Technology*, 246 (2020) 1193–1205.

7 Secondary Ion Mass Spectrometry

Paweł Piotr Michałowski
Łukasiewicz Research Network – Institute of Microelectronics and Photonics

CONTENTS

7.1 INTRODUCTION

Secondary ion mass spectrometry (SIMS) is a very precise surface-sensitive analytical technique. A sample is bombarded with a primary ion beam which leads to the sputtering of the matter from its surface. A small part of the sputtered particles are charged (secondary ions). They are collected and undergo spectral analysis which provides information about their mass to charge ratio. A proper interpretation allows

DOI: 10.1201/9781003299295-7

for determining the sample's elemental and/or isotopic composition. Subsequent layers of the sample are removed during the analysis and thus, it is possible to determine how the composition changes as a function of depth, creating so-called depth profiles. The lateral analysis of the signal allows the creation of 3D images and cross-section views of the sample.

SIMS is particularly well known for its excellent detection limits ranging from parts per million to parts per billion for most elements and materials [1–5]. In some cases, it can even reach a regime of tens of parts per trillion [6]. Such accuracy is invaluable for the semiconductor industry as it allows to control whether the high purity requirements of the front-end-of-line integration approaches are met. Furthermore, it is used to determine the dopant distribution which is critical for the optimization of the CMOS technology and thus, SIMS is considered to be the most important metrology instrument in the semiconductor industry.

The same cannot be said about lithium-ion batteries (LIBs) technology. While there are many examples of successful application of SIMS to LIBs characterization, the instrument is not considered to be a must-have by LIBs manufacturers. It may seem very surprising, as the aforementioned best detection limit was reported for the lithium element. Even though it was registered for a HgCdTe matrix, it is usually only slightly worse for other materials, and much better than for most other elements. Thus, one could have expected that SIMS should have played a more important role in LIBs characterization.

Thus, the aim of this chapter is to explain this paradox. Section 7.2 will provide a very brief introduction to the principles of the SIMS technique. While a comprehensive description can be found elsewhere [7–13], the section will focus on those aspects that are the most relevant for LIBs characterization and may not be valid for other materials. Most of the basic information like primary ions generation, working principles of detectors, and electronic lenses will be omitted.

Section 7.3 will be devoted to the presentation of the challenges and pitfalls of SIMS characterization of LIBs and should be considered the most important part of this chapter for scientists who wish to start applying SIMS for LIBs analysis. Information provided here will help to optimize SIMS measurement procedures and understand why some already published experiments are difficult to reproduce. Furthermore, the presentation of possible measurement artifacts should develop a more critical approach and help differentiate which SIMS results are realistic and which may contain measurement artifacts.

Eventually, the most important examples will be discussed. They will be divided into three categories: lithium distribution analysis, electrode materials characterization, and formation of solid electrolyte interface (SEI) – Sections 7.4–7.6, respectively. It should be noted that these categories are not fully separable. For example, a formation of SEI usually contains some conclusions about both lithium distribution and changes that occurred at the electrodes. However, most articles focus on one of these aspects and thus, this division will increase the clarity of the presented information.

7.2 PRINCIPLES OF THE TECHNIQUE

7.2.1 Basic Phenomena

SIMS is a very precise analytical technique that can provide information about the elemental/isotopic composition of the sample. The basic operation involves a bombardment of the surface of the sample with ions with energy usually in the range of 0.1–30 keV. These primary ions collide with atoms of the sample and transfer part of their energy. If it is higher than the binding energy, then those atoms are also set in motion and further collide with other atoms in the sample, giving rise to a phenomenon called collision cascade. A part of the energy of primary ions is transferred back to the surface region enabling atoms and molecules to be emitted from the sample, provided that the transferred energy is higher than the surface binding energy. Most of the emitted species are neutral; only a small fraction is positively or negatively charged – and they are called secondary ions. The mass spectral analysis of those secondary ions provides information about the composition of the sample surface. During a prolonged bombardment, the material is sputtered and a formation of a crater occurs. It is, therefore, possible to monitor how the composition changes as a function of depth, thus obtaining so-called depth profiles.

Figure 7.1 schematically presents an interaction of the primary beam (marker 1) with a sample. The ion bombardment leads to the collision cascade (marker 2) – the incident ions transfer part of their energy to atoms of the sample and set them in motion. They further collide with other atoms and as long as the transferred energy is higher than the binding energy, new atoms are knocked out from their positions. Most of the primary ions are eventually stopped and trapped in the sample. An obvious

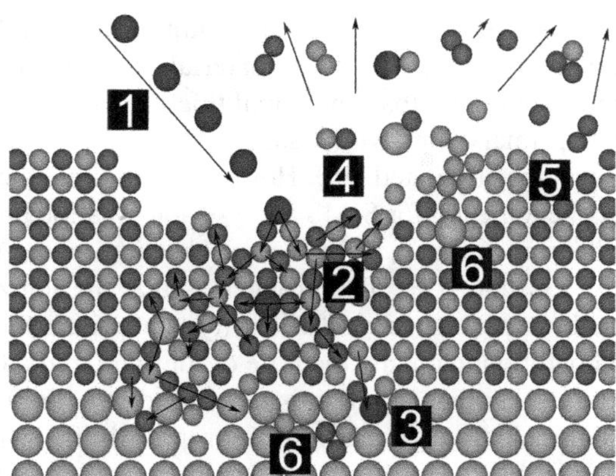

FIGURE 7.1 The principles of the SIMS technique. The primary beam (1) interacts with the atoms from a sample, generating a collision cascade (2). The results of the bombardment are primary ions implantation (3), sputtering and ionization (4), preferential sputtering (5), and mixing (6).

consequence is the change in the chemical composition of the sample as the primary ion species are introduced and mixed with the sample (marker 3). Furthermore, the addition of the primary ions leads to the swelling of the sample as other atoms have to move to make space for incoming ions.

A part of the energy transferred during the collision cascade is brought back to the surface of the sample and atoms and bigger clusters are emitted from the sample – the phenomenon is called sputtering (marker 4). The sputtering process is quantitatively described by the parameter called partial sputter yield, which is defined as the number of sputtered species A per incident ion. This parameter strongly depends on both the bombardment conditions (type, energy, and incident angle of the primary ion, etc.) and the properties of the sample (chemical composition, topography, etc.). The probability of sputtering of one species can be significantly different from the other and thus, a preferential sputtering may occur – a higher probability to sputter one element when compared to the other results in the depletion of the first element in the sample (marker 5).

The direct result of the primary ion bombardment and the collision cascade is the mixing of the sample. Some of the knocked-out atoms and primary ions have sufficiently high energy to penetrate the sample for distances as long as several tens of nanometers. This can lead to a significant alteration of the initial composition of the sample (marker 6). This effect is strongly visible if the sample consists of several thin, initially well-defined layers. The atoms from the upper layers penetrate the deeper layers knocking off other atoms which in turn may travel back to the surface of the sample. The effect is strongly asymmetrical, i.e., more atoms are pushed into a deeper region than brought back toward the surface. This makes it difficult to interpret properly where there is the beginning and the end of a specific layer since the atoms can be detected at the depth where initially they were not present. The penetration depth depends on the experimental condition, especially on the primary ion energy and the type of substrate. For soft materials, the penetration depth can be a few orders of magnitude higher than the initial thickness of the layer.

The interaction of primary ions with a sample surface is in general well understood [14] and can even be simulated [15]. However, a credible theory that can predict another crucial phenomenon for SIMS analysis – ionization probability – is still lacking. As it has been mentioned, only a small fraction of the sputtered species is ionized (typically less than 1%). The problem is that the ionization probability depends not only on the experimental conditions (type of primary ion, impact energy and angle, vacuum condition, etc.) and the type of the sputtered atom but also on its surroundings, i.e. the matrix that the atom is embedded into. For example, a silicon atom may have a significantly different ionization probability when it is sputtered from a silicon substrate than when it is embedded in a gallium nitride matrix. The difference can be up to several orders of magnitude and thus, a direct interpretation of the SIMS result is often not possible. This phenomenon is called a matrix effect. It should be emphasized that a type of matrix may change the sputtering probability as well but the differences are usually much smaller than for the ionization probability.

Thus, when all phenomena are taken into consideration the equation describing the intensity of the registered SIMS signals is given as follows:

$$I_A = I_p C(A) Y(A)\, \alpha(A) \delta, \tag{7.1}$$

where I_A and I_p are secondary and primary ion currents, respectively; C(A) is the concentration of element A; Y(A) is a partial sputter yield (number of sputtered species A per incident primary ion); α(A) is the ionization probability; and δ is the transmission and detection coefficient of the instrument.

Both ion currents are measured, and to calculate the concentration of element A one has to determine partial sputter yield, ionization probability, and the transmission and detection coefficient. However, all three depend heavily on the type of the composition of a sample and the experimental conditions. Thus, direct quantification of SIMS results is not possible and external reference samples are needed for quantification. It should be, however, emphasized that even with such samples the quantification is still a very complex process and prone to errors. There are two widely accepted quantification procedures:

- It has been shown that when cesium ions are used as a primary source it is possible to register secondary ions as CsX+ species. For some simpler compounds, this mode is often considered to be self-quantitative, i.e. the intensity ratio of two signals is proportional to the concentration ratio of these two elements [16–19].
- For dopant/contaminant analysis (i.e. atoms with low concentration, typically much less than 1%) it can be approximated that the presence of these low-concentration atoms does not change the properties of the matrix [20–22]. Thus, the concentration of these atoms can be calculated as:

$$C(A) = \frac{I_A}{I_M} \mathrm{RSF}\left(\frac{A}{M}\right), \tag{7.2}$$

where C(A) is the concentration of element A; I_A and I_M are signal intensities of element A and some matrix element, respectively; $\mathrm{RSF}(A / M)$ is a relative sensitivity factor for these two signals. It can be determined from a sample with a known dopant distribution. However, it should be noted that proper quantification requires a reference sample with the same composition, i.e. a $Si_{0.9}Ge_{0.1}$ sample with a known concentration of boron should not be used as a reference sample for $Si_{0.8}Ge_{0.2}$ sample with unknown B distribution. In some cases, the error introduced when a similar sample is used may be relatively small but in others, there can be even an order of magnitude difference.

Unfortunately, both of these procedures rarely can be applied to materials used for LIBs technology, as some of them are too complex (e.g. $LiNiMnCoO_2$) and even when a simpler material is used (e.g. graphite) the concentration of Li by far exceeds the threshold where RSF approach can be applied.

It can be therefore concluded that while SIMS quantification is generally a complicated process, it may be completely impossible for LIBs.

It is important to mention that SIMS is a destructive technique as it requires the sputtering process to occur. However, two different modes are usually distinguished: static and dynamic SIMS. The former is sometimes called to be semi-nondestructive as it limits the total fluence of the primary ions to 10^{12}–10^{13} ions/cm^2, which means that only about 0.1%–1% of the sample surface has been bombarded and the alteration of its composition can be neglected. This mode is therefore limited to surface analysis. For the dynamic SIMS subsequent layers of a sample are eroded and thus, it is possible to create depth profiles of a sample and analyze how the composition is changing as a function of depth.

7.2.2 An Overview of Different Instruments

SIMS instruments operate under high vacuum conditions (usually in a range of 10^{-6}–10^{-11} mbar) and consist of a primary ion source; primary ion column where the beam is accelerated and focused; analysis chamber, where samples are introduced via a load-lock; extraction lenses that collect secondary ions; secondary ion column, where the secondary beam is focused; mass analyzer, where ions are separated based on their mass to charge ratio; and detectors that count the intensity of the registered signals.

While the working principles of most of these systems are described elsewhere [7–13], it is important to emphasize the differences between various mass analyzers. The most commonly used for LIBs characterization is the Time-of-Flight (ToF) analyzer. Its working principle is based on a measurement of the time that ions need to travel the distance between the sample and the detector. ToF-SIMS instruments use very short pulses of the primary ion beam and thus, it can be approximated that all secondary ions are emitted from the sample surface at the same time. They are collected and accelerated to common kinetic energy in a constant electrical field. Formulas for the velocity and the kinetic energy of an ion are given by:

$$v = \frac{L}{t}, \tag{7.3}$$

$$qU = \frac{mv^2}{2}, \tag{7.4}$$

where v is the velocity, L is the distance between the sample surface and the detector, t is the time needed to travel this distance, q is the charge of the ion, U is the accelerating voltage, and m is the mass of the ion. Thus, a final formula that describes ions separation can be derived as

$$\frac{m}{q} = \frac{2Ut^2}{L^2}. \tag{7.5}$$

It can be immediately noted that mass analyzers (including different types of analyzers) do not determine the mass of ions but rather their mass-to-charge ratio. However, in most SIMS experiments multiple ionized signals are negligible.

The other two most common mass analyzers are called magnetic sector and quadrupole. The former applies a magnetic field that deflects the primary beam and only ions with a specific mass to charge ratio can reach the exit slit. In the latter direct and alternating currents are applied to four parallel cylindrical metal rods and thus, only a resonant ion with a specific mass to charge ratio may pass through the exit slit. In both cases, only one ion can be measured at the same time and if more signals are required, it is necessary to cycle through them.

On the other hand, the ToF analyzer can measure all different masses at the same time and this feature makes it most applicable to the LIBs technology as it enables registering all different signals coming from organic and inorganic materials at the same time. ToF analyzers are less sensitive, especially when compared to magnetic-sector SIMS but this disadvantage is particularly important in the semiconductor industry. Detection limits in the range of 10^{16}–10^{18} atoms/cm^3 offered by ToF instruments are usually more than enough for LIBs characterization.

It should also be noted that since ToF analyzers operate on very short ion pulses they fulfill the static SIMS regime, i.e. are ultra-surface-sensitive which is valuable for LIBs. Depth profiling is also possible because most ToF-SIMS instruments have additional primary columns which can generate so-called sputtering beams which are characterized by high fluence. It is possible to operate in a dual-beam mode where primary ions are alternating between analyzing and sputtering beams: the former is used to register secondary ion signals while the latter is used to remove subsequent layers of the sample.

It is also essential to briefly introduce recent advances in SIMS instrumentation, namely the addition of the ToF-SIMS analyzer to the focused ion beam/scanning electron microscope (FIB/SEM) [23]. As the name suggests, it is a direct implementation of the ToF-SIMS module into FIB/SEM equipment. Electrons are used to obtain the topography of a sample surface (standard SEM operation), whereas the FIB is simultaneously used to etch the sample and as a primary ion source for ToF-SIMS analysis. In this way, fully 3D pictures of a sample may be obtained with additional information about the chemical composition. Such measurements typically have much better lateral resolution than standard ToF-SIMS analysis and better sensitivity than energy-dispersive X-ray spectroscopy (EDX) which is commonly used in SEM instruments. Furthermore, it is possible to detect light elements (starting with hydrogen), which is not accessible for EDX. The complementary use of ToF-SIMS and FIB/SEM techniques is being intensively developed but there are still some major issues with quantification and proper identification of heavy ions. However, lithium analysis is one of the most prominent examples of this improvement and thus this advancement is very useful for LIBs technology.

7.3 CHALLENGES AND PITFALLS OF SIMS CHARACTERIZATION OF LIBs

It is important to emphasize that the SIMS technique is well known for its numerous measurement artifacts. While there are certain ways to mitigate them, they are seldom discussed in recent articles, particularly concerning LIBs technology. Thus, the interpretation of SIMS results is never straightforward, and it is relatively easy to make a mistake.

7.3.1 Matrix Effect

The matrix effect, as it has been already discussed, is related to the fact that the partial sputter yield and particularly the ionization probability of element A depends strongly on its surroundings, i.e. the matrix that the atom is embedded into. Thus, a proper concentration calibration is often not possible. It should be, however, emphasized that a straightforward qualitative interpretation may also be misleading. For example, a lithium signal may have higher intensity in one region of a sample but it does not necessarily mean that its concentration is bigger than in other regions. As mentioned, the ionization probability can be several orders of magnitude different for various matrices.

A particularly problematic moment during depth profiling is an interface between two different layers. Therefore, the ionization probability may abruptly change and a very intense increase/decrease of signal intensity may be registered. This can be falsely interpreted as agglomeration/depletion of the measured species at the interface.

The matrix effect can even affect the isotope analysis and the signals' ratio of various isotopes measured in one matrix is not necessarily the same in another. This can be particularly problematic for LIBs technology as isotopic analysis of ^{6}Li and ^{7}Li distributions is an important issue.

7.3.2 Sputtering Rate

Proper depth calibration is also a non-trivial task. The easiest solution is to measure the depth of the crater and calculate the sputtering rate which enables a direct depth calibration. However, such an approach can only be applied for uniform samples with only one material type. Different materials are sputtered at different rates and thus, only the average sputtering rate can be determined.

Once again, a straightforward interpretation can be misleading, because even if it takes considerably more time to sputter one layer than another, it does not necessarily mean that this layer is thicker. It may simply be more resistant to the sputtering process.

Thus, for multilayer structures, it is necessary to perform several experiments on reference samples and determine the sputtering rate for each material independently. Such an approach is time-consuming and requires the acquisition of well-defined reference samples, which is not always possible, particularly for LIBs technology where there are many layers with varying compositions.

7.3.3 Mass Spectrum Analysis

The mass analyzer separates the secondary ions based on their mass to charge ratio and thus, a mass spectrum can be obtained during the experiment. The registered peaks, however, have some finite widths because no analyzer can perform the separation perfectly. Thus mass interferences may occur, especially because during the SIMS experiments both monoatomic and polyatomic ions are registered. It is relatively common that a nominal mass of a polyatomic ion is the same as the mass of a monoatomic ion (e.g. $^{6}Li_2$ and ^{12}C or ^{31}P $^{12}C_2$ and ^{55}Mn). However, the actual mass of these species is different by a fraction of a unit and in most cases, it is possible to resolve these peaks in a mass spectrum. A parameter called mass resolving power (MRP) is used to evaluate whether two interfering ions can be resolved and it is defined as the ratio of the nominal mass and the actual mass difference.

It should be noted that mass interferences are usually not problematic when LIBs samples are analyzed. When various important elements, namely H, Li, C, N, O, F, Al, Si, P, Ti, Mn, Fe, Co, and Ni, are considered, there are almost 3,000 possible mass interferences. However, only 42 require MRP higher than 3,000 which is not a problem for a vast majority of spectrometers. Most ToF analyzers provide MRP ~10,000 and in such a case only ^{58}Fe and ^{58}Ni cannot be resolved as they require MRP of about 28,000. This is not, however, a problem because a natural abundance of ^{58}Fe is about 0.28% and it is more natural to use a ^{56}Fe signal to monitor iron distribution.

There is, however, a more important issue related to the mass spectrum during the analysis of LIBs samples, namely peak identification. It should be noted that ToF-SIMS usually generate an enormous amount of data: secondary ions are registered on a position-sensitive detector and for each pixel, a complete mass spectrum is registered separately. If 128×128 pixel density is used, then 16,384 mass spectra are registered, each with an accuracy of about 0.0001 u. During the depth profiling, this number is further multiplied by the number of data points which is typically several hundreds or even thousands. During the peak identification procedure, an integrated mass spectrum is presented and the user manually assigns the most important peaks. Thus, it is very easy to overlook some minor peaks which only appear at one specific location within a sample, and they could have provided important information about the properties of the tested material.

7.3.4 Mixing Effect

During the SIMS analysis, a considerable amount of energy is transferred from primary ions to atoms from a sample. As it has been already mentioned, it gives rise to a phenomenon called collision cascade and causes the mixing effect. As a consequence, very sharp interfaces are usually not observed during the SIMS analysis and it is not easy to determine whether these blurred signals are caused by an actual intermixing of layers or just measurement artifacts.

A suitable but time-consuming solution is to perform several measurements with varying primary ion energy. The mixing effect is proportional to the primary ion energy and thus, if the decay length of a signal decreases for lower energies, the intermixing of layers can be excluded. However, if it remains unchanged, it confirms it.

It should also be noted that the mixing effect may cause the formation of bonds between various atoms which were not present before the SIMS analysis. For example, the detection of LiO or Li_2O signals does not necessarily confirm the presence of lithium oxide. It may happen that both types of atoms were in the vicinity before the

SIMS experiment and the formation of a complex ion has been caused by the mixing effect. Thus, any conclusions about the chemical state of the tested samples should be verified by other techniques like X-ray photoelectron spectroscopy (XPS).

7.3.5 Lithium Mobility

Lithium is a very light and mobile element and it can cause a serious problem during the SIMS analysis as Coulomb interaction with primary ions may induce its movement. It should be noted that most primary ions used in SIMS spectrometers are cations (Bi^+, Cs^+, O_2^+, Ga^+, C_{60}^+, etc.) and thus lithium ions are repelled and may propagate into deeper regions of a sample.

Holzer et al. analyzed the migration artifact of alkali metals in the SiO_2 matrix and concluded that an oxygen cluster ion beam is a suitable solution to the problem [24]. The cluster contained about 1,300 molecules and was single charged. Thus, the amount of charge transferred to the sample surface was considerably lower while the erosion rate was similar to a standard O_2^+ primary ions bombardment. In this way, the migration of alkali metals was suppressed.

This approach, however, is not used for the SIMS characterization of LIBs samples. On the other hand, the tested materials are not insulating as SiO_2 and the excess charges at the surface will be quickly dissipated. Nevertheless, the migration effect may still be present and the long decay length of lithium-related signals may be caused by either diffusion of this element before the SIMS experiment or a measurement artifact. It should be emphasized that the effect is typically not considered in LIBs literature and thus, a reader should avoid straightforward interpretation of lithium depth profiles.

7.3.6 Non-Planar Surface

In the semiconductor industry, SIMS experiments are usually performed on flat surfaces and many scientists treat it as a prerequisite for reliable SIMS analysis. However, this condition is typically not fulfilled for LIBs technology, especially for SEI which is porous, with complex morphology and composed of various materials. All the aforementioned problems can be therefore amplified:

- As mentioned the matrix effect is particularly intensive at the interface. Given a complex composition of an SEI one may expect that the ionization probability may fluctuate a lot during the analysis of this region. Furthermore, it is not possible to prepare any reference samples to quantify the SIMS signals.
- The sputtering of a porous material is a very complex process and it is not possible to predict how pores may affect the sputtering rate. In theory, high porosity may cause the material to be less resistant to the etching but on the other hand, large void spaces may impede the collision cascade. Thus, a reader should not immediately accept any conclusion about the thickness of SEI which is based on the total sputtering time.

- For porous materials secondary ions can be ejected from various depths which in theory may influence the mass resolution of the ToF analyzer. However, the distance between the sample and the detector is usually bigger than 1 m so a difference of tens of nanometers will not play a significant role. It should be, however, noted that mass spectra may be additionally convoluted as they may contain signals from varying depths; thus, a proper interpretation will be even more difficult.
- The porosity of the material may significantly influence the amplitude of the mixing effect. However, similar to the sputtering rate, it is difficult to predict whether the mixing effect will be enhanced or diminished. Even worse is the presence of various compounds in close vicinity. For example, at the interface between lithium fluoride and polyethylene oxide strong LiO and Li_2O signals may be detected. However, the conclusion about the presence of lithium oxide in this region would be unfounded. Similarly, lithium migration artifacts may be enhanced or diminished as well.

A complex morphology and composition of the SEI cause an additional problem with reproducibility. In the semiconductor industry, it is customary to perform several measurements in several spots of a sample. This is not feasible for LIBs technology, as the composition of a sample in various places may be significantly different.

7.3.7 Battery Sample Extraction and Transfer

Last but not least, it should be noted that in many cases, lithium batteries have to be dismounted before the SIMS analysis. Most articles describe the procedure and emphasize that it has been performed in a glove box in an inert atmosphere. However, samples have to be transferred from a glove box to the SIMS instrument and in most cases, samples are exposed to environmental factors during the process. This fact is seldom reported and its influence on the samples is not fully investigated. Thus, it may happen that the SIMS analysis is performed on a sample that has already undergone significant changes, particularly oxidation, and the results are not representative.

7.4 LITHIUM DISTRIBUTION ANALYSIS

The SIMS technique is particularly useful for the analysis of lithium distribution because, unlike many other techniques, there are no limitations in the detection of light elements. On the contrary, the detection limits of lithium are one of the best that SIMS has to offer. For this reason, it may be slightly surprising that the number of articles that focus on lithium distribution analysis is not very high. This can be explained by the number of possible measurement artifacts, particularly lithium mobility, that have been previously discussed. It is also important to emphasize that for a vast majority of articles that do focus on this aspect the authors use some non-trivial approaches, namely using isotopically labeled electrolytes [25–27], ToF-SIMS FIB/SEM multimodal microscopy [28,29], and *operando* measurements [30,31]. It illustrates how challenging can it be to obtain reliable information on lithium distribution.

7.4.1 Isotopically Labeled Materials

Lu and Harris use isotopically labeled $^6LiBF_4$ electrolyte to study lithium transport within a $LiClO_4$ SEI grown on copper [25,26]. They show that the penetration depth of the boron profile is about 4–5 nm and is the same for samples immersed in the electrolyte for 30 seconds, 3 minutes, and 15 minutes. Thus, they conclude that the top part of SEI is porous. As shown in Figure 7.2a at the same depth a broad peak of the Li_2O^+ signal is detected. The authors conclude that the presence of densely packed Li_2O or Li_2CO_3 limits the penetration of the electrolyte. At the same time (Figure 7.2b) they observe a much deeper penetration of lithium ions from the isotopically labeled electrolyte. In theory, Li^+ ions can penetrate the SEI through pores or grain boundaries but that would require BF_4^- ions to move along to ensure the charge balance. Since the distribution of B is limited to 5 nm the authors conclude that further Li^+ transport is via ion exchange.

Berthault et al. show a similar approach but focus on the dynamics of the $^6Li/^7Li$ exchange [27]. In the beginning, they form an SEI on a graphite electrode using electrolyte and lithium counter electrodes both in natural lithium abundance (92.7% of ^7Li and 7.3% of ^6Li). Then they perform experiments with materials enriched in ^6Li (up to 95%) and study the dynamics of Li isotopic exchange in the self-diffusion conditions (by dipping in the electrolyte) and galvanostatic cycling. They observe that in both cases all ^7Li ions in the SEI have been exchanged with ^6Li and thus conclude that the SEI acts as an ion-exchange resin. They notice that the process is more than 50 times slower for the cycling experiment than for the free diffusion. This is not found surprising since in these conditions ion exchange is governed not only by diffusion but by ion migration as well. The movement of both cations and anions is

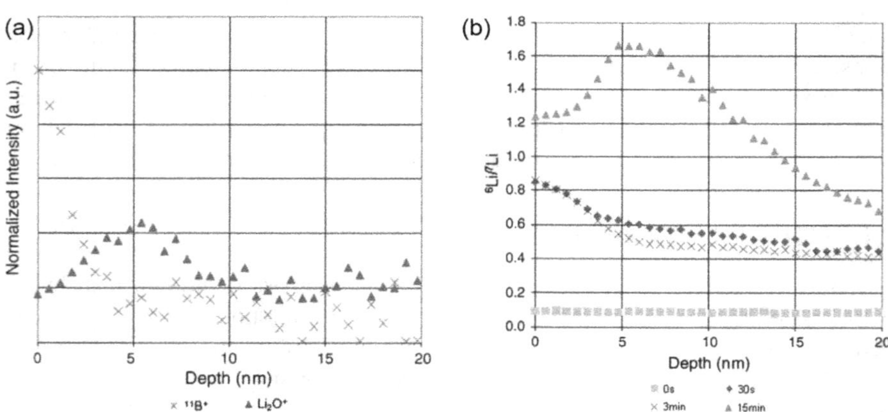

FIGURE 7.2 SIMS profiles of $^7LiClO_4$ SEI immersed in $^6LiBF_4$ electrolyte. A broad peak of the Li_2O^+ signal suggests the formation of a densely packed Li_2O or Li_2CO_3 layer which limits the diffusion of boron to 4–5 nm (a). Lithium penetrates a much bigger depth, which suggests an ion-exchange mechanism (b). (Reprinted from Lu, P.; Harris, S. J., 2011. Lithium transport within the solid electrolyte interphase, *Electrochem. Commun.*, 13 (10), 1035–1037, with permission from Elsevier.)

influenced by the electrical field, and during charging/discharging cycles the impact is the opposite.

It is also worth mentioning that the authors realize that the isotope analysis performed on a freshly formed SEI does not match the natural abundance of ^{6}Li and ^{7}Li isotopes even though it is possible for the lithium electrode. As mentioned, a change in the matrix may even influence the ionization probability of different isotopes of the same element. This visualizes the aforementioned problem of proper calibration of SIMS signals.

7.4.2 ToF-SIMS FIB/SEM Multimodal Microscopy

Sui et al. present a very interesting study of lithium distribution on a cathode material [28]. As presented in Figure 7.3 ToF-SIMS elemental mapping reveals the presence of Li-rich spots. The authors do not formulate a straightforward conclusion as they know that the surface morphology, particularly pores and grain boundaries, may influence the secondary ion yield. Indeed, a correlative comparison of secondary electron images and elemental mapping shows that many Li-rich spots are located at grain boundaries. However, similar spots are also observed for fully charged samples and thus, the authors conclude that not all of these spots can be attributed to measurement artifacts. They are Li trapping sites and the ability to monitor their formation will undoubtedly enable further optimization of cathode materials.

Bessette et al. use the ToF-SIMS FIB/SEM multimodal microscopy as well and intend to establish calibration curves needed for Li distribution quantification [29]. The authors identify many SIMS-related artifacts and problems (inability to apply the relative sensitivity factor approach, matrix effect, edge effect, etc.) and propose a calibration procedure that is based solely on the lithium signal intensity. The results are convincing and many shortcomings of the approach have been identified, for example, the necessity to establish a new calibration curve for different primary ions. However, the article lacks reproducibility tests, and since even small changes in the experimental conditions (vacuum, aged ion source) may significantly influence the outcome the procedure should not be seen as fully completed. The ability to quantify lithium concentration is without any doubt a major achievement but the reproducibility of the method remains unconfirmed.

7.4.3 Operando Measurements

The recent development of the ToF-SIMS instrumentation enables performing *operando* measurements of all-solid-state lithium batteries. This enables a unique ability to obtain lithium distribution maps in the initial, charged and discharge states from the same region. Masuda et al. have used it to visualize the decrease/increase of the lithium concentration during cycling [30]. The cell was composed of $LiCoPO_4$ (LCP), $Li_{1+x}Al_xTi_{2-x}(PO_4)_3$ (LATP), and Pd. For the cathode, LCP, LAPT, and Pd act as active material, solid electrolyte, and conductive additive, respectively. As shown in Figure 7.4 a clear decrease/increase in lithium concentration can be observed for LCP particles during charging/discharging. Lithium removal from LAPT particles after the

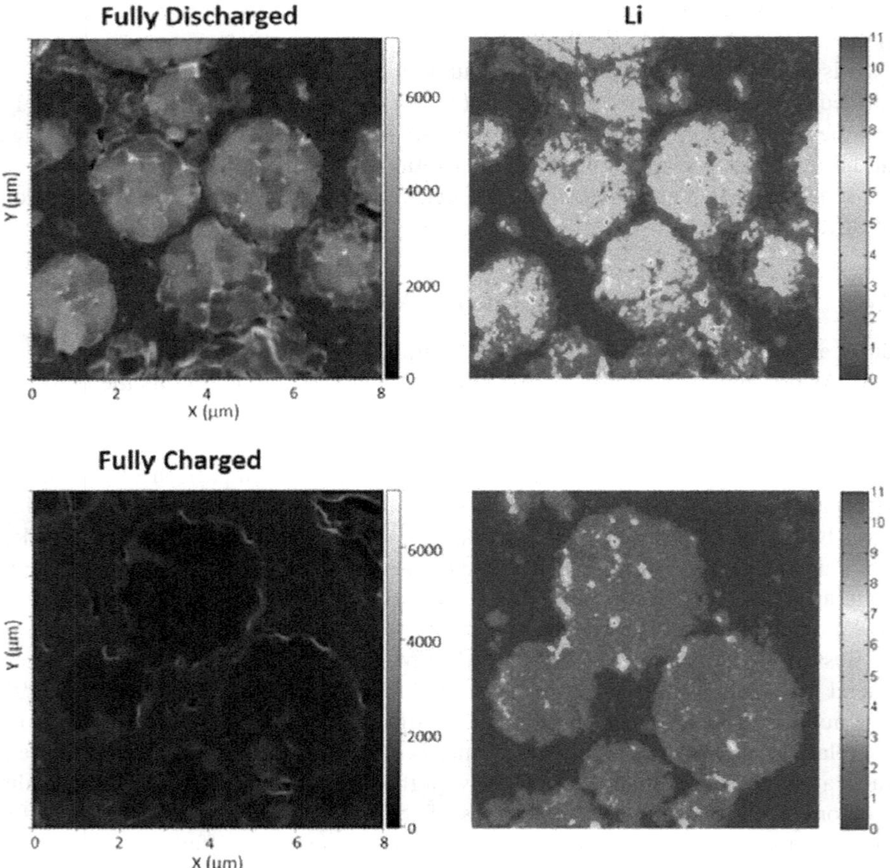

FIGURE 7.3 Structural (FIB/SEM) and chemical mapping (ToF-SIMS) performed on cathode material in fully discharged and charged states. (Reprinted from Sui, T.; Song, B.; Dluhos, J.; Lu, L.; Korsunsky, A. M., 2015. Nanoscale chemical mapping of Li-ion battery cathode material by FIB-SEM and TOF-SIMS multi-modal microscopy, *Nano Energy*, 17, 254–260, with permission from Elsevier.)

first charging is also observed but the authors have realized that more research is required before formulating any definite conclusions.

Yamagishi et al. have used a similar approach and observed the evolution of lithium distribution of a pristine material and after charging and discharging [31]. They also study PO_x^- and SO_x^- signals and visualize the degradation of the electrode during the cycling. Further development of *operando* ToF-SIMS measurements will certainly contribute to the development and optimization of all-solid-state lithium batteries.

7.5 ELECTRODE MATERIALS CHARACTERIZATION

SIMS is commonly used for the characterization of electrodes used in LIBs technology. Naturally, lithium-related signals are also monitored but the focus is shifted toward the electrodes and potential changes that affect the materials they are composed of. There are numerous articles that study the composition of electrodes [32–38], various coatings that are used to enhance their performance [39–44], and degradation mechanism [45–49]. It is important to emphasize that many of these works do not even require an assembly of a full cell and thus, the problem with sample transfer is omitted.

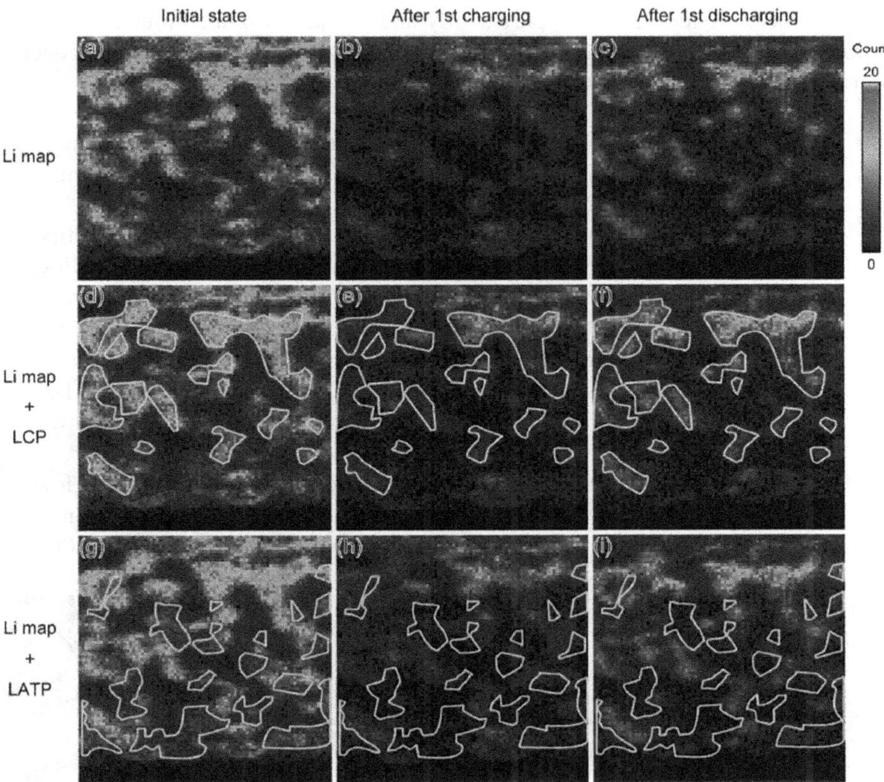

FIGURE 7.4 Lithium distribution maps of the same area (*operando* measurements) in the initial state and after the first charging and discharging (a–c). Panels d–f and g–i show outlines of LCP and LATP particles, respectively. (Reprinted from Masuda, H.; Ishida, N.; Ogata, Y.; Ito, D.; Fujita, D., 2018. In situ visualization of Li concentration in all-solid-state lithium ion batteries using time-of-flight secondary ion mass spectrometry, *J. Power Sources*, 400, 527–532, with permission from Elsevier.)

7.5.1 The Composition of Electrodes

There are many examples of the SIMS technique used for the compositional characterization of electrodes, usually in the form of depth profiles. There are examples of compounds that may potentially act as electrodes like V_2O_5 [32,33], Cr_2O_3 [34], Sn-Co [35], and Sn-Ni alloys [36], and commercially used materials like lithium manganese oxide (LMO)/lithium nickel manganese cobalt oxide (NCM) [37] or metallic lithium [38].

It should be noted that most of these SIMS experiments are qualitative and the results, while certainly very useful, are not easy to interpret. For example, Li et al. compare depth profiles of Sn-Ni electrodes in the initial state and after the first discharging process [36]. As shown in Figure 7.5 the difference is visible but there are numerous problems with these results. The authors claim that the initial Si-Ni alloy has been divided into two parts, fully and partially lithiated (outer and inner regions, respectively). There are, however, many doubts about whether this conclusion is valid, some of them are as follows:

- The reproducibility of measurements is poor, as evidenced by the intensity of the Cu signal coming from the substrate – it is almost one order of magnitude lower for the pristine sample.
- The author claims that a moderate intensity of Cu signal in the pristine sample suggests the presence of pinhole defects and/or cavities in the alloy. In many cases that would be a reasonable assumption but the layer is about 3 μm and the incident angle of the primary ions is about 45°; thus, the extraction of Cu secondary ions from the substrate is highly unlikely.
- Furthermore, the intensity of the Cu signal is very low for the region marked as "lithiated alloy" (discharged sample), and the authors do not acknowledge nor discuss it.
- Sn and Ni signals in the "lithiated alloy" alloy are initially of very low intensity and only later start to increase. The authors also analyze the Sn to Li signals ratio (in such a case a plateau can be observed up to about 2,000

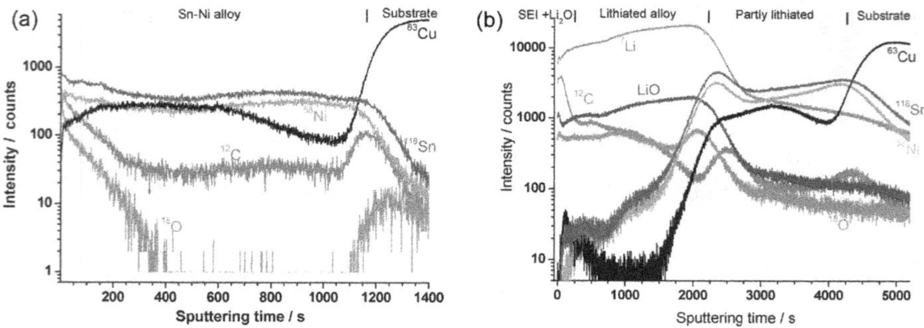

FIGURE 7.5 Depth profile of pristine (a) and discharged (b) Si-Ni alloy electrode. (Reprinted from Li, J. T. et al., 2011. XPS and ToF-SIMS study of electrode processes on Sn-Ni alloy anodes for Li-ion batteries, *J. Phys. Chem. C*, 115 (14), 7012–7018, with permission from American Chemical Society.)

seconds of the sputtering time). Nevertheless, it seems to be unsubstantiated to treat this region as a Si-Ni alloy.

- The LiO signal seems to be perfectly parallel to the Li signal, whereas the O signal is modulating which should somehow affect the shape of the LiO signal. It should be noted that in depth profiling mode it is not possible to distinguish oxygen coming from various lithium oxides and other components (e.g. Ni_xO_y, Sn_xO_y, Cu_xO_y, and organic materials). Thus, the shape of the oxygen signal should still affect the LiO signal.
- The decay length of the Li signal is very long and will most probably expand deep into the substrate region (probably caused by electrostatic repulsion during the SIMS experiment).

This example emphasizes the aforementioned possibilities of various measurement artifacts during the analysis of materials used in LIBs technology. It can be therefore concluded that it is not clear whether the interpretation provided by the authors is valid.

Another problem is that these conclusions are based on a few selected signals that have been used in depth profiling mode. As mentioned, during ToF-SIMS experiments a full mass spectrum is registered for each data point, and in some cases, it is better to perform a comparative analysis of these spectra, not just a few signals. This is exactly what Yabuuchi et al. have performed during the analysis of LMO/NCM electrodes [37]. Only the mass spectrum obtained for discharged electrodes contains very strong Li_2O^+ and LiC_2^+ peaks and thus, the authors have concluded that Li_2CO_3 is formed at the surface of the electrode material. Perhaps the same conclusion may be valid for the work of Li et al.

As mentioned before, the compositional analysis of electrodes in some cases can be performed even without the assembly of a full cell. The work of Otto et al. is a prime example [38]. The authors have focused on a lithium electrode and used the ToF-SIMS technique to analyze the passivation layer at its surface. They have shown that the layer is several nanometers thick and its outer region is predominantly composed of LiOH and Li_2CO_3, while at the interface with the bulk material the formation of Li_2O has been confirmed. The work is particularly important since the presence of passivation layers at electrodes before the assembly of a full cell is often omitted.

7.5.2 Coatings

A relatively simple application of the SIMS technique in the LIBs technology is the analysis of various coatings applied to the electrode materials. In most cases, the aim is to confirm the successful deposition of coating layers and their homogeneity. In some cases, authors also analyze signals which are related to the degradation of electrodes and show that their intensities are lower for coated samples.

For the organic coatings, Fedorková et al. confirm the deposition of polyethylene glycol and polypyrrole polymer blend coating on particles of $LiFePO_4$ which can be used as a cathode material [39,40], whereas Neudeck et al. use organophosphates to coat NCM particles [45]. The latter work is particularly interesting as the authors have performed a full 3D reconstruction of the ToF-SIMS analysis, as presented in Figure 7.6.

FIGURE 7.6 ToF-SIMS 3D reconstruction (10 × 20 × 20 μm³) of NiO_2^- and combined PO_2^-/PO_3^- signals of NCM particles coated with tris (4 -nitrophenyl) phosphate (left) and tris (trimethylsilyl) phosphate (right). (Reprinted from Neudeck, S. et al., 2018. Molecular surface modification of NCM622 cathode material using organophosphates for improved Li-ion battery full-cells, *ACS Appl. Mater. Interfaces*, 10 (24), 20487–20498, with permission from American Chemical Society.)

For inorganic coatings, ToF-SIMS analyses confirming a successful formation of the Li_3PO_4 layer on NCM particles are discussed by Jo et al. [42] and Lee et al. [44]. Hong et al. coat the same material with a thin layer of $[B, Al]_2O_3$ [41]. The work is notable because it is one of the very rare examples where magnetic-sector SIMS has been used to characterize LIBs samples. Figure 7.7 shows the distribution of B, Al, Mn, and Ni on the surface (i) and the cross-section (ii) of the coated particles. Visbal et al. show the deposition of diamond-like carbon coating and confirms with the ToF-SIMS technique that its presence prevents the interfacial reaction [43].

7.5.3 Degradation Analysis

SIMS technique can be used to monitor the degradation process of the LIB samples. It can be achieved either by monitoring a redistribution of elements that the electrodes are composed of or by analyzing the byproducts of reactions that occur during cycling.

Börner et al. present a correlative application of several analytical techniques to study the degradation mechanism of NCM electrodes [46]. Notably, SEM has been used to confirm the cracking of aged NCM particles, whereas ToF-SIMS has been employed to study transition metal dissolution. The authors acknowledge that the topography of the sample has an impact on the intensity of various signals but this effect is similar for all registered elements. As shown in Figure 7.8 the distribution

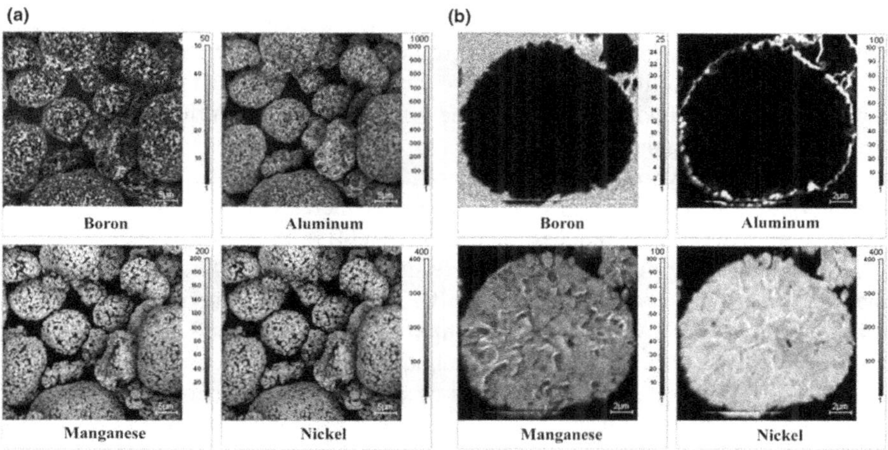

FIGURE 7.7 Elemental distribution maps of $[B, Al]_2O_3$-coated NCM particles on the surface (a) and the cross-section (b). (Reprinted from Hong, T. E. et al., 2012. Nano SIMS characterization of boron- and aluminum-coated $LiNi_{1/3}Co_{1/3}Mn_{1/3}O_2$ cathode materials for lithium secondary ion batteries, *J. Appl. Electrochem.*, 42 (1), 41–46, with permission from Springer Nature.)

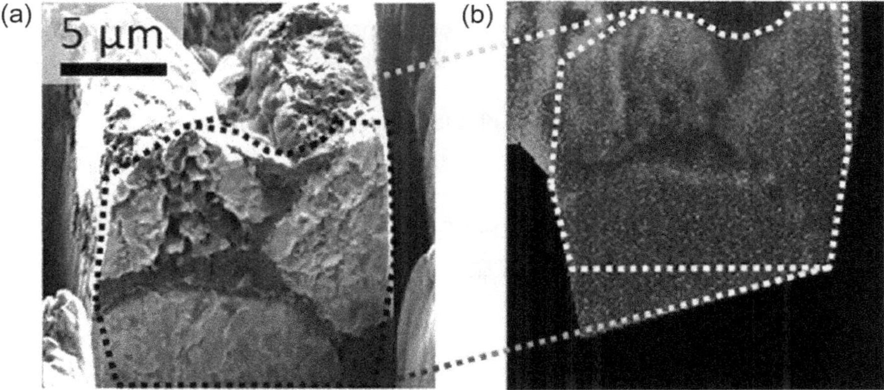

FIGURE 7.8 Elemental distribution maps of a cracked NMC particle. (a) secondary electron image, (b) colored overlay of Li (gray), Mn (red), Ni (green), and Co (blue). A clear agglomeration of manganese at the surface of a cracked particle is visible, whereas other elements are uniformly distributed. (Reprinted from Börner, M. et al., 2016. Degradation effects on the surface of commercial $LiNi_{0.5}Co_{0.2}Mn_{0.3}O_2$ electrodes, *J. Power Sources*, 335, 45–55, with permission from Elsevier.)

of lithium, nickel, and cobalt is rather homogenous, whereas the manganese signal is significantly higher on the surface of a cracked particle. This observation confirms a preferential dissolution of manganese and its subsequent re-deposition on the surface of cracked particles.

Gilbert et al. have performed degradation tests on NCM/graphite cells [48]. The authors have used SIMS to obtain depth profiles of graphite electrode and found that after the formation of the cell and intense cycling, Ni and Mn atoms can be found deposited on the cathode. Similarly to the previous study, the effect is particularly pronounced for manganese.

The analysis of byproducts is usually limited to monitoring sulfur, phosphorous, and fluorine-related signals and showing how they increase during the cell operation [48,49] or that the coating reduces their formation [41,42].

As mentioned, a typical ToF-SIMS analysis provides a very large amount of data and it is not feasible to analyze it all manually. Thus, a statistical approach called multivariate analysis (MVA) is increasingly more commonly applied in the processing of ToF-SIMS data. Detailed information about this approach is presented elsewhere [50]. Briefly, MVA is a set of tools that is based on the simultaneous observation of more than one outcome variable. The analysis helps to understand the relationships between variables and is used to represent the distributions of observed data. Among its other advantages, MVA can be used to identify peaks in a mass spectrum, even if only limited knowledge about the sample is available. In this light, the article by Heller et al. is particularly interesting for the degradation studies [47]. The authors use the MVA approach to obtain information about various degradation products occurring at the aged electrodes and determine the layer structure. It is important to emphasize that even with this approach not all unknown peaks from a mass spectrum have been identified, which may help to realize how complex peak identification can be. The authors show that from a group of previously unknown peaks they are able to determine the chemical nature of up to about 76% of 170 major peaks. With this approach, it is also possible to analyze small differences between degradation mechanisms depending on the type of additives (vinylene carbonate, fluoroethylene carbonate, and ethylene sulfite).

7.6 FORMATION OF SOLID ELECTROLYTE INTERFACE

The SIMS analysis focused on the formation of an SEI is a very non-trivial task. SEI is known to be porous, inhomogeneous, and composed of many different compounds that are packed in close vicinity to each other. Thus, all the aforementioned challenges and pitfalls are particularly pronounced during the analysis of SEI. For this reason, in most cases, SIMS measurements are used to obtain some basic qualitative information about SEI, and the results are used to complement other analytical techniques, in most cases XPS.

It does not, however, diminish the importance of the SIMS results. Peled et al. show that the composition of SEI on highly ordered pyrolytic graphite (HOPG) is inhomogeneous, and basal planes contain significantly more organic materials than edge planes [51]. Ota et al. analyze mass spectra of SEI formed on HOPG electrode and shows a significant difference between samples that have been immersed in electrolyte with and without vinylene carbonate additive [52].

The most common application of the SIMS technique for SEI characterization is depth profiling [53–57]. By choosing suitable signals and complementing XPS results many scientists have confirmed the formation of two-layered SEI where the outer and inner layers are predominantly composed of organic and inorganic materials, respectively. Interestingly, Zhou et al. used liquid-SIMS analysis on a sample that still contained the electrolyte [57]. In such measurements, a protective membrane (Si_3N_4 in this case) is used to separate the liquid from the vacuum and the crater size is very small (1–2 μm) so that the surface tension of the liquid holds it in the cell.

Among these experiments, the work of Liu et al. stands out as they used isotopically labeled materials to study the formation of the SEI on a copper electrode [56]. In this way, the authors not only have confirmed the division of SEI into organic/inorganic parts but also shown a bottom-up formation mechanism. In their experiments, the electrolyte has been ^{6}Li-enriched, whereas the counter/reference electrode is pure ^{7}Li metal. During the formation of an SEI on the copper electrode, more and more ^{7}Li is dissolved into the electrolyte. Thus, a decrease in the ^{6}Li/^{7}Li ratio can be directly related to the time scale of the SEI formation. As shown in Figure 7.9 the ratio is the highest at the beginning of the depth profile, which indicates that this part has been formed first. The knowledge about the growth kinetics has allowed the authors to formulate conclusions about the dynamics of one-electron and two-electron reduction reactions which in turn corroborate the division of the SEI into organic/inorganic layers.

Otto et al. present a very interesting approach where they have performed in situ deposition of lithium on solid electrolyte pellets within the ToF-SIMS instrument [58]. In this way, they avoided the problem of the exposure of a sample to environmental factors. Further optimization of the process (there are problems with film

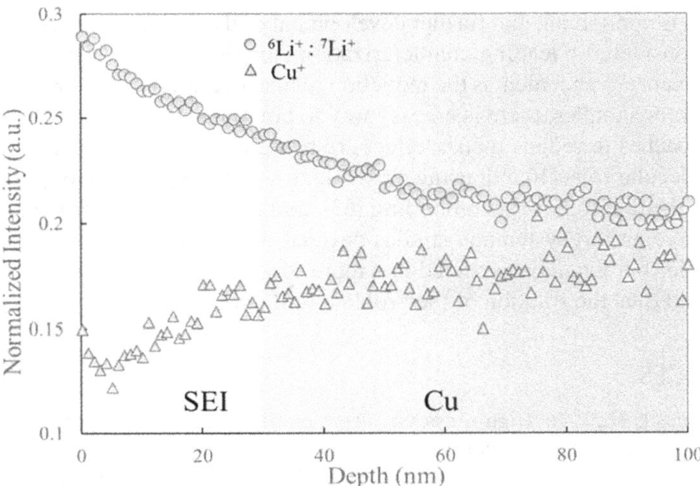

FIGURE 7.9 ToF-SIMS depth profile of SEI formed on a copper electrode. The analysis of the ^{6}Li/^{7}Li ratio indicates that the region close to the surface has been formed first. (Reprinted from Liu, Z.; Lu, P.; Zhang, Q.; Xiao, X.; Qi, Y.; Chen, L. Q., 2018. A bottom-up formation mechanism of solid electrolyte interphase revealed by isotope-assisted time-of-flight secondary ion mass spectrometry, *J. Phys. Chem. Lett.*, 9 (18), 5508–5514, with permission from American Chemical Society.)

homogeneity and roughness) will undoubtedly yield important information about SEI formation.

7.7 SUMMARY

The SIMS is a very powerful analytical technique, particularly known for its excellent detection limits and no impediment to the detection of light elements. However, due to numerous potential measurement artifacts, its utilization for the LIBs technology is a very non-trivial task. What is even more problematic is the fact that these shortcomings are rarely discussed in the literature, and the procedures to overcome these problems and verify the validity of the results are seldom discussed. Thus, it may be challenging to reproduce some SIMS experiments described in the literature.

Nevertheless, there are many examples of successful utilization of the SIMS technique for the characterization of LIBs samples. In some cases, SIMS is used as a complementary technique and the measurements are used to validate and better understand the results of other analytical techniques, most notably SEM and XPS. There are also some examples, where SIMS is the leading analytical technique and the conclusions are based almost exclusively on the SIMS results. It is important to emphasize that these exceptional results are often obtained when the authors have carefully designed the experiments and not simply tried to use the technique as is. The most notable example is the usage of isotopically labeled materials. Given that the SIMS technique can infallibly distinguish ^{6}Li and ^{7}Li isotopes, it is possible to precisely study processes that occur in the LIBs samples.

The fact, however, remains that the application of the SIMS technique in the LIBs technology is non-trivial, and further development of the technique is required before it can be considered a leading characterization technique in the LIBs industry. The most important advancement is the reduction of matrix effect because, contrary to the semiconductor samples, there is no easy way to circumvent it. It should be noted that some approaches to reduce matrix effect are being developed, for example registering the molecular ions [16,59], using cluster ions as primary source [60,61], dynamic reactive ionization [62], oxygen flooding [63], and the storing matter technique [64]. To date, there are no systematic studies devoted to testing these approaches on the LIBs samples but it can be expected that more SIMS specialist will focus on lithium batteries and thus the solution of the problem will emerge.

REFERENCES

[1] Wittmaack, K., 1976. High-sensitivity depth profiling of arsenic and phosphorus in silicon by means of SIMS, *Appl. Phys. Lett.*, 29 (9), 552–554.

[2] Ber, B. Y. et al., 1998. Secondary ion mass spectroscopy investigations of magnesium and carbon doped gallium nitride films grown by molecular beam epitaxy, *Semicond. Sci. Technol.*, 13 (1), 71–74.

[3] Matsunaga, T.; Yoshikawa, S.; Tsukamoto, K., 2002. Secondary ion yields of C, Si, and Ge in InP, and Cs surface density and concentration, in SIMS, *Surf. Sci.*, 515 (2–3), 390–402.

[4] Chiou, C. Y.; Wang, C. C.; Ling, Y. C.; Chiang, C. I., 2003. Secondary ion mass spectrometry analysis of In-doped p-type GaN films, *Appl. Surf. Sci.*, 203–204, 482–485.

[5] Emziane, M.; Durose, K.; Halliday, D. P.; Bosio, A.; Romeo, N., 2006. In situ oxygen incorporation and related issues in CdTe/CdS photovoltaic devices, *J. Appl. Phys.*, 100 (1), 013513.

[6] Gnaser, H., 1997. SIMS detection in the 1012 atoms cm-3 range, *Surf. Interface Anal.*, 25 (10), 737–740.

[7] "Benninghoven, A.; Rudenauer, F. G.; Werner, H. W., 1987. Secondary Ion Mass Spectrometry: Basic Concepts, Instrumental Aspects, Applications and Trends (Chemical Analysis: A Series of Monographs on Analytical Chemistry and Its Applications) Wiley-Interscience; 1st edition".

[8] Chatzitheodoridis, E.; Kiriakidis, G.; Lyon, I., 2002. Secondary Ion Mass Spectrometry and Its Application to Thin Film Characterization, in *Handbook of Thin Films.*

[9] Hutter, H., 2011. Dynamic Secondary Ion Mass Spectrometry (SIMS), in *Surface and Thin Film Analysis: A Compendium of Principles, Instrumentation, and Applications*, Second Edition.

[10] Leng, Y., 2013. Secondary Ion Mass Spectrometry for Surface Analysis, in *Materials Characterization.*

[11] Van Der Heide, P., 2014. *Secondary Ion Mass Spectrometry: An Introduction to Principles and Practices.* 9781118480489.

[12] Baker, J. E., 2014. Secondary Ion Mass Spectrometry, in *Practical Materials Characterization.*

[13] Fearn, S., 2015. *An Introduction to Time-of-Flight Secondary Ion Mass Spectrometry (ToF-SIMS) and Its Application to Materials Science.*

[14] Möller, W.; Resume, S., 2002. Fundamentals of ion-surface interaction, a Lect. held Tech. Univ. Dresden.

[15] Ziegler, J. F.; Ziegler, M. D.; Biersack, J. P., 2010. SRIM - The stopping and range of ions in matter (2010), *Nucl. Instruments Methods Phys. Res. Sect. B Beam Interact. Mater. Atoms*, 268 (11–12), 1818–1823.

[16] Gao, Y., 1988. A new secondary ion mass spectrometry technique for III-V semiconductor compounds using the molecular ions CsM+, *J. Appl. Phys.*, 64 (7), 3760–3762.

[17] Magee, C. W.; Harrington, W. L.; Botnick, E. M., 1990. On the use of CsX+ cluster ions for major element depth profiling in secondary ion mass spectrometry, *Int. J. Mass Spectrom. Ion Process.*, 103 (1), 45–56.

[18] Brison, J., 2004. Cesium/xenon dual beam depth profiling with TOF-SIMS: measurement and modeling of M+, MCs+, and M2Cs2+ yields, *Appl. Surf. Sci.*, 231–232, 749–753.

[19] Miyamoto, T.; Numao, S.; Hasegawa, T.; Karen, A., 2013. Origin of differences between MCs+ and MCs2+ SIMS depth profiles, *Surf. Interface Anal.*, 45 (1), 101–102.

[20] Wilson, R. G.; Novak, S. W., 1991 Systematics of secondary-ion-mass spectrometry relative sensitivity factors versus electron affinity and ionization potential for a variety of matrices determined from implanted standards of more than 70 elements, *J. Appl. Phys*, 69 (1), 466–474.

[21] Wilson, R. G., 1995. SIMS quantification in Si, GaAs, and diamond - an update, *Int. J. Mass Spectrom. Ion Process.*, 143 (C), 43–49.

[22] Douglas, M. A.; Chen, P. J., 1998. Quantitative trace metal analysis of silicon surfaces by ToF-SIMS, *Surf. Interface Anal.*, 26 (13), 984–994.

[23] Stevie, F. A., 2005. Focused ion beam secondary ion mass spectrometry (FIB-SIMS), in *Introduction to Focused Ion Beams: Instrumentation, Theory, Techniques and Practice.*

[24] Holzer, S.; Krivec, S.; Kayser, S.; Zakel, J.; Hutter, H., 2017. Large O_2 cluster ions as sputter beam for ToF-SIMS depth profiling of alkali metals in thin SiO_2 films, *Anal. Chem.*, 89 (4), 2377–2382.

[25] Lu, P.; Harris, S. J., 2011. Lithium transport within the solid electrolyte interphase, *Electrochem. Commun.*, 13 (10), 1035–1037.

[26] Harris, S. J.; Lu, P., 2013. Effects of inhomogeneities - nanoscale to mesoscale -on the durability of Li-ion batteries, *J. Phys. Chem. C*, 117 (13), 6481–6492.

[27] Berthault, M.; Santos-Pena, J.; Lemordant, D.; De Vito, E., 2021. Dynamics of the 6Li/7Li exchange at a graphite-solid electrolyte interphase: a time of flight-secondary ion mass spectrometry study, *J. Phys. Chem. C*, 125 (11), 6026–6033.

[28] Sui, T.; Song, B.; Dluhos, J.; Lu, L.; Korsunsky, A. M., 2015. Nanoscale chemical mapping of Li-ion battery cathode material by FIB-SEM and TOF-SIMS multi-modal microscopy, *Nano Energy*, 17, 254–260.

[29] Bessette, S. et al., 2018. Nanoscale lithium quantification in LiXNiyCowMnZO2 as cathode for rechargeable batteries, *Sci. Rep.*, 8 (1), 17575.

[30] Masuda, H.; Ishida, N.; Ogata, Y.; Ito, D.; Fujita, D., 2018. In situ visualization of Li concentration in all-solid-state lithium ion batteries using time-of-flight secondary ion mass spectrometry, *J. Power Sources*, 400, 527–532.

[31] Yamagishi, Y.; Morita, H.; Nomura, Y.; Igaki, E., 2021. Visualizing lithium distribution and degradation of composite electrodes in sulfide-based all-solid-state batteries using operando time-of-flight secondary ion mass spectrometry, *ACS Appl. Mater. Interfaces*, 13 (1), 580–586.

[32] Castle, J. E. et al., 2008. XPS and TOF-SIMS study of the distribution of Li ions in thin films of vanadium pentoxide after electrochemical intercalation, *Surf. Interface Anal.*, 40 (3–4), 746–750.

[33] Światowska-Mrowiecka, J. et al., 2008. The distribution of lithium intercalated in V2O5 thin films studied by XPS and ToF-SIMS, *Electrochim. Acta*, 53 (12), 4257–4266.

[34] Li, J. T. et al., 2009. XPS, time-of-flight-SIMS and polarization modulation IRRAS study of Cr2O3 thin film materials as anode for lithium ion battery, *Electrochim. Acta*, 54 (14), 3700–3707.

[35] Li, J. T. et al., 2010. XPS and ToF-SIMS study of Sn-Co alloy thin films as anode for lithium ion battery, *J. Power Sources*, 195 (24), 8251–8257.

[36] Li, J. T. et al., 2011. XPS and ToF-SIMS study of electrode processes on Sn-Ni alloy anodes for Li-ion batteries, *J. Phys. Chem. C*, 115 (14), 7012–7018.

[37] Yabuuchi, N.; Yoshii, K.; Myung, S. T.; Nakai, I.; Komaba, S., 2011. Detailed studies of a high-capacity electrode material for rechargeable batteries, Li2MnO3-LiCo1/3Ni 1/3Mn1/3O_2, *J. Am. Chem. Soc.*, 133 (12), 4404–4419.

[38] Otto, S. K. et al., 2021. In-depth characterization of lithium-metal surfaces with XPS and ToF-SIMS: toward better understanding of the passivation layer, *Chem. Mater.*, 33 (3), 859–867.

[39] Fedorková, A. et al., 2010. PPy doped PEG conducting polymer films synthesized on LiFePO4 particles, *J. Power Sources*, 195 (12), 3907–3912.

[40] Fedorková, A.; Oriňáková, R.; Oriňák, A.; Heile, A.; Wiemhöfer, H. D.; Arlinghaus, H. F., 2011. Electrochemical and TOF-SIMS investigations of PPy/PEG-modified LiFePO 4 composite electrodes for Li-ion batteries, *Solid State Sci.*, 13 (5), 824–830.

[41] Hong, T. E. et al., 2012. Nano SIMS characterization of boron- and aluminum-coated LiNi 1/3Co 1/3Mn 1/3O 2 cathode materials for lithium secondary ion batteries, *J. Appl. Electrochem.*, 42 (1), 41–46.

[42] Jo, C. H.; Cho, D. H.; Noh, H. J.; Yashiro, H.; Sun, Y. K.; Myung, S. T., 2015. An effective method to reduce residual lithium compounds on Ni-rich Li[Ni0.6Co0.2Mn0.2]O_2 active material using a phosphoric acid derived Li3PO4 nanolayer, *Nano Res.*, 8 (5), 1464–1479.

[43] Visbal, H.; Aihara, Y.; Ito, S.; Watanabe, T.; Park, Y.; Doo, S., 2016. The effect of diamond-like carbon coating on LiNi0.8Co0.15Al0.05O2 particles for all solid-state lithium-ion batteries based on Li2S-P2S5 glass-ceramics, *J. Power Sources*, 314, 85–92.

[44] Lee, S. W. et al., 2017. Li3PO4 surface coating on Ni-rich LiNi0.6Co0.2Mn0.2O2 by a citric acid assisted sol-gel method: Improved thermal stability and high-voltage performance, *J. Power Sources*, 360, 206–214.
[45] Neudeck, S. et al., 2018. Molecular surface modification of NCM622 cathode material using organophosphates for improved Li-ion battery full-cells, *ACS Appl. Mater. Interfaces*, 10 (24), 20487–20498.
[46] Börner, M. et al., 2016. Degradation effects on the surface of commercial LiNi0.5Co0.2 Mn0.3O2 electrodes, *J. Power Sources*, 335, 45–55.
[47] Heller, D.; Hagenhoff, B.; Engelhard, C., 2016. Time-of-flight secondary ion mass spectrometry as a screening method for the identification of degradation products in lithium-ion batteries—A multivariate data analysis approach, *J. Vac. Sci. Technol. B, Nanotechnol. Microelectron. Mater. Process. Meas. Phenom.*, 34 (3), 03H138.
[48] Gilbert, J. A. et al., 2017. Cycling behavior of NCM523/graphite lithium-ion cells in the 3–4.4 V range: diagnostic studies of full cells and harvested electrodes, *J. Electrochem. Soc.*, 164 (1), A6054–A6065.
[49] Walther, F. et al., 2019. Visualization of the interfacial decomposition of composite cathodes in argyrodite-based all-solid-state batteries using time-of-flight secondary-ion mass spectrometry, *Chem. Mater.*, 31 (10), 3745–3755.
[50] Graham, D. J.; Castner, D. G., 2012. Multivariate analysis of ToF-SIMS data from multicomponent systems: the why, when, and how, *Biointerphases*, 7 (1–4), 49.
[51] Peled, E.; Bar Tow, D.; Merson, A.; Gladkich, A.; Burstein, L.; Golodnitsky, D., 2001. Composition, depth profiles and lateral distribution of materials in the SEI built on HOPG-TOF SIMS and XPS studies, *J. Power Sources*, 97–98, 52–57.
[52] Ota, H.; Sakata, Y.; Inoue, A.; Yamaguchi, S., 2004. Analysis of vinylene carbonate derived SEI layers on graphite anode, *J. Electrochem. Soc.*, 151 (10), A1659–A1669.
[53] Lu, P.; Li, C.; Schneider, E. W.; Harris, S. J., 2014. Chemistry, impedance, and morphology evolution in solid electrolyte interphase films during formation in lithium ion batteries, *J. Phys. Chem. C*, 118 (2), 896–903.
[54] Lee, J. T.; Nitta, N.; Benson, J.; Magasinski, A.; Fuller, T. F.; Yushin, G., 2013. Comparative study of the solid electrolyte interphase on graphite in full Li-ion battery cells using X-ray photoelectron spectroscopy, secondary ion mass spectrometry, and electron microscopy, *Carbon N. Y.*, 52, 388–397.
[55] Li, W. et al., 2017. Dynamic behaviour of interphases and its implication on high-energy-density cathode materials in lithium-ion batteries, *Nat. Commun.*, 8, 14589.
[56] Liu, Z.; Lu, P.; Zhang, Q.; Xiao, X.; Qi, Y.; Chen, L. Q., 2018. A bottom-up formation mechanism of solid electrolyte interphase revealed by isotope-assisted time-of-flight secondary ion mass spectrometry, *J. Phys. Chem. Lett.*, 9 (18), 5508–5514.
[57] Zhou, Y. et al., 2020. Real-time mass spectrometric characterization of the solid–electrolyte interphase of a lithium-ion battery, *Nat. Nanotechnol.*, 15 (3), 224–230.
[58] Otto, S. K. et al., 2022. In situ investigation of lithium metal–solid electrolyte anode interfaces with ToF-SIMS, *Adv. Mater. Interfaces*, 9 (13), 2102387.
[59] Wittmaack, K., 1992. Basic requirements for quantitative SIMS analysis using cesium bombardment and detection of MCs+ secondary ions, *Nucl. Instruments Methods Phys. Res. Sect. B Beam Interact. Mater. Atoms*, 64 (1–4), 621–625.
[60] Shard, A. G.; Spencer, S. J.; Smith, S. A.; Havelund, R.; Gilmore, I. S., 2015. The matrix effect in organic secondary ion mass spectrometry, *Int. J. Mass Spectrom.*, 377, 599–609.
[61] Alnajeebi, A. M.; Sheraz née Rabbani, S.; Vickerman, J. C.; Lockyer, N. P., 2016. SIMS Matrix Effects of Biological Molecules under Cluster Ion Beam Bombardment, in *Proceedings of the Eighth Saudi Students Conference in the UK*, Feb. 2016, 437–444.

[62] Tian, H.; Wucher, A.; Winograd, N., 2016. Reducing the matrix effect in organic cluster SIMS using dynamic reactive ionization, *J. Am. Soc. Mass Spectrom.*, 27 (12), 2014–2024.
[63] Waddilove, A.; Chew, A.; Sykes, D., 1993. The effect of oxygen flooding on the secondary ion yields of matrix and impurity species in InP and InGaAs, *Vacuum*, 44 (3–4), 389–395.
[64] Kasel, B.; Wirtz, T., 2014. Reduction of the SIMS matrix effect using the storing matter technique: a case study on Ti in different matrices, *Anal. Chem.*, 86 (8), 3750–3755.

8 Nuclear Magnetic Resonance Microscopy

Atom to Micrometer

Chenjie Lou, Jie Liu, and Mingxue Tang
Center for High Pressure Science & Technology Advanced Research (HPSTAR)

CONTENTS

8.1 INTRODUCTION

Energy storage innovations for advanced technology are explosively speeding up, and rechargeable metal-ion batteries (such as lithium- [LIBs] and sodium-ion batteries [NIBs]) with fast cycling performance, long life span, high safety, and low cost are thus considered to be effective and reliable energy carriers.[1–3] Optimizations and breakthroughs for promoting metal-ion batteries need to unveil the components assignment, structure evolution, and reacting mechanisms under electrochemical cycling. The insights obtained for the structure and functioning properties at an atomic level are pivotal to identifying the key factors that restrict battery performance, which will, in turn, improve the material design and the fabrication of batteries. Therefore, characterization techniques play a significant role in the battery community. The widely claimed and reported analysis approaches are diverse and complementary,[3–5] such as infrared (IR),[6,7] Raman,[8–12] X-ray diffraction (XRD),[13–15] neutron techniques,[16–19] scanning electron microscopy (SEM),[20–22] transmission electron microscopy (TEM),[23–27] atomic force microscopy (AFM),[28–32] electron paramagnetic resonance (EPR),[33–36] and nuclear magnetic resonance (NMR).[37–41] Among these, EPR and NMR are the

DOI: 10.1201/9781003299295-8

usually considered approaches for the local structure determination at atomic scale. In addition, EPR and NMR reveal information on the electron cloud and provide complementary information without destroying the samples under investigation.

In energy batteries, many NMR-active isotopes like ^{1}H, $^{6,7}Li$, ^{11}B, ^{13}C, ^{15}N, ^{17}O, ^{19}F, ^{23}Na, ^{25}Mg, ^{27}Al, ^{29}Si, ^{31}P, ^{33}S, ^{39}K, ^{43}Ca, ^{51}V, ^{67}Zn, and ^{119}Sn can be measured and analyzed to provide structure and dynamics information.[37–41] The most used isotopes in metal-ion batteries are summarized in Table 8.1. These nuclei might be contained in different types of batteries or in different parts of the electrochemical cell batteries, such as anode, cathode, and electrolyte materials used in LIBs, NIBs, and K-ion batteries (KIBs). Local structure insights can be deduced by the NMR chemical shift, signal shape, and intensity, as well as the dynamic information that is drawn from relaxation, exchange, and diffusion measurement and analysis. Thus, it is possible to distinguish the diamagnetic and paramagnetic contributions from the electrodes and electrolyte, enabling ex situ or in situ[37–45] investigation without separating or unpacking batteries. Besides, the information on ion dynamics, e.g., activation energies and diffusion, from NMR investigations can be further correlated with theoretical calculations and other experimental techniques such as impedance spectroscopy. NMR of LIBs and NIBs are briefly summarized in Schematic 8.1, from which structure, dynamics, and distribution could be extracted for all components contained in the electrochemical cells.

This chapter focuses on NMR characterizations and applications to metal-ion batteries from ex situ and in situ aspects. First, the concept of NMR is briefly discussed with the possible shifts caused by chemical, Knight, and Fermi contact interactions, together with dipolar and quadrupole coupling influences. Second, examples and applications of NMR to cathodes, anodes, electrolyte, and solid electrolyte interface (SEI) are demonstrated, respectively. Third, the set-up of in situ NMR is given, and its applications to batteries are discussed. Furthermore, in situ magnetic resonance imaging (MRI) with the aid of pulsed field gradient (PFG) is extended to follow the evolution of ion distribution within electrolytes and electrodes upon electrochemical cycling. Finally, NMR from an energy storage perspective is summarized.

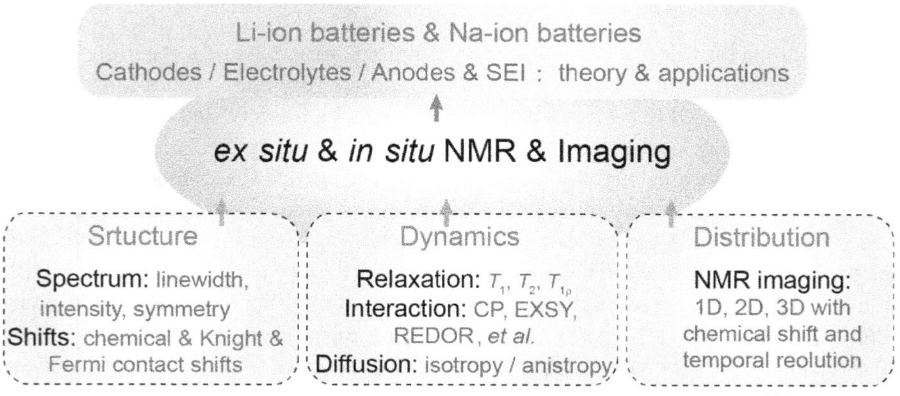

SCHEMATIC 8.1 Information deduced from NMR for metal-ion batteries. CP stands for cross polarization, EXSY for exchange spectroscopy, and REDOR is rotational-echo double-resonance.

TABLE 8.1
NMR Nuclei for Various Energy Storage Materials[a]

Isotope	I	N.A. (%)	Q (fm^2)	F.R. (%)	Sensitivity	Major Challenges	Information and Applications
^{1}H	1/2	99.99	–	100.00	1	Signal background	Electrolytes (polymer), supercapacitors, surface chemistry
^{2}H	1	0.01	0.29	15.35	1.11e−6	Low N.A. (enrichment)	Same to ^{1}H; better suited to study dynamics
^{6}Li	1	7.59	−0.08	14.72	6.45e−4	Low N.A. (enrichment)	Same to ^{7}Li; improved resolution vs ^{7}Li (smaller quadrupolar and dipolar broadening); provides some complementary information when combined with ^{7}Li
^{7}Li	3/2	92.41	−4.01	38.86	0.27	Comparably small diamagnetic shift range	Electrolytes (liquid/solid), electrodes, dynamics, SEI
^{11}B	3/2	80.10	4.06	32.08	0.13		Electrolytes, electrodes
^{13}C	1/2	1.07	–	25.15	1.70e−4	Low N.A. (enrichment); partial enrichment allows detection of specific functional groups	Electrolytes, supercapacitors, carbon anodes, SEI
^{17}O	5/2	0.04	−2.56	13.56	1.11e−5	Low N.A. (enrichment); enrichment readily available but expensive	Metal-air batteries, electrolytes (liquid/solid), electrodes, oxygen dynamics in oxide materials
^{19}F	1/2	100	–	94.09	0.83	Signal background	Electrolytes (liquid/solid), supercapacitors, cathode coatings, SEI
^{23}Na	3/2	100	10.40	26.45	0.09	None	Electrolytes (liquid/solid), electrodes, dynamics, SEI
^{25}Mg	5/2	10.00	19.94	6.12	2.68e−4	Low FR, low N.A., large quadrupole moment; low sensitivity without expensive enrichment	Mg^{2+} batteries; solid electrolytes
^{27}Al	5/2	100	14.66	26.06	0.21	Large quadrupole moment	Al^{3+} batteries, Al-S batteries; cathode coatings, solid electrolytes
^{29}Si	1/2	4.68	–	19.87	3.68e−4	Low N.A. (enrichment)	Si anodes; solid electrolytes

(*Continued*)

TABLE 8.1 (*Continued*)
NMR Nuclei for Various Energy Storage Materials[a]

Isotope	I	N.A. (%)	Q (fm^2)	F.R. (%)	Sensitivity	Major Challenges	Information and Applications
^{31}P	1/2	100	–	40.48	0.07	Slow relaxation can cause long experiment times	Phosphide and phosphate electrodes, electrolytes (liquid/solid), SEI
^{33}S	3/2	0.76	−6.78	7.68	1.72e−5	Low N.A. (enrichment); expensive enrichment; very broad signals	Li-S batteries; solid electrolytes
^{35}Cl	3/2	75.78	−8.17	9.80	3.58e−3	Low sensitivity	Electrolytes (liquid/solid)
^{39}K	3/2	93.26	5.85	4.67	4.76e−4	Low sensitivity and slow relaxation	K^+ batteries
^{43}Ca	7/2	0.14	−4.08	6.73	8.68e−6	Low N.A. (enrichment); expensive enrichment	Ca^{2+} batteries; solid electrolytes
^{45}Sc	7/2	100	−22.00	24.29	0.30	Large quadrupole moment	Electrodes
^{51}V	7/2	99.75	−5.20	26.30	0.38	Large quadrupole moment	Electrodes; redox flow batteries
^{57}Fe	1/2	2.12	–	3.24	7.24e−7		Cathodes
^{59}Co	7/2	100	42.00	23.73	0.28	Large quadrupole moment	Electrodes
^{63}Cu	3/2	69.17	−22.00	26.52	0.06	Low sensitivity; large quadrupole moment	Electrodes
^{67}Zn	5/2	1.68	15.00	6.26	1.18e−4	Low N.A. (enrichment); large quadrupole moment	Zn-air batteries, redox flow batteries
^{71}Ga	3/2	39.89	10.70	30.50	0.06	Low sensitivity; large quadrupole moment	Solid electrolytes
^{73}Ge	9/2	7.73	−19.60	3.49	1.09e−4	Low N.A. (enrichment); large quadrupole moment	Solid electrolytes, electrodes
^{79}Br	3/2	50.69	32.30	25.05	0.04	Low sensitivity; large quadrupole moment	Solid electrolytes
^{119}Sn	1/2	8.59	–	37.29	4.53e−3	Comparably low N.A.	Sn metal and Sn compound anodes
^{127}I	5/2	100	−71.00	20.01	0.10	Large quadrupole moment	Solid electrolytes
^{139}La	7/2	99.91	20.00	14.13	0.06	Large quadrupole moment	Solid electrolytes, electrodes

[a] Isotopes, nuclear spin (I), natural abundance (N.A.), quadrupole moment (Q), frequency ratio (F.R.), and sensitivity (relative to 1H) are given, together with experimental challenges and applications.

8.2 PRINCIPLE OF NMR

Atoms or molecules with a nuclear spin $I \neq 0$ possess a magnetic moment and are NMR-active. The nuclear spins can be considered tiny compass needles and homogeneously direct any position,[43] summing up with zero magnetization if there is no external magnetic field present (Figure 8.1a). However, if the spins are located in an external magnetic field, the distribution is relocated and slightly more spins are aligned in the B_0 direction to produce a net magnetic moment along the z direction (Figure 8.1b). For a spin of 1/2, the energies in the lower and the higher states are $E_\alpha = -\gamma h B_0/4\pi$ and $E_\beta = \gamma h B_0/4\pi$, respectively. The energy difference is thus $\Delta E = \gamma h B_0/2\pi$, and the spin population ratio is calculated as Boltzmann distribution (Equation 8.1). The net magnetization is determined by the population of different energy states (Figure 8.1c), the external magnetic field, and the absolute temperature. After radio-frequency (RF) excitation for NMR signal acquisition, the process of bringing the net magnetization from any nonequilibrium state back to its equilibrium state (relaxation mechanism) is typically fast for paramagnetic materials and slow for diamagnetic materials. As a matter of fact, NMR determines and analyzes the nuclear spins' Larmor frequency ω_0 or ν_0 (Equation 8.2) with respect to B_0, which is closely related to the transitions between the neighboring energy levels, e.g., $-1/2 \leftrightarrow +1/2$. Interactions within materials cause perturbations to these distinct energy levels or frequencies that NMR acquired. The net magnetization is further rotated by RF pulses to be transferred into the x-y plane viewed from the so-called rotating frame

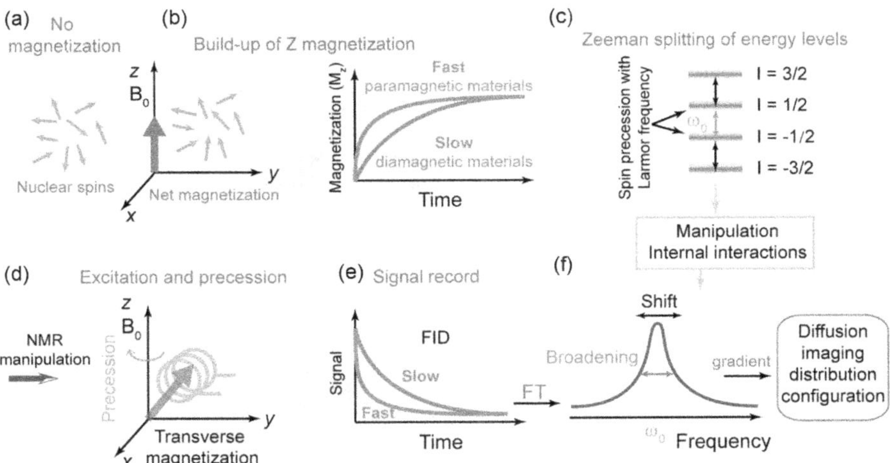

FIGURE 8.1 Principle of NMR spectroscopy. (a) The randomly orientated nuclear spins without external magnetic field. (b) Net magnetization along the z-axis produced in an external magnetic field B_0. (c) The Larmor frequency related to the Zeeman splitting of the energy levels. (d) RF pulse rotates the net magnetization into the x–y plane, as described in the rotating frame, followed by a precession along z-axis to induce current inside the NMR coil for further signal recording according to free induction decay (FID). (f) Fourier transformation (FT) of the time-domain signal to frequency NMR signal.[43] (Figure partially adapted from and inspired by Ref. 43 with permission from the American Chemical Society, Copyright [2017].)

(transverse magnetization) for NMR signal recording during their precession about the *z*-axis (Figure 8.1d). The collected NMR signal is, in fact, from the induced voltage by the precession of the magnetic moment inside the detecting coil. Due to the influence from the hyperfine interaction caused by unpaired or delocalized electrons, a relatively fast free induction decay (FID) is observed for the paramagnetic materials, but a slow recovery for the diamagnetic materials (Figure 8.1e). The materials with fast decay usually need special signal acquisition techniques, such as Hahn-echo (Schematic 8.2b) for recalling the signal buried during the dead time, which cannot be avoided due to the nature of NMR detection from hardware aspect by using single-pulse excitation (Schematic 8.2a). Herein, the same coil is used for NMR magnetization excitation and signal acquisition, and the acquisition can only be started after the residual excitation current disappears completely. The so-called dead time is the period between turning off the excitation pulse and starting the acquisition. For the dilute nuclei, such as ^{13}C and ^{6}Li, cross polarization (CP) is usually employed to enhance NMR signal from the surrounding abundant nuclei (^{1}H, ^{19}F) (Schematic 8.2c). CP works well for the organic electrodes, polymer electrolytes, and SEI. Via Fourier transformation (FT), the time-domain signal is finally translated into an intensity–frequency signal (Figure 8.1f and Equation 8.3), where the shift and line shape can be further used to decrypt the component assignment, local structure, and internal interactions. In the advanced NMR community, a PFG is further introduced to map the Larmor frequency correlated to the strength of fields to obtain additional

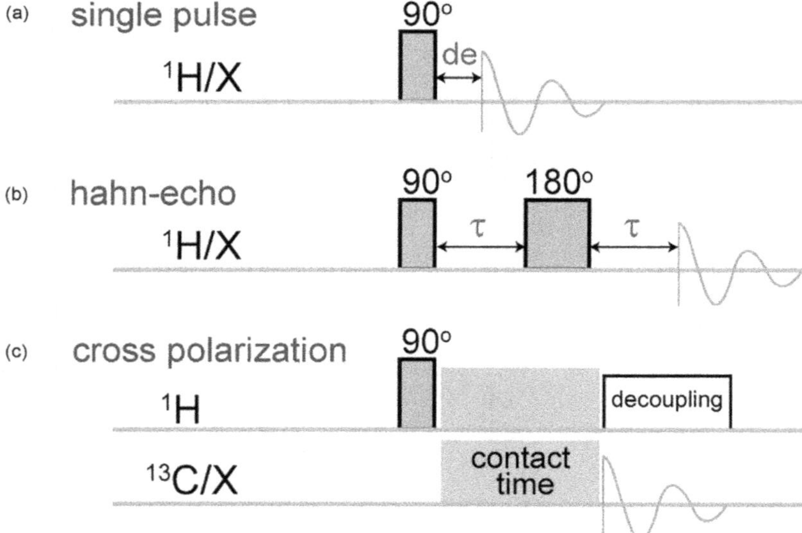

SCHEMATIC 8.2 Essential pulse sequences used in solid-state NMR. (a) Single pulse: the acquisition takes place after 90° excitation; *de* stands for dead time that cannot be avoided due to the nature of NMR coil. (b) Hahn-echo: one 180° pulse is inserted at the middle of 90° excitation and acquisition to recall signal decayed during dead time as shown in (a). (c) Cross polarization: the magnetization of abundant nuclear (^{1}H) is excited and followed by cross polarization via contact time to the surrounding dilute nuclei ($^{13}C/X$) to enhance the signal of $^{13}C/X$.

information, such as the thermal diffusion coefficient, imaging, and concentration dispersion. In words, local information such as structural defects, atomic disorder, and dynamic information can be powerfully distinguished on an atomic level.

$$N_\alpha / N_\beta = \exp(\Delta E / (kT)) = \exp(\gamma h B_0 / (2\pi kT)) \quad (8.1)$$

$$\omega_0 = -\gamma B_0 \text{ (in radians) or } \nu_0 = -\gamma B_0/(2\pi) \text{ (in Hz)} \quad (8.2)$$

$$S(t) = S_0 \exp(i\Omega t)\ \exp(-t/T_2) \quad (8.3)$$

Where, γ is gyromagnetic ratio, h is Plank constant, B_0 stands for the strength of magnetic field, k is Boltzmann constant, T is absolute temperature in K, Ω is the phase for signal detection, and T_2 is the transverse or spin–spin relaxation time.

The NMR of battery materials is complicated because all components containing active isotopes give resonance signal. Therefore, it is essential to extract various coupling parameters to obtain a reliable analysis and quantification of the various internal interactions, as shown in Figure 8.2. The chemical shift originates from the shielding influence by surrounding valence electrons. For the commonly used nuclei, such as ^{6,7}Li and ^{23}Na, the chemical shift shows a minor effect on the battery materials, making NMR analysis challenging. Besides, two additional interactions from the spin magnetic moment of unpaired electrons, namely Knight and Fermi contact shifts (Figure 8.2a), make the NMR spectrum analysis even more complex. The shift caused by the conduction electrons for metals or conductive materials is a Knight shift. For example, the metallic Li has a Knight shift at around 260 ppm, which is far from the typical diamagnetic shift near 0 ppm (e.g., Li_2S, Li_2CO_3, and LiF). Another origin is the hyperfine interaction caused by unpaired electrons for paramagnetic compounds via the bond interaction of nuclear spins with magnetic moments causing a Fermi contact shift. For example, Li in the transition metal (TM) Mn layer shows a larger Fermi shift when compared to the shift for the Li in the Li metal layer.[46–48] Dipole–dipole interaction may provide a direct spectroscopic route to determine

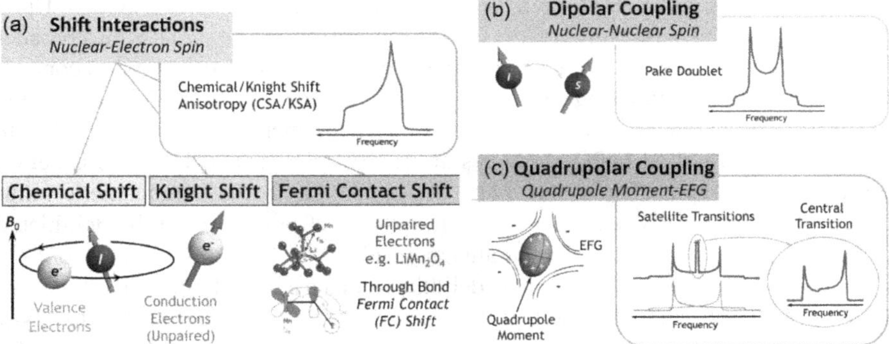

FIGURE 8.2 NMR shifts caused by various interactions. (a) The origins of chemical, Knight, and Fermi contact shifts. (b) Dipolar and (c) quadrupole coupling. Characteristic NMR signal shapes caused by the anisotropy and couplings are depicted in the boxes.[43] (Reproduced from Ref. 43 with permission from the American Chemical Society, Copyright [2017].)

interatomic/internuclear distances since its amplitude is proportional to the distance item of $1/r^3$ (Figure 8.2b). The dipole–dipole coupling can occur between the same NMR-active nuclei (homonuclear) or different nuclei (heteronuclear). Furthermore, the surrounding interaction produced by the nonspherical quadrupole moment Q ($I > 1/2$) with the electric field gradient (EFG) is known as quadrupole coupling (Figure 8.2c), with possible information on the structural distortion. The above interactions are all "anisotropic" origins. Reliable assignments could be buried by these anisotropies if the NMR spectrum is not well resolved. However, magic angle spinning (MAS) is an essential technique to gain high spectral resolution for powders by spinning the samples at an angle of 54.7° relative to the external magnetic field. At an angle of 54.7°, most interactions collapse due to the item $[3/2 \times (\cos\theta)^2 - 1]$, where θ is to the orientation of the internuclear vector with respect to the static B_0 field, contained in different interactions turning to zero, and therefore, averaging out many anisotropic interactions and leading to well-resolved isotropic signals.

Next, NMR investigations on cathodes, anodes, electrolytes, and SEI are discussed individually with presentative materials. In addition, in situ NMR and imaging (MRI) are also provided to extend the spatial and temporal resolution for tracking the evolution of functioning batteries.

8.3 NMR OF CATHODE MATERIALS

The local sensitive NMR probe is suitably used to determine the structural evolution of cathodes upon electrochemical cycling, even though the materials become amorphous.[44,49–60] Menetrier and coworkers[61] reported the detailed ^{7}Li NMR study of $LiCoO_2$. As shown in Figure 8.3a, the pristine Li_xCoO_2 ($x = 1.0$) displays a narrow symmetric ^{7}Li signal at 0 ppm due to the diamagnetic low-spin Co^{3+}. Upon delithiation, the Li^+ ion is extracted from the cathode, and a reduced signal is obtained[61] since the NMR signal is proportional to the number of studied isotope (^{7}Li) if measurements and analysis are properly carried out. Only one peak is observed for the delithiation until $x = 0.94$. In the range of $x = 0.90$ to $x = 0.75$, an additional peak is detected in the range of 50–90 ppm, revealing a biphasic reaction for Li_xCoO_2. Upon deeper delithiation in the range of $x = 0.70$ to $x = 0.50$, only one signal with a broader linewidth appears at 80–130 ppm due to the Knight shift (electron orbital to conduction bonds) that evolves from the paramagnetic interaction. $LiFePO_4$ has become a popular cathode material and is widely used in commercial batteries. Solid-state ^{31}P NMR spectra of Li_xFePO_4 cathodes are shown in Figure 8.3b, where $LiFePO_4$ presents an isotropic ^{31}P resonance at 3,800 ppm[62] due to the paramagnetic influence $t_{2g}{}^4e_g{}^2$ from Fe^{2+}. The resonance shifts up to 5,800 ppm for the fully delithiated cathode $FePO_4$ due to its stronger paramagnetic electron configuration $t_{2g}{}^3e_g{}^2$ of Fe^{3+}. Two phases are observed for the half-delithiated material $Li_{0.5}FePO_4$, in which two isotopic resonances at 3,800 and 5,800 ppm are detected.[62] In addition, the ^{7}Li NMR spectrum of $LiFePO_4$ shows an isotropic resonance at −11 ppm, but at 75 ppm for $LiMnPO_4$ (Figure 8.3c), where the latter displays a larger paramagnetic shift due to its $t_{2g}{}^3e_g{}^2$ electron configuration of Mn^{2+}, and the narrower signal is mainly due to the short spin-lattice relaxation time T_{1e} term.[62] Furthermore, high-resolution ^{7}Li and ^{31}P NMR have been widely used to determine the $LiFePO_4$-based cathodes.[63–74] Salager

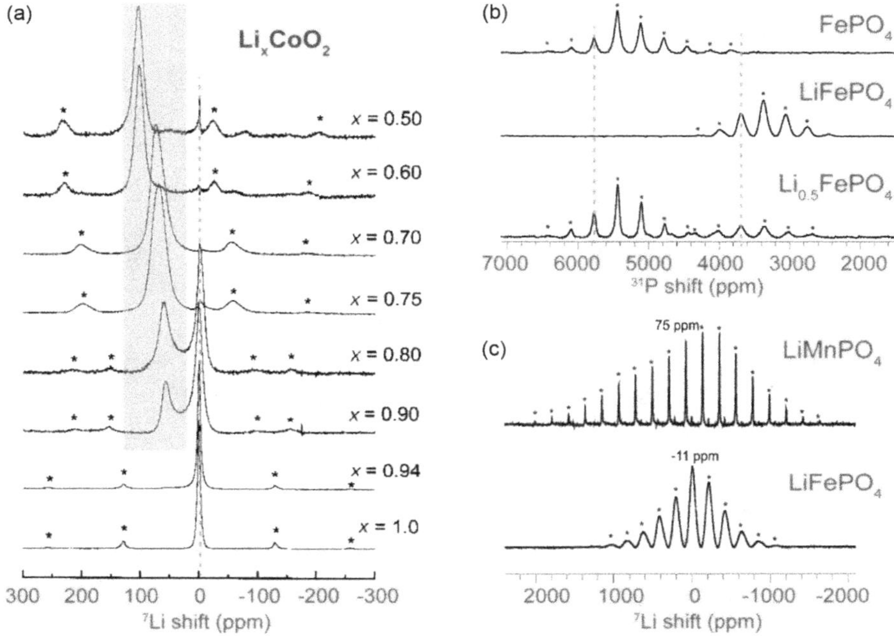

FIGURE 8.3 (a) Solid-state ^{7}Li NMR spectra of Li$_x$CoO$_2$ at various cycling states. The sharp signals at 0 ppm for the sample $x = 0.50$ and $x = 0.60$ are artifacts from the irradiation pulse.[61] (b) ^{31}P MAS NMR of FePO$_4$, LiFePO$_4$, and biphasic Li$_{0.5}$FePO$_4$. (c) ^{7}Li MAS NMR spectra of LiMnPO$_4$ and LiFePO$_4$.[62] Asterisks denote spinning sidebands. (Reproduced from Ref. 61 with permission from the Royal Society of Chemistry, Copyright [1999]. Reproduced from Ref. 62 with permission from the Royal Society of Chemistry, Copyright [2011].)

and coworkers employed ^{7}Li NMR and theoretical calculations to reveal the lithium environments with changing Ru and Sn concentrations, together with the delithiation process of $Li_2Ru_{0.75}Sn_{0.25}O_3$.[56]

Due to powerful hyperfine interactions, the NMR spectra become extremely broad up to ~MHz for some paramagnetic materials. As such, the fastest spinning rate of ~120 kHz to date cannot completely remove the internal interactions as aforementioned. Fortunately, Gan and coworkers developed projection-Magic-Angle-Turning and Phase-Adjusted Sideband Separation (pjMATPASS) to obtain clean and well-resolved spectra for the anisotropy beyond 1 MHz.[75] This technique was extended[44] to study the Mn-based cathodes, as shown in Figure 8.4. The conventional stimulated echo only gives a great many spinning sidebands (SSBs) for Li_2MnO_3 (LMO), as shown on the top of Figure 8.4a. By employing pjMATPASS, a pseudo-two-dimensional spectrum modulated by phase is recorded (Figure 8.4a), followed by shearing along the F2 dimension to produce aligned spectra, and the row projection sums up a pure isotropic ^{7}Li spectrum with two major resonances shown on the top of Figure 8.4b, which are further assigned to Li_{TM} (with a Fermi contact shift of 1,510 ppm) and Li_{Li} (750 ppm). The pjMATPASS method was also applied to the Ni- and Co-doped Li_2MnO_3 and their delithiated cathodes.[44] Upon the introduction

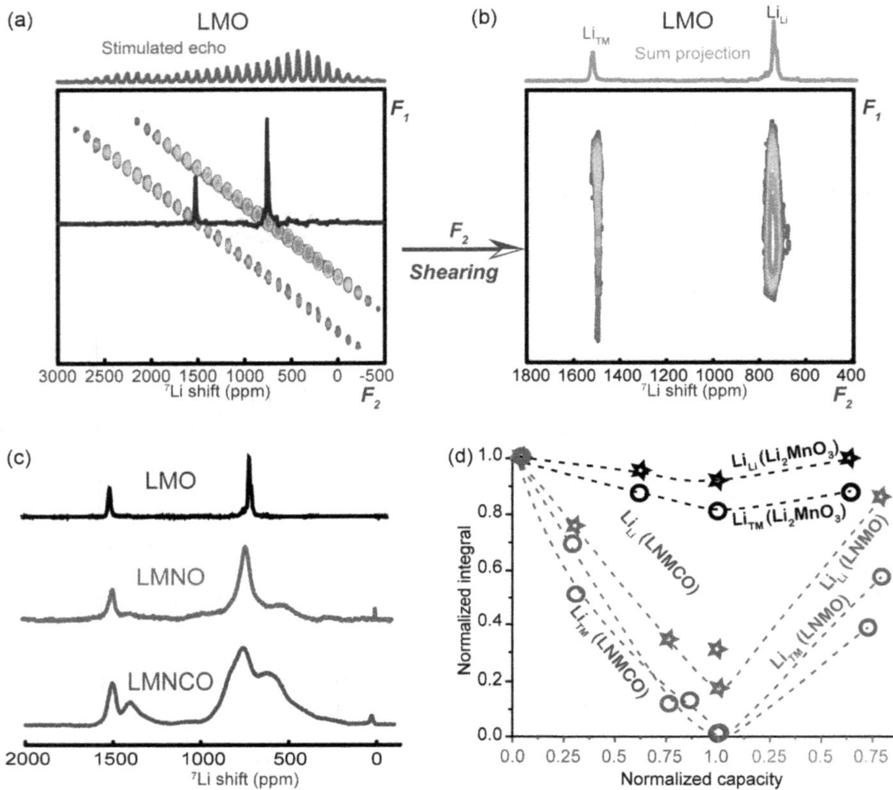

FIGURE 8.4 (a and b) pjMATPASS NMR of Li_2MnO_3 (LMO) to achieve a pure isotropic spectrum. (c) Comparison of pjMATPASS NMR spectra of LMO, LMNO, and LMNCO. (d) Normalized intensities of the three electrodes at different cycling states.[44] (Reproduced from Ref. 44 with permission from the American Chemical Society, Copyright [2017].)

of Ni and Co, ^{7}Li NMR spectra become more complex due to the changes in spin density transfer between Ni, Co, and Mn to Li (Figure 8.4c). The large peak width of the resonances originates from the heterogeneity of Ni, Co, and Mn coordination to Li.[44] As shown in Figure 8.4d, delithiation quantification indicates that Li^+ is first extracted from the TM layer and then follows a relatively constant ratio from both the TM and Li layers.

In conclusion, although metal ions in cathode materials have intensive hyperfine interactions, various NMR-active isotopes, such as ^{6,7}Li, ^{17}O, ^{19}F, ^{23}Na, ^{27}Al, and ^{31}P, allow us to determine the local structure and cycling dynamics at an atomic level in short range.[57–60,76–90]

8.4 NMR OF ANODE MATERIALS

$Li_4Ti_5O_{12}$ (LTO) is a widely studied anode material due to its extremely stable structure, fast rate performance, and large capacity since a maximum of five Li^+ can be intercalated if the discharge voltage extends to ~0 V. The LTO synthesized with CH_3COOLi (LiAc) shows an improved rate performance due to the formed bridged grain boundaries and percolated structural networks.[91] High-resolution ^{7}Li NMR spectra of the cycled LTO under different states (Figure 8.5a) reveal the clear Li^+ environments at 8a, 16d, and 16c sites (Figure 8.5b). The pristine LTO displays Li^+ at 8a and 16d sites. Upon discharge, the lithium is stripped from Li metal electrode and intercalated into 16c site in LTO and exchanges with Li^+ at 8a sites to prevent phase separation, which is beneficial for good electrochemical performance. NMR spectra in Figure 8.5c reflect that slow discharge allows more Li^+ occupying at 16c site.[91] The Li^+ pathway is probed by ^{6}Li NMR for LTO as well.[92] Recently, Louvain and coworkers employed both ^{7}Li and ^{19}F NMR to determine the mechanism for the fluorine-coated LTO with improved performance.[49]

Graphite is a widely-used commercial anode material, and solid-state ^{13}C and ^{7}Li NMR were employed to determine the mechanism of Li^+ intercalation.[93–96] As shown in Figure 8.6, LiC_x and NaC_x show similar ^{13}C NMR evolutions[97] for both LIBs and NIBs. The pristine graphite shows a broad ^{13}C resonance at around 100 ppm, which can be further deconvoluted into two peaks at 73 and 114 ppm (Figure 8.6a). Upon discharge to 0.25V, the signal at 114 ppm shifts to 133 ppm, and the linewidth narrows down to 17 ppm from 74 ppm. The intensive changes in the ^{13}C position and line shape are possibly caused by the increased Knight shift, which is related to a nonzero Fermi density of states and becomes dominant in alkali-metal-intercalated graphite. The opposite trend is observed during the subsequent charge process, indicating excellent reversibility for the Li^+ intercalation and extraction for graphite. However, for the NIB shown in Figure 8.6b, the peak (B-line) increases in intensity and shifts from 107 to 124 ppm, and the A-line signal decreases upon discharge are due to a relatively small Knight shift when compared to those of the LIB. The signal does not recover to its pristine state because the NIB presents less reversibility, possibly due

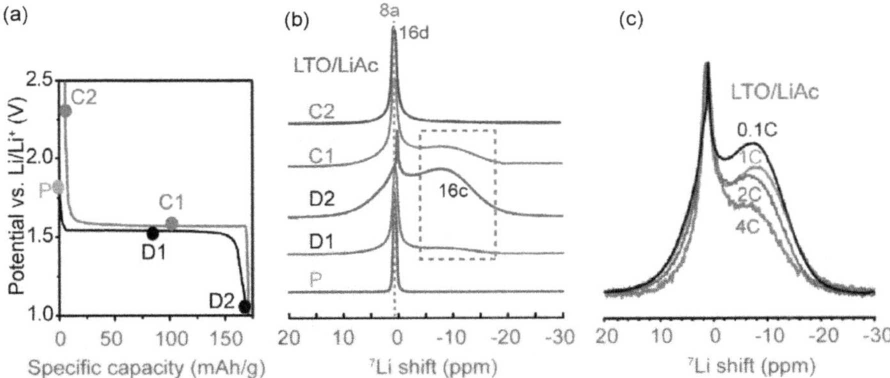

FIGURE 8.5 (a) Electrochemical profile of a typical LTO//Li half-cell battery; (b) Ex situ high-resolution ^{7}Li NMR of LTO electrodes cycling at different states as indicated in (a). P stands for the pristine LTO electrode. (c) ^{7}Li NMR of LiAc-treated LTO electrodes discharged to 1 V at cycling rates of 0.1 C, 1 C, 2 C, and 4 C, respectively.[91] (Reproduced from Ref. 91 with permission from Elsevier, Copyright [2017].)

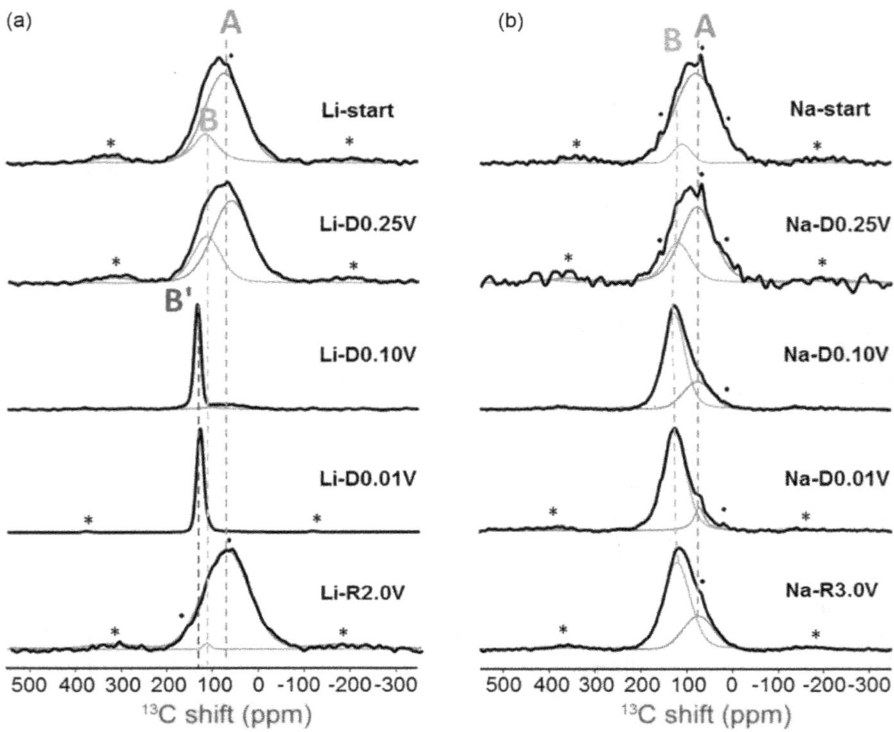

FIGURE 8.6 ^{13}C MAS NMR spectra of anodes at various electrochemical cycling states for (a) Li- and (b) Na-ion batteries. Asterisks represent spinning sidebands (SSBs), and black dots are the residual signals from the used electrolyte.[97] (Reproduced from Ref. 97 with permission from the American Chemical Society, Copyright [2019].)

to its larger radius compared to Li^+. Furthermore, the ^{13}C signal integration changes upon discharge, and the charge process is possibly due to the changing demagnetizing fields for the graphitic regions, which in turn affects the polarization of ^{13}C spins.

In addition to ^{13}C NMR, high-resolution ^{7}Li and ^{23}Na NMR were used to probe the reactions occurring in the graphite anode.[97] As shown in Figure 8.7a, the signal centered at ~0ppm is mainly from the SEI or decomposed components. The resonances at 2, 6, 10, and 46ppm are attributed to LiC_{36}, LiC_{27}, LiC_{18}, and LiC_{12}, respectively. Detailed quantification is further analyzed to track the evolution along the electrochemical cycling, as shown in Figure 8.7b. Turning to the NIB, the ^{23}Na NMR spectrum (Figure 8.7c) reveals the presence of Na located on the surface (−10ppm) and the formation of the NaBr phase (~5 ppm), with the Br coming from the fine-grained graphite. The electrolyte shows a strong signal at around 11 ppm, and the SEI is observed at −29ppm. The evolution of each component is also quantified and summarized in Figure 8.7d.

Alloying is an important reaction for the anodes. Long-range sensitive techniques, such as XRD, may not always be available for alloy structure determination. Well-resolved NMR spectra, alternately obtained under MAS or static conditions, provide access to monitor the structural evolution. As shown in Figure 8.8,

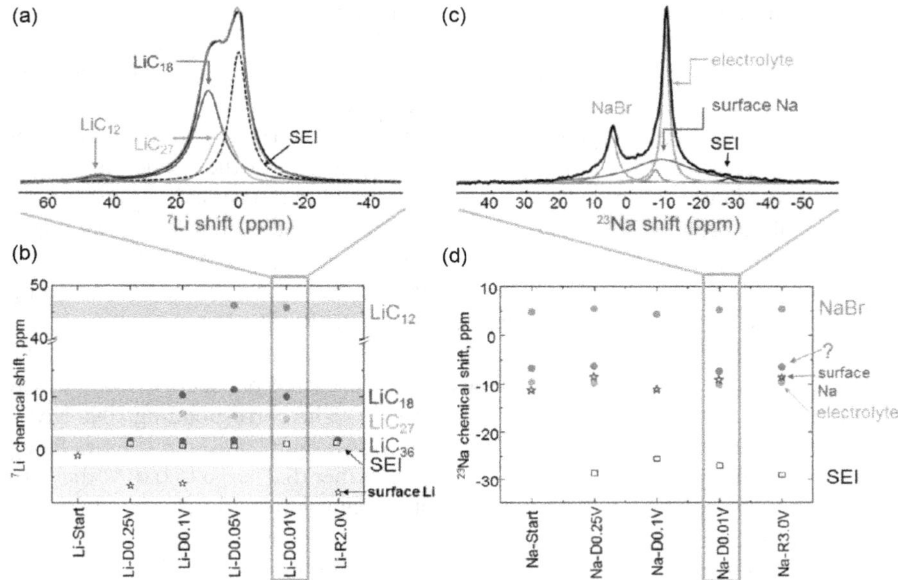

FIGURE 8.7 (a) ^{7}Li MAS NMR spectrum of the anode discharged to 0.01V (Li-D0.01V) and its simulation. (b) Deconvoluted ^{7}Li chemical shifts for the cycled anodes. (c) ^{23}Na NMR spectrum of Na-D0.01V and its deconvolution. (d) Deconvoluted ^{23}Na chemical shifts for various cycled anodes in Na-ion batteries.[97] (Reproduced from Ref. 97 with permission from the American Chemical Society, Copyright [2019].)

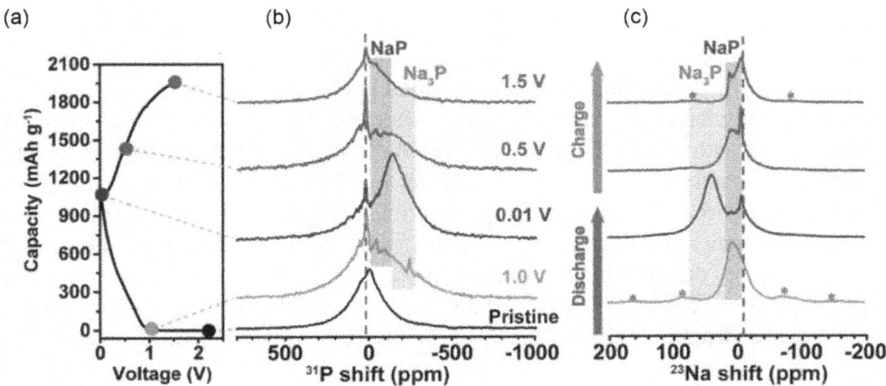

FIGURE 8.8 The electrochemical evolution of P anodes probed by NMR. (a) The charge/discharge curve with selected stages for ex situ NMR. (b) ^{31}P and (c) ^{23}Na NMR spectra of Na_xP anode at various stages marked in (a). The peaks labeled with asterisks are SSBs.[98] (Reproduced from Ref. 98 with permission from the American Association for the Advancement of Science, Copyright [2021].)

the sodiation/desodiation of Na_xP phases, which is not accessible to XRD, possesses various components in the ^{23}Na and ^{31}P spectra.[98] Figure 8.8a displays the data points where cycling stopped for the ex situ NMR investigation. As shown in Figure 8.8b, the pristine phosphorus P anode shows a single ^{31}P resonance centering at 0 ppm and spreading from −500 to 500 ppm. The broad and asymmetric peak reflects different chemical environments due to its non-crystalline nature. For the anode discharged to 1.0 V, a broad tail is observed at the right shoulder of the ^{31}P signal, suggesting the formation of amorphous Na_xP. The minor sharp peak at −245 ppm is from an impurity. Upon further discharging to 0.01 V, a variety of Na_xP alloys is formed. The overlapping signals of crystalline NaP and Na_3P are reasonably assigned according to the Ab initio calculation given to predict the ^{31}P NMR of Li_xP and Na_xP anodes.[99] The purple and green shadows represent the NaP and Na_3P phases, respectively. The sharp peak at 14 ppm likely comes from the unreacted phosphorus, which will decrease during the alloying reaction to form Na_xP. The corresponding ^{23}Na NMR spectrum for the anode discharged to 1.0 V presents a signal at 6 ppm (Figure 8.8c), which consolidates the formation of NaP. Upon further discharge to 0.01V, the NaP signal almost disappears, and a new peak at 50 ppm noticeably grows due to the generation of Na_3P. Upon the subsequent charging process, reversible trends are observed from both the ^{31}P and ^{23}Na NMR spectra. In the spectra of 0.5 and 1.5 V, resonances of ^{31}P show a downfield shift, whereas ^{23}Na moves toward the upfield. The sodium is mostly extracted after one cycle to 1.5V, which results in a dominant signal at −11 ppm for Na in the residual electrolyte. In summary, the amorphous sodiation/desodiation of P/Sn-based anode material is clearly characterized by the combination of solid-state ^{23}Na and ^{31}P NMR spectroscopy.

The reaction mechanism for a complex anode is also determined by solid-state ^{7}Li NMR.[98,100–103] All the FeCo-NiS@NC cycled samples show an isotropic resonance located at ~0 ppm with multiple SSBs (Figure 8.9a).[100] The electrode discharged to 1.52 V displays a large anisotropy of ~2,000 ppm due to the significant anisotropy caused by strong electron-nuclear dipolar coupling raised by the unpaired electron coming from metallic elements (Fe, Co, and Ni). The width of the full spectra expands upon further lithiation. A reverse trend is observed for these spectra during the subsequent charge process. The anode after one cycle shows a sharp intensive isotropic peak with little SSBs, which is possibly from the residual electrolyte and the incomplete capacity recovery, such as SEI formation. The isotropic resonances are enlarged in Figure 8.9b. All spectra localized near 0 ppm with a linewidth of more than 20 ppm, indicating the formation of Li_2S.[104] Upon further discharge, the amount of Li_2S increases due to lithiation into the anode. The isotropic linewidth grows, and the peak position seems to shift toward downfield, possibly due to the change of redox states and the formation of metallic Fe, Co and Ni, resulting in a paramagnetic shift.

8.5 NMR OF ELECTROLYTES IN BATTERIES

Solid electrolytes, such as $Li_7La_3Zr_2O_{12}$ (LLZO) and $Li_{10}GeP_2S_{12}$ (LGPS), used in LIBs can significantly improve the safety and energy density of solid-state batteries and attract intensive attention in both academic and industrial developments. NMR has been introduced to determine the structural evolution of both liquid and solid

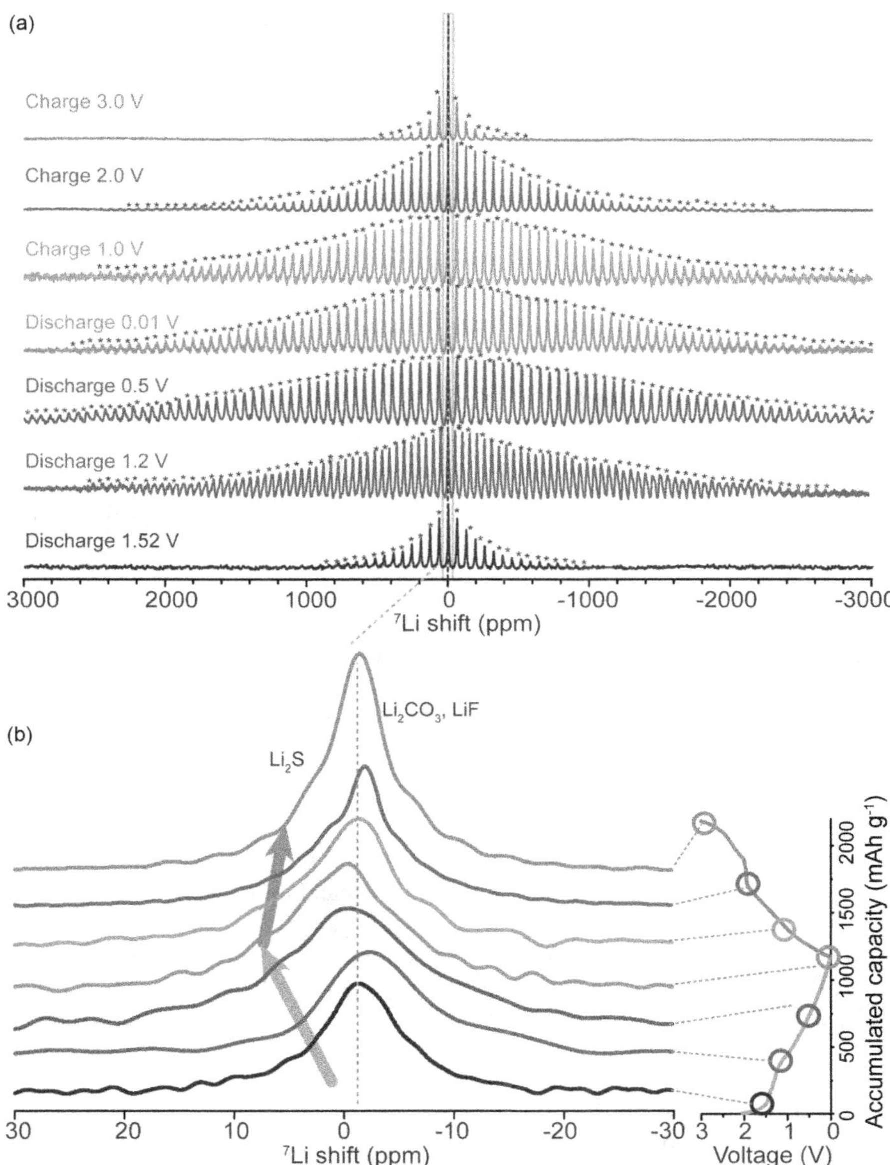

FIGURE 8.9 (a) Solid-state ^{7}Li NMR spectra for the FeCo-NiS@NC electrodes obtained at various charge-discharge states, with the isotropic peaks shown in (b).[100] SSBs are marked with asterisks. (Reproduced from Ref. 100 with permission from Elsevier, Copyright [2022].)

electrolytes.[105–123] A solid electrolyte composed of ceramic LLZO and polyethylene oxide (PEO) has high conductivity from LLZO and mechanical properties from PEO. The local correlation at the interface between LLZO and PEO is soft and invisible with many conventional analytical methods. In addition, determining the pathway

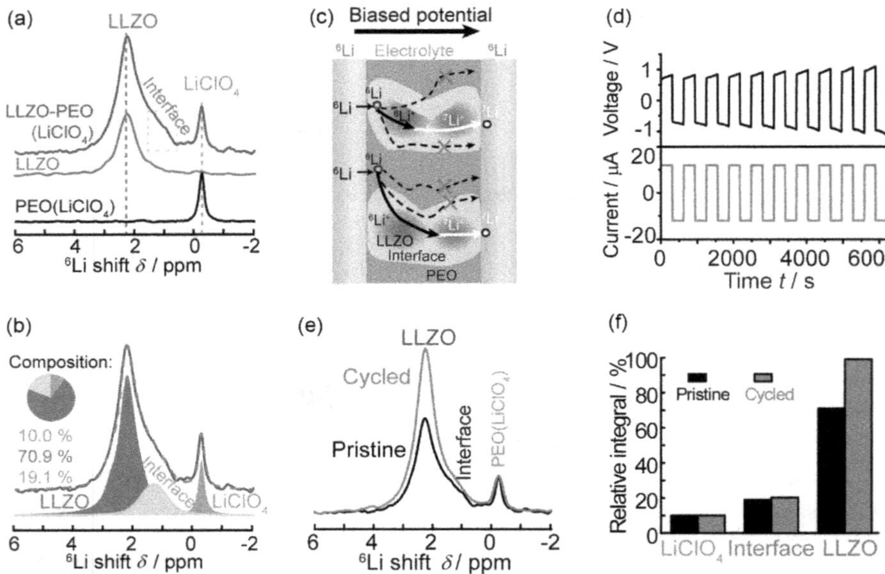

FIGURE 8.10 (a) ^{6}Li NMR of $LiClO_4$ in PEO, pure LLZO, and LLZO-PEO composite. (b) ^{6}Li CP NMR spectrum and quantification for the LLZO-PEO electrolyte. (c) Illustration of the symmetric ^{6}Li/LLZO-PEO/^{6}Li battery and possible Li^+ transport pathways within the composite electrolyte upon cycling. (d) Electrochemical profile of the symmetric battery cycled with a constant current. (e) Comparison of the ^{6}Li NMR spectra of the pristine and the cycled LLZO-PEO. (f) Quantitative analysis of ^{6}Li amount in PEO, interface, and LLZO of the LLZO-PEO, as shown in (e).[105] (Reproduced from Ref. 105 with permission from John Wiley and Sons, Copyright [2016].)

of Li^+ inside the electrolyte is a vital key to understanding its working mechanism and further optimizing electrolyte design from an atomic perspective. Solid-state NMR is isotope-orientated and sensitive to atomic configurations. Highly resolved ^{6}Li NMR spectra (shown in Figure 8.10a) clearly distinguish the signal of $LiClO_4$ in PEO appearing at −0.2 ppm and the intensive resonance at 2.2 ppm from the ceramic LLZO. A shoulder on the right side of LLZO is observed for the interface Li^+ in the LLZO-PEO composite electrolyte.[105] The ^{1}H-^{6}Li CP NMR spectrum is plotted and quantified in Figure 8.10b, from which the LLZO takes up 71%, and the interface concentration is up to 19%. ^{6}Li-riched Li metal is employed as two electrodes to determine the Li^+ pathway within LLZO-PEO (Figure 8.10c). Under potential polarization, the ^{6}Li is stripped from the electrodes and replaces the ^{7}Li on its way across the LLZO-PEO electrolyte. A constant current with an alternative direction is used for back-and-forth polarization for days to obtain a reasonable replacement, with the initial cycles shown in Figure 8.10d. The comparison of ^{6}Li NMR for the pristine and the cycled LLZO-PEO shows an obvious increase in intensity for the LLZO (Figure 8.10e and f), indicating the Li^+ pathway is mainly within LLZO rather than the interface predicted by theory simulations.[124,125] By varying the composition and introducing the additives, the Li^+ pathway is strongly influenced.[126]

In addition, many NMR works on solid electrolytes have been claimed.[122,127–132] For example, ^{7}Li and ^{19}F were employed to determine the local correlation in poly(ethylene oxide)-lithium aluminum titanium phosphate (PEO-LICGC) composite electrolyte.[133] Two-dimensional ^{7}Li exchange spectroscopy (EXSY) measurements were performed to probe the Li^+ transport across the interface in a Li_6PS_5Cl-Li_2S solid electrolyte-electrode.[134] Solid-state $^{6,7}Li$ and ^{31}P NMR spectra were analyzed for the reactions occurring at the $Li_6PS_5Cl_{0.5}Br_{0.5}$ (LPSCB)-based composite in the solid-state battery (InLi/LPSCB/LPSCB-MWCNTs).[108] Solid-state ^{7}Li, ^{45}Sc, and ^{71}Ga NMR, together with theoretical calculations, were carried out for Ga/Sc co-doped LLZO.[135] ^{7}Li NMR imaging of LLZO and LGPS was also investigated and will be discussed in detail in Section 8.8.

8.6 NMR OF SOLID ELECTROLYTE INTERFACE

The SEI layer formed on electrodes during electrochemical cycling is critical to maintaining their long-term capacity. However, it is inherently difficult to investigate due to its very thin thickness and amorphous characteristics. NMR of ^{1}H, ^{7}Li, ^{13}C, ^{19}F, ^{23}Na, and ^{31}P isotopes was carried out to gain insights into the decomposition products in the SEI.[93,94,136–139]

The SEI formation on the C/Si electrode strongly depends on the cycling voltage windows,[136] and different spectra were acquired for the materials stopped at different voltages, as shown in Figure 8.11a. The species become more distinguished when the cut-off voltage is 1 mV, with the resonance at 4.6 ppm for ethylene carbonate (EC) and the peak at 1.3 ppm for the CH_3R or $R'CH_2R$. The signal at 3.8 ppm is not clear but could be finally sorted out to be CH_3OX group, such as CH_3OCO_2Li, with further aid from heteronuclear spectra and ^{13}C-labed correlation analysis.[136] As shown in Figure 8.11b, ^{7}Li NMR spectrum shows an asymmetric signal centered at 0 ppm, which can be further deconvoluted as two signals at −0.4 and −2.5 ppm from ^{1}H-^{7}Li heteronuclear spectra displayed in Figure 8.11c and d. The signal at −0.4 ppm is basically from lithium salt contained in SEI (such as LiF and Li_2CO_3), and the peak at −2.5 ppm is assigned to $LiPF_6$. Meanwhile, the ^{19}F spectrum confirms the presence of $LiPF_6$, and a small amount of LiF is also observed in the SEI (Figure 8.11b).

In addition to LIBs, NMR was also extended to determine the components of electrolytes used in NIBs.[138,139] As shown in Figure 8.12, both ^{23}Na and ^{11}B NMR spectra reveal that the components are involved in the evolution during cycling. Typically, NaOH, $Na_4B_2O_5$, and NaF are formed on the surface of the Na electrode, together with oligomeric borate, $NaBF_4$, and NaDFOB.[139]

8.7 EX SITU AND IN SITU NMR

NMR investigations of battery materials are typically sorted into two categories: ex situ and in situ, depending on the sample states under NMR measurements. Generally, ex situ NMR means that the measured electrodes/electrolytes are extracted from the cycled batteries that were stopped at certain potentials. The electrode materials are extracted from the cycled batteries with potential extra treatments, such as being washed with a solvent, dried and ground to be filled into a MAS rotor. As such,

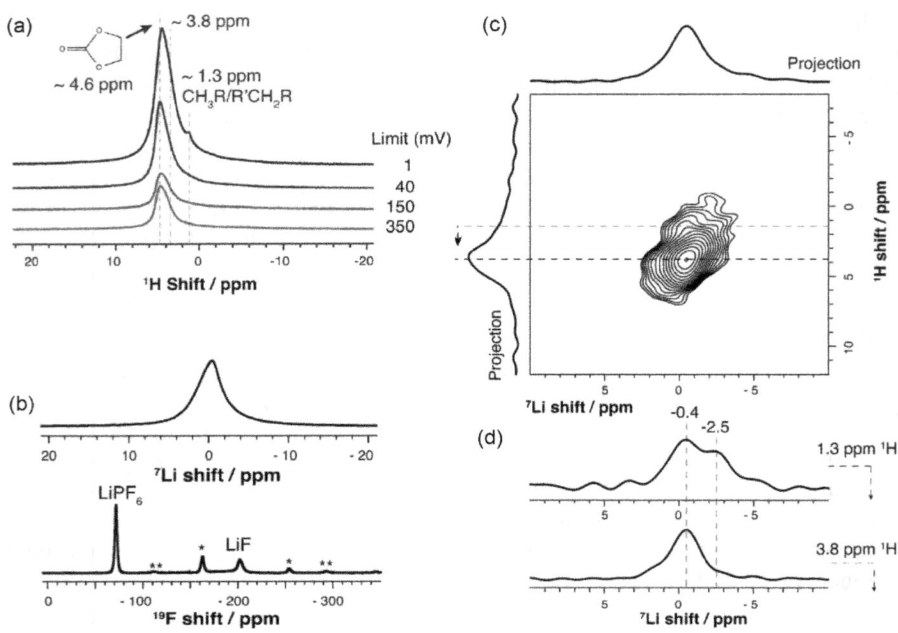

FIGURE 8.11 (a) ^{1}H Hahn-echo NMR spectra of the unwashed C/Si electrode samples, with their limited cycling voltages marked to the right. The unwashed C/Si electrode sample discharged to a 1 mV limit was investigated using (b) ^{7}Li single-pulse and ^{19}F Hahn echo and (c) ^{1}H-^{7}Li heteronuclear correlation NMR experiments. (d) Slices of the 2D spectra extracted from (c). SSBs are indicated with asterisks.[136] (Reproduced from Ref. 136 with permission from the American Chemical Society, Copyright [2016].)

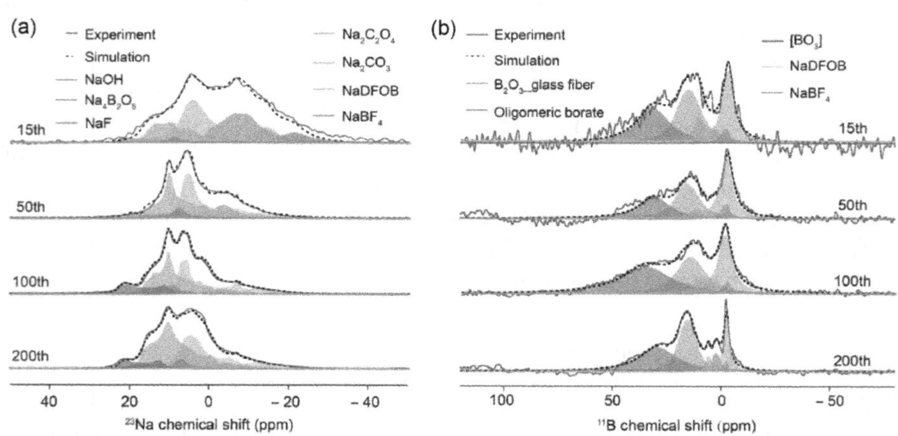

FIGURE 8.12 Solid-state (a) ^{23}Na and (b) ^{11}B NMR spectra and simulations of SEI species harvested from the Na metal surface after the 15th, 50th, 100th, and 200th cycles.[139] (Reproduced from Ref. 139 with permission from the American Association for the Advancement of Science, Copyright [2022].)

the measured samples are fully relaxed to their equilibrium state, and the composition may be changed by a series of post-treatments.[140] The lost information could be alternately complemented by in situ NMR. In situ investigation means that the NMR spectra are acquired on the unperturbed active components inside batteries.

The in situ NMR of batteries was initially proposed by Rathke in 1997, by mapping the first ^{19}F concentration of triflate ($CF_3SO_3^-$) electrolyte.[37,141] Later, Letellier investigated lithium-metal/hard carbon batteries using in situ ^7Li NMR spectroscopy. Plenty of in situ NMR works have been claimed in the past decades by various groups, such as the Grey and coworkers at Cambridge University,[41,142–144] Yan-Yan Hu's group in the United States,[44,145,146] the Goward team in Canada,[117,147–149] and Salager et al. in France.[42] The in situ NMR of batteries were also claimed by Hu's and Zhong's team.[150,151] Furthermore, Goward and coworkers prepared a spinning battery assembled inside a 4-mm rotor to achieve improved spectral resolution, which can record MAS-NMR spectra without unpacking the battery.[152]

Since the used RF in NMR excitation is significantly attenuated when it penetrates the metal package used for commercial batteries, it results in very weak signal for analysis. Therefore, performing an in situ NMR investigation needs a special cell design to optimize the NMR signal in a reasonable temporal resolution. To date, most cells used for in situ NMR are made of plastic, including bag cells and cylindrical cells. The trend of home-made cells for in situ NMR is portrayed in Figure 8.13.[153] Due to the complex assembly and lack of fine sealing, the coin-type cell shown in Figure 8.13a is not widely employed. The bag cell demonstrated in Figure 8.13b is easily fabricated, and the size and shape can be adjusted to the configuration of the NMR coil. However, the bag cell is hard for nuclear magnetic imaging since a slight tilt in locating the cell in the coil would cause a big deviation in analyzing the MRI data. The cylindrical cells shown in Figure 8.13c to f are mostly adapted to use. Other designs, such as the pouch cell (Figure 8.13g) and jelly roll cell (Figure 8.13h), are specially designed for MRI and MAS signal acquisition. In addition, a three-dimensional (3D) printed cell was also introduced into in situ NMR for long-time cycling capability.[160]

The typical in situ set-up is demonstrated in Figure 8.14. The home-made cylindrical cell or bag cell (Figure 8.14a) is mounted into the coil at the top of the NMR probe (Figure 8.14b). Cables from cell are conducted out to the galvanostat for the battery cycling. The entire probe set-up with battery is inserted into the magnetic field (Figure 8.14c). The battery is located at the position with the strongest and the most homogenous magnetic flux. It is worth noting that the gradient is not needed if only acquiring conventional NMR spectra. The gradient coil shown in Figure 8.14b works for the diffusion and imaging investigations discussed in detail in Section 8.8.

Phosphides work as promising anodes because of their high capacity and low potential. The electrochemical reaction for most phosphide materials is the formation of alloys with a typically disordered structure, which is the right system for NMR investigation. The conversion mechanism of Cu_3P was successfully probed by in situ NMR.[154] Nano-sheets Ni_5P_4 were prepared by a chemical vapor deposition (CVD) approach with Ni foam as a substrate, resulting in a short electron pathway and enhanced rate performance when used as an anode.[146] Five types of coordination are analyzed for the pristine Ni_5P_4 (Figure 8.15a). A Ni_5P_4//Li bag cell battery was

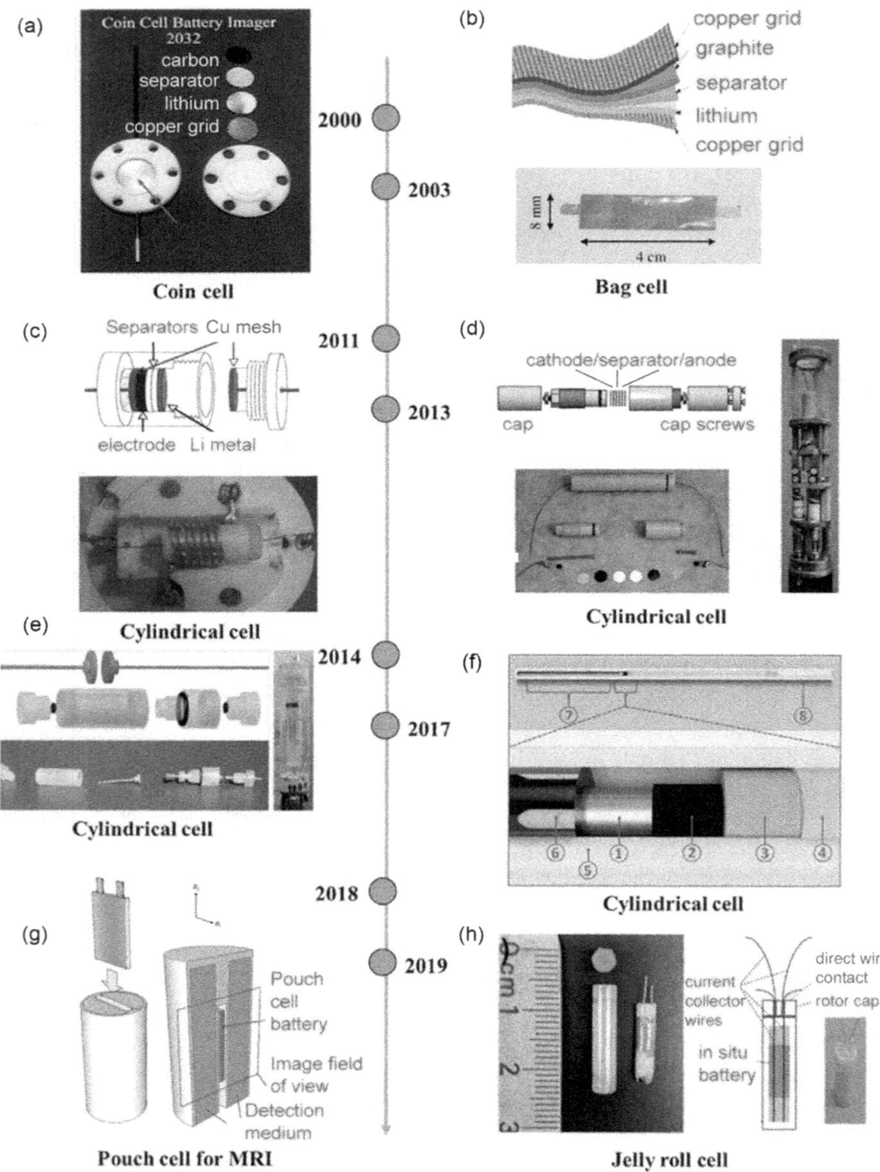

FIGURE 8.13 The evolution process of an in situ cell used for NMR study: (a) coin cell; (b) bag cell; (c–e) different configurations of cylindrical cells; (f) cylindrical cell for detecting liquid electrolyte; (g) pouch cell for MRI; (h) jelly roll cell for in situ MAS NMR.[153] (Reproduced with permission from Ref. 153 with permission from Wiley-Blackwell, Copyright [2021].)

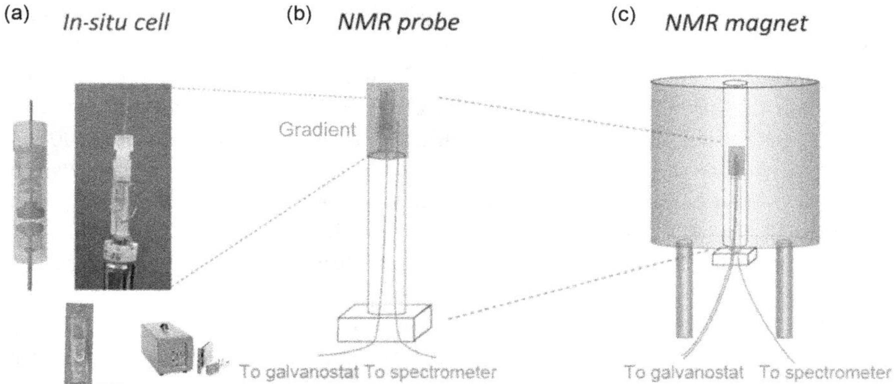

FIGURE 8.14 Experimental set-up for in situ NMR measurement. (a) The home-made in situ cylindrical cell or bag cell is located in the center of the radio-frequency and gradient-field coils of NMR probe (b), connecting to a galvanostat for (dis)charging and a spectrometer. The entire probe is inserted into the magnetic field as shown in (c).

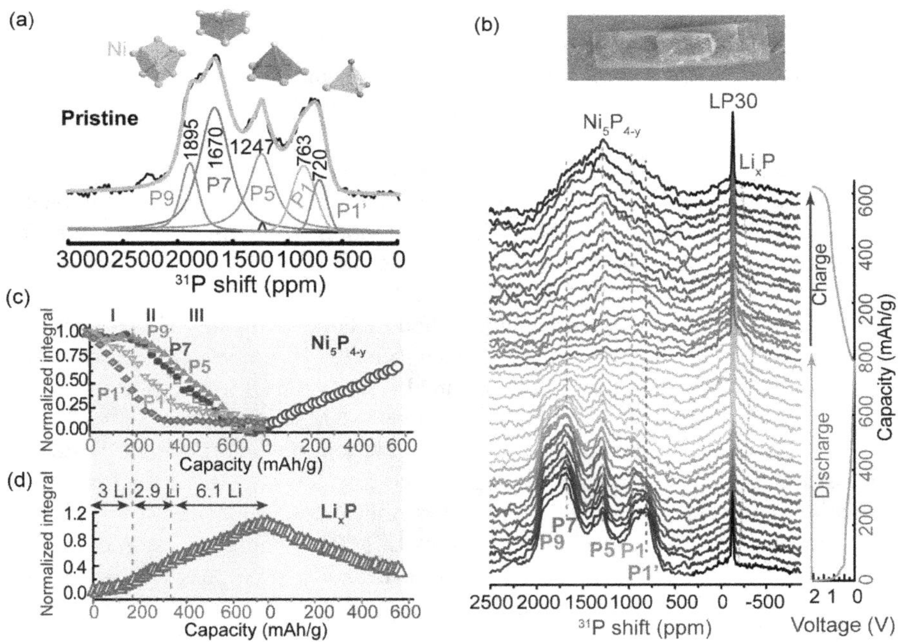

FIGURE 8.15 (a) Static ^{31}P NMR spectrum and deconvolution for the as-prepared Ni_5P_4. P1, P5, P7, and P9 are phosphorus coordinated with 1, 5, 7, and 9 Ni atoms, respectively. (b) In situ ^{31}P NMR during the first cycle. Normalized ^{31}P signal evolutions extracted from (b) for Ni_5P_{4-y} (c) and Li_xP (d).[146] (Reproduced from Ref. 146 with permission from the Royal Society of Chemistry, Copyright [2018].)

assembled (top of Figure 8.15b), and the in situ ^{31}P NMR spectra during the initial discharge and charge process are plotted in Figure 8.15b. The broad resonances ranging from 500 to 2,200 ppm belong to the Ni_5P_4, and the sharp signal at −140 ppm is assigned to the PF_6^- anion in the LP30 electrolyte. Upon discharge, the signal of the Ni_5P_4 decreases because of the formation of Li_xP appearing between 200 and −500 ppm. A reversible evolution is observed during the subsequent charge process, except the peaks of Ni_5P_4 merge into a broad one without spectral resolution due to the increased disorder. Quantitative analysis from ^{31}P signals reveals that the phosphorus atoms surrounded by less nickel react first (Figure 8.15c). As expected, the in situ ^{31}P resonance for the $Li_xP(Ni)$ presents a linear evolution upon cycling (Figure 8.15d).[146] However, the ^7Li signal of the metallic Li electrode remains almost constant during the first discharge due to the skin depth effect and the homogenously stripped Li^+ from the Li surface. Upon charge, the Li^+ is extracted from $Li_xP(Ni)$ and loosely deposited onto Li metal, allowing RF penetration to excite all spins in the newly formed moss, further resulting in an increased linear signal.[146]

8.8 NUCLEAR MAGNETIC RESONANCE IMAGING

In addition to the conventional NMR, introducing an extra field gradient surrounding the NMR coil (refer to Figure 8.14) will differentiate the spin densities in space. The thermal diffusion coefficient could be obtained to evaluate the Li^+ transport properties by fitting the signal attenuation as a function of the diffusion time or the applied gradient.[118,155–159] Diffusion constants at different temperatures would provide the activation energy of the studied ions. Here, we mainly focus on the MRI of batteries with the aid of PFGs. In situ MRI was developed to map various active components within batteries, such as the formation and distribution of lithium microstructures on the electrodes,[41,143] the evolution of lithium concentration in liquid electrolyte upon cycling,[118,147,148] stray-field MRI (STRAFI) tracking the distribution of lithium metal electrodes, and imaging of paramagnetic cathode $LiFePO_4$ with high spatial resolution due to the strong gradient,[160] scanning image-selected in situ spectroscopy (S-ISIS) capturing solid electrodes $LiCoO_2$ (LCO) and $Li_4Ti_5O_{12}$ (LTO) with a short transverse relaxation time T_2,[42] and the evolution of Li^+ ions inside solid electrolytes.[113,145]

MRI can visualize the distribution, configuration, and morphology of lithium microstructures. In situ ^7Li MRI was proposed for mapping the Li microstructure formation into a carbon electrode for the first time.[38] Modification of Li upon electrochemical cycling was mapped by in situ ^7Li MRI.[41] The Li//Li symmetric bag cell (Figure 8.16a) was located parallel with B_0 field. The pristine cell shows a sharp peak centered at 275 ppm due to the Knight shift caused by the conductive electrons (Figure 8.16b). Upon polarization, the lithium is stripped and loosely deposited onto the other electrode, resulting in a wide resonance spreading to 250 ppm because of the broad distribution of lithium environments inside the microstructures. The large shift change here is mainly raised from the bulk magnetic susceptibility (BMS).[161] The Li surface becomes rough after polarization, as demonstrated by the 3D in situ images (Figure 8.16c). Chemical shift imaging (CSI) spectra extracted from the x–y plane for the battery reflect the growth of dendrites on one electrode after polarization (Figure 8.16d).

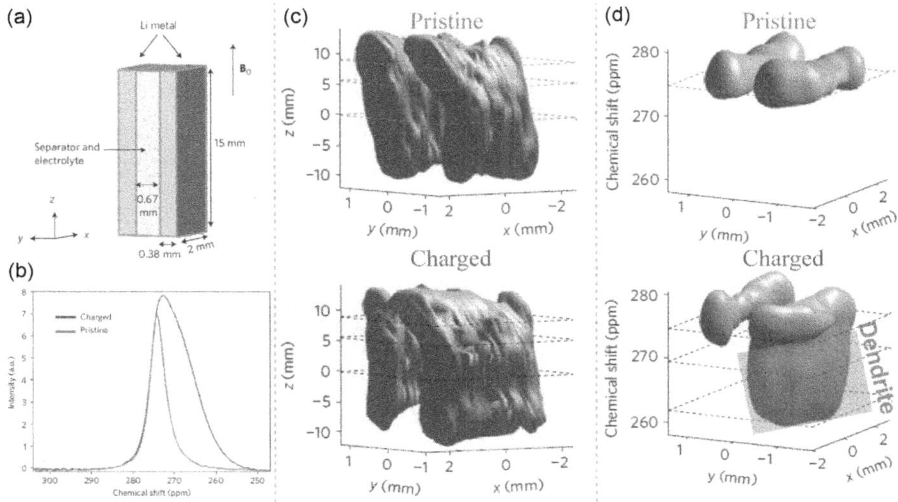

FIGURE 8.16 In situ ^{7}Li NMR study of a Li//Li symmetric cell.[41] (a) Schematic of the Li//Li cell. (b) ^{7}Li NMR spectra of the pristine and polarized battery. (c) 3D ^{7}Li MRI. (d) CSI extracted from *x*–*y* plane. (Adapted from Ref. 41 with permission from Springer Nature, Copyright [2012].)

MRI is conventionally applied to map the battery components with narrow NMR signals and a long transverse relaxation time T_2, such as liquid electrolytes and metallic Li or Na electrodes. By advancing methodology, Li^+ in paramagnetic $LiFePO_4$ was successfully mapped with the STRAFI.[160] However, STRAFI and conventional CSI cannot simultaneously acquire both spatial and chemical information for solids with short T_2. Solid electrodes, except for metallic Li, usually have T_2 shorter than 300 μs, which is insufficient for the CSI acquisition (a minimum of ~450 μs is needed for CSI).[42] To overcome this challenge, the longer longitudinal time T_1 is used to store the information on the position of spins rather than T_2.[42] By using the scanning image-selected in-situ spectroscopy (S-ISIS), all components in the battery (Figure 8.17a) are detected with both position and chemical shift information (Figure 8.17b), whereas $LiCoO_2$ (LCO) and $Li_4Ti_5O_{12}$ (LTO) are not detected by conventional CSI (Figure 8.17c). The S-ISIS images provide direct evidence that the region next to the separator reacts first for the LTO electrodes upon charge and discharge, as they display poor lithium diffusion properties (Figure 8.17d–i),[42] with biphasic and solid-solution reaction mechanisms for LTO and LCO, respectively.

MRI is capable of nondestructively visualizing the evolution of solid electrolytes (Figure 8.18). $Li_{10}GeP_2S_{12}$ (LGPS) is an attractive solid electrolyte due to high ion conduction. However, its poor chemical stability could be improved by the coating of PEO (PEO-LGPS) (Figure 8.18c). The origin of the increased impedance upon polarization is unclear, and there are no conventional tools that can noninvasively access the underlying cause. A cylindrical Li/LGPS/Li symmetric cell[145] was assembled (Figure 8.18a and b). LGPS was pressed between two lithium disks on both sides, and two screws and three O-rings were used to keep it airtight. The ^{7}Li concentration

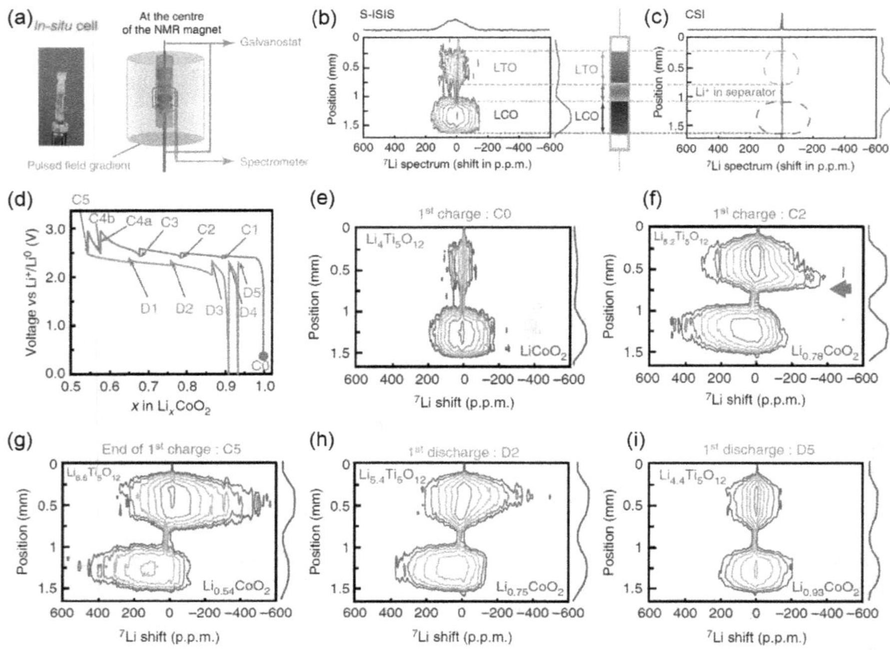

FIGURE 8.17 (a) Cylindrical cell for in situ NMR study with gradient for imaging. (b) S-ISIS imaging and (c) in situ conventional CSI. The vertical axis measures the position along the battery and the horizontal dimension for the chemical shift. (d) Electrochemical control. (e–i) S-ISIS images recorded at different cycling states are shown in (d).[42] (Reproduced from Ref. 42 with permission from Springer Nature, Copyright [2016].)

across the electrolyte pellet is homogenous for the pristine LGPS and PEO-LGPS (Figure 8.18d and e) as observed from ^7Li MRI. Many cracks are observed for the cycled LGPS (to the right of Figure 8.18d). The resulting disconnections between LGPS fragments prevent Li^+ transport, resulting in high internal resistance. For the PEO-LGPS, the Li^+ distribution is almost unchanged after 15 days of cycling (Figure 8.18e). The quantification deduces that the inhomogeneity of Li^+ concentration across the electrolyte and the deficiency at the electrodes/electrolyte interface is the main reason for the increased resistance.

8.9 SUMMARY AND PERSPECTIVES

NMR and MRI can powerfully unveil the atomic structure, dynamic interactions, and spin distribution for crystalline or amorphous materials. The chemical shift and resonance configuration undoubtedly reveal the coordination of investigated isotopes. The NMR signal ratio is proportional to the number of relevant atoms if all signals are properly recorded. Advanced techniques, such as relaxation, diffusion, exchange, spin diffusion, homonuclear and heteronuclear correlation, would provide additional power to determine more detailed structure dynamics information for materials.

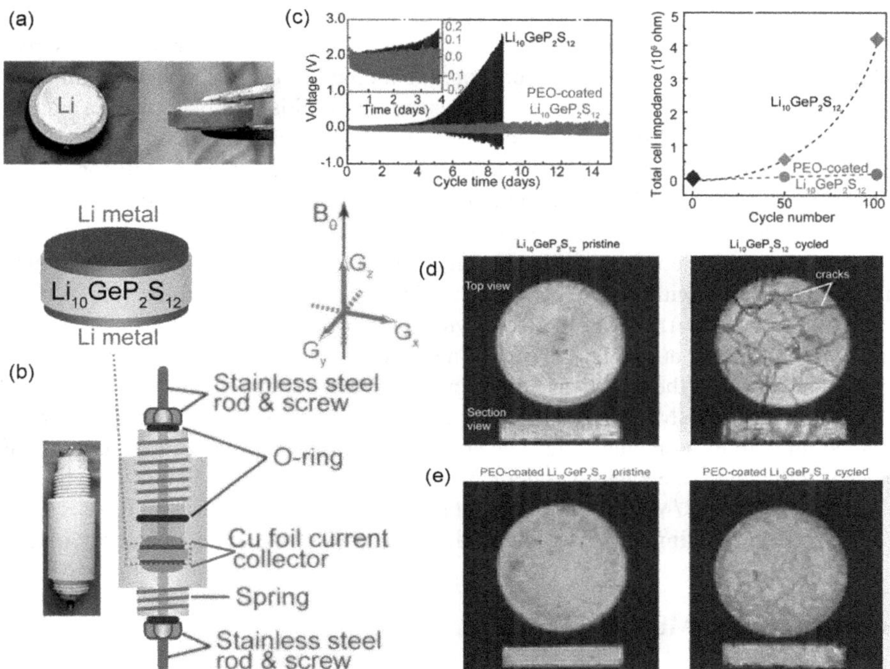

FIGURE 8.18 In situ ^{7}Li MRI of LGPS correlating to its failure in electrochemical cycling.[145] (a) Picture diagram of Li/LGPS/Li battery located in a home-made cylindrical cell (b). (c) Cycling performance of the LGPS and PEO-coated LGPS (PEO-LGPS). Top and section views from the in situ ^{7}Li MRI of (d) the pristine and the cycled LGPS, and (e) the pristine and cycled PEO-LGPS, respectively. (Reproduced from Ref. 145 with permission from the American Chemical Society, Copyright [2018].)

With the aid of a PFG, spatial information can be easily obtained to analyze the concentration distribution, with resolutions ranging from μm to cm depending on the used gradient, the nature of the studied materials and the isotopes under investigation. In addition, due to their inherently noninvasive characteristics, NMR/MRI could be extended to in situ research of working batteries in real time, thus providing an extra temporal resolution. The plastic bag cell and cylindrical cell have been well adapted to in situ NMR/MRI. Isotopes such as ^{1}H, ^{6,7}Li, ^{11}B, ^{13}C, ^{15}N, ^{17}O, ^{19}F, ^{23}Na, ^{23}Al, ^{29}Si, ^{31}P, ^{35}Cl, ^{63}Cu, ^{71}Ga, ^{79}Br, and ^{119}Sn have been successful in revealing the structure and reaction mechanisms of electrolyte and electrodes in batteries. The causes of battery failure are also evidenced by in situ NMR/MRI, such as self-discharge, dendrite formation, and deficiency.

Although NMR has demonstrated many merits, it has suffered from a low signal-to-noise (S/N) ratio due to Boltzmann distribution. Improving sensitivity is one of the key directions worth paying intensive attention to, from both hardware and software aspects. From a hardware perspective, building up a high magnetic field, designing a special coil with a high filling factor, fabricating a stronger PFG, and adding a filter for the NMR signal manipulations are the main focus of development.

Turning to the software aspect, optimizing and developing pulse programming and the post-data process should be conducted to improve the spectral resolution, signal reliability, and results analysis. In addition, the method of transition spin signal from the surrounding unpaired electron to the measured nuclei is promising with a significantly enhanced signal or so-called dynamic nuclear polarization (DNP).[116,162–165] An improved spectral resolution could be achieved by using a spinning cell for in situ NMR investigations.[152] Extending data analysis could be achieved by distinguishing time-independent and -dependent signals to improve spectral resolution by a derivative *operando* (dOp) approach.[166] In situ MRI serves as a noninvasive approach to map the ionic evolution within the electrode and electrolyte. It paves a direct way to track mass transport properties within electrolytes under operation conditions. The spatial and temporal information obtained from in situ MRI could help understand the working mechanism and the causes of failure in the batteries.

In conclusion, NMR/MRI can comprehensively investigate battery materials from atomic to micrometer scale. The correlation between structure and function of battery, and dynamic evolution under electrochemical operation can be studied using a noninvasive NMR/MRI probe. The obtained insights on batteries via NMR shed vital light on achieving better structural design and electrochemical performance.

ACKNOWLEDGMENTS

This work is supported by the National Natural Science Foundation of China under Grant Nos. 21974007 and 22090043.

ABBREVIATIONS

NMR, nuclear magnetic resonance; MRI, magnetic resonance imaging; SSBs, spinning sidebands; BMS, bulk magnetic susceptibility; S/N, signal-to-noise; MAS, magic angle spinning; pjMATPASS, projection-Magic-Angle-Turning and Phase-Adjusted Sideband Separation; CP, cross polarization; REDOR, rotational-echo double-resonance; S-ISIS, scanning image-selected in situ spectroscopy; CSI, chemical shift imaging; STRAFI, stray-field imaging; DNP, dynamic nuclear polarization; PFG, pulsed field gradient; RF, radio frequency; LIB, lithium ion battery; SEI, solid electrolyte interphase; CVD, chemical vapor deposition; LFP, $LiFePO_4$; LCO, $LiCoO_2$; LTO, $Li_4Ti_5O_{12}$; LLZO, $Li_7La_3Zr_2O_{12}$; PEO, polyethylene oxide; LGPS, $Li_{10}GeP_2S_{12}$.

REFERENCES

[1] Tarascon, J.-M.; Armand, M. 2001. Issues and Challenges Facing Rechargeable Lithium Batteries. *Nature*, *414* (6861), 359–367.

[2] Armand, M.; Tarascon, J.-M. 2008. Building Better Batteries. *Nature*, *451* (7179), 652–657.

[3] Grey, C. P.; Tarascon, J. M. 2017. Sustainability and In Situ Monitoring in Battery Development. *Nat. Mater.*, *16* (1), 45–56.

[4] Harks, P. P. R. M. L.; Mulder, F. M.; Notten, P. H. L. 2015. In Situ Methods for Li-Ion Battery Research: A Review of Recent Developments. *J. Power Sources*, *288*, 92–105.

[5] Yang, J.; Muhammad, S.; Ru Jo, M.; Kim, H.; Song, K.; Adjei Agyeman, D.; Kim, Y.-I.; Yoon, W.-S.; Kang, Y.-M. 2016. In Situ Analyses for Ion Storage Materials. *Chem. Soc. Rev.*, *45* (20), 5717–5770.

[6] Li, J.-T.; Chen, S.-R.; Ke, F.-S.; Wei, G.-Z.; Huang, L.; Sun, S.-G. 2010. In Situ Microscope FTIR Spectroscopic Studies of Interfacial Reactions of Sn–Co Alloy Film Anode of Lithium Ion Battery. *J. Electroanal. Chem.*, *649* (1–2), 171–176.

[7] Li, J.-T.; Zhou, Z.-Y.; Broadwell, I.; Sun, S.-G. 2012. In-Situ Infrared Spectroscopic Studies of Electrochemical Energy Conversion and Storage. *Acc. Chem. Res.*, *45* (4), 485–494.

[8] Panitz, J.-C.; Joho, F.; Novak, P. 1999. In Situ Characterization of a Graphite Electrode in a Secondary Lithium-Ion Battery Using Raman Microscopy. *Appl. Spectrosc.*, *53* (10), 1188–1199.

[9] Dokko, K.; Shi, Q.; Stefan, I. C.; Scherson, D. A. 2003. In Situ Raman Spectroscopy of Single Microparticle Li^+–Intercalation Electrodes. *J. Phys. Chem. B*, *107* (46), 12549–12554.

[10] Hy, S.; Felix; Chen, Y.-H.; Liu, J.; Rick, J.; Hwang, B.-J. 2014. In Situ Surface Enhanced Raman Spectroscopic Studies of Solid Electrolyte Interphase Formation in Lithium Ion Battery Electrodes. *J. Power Sources*, *256*, 324–328.

[11] Stancovski, V.; Badilescu, S. 2014. In Situ Raman Spectroscopic–Electrochemical Studies of Lithium-Ion Battery Materials: A Historical Overview. *J. Appl. Electrochem.*, *44* (1), 23–43.

[12] Venkateswara Rao, C.; Soler, J.; Katiyar, R.; Shojan, J.; West, W. C.; Katiyar, R. S. 2014. Investigations on Electrochemical Behavior and Structural Stability of $Li_{1.2}Mn_{0.54}Ni_{0.13}Co_{0.13}O_2$ Lithium-Ion Cathodes via In-Situ and Ex-Situ Raman Spectroscopy. *J. Phys. Chem. C*, *118* (26), 14133–14141.

[13] Zhou, H.; Einarsrud, M.-A.; Vullum-Bruer, F. 2013. In Situ X-Ray Diffraction and Electrochemical Impedance Spectroscopy of a Nanoporous Li_2FeSiO_4/C Cathode during the Initial Charge/Discharge Cycle of a Li-Ion Battery. *J. Power Sources*, *238*, 478–484.

[14] Placke, T.; Schmuelling, G.; Kloepsch, R.; Meister, P.; Fromm, O.; Hilbig, P.; Meyer, H.-W.; Winter, M. 2014. In Situ X-Ray Diffraction Studies of Cation and Anion Intercalation into Graphitic Carbons for Electrochemical Energy Storage Applications. *Z. Anorg. Allg. Chem.*, *640* (10), 1996–2006.

[15] Kirshenbaum, K.; Bock, D. C.; Lee, C.-Y.; Zhong, Z.; Takeuchi, K. J.; Marschilok, A. C.; Takeuchi, E. S. 2015. In Situ Visualization of $Li/Ag_2VP_2O_8$ Batteries Revealing Rate-Dependent Discharge Mechanism. *Science*, *347* (6218), 149–154.

[16] Le Toquin, R.; Paulus, W.; Cousson, A.; Prestipino, C.; Lamberti, C. 2006. Time-Resolved In Situ Studies of Oxygen Intercalation into $SrCoO_{2.5}$, Performed by Neutron Diffraction and X-Ray Absorption Spectroscopy. *J. Am. Chem. Soc.*, *128* (40), 13161–13174.

[17] Senyshyn, A.; Mühlbauer, M. J.; Nikolowski, K.; Pirling, T.; Ehrenberg, H. 2012. "In-Operando" Neutron Scattering Studies on Li-Ion Batteries. *J. Power Sources*, *203*, 126–129.

[18] Hu, C.-W.; Sharma, N.; Chiang, C.-Y.; Su, H.-C.; Peterson, V. K.; Hsieh, H.-W.; Lin, Y.-F.; Chou, W.-C.; Shew, B.-Y.; Lee, C.-H. 2013. Real-Time Investigation of the Structural Evolution of Electrodes in a Commercial Lithium-Ion Battery Containing a V-Added $LiFePO_4$ Cathode Using In-Situ Neutron Powder Diffraction. *J. Power Sources*, *244*, 158–163.

[19] Bobrikov, I. A.; Balagurov, A. M.; Hu, C.-W.; Lee, C.-H.; Chen, T.-Y.; Deleg, S.; Balagurov, D. A. 2014. Structural Evolution in $LiFePO_4$-Based Battery Materials: In-Situ and Ex-Situ Time-of-Flight Neutron Diffraction Study. *J. Power Sources*, *258*, 356–364.

[20] Orsini, F.; Du Pasquier, A.; Beaudoin, B.; Tarascon, J. M.; Trentin, M.; Langenhuizen, N.; De Beer, E.; Notten, P. 1998. In Situ Scanning Electron Microscopy (SEM) Observation of Interfaces within Plastic Lithium Batteries. *J. Power Sources*, *76* (1), 19–29.

[21] Raimann, P. R.; Hochgatterer, N. S.; Korepp, C.; Möller, K. C.; Winter, M.; Schröttner, H.; Hofer, F.; Besenhard, J. O. 2006. Monitoring Dynamics of Electrode Reactions in Li-Ion Batteries by In Situ ESEM. *Ionics*, *12* (4–5), 253–255.

[22] Hovington, P.; Dontigny, M.; Guerfi, A.; Trottier, J.; Lagacé, M.; Mauger, A.; Julien, C. M.; Zaghib, K. 2014. In Situ Scanning Electron Microscope Study and Microstructural Evolution of Nano Silicon Anode for High Energy Li-Ion Batteries. *J. Power Sources*, *248*, 457–464.

[23] Orsini, F.; du Pasquier, A.; Beaudouin, B.; Tarascon, J. M.; Trentin, M.; Langenhuizen, N.; de Beer, E.; Notten, P. 1999. In Situ SEM Study of the Interfaces in Plastic Lithium Cells. *J. Power Sources*, *81–82*, 918–921.

[24] Abellan, P.; Mehdi, B. L.; Parent, L. R.; Gu, M.; Park, C.; Xu, W.; Zhang, Y.; Arslan, I.; Zhang, J.-G.; Wang, C.-M.; et al. 2014. Probing the Degradation Mechanisms in Electrolyte Solutions for Li-Ion Batteries by In Situ Transmission Electron Microscopy. *Nano Lett.*, *14* (3), 1293–1299.

[25] Layla Mehdi, B.; Gu, M.; Parent, L. R.; Xu, W.; Nasybulin, E. N.; Chen, X.; Unocic, R. R.; Xu, P.; Welch, D. A.; Abellan, P.; et al. 2014. In-Situ Electrochemical Transmission Electron Microscopy for Battery Research. *Microsc. Microanal.*, *20* (02), 484–492.

[26] Liu, S.; Xie, J.; Su, Q.; Du, G.; Zhang, S.; Cao, G.; Zhu, T.; Zhao, X. 2014. Understanding Li-Storage Mechanism and Performance of $MnFe_2O_4$ by in Situ TEM Observation on Its Electrochemical Process in Nano Lithium Battery. *Nano Energy*, *8*, 84–94.

[27] Wu, F.; Yao, N. 2015. Advances in Sealed Liquid Cells for In-Situ TEM Electrochemial Investigation of Lithium-Ion Battery. *Nano Energy*, 11, 196–210.

[28] Cohen, Y.; Aurbach, D. 1999. The Use of a Special Work Station for In Situ Measurements of Highly Reactive Electrochemical Systems by Atomic Force and Scanning Tunneling Microscopes. *Rev. Sci. Instrum.*, *70* (12), 4668–4675.

[29] Cohen, Y. S.; Aurbach, D. 2004. Surface Films Phenomena on Vanadium-Pentoxide Cathodes for Li and Li-Ion Batteries: In Situ AFM Imaging. *Electrochem. Commun.*, *6* (6), 536–542.

[30] Zhang, J.; Wang, R.; Yang, X.; Lu, W.; Wu, X.; Wang, X.; Li, H.; Chen, L. 2012. Direct Observation of Inhomogeneous Solid Electrolyte Interphase on MnO Anode with Atomic Force Microscopy and Spectroscopy. *Nano Lett.*, *12* (4), 2153–2157.

[31] Ramdon, S.; Bhushan, B.; Nagpure, S. C. 2014. In Situ Electrochemical Studies of Lithium-Ion Battery Cathodes Using Atomic Force Microscopy. *J. Power Sources*, *249*, 373–384.

[32] Wang, L.; Deng, D.; Lev, L. C.; Ng, S. 2014. In-Situ Investigation of Solid-Electrolyte Interphase Formation on the Anode of Li-Ion Batteries with Atomic Force Microscopy. *J. Power Sources*, *265*, 140–148.

[33] Sathiya, M.; Leriche, J.-B.; Salager, E.; Gourier, D.; Tarascon, J.-M.; Vezin, H. 2015. Electron Paramagnetic Resonance Imaging for Real-Time Monitoring of Li-Ion Batteries. *Nat. Commun.*, *6*(1), 1–7.

[34] Wandt, J.; Marino, C.; Gasteiger, H. A.; Jakes, P.; Eichel, R.-A.; Granwehr, J. 2015. Operando Electron Paramagnetic Resonance Spectroscopy – Formation of Mossy Lithium on Lithium Anodes during Charge–Discharge Cycling. *Energy Environ. Sci.*, *8* (4), 1358–1367.

[35] Wang, Q.; Zheng, J.; Walter, E.; Pan, H.; Lv, D.; Zuo, P.; Chen, H.; Deng, Z. D.; Liaw, B. Y.; Yu, X.; et al. 2015. Direct Observation of Sulfur Radicals as Reaction Media in Lithium Sulfur Batteries. *J. Electrochem. Soc.*, *162* (3), A474–A478.

[36] Niemöller, A.; Jakes, P.; Eichel, R.-A.; Granwehr, J. 2018. EPR Imaging of Metallic Lithium and Its Application to Dendrite Localisation in Battery Separators. *Sci. Rep.*, *8* (1), 1–7.

[37] Gerald, R. E.; Klingler, R. J.; Rathke, J. W.; Sand, G.; Woelk, K. 1998. In Situ Imaging of Charge Carriers in an Electrochemical Cell. In *Spatially Resolved Magnetic Resonance*; Blmler, P., Blmich, B., Botto, R., Fukushima, E., Eds.; Wiley-VCH Verlag GmbH: Weinheim, Germany, pp 111–119.

[38] Ii, R. E. G.; Klingler, R. J.; Sandı, G.; Johnson, C. S.; Scanlon, L. G.; Rathke, J. W. 2000. ^{7}Li NMR Study of Intercalated Lithium in Curved Carbon Lattices. *J. Power Sources*, *89*, 237–243.

[39] Gerald, R. E.; Sanchez, J.; Johnson, C. S.; Klingler, R. J.; Rathke, J. W. 2001. *In Situ* Nuclear Magnetic Resonance Investigations of Lithium Ions in Carbon Electrode Materials Using a Novel Detector. *J. Phys. Condens. Matter*, *13* (36), 8269–8285.

[40] Blanc, F.; Leskes, M.; Grey, C. P. 2013. In Situ Solid-State NMR Spectroscopy of Electrochemical Cells: Batteries, Supercapacitors, and Fuel Cells. *Acc. Chem. Res.*, *46* (9), 1952–1963.

[41] Chandrashekar, S.; Trease, N. M.; Chang, H. J.; Du, L.-S.; Grey, C. P.; Jerschow, A. 2012.^{7}Li MRI of Li Batteries Reveals Location of Microstructural Lithium. *Nat. Mater.*, *11* (4), 311–315.

[42] Tang, M.; Sarou-Kanian, V.; Melin, P.; Leriche, J.-B.; Ménétrier, M.; Tarascon, J.-M.; Deschamps, M.; Salager, E. 2016. Following Lithiation Fronts in Paramagnetic Electrodes with in Situ Magnetic Resonance Spectroscopic Imaging. *Nat. Commun.*, *7* (1), 1–8.

[43] Pecher, O.; Carretero-González, J.; Griffith, K. J.; Grey, C. P. 2017. Materials' Methods: NMR in Battery Research. *Chem. Mater.*, *29* (1), 213–242.

[44] Li, X.; Tang, M.; Feng, X.; Hung, I.; Rose, A.; Chien, P.-H.; Gan, Z.; Hu, Y.-Y. 2017. Lithiation and Delithiation Dynamics of Different Li Sites in Li-Rich Battery Cathodes Studied by Operando Nuclear Magnetic Resonance. *Chem. Mater.*, *29* (19), 8282–8291.

[45] Romanenko, K.; Jerschow, A. 2019. Distortion-Free Inside-Out Imaging for Rapid Diagnostics of Rechargeable Li-Ion Cells. *Proc. Natl. Acad. Sci.*, *116* (38), 18783–18789.

[46] Lebens-Higgins, Z. W.; Faenza, N. V.; Radin, M. D.; Liu, H.; Sallis, S.; Rana, J.; Vinckeviciute, J.; Reeves, P. J.; Zuba, M. J.; Badway, F.; et al. 2019. Revisiting the Charge Compensation Mechanisms in $LiNi_{0.8}Co_{0.2-y}Al_yO_2$ Systems. *Mater. Horiz.*, *6* (10), 2112–2123.

[47] Rosina, K. J.; Jiang, M.; Zeng, D.; Salager, E.; Best, A. S.; Grey, C. P. 2012. Structure of Aluminum Fluoride Coated $Li[Li_{1/9}Ni_{1/3}Mn_{5/9}]O_2$ Cathodes for Secondary Lithium-Ion Batteries. *J. Mater. Chem.*, *22* (38), 20602–20610.

[48] Armstrong, A. R.; Paterson, A. J.; Dupré, N.; Grey, C. P.; Bruce, P. G. 2007. Structural Evolution of Layered $Li_xMn_yO_2$: Combined Neutron, NMR, and Electrochemical Study. *Chem. Mater.*, *19* (5), 1016–1023.

[49] Charles-Blin, Y.; Flahaut, D.; Ledeuil, J.-B.; Guérin, K.; Dubois, M.; Deschamps, M.; Perbost, A.-M.; Monconduit, L.; Martinez, H.; Louvain, N. 2019. Atomic Layer Fluorination of the $Li_4Ti_5O_{12}$ Surface: A Multiprobing Survey. *ACS Appl. Energy Mater.*, *2* (9), 6681–6692.

[50] Märker, K.; Reeves, P. J.; Xu, C.; Griffith, K. J.; Grey, C. P. 2019. Evolution of Structure and Lithium Dynamics in $LiNi_{0.8}Mn_{0.1}Co_{0.1}O_2$(NMC811) Cathodes during Electrochemical Cycling. *Chem. Mater.*, *31* (7), 2545–2554.

[51] Jung, H.; Allan, P. K.; Hu, Y.-Y.; Borkiewicz, O. J.; Wang, X.-L.; Han, W.-Q.; Du, L.-S.; Pickard, C. J.; Chupas, P. J.; Chapman, K. W.; et al. 2015. Elucidation of the Local and Long-Range Structural Changes That Occur in Germanium Anodes in Lithium-Ion Batteries. *Chem. Mater.*, *27* (3), 1031–1041.

[52] Buzlukov, A.; Mouesca, J.-M.; Buannic, L.; Hediger, S.; Simonin, L.; Canevet, E.; Colin, J.-F.; Gutel, T.; Bardet, M. 2016. Li-Rich Mn/Ni Layered Oxide as Electrode Material for Lithium Batteries: A ^{7}Li MAS NMR Study Revealing Segregation into (Nanoscale) Domains with Highly Different Electrochemical Behaviors. *J. Phys. Chem. C*, *120* (34), 19049–19063.

[53] Murakami, M.; Noda, Y.; Koyama, Y.; Takegoshi, K.; Arai, H.; Uchimoto, Y.; Ogumi, Z. 2014. Local Structure and Spin State of Cobalt Ion at Defect in Lithium Overstoichiometric $LiCoO_2$ As Studied by $^{6/7}$Li Solid-State NMR Spectroscopy. *J. Phys. Chem. C*, *118* (28), 15375–15385.

[54] Ilott, A. J.; Mohammadi, M.; Schauerman, C. M.; Ganter, M. J.; Jerschow, A. 2018. Rechargeable Lithium-Ion Cell State of Charge and Defect Detection by In-Situ Inside-Out Magnetic Resonance Imaging. *Nat. Commun.*, *9* (1), 1776.

[55] Guo, L.; Zhang, Y.; Wang, J.; Ma, L.; Ma, S.; Zhang, Y.; Wang, E.; Bi, Y.; Wang, D.; McKee, W. C.; et al. 2015. Unlocking the Energy Capabilities of Micron-Sized $LiFePO_4$. *Nat. Commun.*, *6*, 7898.

[56] Salager, E.; Sarou-Kanian, V.; Sathiya, M.; Tang, M.; Leriche, J.-B.; Melin, P.; Wang, Z.; Vezin, H.; Bessada, C.; Deschamps, M.; et al. 2014. Solid-State NMR of the Family of Positive Electrode Materials $Li_2Ru_{1-y}Sn_yO_3$ for Lithium-Ion Batteries. *Chem. Mater.*, *26* (24), 7009–7019.

[57] Song, B.; Tang, M.; Hu, E.; Borkiewicz, O. J.; Wiaderek, K. M.; Zhang, Y.; Phillip, N. D.; Liu, X.; Shadike, Z.; Li, C.; et al. 2019. Understanding the Low-Voltage Hysteresis of Anionic Redox in $Na_2Mn_3O_7$. *Chem. Mater.*, *31* (10), 3756–3765.

[58] Hou, J.; Hadouchi, M.; Sui, L.; Liu, J.; Tang, M.; Kan, W. H.; Avdeev, M.; Zhong, G.; Liao, Y.-K.; Lai, Y.-H.; et al. 2021. Unlocking Fast and Reversible Sodium Intercalation in NASICON $Na_4MnV(PO_4)_3$ by Fluorine Substitution. *Energy Storage Mater.*, *42*, 307–316.

[59] Li, X.; Li, X.; Monluc, L.; Chen, B.; Tang, M.; Chien, P.-H.; Feng, X.; Hung, I.; Gan, Z.; Urban, A.; et al. 2022. Stacking-Fault Enhanced Oxygen Redox in Li_2MnO_3. *Adv. Energy Mater.*, 12(18), 2200427.

[60] Hadouchi, M.; Yaqoob, N.; Kaghazchi, P.; Tang, M.; Liu, J.; Sang, P.; Fu, Y.; Huang, Y.; Ma, J. 2021. Fast Sodium Intercalation in $Na_{3.41}£_{0.59}FeV(PO_4)_3$: A Novel Sodium-Deficient NASICON Cathode for Sodium-Ion Batteries. *Energy Storage Mater.*, *35*, 192–202.

[61] Ménétrier, M.; Saadoune, I.; Levasseur, S.; Delmas, C. 1999. The Insulator-Metal Transition upon Lithium Deintercalation from $LiCoO_2$: Electronic Properties and 7Li NMR Study. *J. Mater. Chem.*, *9* (5), 1135–1140.

[62] Davis, L. J. M.; Heinmaa, I.; Ellis, B. L.; Nazar, L. F.; Goward, G. R. 2011. Influence of Particle Size on Solid Solution Formation and Phase Interfaces in $Li_{0.5}FePO_4$ Revealed by ^{31}P and^7Li Solid State NMR Spectroscopy. *Phys. Chem. Chem. Phys.*, *13* (11), 5171–5177.

[63] Tucker, M. C.; Doeff, M. M.; Richardson, T. J.; Fiñones, R.; Reimer, J. A.; Cairns, E. J. 2002. ^7Li and ^{31}P Magic Angle Spinning Nuclear Magnetic Resonance of $LiFePO_4$-Type Materials. *Electrochem. Solid-State Lett.*, *5* (5), A95–A98.

[64] Aron, D.; Zorko, A.; Dominko, R.; Jaglii, Z. 2004. A Comparative Study of Magnetic Properties of $LiFePO_4$ and $LiMnPO_4$. *J. Phys. Condens. Matter*, *16* (30), 5531–5548.

[65] Wilcke, S. L.; Lee, Y.-J.; Cairns, E. J.; Reimer, J. A. 2007. Covalency Measurements via NMR in Lithium Metal Phosphates. *Appl. Magn. Reson.*, *32* (4), 547–563.

[66] Cabana, J.; Shirakawa, J.; Chen, G.; Richardson, T. J.; Grey, C. P. 2010. MAS NMR Study of the Metastable Solid Solutions Found in the $LiFePO_4/FePO_4$ System. *Chem. Mater.*, *22* (3), 1249–1262.

[67] Castets, A.; Carlier, D.; Zhang, Y.; Boucher, F.; Marx, N.; Croguennec, L.; Ménétrier, M. 2011. Multinuclear NMR and DFT Calculations on the $LiFePO_4{\cdot}OH$ and $FePO_4{\cdot}H_2O$ Homeotypic Phases. *J. Phys. Chem. C*, *115* (32), 16234–16241.

[68] Castets, A.; Carlier, D.; Zhang, Y.; Boucher, F.; Ménétrier, M. 2012. A DFT-Based Analysis of the NMR Fermi Contact Shifts in Tavorite-like $LiMPO_4{\cdot}OH$ and $MPO_4{\cdot}H_2O$ (M = Fe, Mn, V). *J. Phys. Chem. C*, *116* (34), 18002–18014.

[69] Nagpure, S. C.; Bhushan, B.; Babu, S. S. 2012. Raman and NMR Studies of Aged $LiFePO_4$ Cathode. *Appl. Surf. Sci.*, *259*, 49–54.

[70] Cuisinier, M.; Dupré, N.; Martin, J.-F.; Kanno, R.; Guyomard, D. 2013. Evolution of the $LiFePO_4$ Positive Electrode Interface along Cycling Monitored by MAS NMR. *J. Power Sources*, *224*, 50–58.

[71] Rudisch, C.; Grafe, H.-J.; Geck, J.; Partzsch, S.; Zimmermann, M. V.; Wizent, N.; Klingeler, R.; Büchner, B. 2013. Coupling of Li Motion and Structural Distortions in Olivine $LiMnPO_4$ from ^7Li and ^{31}P NMR. *Phys. Rev. B*, *88* (5), 054303.

[72] Dupré, N.; Cuisinier, M.; Martin, J.-F.; Guyomard, D. 2014. Interphase Evolution at Two Promising Electrode Materials for Li-Ion Batteries: $LiFePO_4$ and $LiNi_{1/2}Mn_{1/2}O_2$. *ChemPhysChem*, *15* (10), 1922–1938.

[73] Kaus, M.; Issac, I.; Heinzmann, R.; Doyle, S.; Mangold, S.; Hahn, H.; Chakravadhanula, V. S. K.; Kübel, C.; Ehrenberg, H.; Indris, S. 2014. Electrochemical Delithiation/Relithiation of $LiCoPO_4$: A Two-Step Reaction Mechanism Investigated by In Situ X-Ray Diffraction, In Situ X-Ray Absorption Spectroscopy, and Ex Situ $^7Li/^{31}P$ NMR Spectroscopy. *J. Phys. Chem. C*, *118* (31), 17279–17290.

[74] Shimoda, K.; Sugaya, H.; Murakami, M.; Arai, H.; Uchimoto, Y.; Ogumi, Z. 2014. Characterization of Bulk and Surface Chemical States on Electrochemically Cycled $LiFePO_4$: A Solid State NMR Study. *J. Electrochem. Soc.*, *161* (6), A1012–A1018.

[75] Hung, I.; Zhou, L.; Pourpoint, F.; Grey, C. P.; Gan, Z. 2012. Isotropic High Field NMR Spectra of Li-Ion Battery Materials with Anisotropy >1 MHz. *J. Am. Chem. Soc.*, *134* (4), 1898–1901.

[76] Lee, Y. J.; Park, S.-H.; Eng, C.; Parise, J. B.; Grey, C. P. 2002. Cation Ordering and Electrochemical Properties of the Cathode Materials $LiZn_xMn_{2-x}O_4$, $0 < x \leq 0.5$: A ^6Li Magic-Angle Spinning NMR Spectroscopy and Diffraction Study. *Chem. Mater.*, *14* (1), 194–205.

[77] Cabana, J.; Dupré, N.; Rousse, G.; Grey, C. P.; Palacín, M. R. 2005. Ex Situ NMR and Neutron Diffraction Study of Structure and Lithium Motion in Li_7MnN_4. *Solid State Ion.*, *176* (29–30), 2205–2218.

[78] Holmes, L.; Peng, L.; Heinmaa, I.; O'Dell, L. A.; Smith, M. E.; Vannier, R.-N.; Grey, C. P. 2008. Variable-Temperature ^{17}O NMR Study of Oxygen Motion in the Anionic Conductor $Bi_{26}Mo_{10}O_{69}$. *Chem. Mater.*, *20* (11), 3638–3648.

[79] Zeng, D.; Cabana, J.; Yoon, W.-S.; Grey, C. P. 2010. Investigation of the Structural Changes in $Li[Ni_yMn_yCo_{(1-2y)}]O_2$ (y=0.05) upon Electrochemical Lithium Deintercalation. *Chem. Mater.*, *22* (3), 1209–1219.

[80] Billaud, J.; Clément, R. J.; Armstrong, A. R.; Canales-Vázquez, J.; Rozier, P.; Grey, C. P.; Bruce, P. G. 2014. β-$NaMnO_2$: A High-Performance Cathode for Sodium-Ion Batteries. *J. Am. Chem. Soc.*, *136* (49), 17243–17248.

[81] Dervişoğlu, R.; Middlemiss, D. S.; Blanc, F.; Lee, Y.-L.; Morgan, D.; Grey, C. P. 2015. Joint Experimental and Computational ^{17}O and ^1H Solid State NMR Study of $Ba_2In_2O_4(OH)_2$ Structure and Dynamics. *Chem. Mater.*, *27* (11), 3861–3873.

[82] Kim, J.; Ilott, A. J.; Middlemiss, D. S.; Chernova, N. A.; Pinney, N.; Morgan, D.; Grey, C. P. 2015. ^2H and ^{27}Al Solid-State NMR Study of the Local Environments in Al-Doped 2-Line Ferrihydrite, Goethite, and Lepidocrocite. *Chem. Mater.*, *27* (11), 3966–3978.

[83] Ma, J.; Bo, S.-H.; Wu, L.; Zhu, Y.; Grey, C. P.; Khalifah, P. G. 2015. Ordered and Disordered Polymorphs of $Na(Ni_{2/3}Sb_{1/3})O_2$: Honeycomb-Ordered Cathodes for Na-Ion Batteries. *Chem. Mater.*, *27* (7), 2387–2399.

[84] Trease, N. M.; Seymour, I. D.; Radin, M. D.; Liu, H.; Liu, H.; Hy, S.; Chernova, N.; Parikh, P.; Devaraj, A.; Wiaderek, K. M.; et al. 2016. Identifying the Distribution of Al^{3+} in $LiNi_{0.8}Co_{0.15}Al_{0.05}O_2$. *Chem. Mater.*, *28* (22), 8170–8180.

[85] Li, Y.; Wu, X.-P.; Jiang, N.; Lin, M.; Shen, L.; Sun, H.; Wang, Y.; Wang, M.; Ke, X.; Yu, Z.; et al. 2017. Distinguishing Faceted Oxide Nanocrystals with ^{17}O Solid-State NMR Spectroscopy. *Nat. Commun.*, *8* (1), 581.

[86] Liu, H.; Choe, M.-J.; Enrique, R. A.; Orvañanos, B.; Zhou, L.; Liu, T.; Thornton, K.; Grey, C. P. 2017. Effects of Antisite Defects on Li Diffusion in $LiFePO_4$ Revealed by Li Isotope Exchange. *J. Phys. Chem. C*, *121* (22), 12025–12036.

[87] Groh, M. F.; Sullivan, M. J.; Gaultois, M. W.; Pecher, O.; Griffith, K. J.; Grey, C. P. 2018. Interface Instability in $LiFePO_4$–$Li_{3+x}P_{1-x}Si_xO_4$ All-Solid-State Batteries. *Chem. Mater.*, *30* (17), 5886–5895.

[88] Li, C.; Shen, M.; Hu, B.; Lou, X.; Zhang, X.; Tong, W.; Hu, B. 2018. High-Energy Nanostructured $Na_3V_2(PO_4)_2O_{1.6}F_{1.4}$ Cathodes for Sodium-Ion Batteries and a New Insight into Their Redox Chemistry. *J. Mater. Chem. A*, *6* (18), 8340–8348.

[89] Liao, Y.; Li, C.; Lou, X.; Hu, X.; Ning, Y.; Yuan, F.; Chen, B.; Shen, M.; Hu, B. 2018. Carbon-Coated $Li_3V_2(PO_4)_3$ Derived from Metal-Organic Framework as Cathode for Lithium-Ion Batteries with High Stability. *Electrochimica Acta*, *271*, 608–616.

[90] Wang, Q.; Huo, H.; Lin, Z.; Zhang, Z.; Wang, S.; Yin, X.; Ma, Y.; Zuo, P.; Wang, J.; Cheng, X.; et al. 2020. Unraveling the Relationship between Ti^{4+} Doping and Li^+ Mobility Enhancement in Ti^{4+} Doped $Li_3V_2(PO_4)_3$. *ACS Appl. Energy Mater.*, *3* (1), 715–722.

[91] Feng, X.-Y.; Li, X.; Tang, M.; Gan, A.; Hu, Y.-Y. 2017. Enhanced Rate Performance of $Li_4Ti_5O_{12}$ Anodes with Bridged Grain Boundaries. *J. Power Sources*, *354*, 172–178.

[92] Schmidt, W.; Wilkening, M. 2016. Discriminating the Mobile Ions from the Immobile Ones in $Li_{4+x}Ti_5O_{12}$: ^{6}Li NMR Reveals the Main Li^+ Diffusion Pathway and Proposes a Refined Lithiation Mechanism. *J. Phys. Chem. C*, *120* (21), 11372–11381.

[93] Huff, L. A.; Tavassol, H.; Esbenshade, J. L.; Xing, W.; Chiang, Y.-M.; Gewirth, A. A. 2016. Identification of Li-Ion Battery SEI Compounds through ^{7}Li and ^{13}C Solid-State MAS NMR Spectroscopy and MALDI-TOF Mass Spectrometry. *ACS Appl. Mater. Interfaces*, *8* (1), 371–380.

[94] Li, Q.; Liu, X.; Han, X.; Xiang, Y.; Zhong, G.; Wang, J.; Zheng, B.; Zhou, J.; Yang, Y. 2019. Identification of the Solid Electrolyte Interface on the Si/C Composite Anode with FEC as the Additive. *ACS Appl. Mater. Interfaces*, *11* (15), 14066–14075.

[95] Harris, K. J.; Reeve, Z. E. M.; Wang, D.; Li, X.; Sun, X.; Goward, G. R. 2015. Electrochemical Changes in Lithium-Battery Electrodes Studied Using ^{7}Li NMR and Enhanced ^{13}C NMR of Graphene and Graphitic Carbons. *Chem. Mater.*, *27* (9), 3299–3305.

[96] Hayes, S.; van Wüllen, L.; Eckert, H.; Even, W. R.; Crocker, R. W.; Zhang, Z. 1997. Solid-State NMR Strategies for the Structural Investigation of Carbon-Based Anode Materials. *Chem. Mater.*, *9* (4), 901–911.

[97] Vyalikh, A.; Koroteev, V. O.; Münchgesang, W.; Köhler, T.; Röder, C.; Brendler, E.; Okotrub, A. V.; Bulusheva, L. G.; Meyer, D. C. 2019. Effect of Charge Transfer upon Li- and Na-Ion Insertion in Fine-Grained Graphitic Material as Probed by NMR. *ACS Appl. Mater. Interfaces*, *11* (9), 9291–9300.

[98] Chen, B.; Yang, Y.; Chen, A.; Zhang, X.; Saddique, J.; Tang, M.; Yu, H. 2021. Sodium-Ion Battery Anode Construction with SnP_x Crystal Domain in Amorphous Phosphorus Matrix. *Energy Mater. Adv.*, 2021, 1–11.

[99] Mayo, M.; Griffith, K. J.; Pickard, C. J.; Morris, A. J. 2016. Ab Initio Study of Phosphorus Anodes for Lithium- and Sodium-Ion Batteries. *Chem. Mater.*, *28* (7), 2011–2021.

[100] Liu, J.; Lou, C.; Fu, J.; Sun, X.; Hou, J.; Ma, J.; Chen, Y.; Gao, X.; Xu, L.; Wei, Q.; Tang, M. 2022. Multiple Transition Metals Modulated Hierarchical Networks for High Performance of Metal-Ion Batteries. *J. Energy Chem.*, *70*, 604–613.

[101] Hui, K.; Fu, J.; Liu, J.; Chen, Y.; Gao, X.; Gao, T.; Wei, Q.; Li, C.; Zhang, H.; Tang, M. 2021. Yolk–Shell Nanoarchitecture for Stabilizing a Ce_2S_3 Anode. *Carbon Energy*, *3* (5), 709–720.

[102] Shi, Y.; Fu, J.; Hui, K.; Liu, J.; Gao, C.; Chang, S.; Chen, Y.; Gao, X.; Gao, T.; Xu, L.; Wei, Q.; Tang, M. 2021. Promoting the Electrochemical Properties of Yolk-Shell-Structured CeO_2 Composites for Lithium-Ion Batteries. *Microstructures*, *1* (1), 2021005.

[103] Frerichs, J. E.; Ruttert, M.; Koppe, J.; Radzieowski, M.; Winter, M.; Placke, T.; Hansen, M. R. 2021. Lithiation Mechanism and Improved Electrochemical Performance of TiSnSb-Based Negative Electrodes for Lithium-Ion Batteries. *Chem. Mater.*, *33* (21), 8173–8182.

[104] Wang, H.; Sa, N.; He, M.; Liang, X.; Nazar, L. F.; Balasubramanian, M.; Gallagher, K. G.; Key, B. 2017. In Situ NMR Observation of the Temporal Speciation of Lithium Sulfur Batteries during Electrochemical Cycling. *J. Phys. Chem. C*, *121* (11), 6011–6017.

[105] Zheng, J.; Tang, M.; Hu, Y.-Y. 2016. Lithium Ion Pathway within $Li_7La_3Zr_2O_{12}$-Polyethylene Oxide Composite Electrolytes. *Angew. Chem. Int. Ed.*, *55* (40), 12538–12542.

[106] Lin, Z.; Guo, X.; Yang, Y.; Tang, M.; Wei, Q.; Yu, H. 2021. Block Copolymer Electrolyte with Adjustable Functional Units for Solid Polymer Lithium Metal Battery. *J. Energy Chem.*, *52*, 67–74.

[107] Ao, X.; Wang, X.; Tan, J.; Zhang, S.; Su, C.; Dong, L.; Tang, M.; Wang, Z.; Tian, B.; Wang, H. 2021. Nanocomposite with Fast Li^+ Conducting Percolation Network: Solid Polymer Electrolyte with Li^+ Non-Conducting Filler. *Nano Energy*, *79*, 105475.

[108] Wang, S.; Tang, M.; Zhang, Q.; Li, B.; Ohno, S.; Walther, F.; Pan, R.; Xu, X.; Xin, C.; Zhang, W.; et al. 2021. Lithium Argyrodite as Solid Electrolyte and Cathode Precursor for Solid-State Batteries with Long Cycle Life. *Adv. Energy Mater.*, *11* (31), 2101370.

[109] Yang, D.; Liang, Z.; Tang, P.; Zhang, C.; Tang, M.; Li, Q.; Biendicho, J. J.; Li, J.; Heggen, M.; Dunin-Borkowski, R. E.; et al. 2022. A High Conductivity 1D π–d Conjugated Metal–Organic Framework with Efficient Polysulfide Trapping-Diffusion-Catalysis in Lithium–Sulfur Batteries. *Adv. Mater.*, *34*, 2108835–2108845.

[110] Zheng, J.; Wang, P.; Liu, H.; Hu, Y.-Y. 2019. Interface-Enabled Ion Conduction in $Li_{10}GeP_2S_{12}$–Poly(Ethylene Oxide) Hybrid Electrolytes. *ACS Appl. Energy Mater.*, *2* (2), 1452–1459.

[111] Xu, H.; Chien, P.-H.; Shi, J.; Li, Y.; Wu, N.; Liu, Y.; Hu, Y.-Y.; Goodenough, J. B. 2019. High-Performance All-Solid-State Batteries Enabled by Salt Bonding to Perovskite in Poly(Ethylene Oxide). *Proc. Natl. Acad. Sci.*, *116* (38), 18815–18821.

[112] Karasulu, B.; Emge, S. P.; Groh, M. F.; Grey, C. P.; Morris, A. J. 2020. Al/Ga-Doped $Li_7La_3Zr_2O_{12}$ Garnets as Li-Ion Solid-State Battery Electrolytes: Atomistic Insights into Local Coordination Environments and Their Influence on ^{17}O, ^{27}Al, and ^{71}Ga NMR Spectra. *J. Am. Chem. Soc.*, *142* (6), 3132–3148.

[113] Marbella, L. E.; Zekoll, S.; Kasemchainan, J.; Emge, S. P.; Bruce, P. G.; Grey, C. P. 2019. ^{7}Li NMR Chemical Shift Imaging to Detect Microstructural Growth of Lithium in All-Solid-State Batteries. *Chem. Mater.*, *31* (8), 2762–2769.

[114] MacFarlane, D. R.; Forsyth, M.; Howlett, P. C.; Kar, M.; Passerini, S.; Pringle, J. M.; Ohno, H.; Watanabe, M.; Yan, F.; Zheng, W.; et al. 2016. Ionic Liquids and Their Solid-State Analogues as Materials for Energy Generation and Storage. *Nat. Rev. Mater.*, *1* (2), 15005.

[115] Wang, X.; Zhu, H.; Greene, G. W.; Zhou, Y.; Yoshizawa-Fujita, M.; Miyachi, Y.; Armand, M.; Forsyth, M.; Pringle, J. M.; Howlett, P. C. 2017. Organic Ionic Plastic Crystal-Based Composite Electrolyte with Surface Enhanced Ion Transport and Its Use in All-Solid-State Lithium Batteries. *Adv. Mater. Technol.*, *2* (7), 1700046.

[116] Sani, M.-A.; Martin, P.-A.; Yunis, R.; Chen, F.; Forsyth, M.; Deschamps, M.; O'Dell, L. A. 2018. Probing Ionic Liquid Electrolyte Structure via the Glassy State by Dynamic Nuclear Polarization NMR Spectroscopy. *J. Phys. Chem. Lett.*, *9* (5), 1007–1011.

[117] Krachkovskiy, S. A.; Bazak, J. D.; Werhun, P.; Balcom, B. J.; Halalay, I. C.; Goward, G. R. 2016. Visualization of Steady-State Ionic Concentration Profiles Formed in Electrolytes during Li-Ion Battery Operation and Determination of Mass-Transport Properties by In Situ Magnetic Resonance Imaging. *J. Am. Chem. Soc.*, *138* (25), 7992–7999.

[118] Krachkovskiy, S. A.; Pauric, A. D.; Halalay, I. C.; Goward, G. R. 2013. Slice-Selective NMR Diffusion Measurements: A Robust and Reliable Tool for In Situ Characterization of Ion-Transport Properties in Lithium-Ion Battery Electrolytes. *J. Phys. Chem. Lett.*, *4* (22), 3940–3944.

[119] Shao, Y.; Zhong, G.; Lu, Y.; Liu, L.; Zhao, C.; Zhang, Q.; Hu, Y.-S.; Yang, Y.; Chen, L. 2019. A Novel NASICON-Based Glass-Ceramic Composite Electrolyte with Enhanced Na-Ion Conductivity. *Energy Storage Mater.*, *23*, 514–521.

[120] Xiang, Y.-X.; Zheng, G.; Zhong, G.; Wang, D.; Fu, R.; Yang, Y. 2018. Toward Understanding of Ion Dynamics in Highly Conductive Lithium Ion Conductors: Some Perspectives by Solid State NMR Techniques. *Solid State Ion.*, *318*, 19–26.

[121] Wang, D.; Zhong, G.; Pang, W. K.; Guo, Z.; Li, Y.; McDonald, M. J.; Fu, R.; Mi, J.-X.; Yang, Y. 2015. Toward Understanding the Lithium Transport Mechanism in Garnet-Type Solid Electrolytes: Li^+ Ion Exchanges and Their Mobility at Octahedral/Tetrahedral Sites. *Chem. Mater.*, *27* (19), 6650–6659.

[122] Li, X.; Wang, D.; Wang, H.; Yan, H.; Gong, Z.; Yang, Y. 2019. Poly(Ethylene Oxide)–$Li_{10}SnP_2S_{12}$ Composite Polymer Electrolyte Enables High-Performance All-Solid-State Lithium Sulfur Battery. *ACS Appl. Mater. Interfaces*, *11* (25), 22745–22753.

[123] Dewing, B. L.; Bible, N. G.; Ellison, C. J.; Mahanthappa, M. K. 2020. Electrochemically Stable, High Transference Number Lithium Bis(Malonato)Borate Polymer Solution Electrolytes. *Chem. Mater.*, *32* (9), 3794–3804.

[124] Kalnaus, S.; Sabau, A. S.; Tenhaeff, W. E.; Dudney, N. J.; Daniel, C. 2012. Design of Composite Polymer Electrolytes for Li Ion Batteries Based on Mechanical Stability Criteria. *J. Power Sources*, *201*, 280–287.

[125] Kalnaus, S.; Tenhaeff, W. E.; Sakamoto, J.; Sabau, A. S.; Daniel, C.; Dudney, N. J. 2013. Analysis of Composite Electrolytes with Sintered Reinforcement Structure for Energy Storage Applications. *J. Power Sources*, *241*, 178–185.

[126] Zheng, J.; Dang, H.; Feng, X.; Chien, P.-H.; Hu, Y.-Y. 2017. Li-Ion Transport in a Representative Ceramic–Polymer–Plasticizer Composite Electrolyte: $Li_7La_3Zr_2O_{12}$–Polyethylene Oxide–Tetraethylene Glycol Dimethyl Ether. *J. Mater. Chem. A*, *5* (35), 18457–18463.

[127] Strangmüller, S.; Eickhoff, H.; Müller, D.; Klein, W.; Raudaschl-Sieber, G.; Kirchhain, H.; Sedlmeier, C.; Baran, V.; Senyshyn, A.; Deringer, V. L.; et al. 2019. Fast Ionic Conductivity in the Most Lithium-Rich Phosphidosilicate $Li_{14}SiP_6$. *J. Am. Chem. Soc.*, *141* (36), 14200–14209.

[128] Restle, T. M. F.; Sedlmeier, C.; Kirchhain, H.; Klein, W.; Raudaschl-Sieber, G.; van Wüllen, L.; Fässler, T. F. 2021. Fast Lithium-Ion Conduction in Phosphide Li_9GaP_4. *Chem. Mater.*, *33* (8), 2957–2966.

[129] Wu, J.-F.; Chen, E.-Y.; Yu, Y.; Liu, L.; Wu, Y.; Pang, W. K.; Peterson, V. K.; Guo, X. 2017. Gallium-Doped $Li_7La_3Zr_2O_{12}$ Garnet-Type Electrolytes with High Lithium-Ion Conductivity. *ACS Appl. Mater. Interfaces*, *9* (2), 1542–1552.

[130] Jiang, Y.; Yan, X.; Ma, Z.; Mei, P.; Xiao, W.; You, Q.; Zhang, Y. 2018. Development of the PEO Based Solid Polymer Electrolytes for All-Solid State Lithium Ion Batteries. *Polymers*, *10* (11), 1237.

[131] Wei, L.; Liu, Q.; Gao, Y.; Yao, Y.; Hu, B.; Chen, Q. 2013. Phase Structure and Helical Jump Motion of Poly(Ethylene Oxide)/$LiCF_3SO_3$ Crystalline Complex: A High-Resolution Solid-State ^{13}C NMR Approach. *Macromolecules*, *46* (11), 4447–4453.

[132] Liu, Z.; Zinkevich, T.; Indris, S.; He, X.; Liu, J.; Xu, W.; Bai, J.; Xiong, S.; Mo, Y.; Chen, H. 2020. $Li_{15}P_4S_{16}Cl_3$, a Lithium Chlorothiophosphate as a Solid-State Ionic Conductor. *Inorg. Chem.*, *59* (1), 226–234.

[133] Peng, J.; Xiao, Y.; Clarkson, D. A.; Greenbaum, S. G.; Zawodzinski, T. A.; Chen, X. C. 2020. A Nuclear Magnetic Resonance Study of Cation and Anion Dynamics in Polymer–Ceramic Composite Solid Electrolytes. *ACS Appl. Polym. Mater.*, *2* (3), 1180–1189.

[134] Yu, C.; Ganapathy, S.; van Eck, E. R. H.; Wang, H.; Basak, S.; Li, Z.; Wagemaker, M. 2017. Accessing the Bottleneck in All-Solid State Batteries, Lithium-Ion Transport over the Solid-Electrolyte-Electrode Interface. *Nat. Commun.*, *8* (1), 1086.

[135] Buannic, L.; Orayech, B.; López Del Amo, J.-M.; Carrasco, J.; Katcho, N. A.; Aguesse, F.; Manalastas, W.; Zhang, W.; Kilner, J.; Llordés, A. 2017. Dual Substitution Strategy to Enhance Li^+ Ionic Conductivity in $Li_7La_3Zr_2O_{12}$ Solid Electrolyte. *Chem. Mater.*, *29* (4), 1769–1778.

[136] Michan, A. L.; Leskes, M.; Grey, C. P. 2016. Voltage Dependent Solid Electrolyte Interphase Formation in Silicon Electrodes: Monitoring the Formation of Organic Decomposition Products. *Chem. Mater.*, *28* (1), 385–398.

[137] Rinkel, B. L. D.; Hall, D. S.; Temprano, I.; Grey, C. P. 2020. Electrolyte Oxidation Pathways in Lithium-Ion Batteries. *J. Am. Chem. Soc.*, *142* (35), 15058–15074.

[138] Gao, L.; Chen, J.; Liu, Y.; Yamauchi, Y.; Huang, Z.; Kong, X. 2018. Revealing the Chemistry of an Anode-Passivating Electrolyte Salt for High Rate and Stable Sodium Metal Batteries. *J. Mater. Chem. A*, *6* (25), 12012–12017.

[139] Gao, L.; Chen, J.; Chen, Q.; Kong, X. 2022. The Chemical Evolution of Solid Electrolyte Interface in Sodium Metal Batteries. *Sci. Adv.*, *8* (6), 4606.

[140] Guérin, K. 2000. A ^{7}Li NMR Study of a Hard Carbon for Lithium–Ion Rechargeable Batteries. *Solid State Ion.*, *127* (3–4), 187–198.

[141] Rathke, J. W.; Klingler, R. J.; Ii, R. E. G.; Kramarz, K. W.; Woelk, K. 1997. Toroids in NMR Spectroscopy1,2. *Prog. Nucl. Magn. Reson. Spectrosc.*, *30*, 209–253.

[142] Key, B.; Bhattacharyya, R.; Morcrette, M.; Seznéc, V.; Tarascon, J.-M.; Grey, C. P. 2009. Real-Time NMR Investigations of Structural Changes in Silicon Electrodes for Lithium-Ion Batteries. *J. Am. Chem. Soc.*, *131* (26), 9239–9249.

[143] Bhattacharyya, R.; Key, B.; Chen, H.; Best, A. S.; Hollenkamp, A. F.; Grey, C. P. 2010. In Situ NMR Observation of the Formation of Metallic Lithium Microstructures in Lithium Batteries. *Nat. Mater.*, *9* (6), 504–510.

[144] Wang, H.; Köster, T. K.-J.; Trease, N. M.; Ségalini, J.; Taberna, P.-L.; Simon, P.; Gogotsi, Y.; Grey, C. P. 2011. Real-Time NMR Studies of Electrochemical Double-Layer Capacitors. *J. Am. Chem. Soc.*, *133* (48), 19270–19273.

[145] Chien, P.-H.; Feng, X.; Tang, M.; Rosenberg, J. T.; O'Neill, S.; Zheng, J.; Grant, S. C.; Hu, Y.-Y. 2018. Li Distribution Heterogeneity in Solid Electrolyte $Li_{10}GeP_2S_{12}$ upon Electrochemical Cycling Probed by ^{7}Li MRI. *J. Phys. Chem. Lett.*, *9* (8), 1990–1998.

[146] Feng, X.; Tang, M.; O'Neill, S.; Hu, Y.-Y. 2018. *In Situ* Synthesis and *in Operando* NMR Studies of a High-Performance Ni_5P_4-Nanosheet Anode. *J. Mater. Chem. A*, *6* (44), 22240–22247.

[147] Leskes, M.; Moore, A. J.; Goward, G. R.; Grey, C. P. 2013. Monitoring the Electrochemical Processes in the Lithium–Air Battery by Solid State NMR Spectroscopy. *J. Phys. Chem. C*, *117* (51), 26929–26939.

[148] Sethurajan, A. K.; Krachkovskiy, S. A.; Halalay, I. C.; Goward, G. R.; Protas, B. 2015. Accurate Characterization of Ion Transport Properties in Binary Symmetric Electrolytes Using In Situ NMR Imaging and Inverse Modeling. *J. Phys. Chem. B*, *119* (37), 12238–12248.

[149] Krachkovskiy, S. A.; Foster, J. M.; Bazak, J. D.; Balcom, B. J.; Goward, G. R. 2018. Operando Mapping of Li Concentration Profiles and Phase Transformations in Graphite Electrodes by Magnetic Resonance Imaging and Nuclear Magnetic Resonance Spectroscopy. *J. Phys. Chem. C*, *122* (38), 21784–21791.

[150] Li, C.; Shen, M.; Lou, X.; Hu, B. 2018. Unraveling the Redox Couples of V^{III}/V^{IV} Mixed-Valent $Na_3V_2(PO_4)_2O_{1.6}F_{1.4}$ Cathode by Parallel-Mode EPR and In Situ/Ex Situ NMR. *J. Phys. Chem. C*, *122* (48), 27224–27232.

[151] Xiang, Y.; Zheng, G.; Liang, Z.; Jin, Y.; Liu, X.; Chen, S.; Zhou, K.; Zhu, J.; Lin, M.; He, H.; et al. 2020. Visualizing the Growth Process of Sodium Microstructures in Sodium Batteries by In-Situ ^{23}Na MRI and NMR Spectroscopy. *Nat. Nanotechnol.*, *15*, 883–890.

[152] Freytag, A. I.; Pauric, A. D.; Krachkovskiy, S. A.; Goward, G. R. 2019. In Situ Magic-Angle Spinning ^{7}Li NMR Analysis of a Full Electrochemical Lithium-Ion Battery Using a Jelly Roll Cell Design. *J. Am. Chem. Soc.*, *141* (35), 13758–13761.

[153] Liu, X.; Liang, Z.; Xiang, Y.; Lin, M.; Li, Q.; Liu, Z.; Zhong, G.; Fu, R.; Yang, Y. 2021. Solid-State NMR and MRI Spectroscopy for Li/Na Batteries: Materials, Interface, and In Situ Characterization. *Adv. Mater.*, *33* (50), 2005878.

[154] Poli, F.; Kshetrimayum, J. S.; Monconduit, L.; Letellier, M. 2011. New Cell Design for In-Situ NMR Studies of Lithium-Ion Batteries. *Electrochem. Commun.*, *13* (12), 1293–1295.

[155] Saito, Y.; Yamamoto, H.; Nakamura, O.; Kageyama, H.; Ishikawa, H.; Miyoshi, T.; Matsuoka, M. 1999. Determination of Ionic Self-Diffusion Coefficients of Lithium Electrolytes Using the Pulsed Field Gradient NMR. *J. Power Sources*, *81–82*, 772–776.

[156] Capiglia, C.; Saito, Y.; Kageyama, H.; Mustarelli, P.; Iwamoto, T.; Tabuchi, T.; Tukamoto, H. 1999. ^{7}Li and ^{19}F Diffusion Coefficients and Thermal Properties of Non-Aqueous Electrolyte Solutions for Rechargeable Lithium Batteries. *J. Power Sources*, *81–82*, 859–862.

[157] Kuhn, A.; Narayanan, S.; Spencer, L.; Goward, G.; Thangadurai, V.; Wilkening, M. 2011. Li Self-Diffusion in Garnet-Type $Li_7La_3Zr_2O_{12}$ as Probed Directly by Diffusion-Induced ^{7}Li Spin-Lattice Relaxation NMR Spectroscopy. *Phys. Rev. B*, *83* (9), 094302.

[158] Lutz, L.; Yin, W.; Grimaud, A.; Alves Dalla Corte, D.; Tang, M.; Johnson, L.; Azaceta, E.; Sarou-Kanian, V.; Naylor, A. J.; Hamad, S.; et al. 2016. High Capacity $Na–O_2$ Batteries: Key Parameters for Solution-Mediated Discharge. *J. Phys. Chem. C*, *120* (36), 20068–20076.

[159] Hrabe, J.; Kaur, G.; Guilfoyle, D. N. 2007. Principles and Limitations of NMR Diffusion Measurements. *J. Med. Phys. Assoc. Med. Phys. India*, *32* (1), 34–42.

[160] Tang, J. A.; Dugar, S.; Zhong, G.; Dalal, N. S.; Zheng, J. P.; Yang, Y.; Fu, R. 2013. Non-Destructive Monitoring of Charge-Discharge Cycles on Lithium Ion Batteries Using ^{7}Li Stray-Field Imaging. *Sci. Rep.*, *3* (1), 1–6.

[161] Trease, N. M.; Zhou, L.; Chang, H. J.; Zhu, B. Y.; Grey, C. P. 2012. In Situ NMR of Lithium Ion Batteries: Bulk Susceptibility Effects and Practical Considerations. *Solid State Nucl. Magn. Reson.*, *42*, 62–70.

[162] Jin, Y.; Kneusels, N.-J. H.; Marbella, L. E.; Castillo-Martínez, E.; Magusin, P. C. M. M.; Weatherup, R. S.; Jónsson, E.; Liu, T.; Paul, S.; Grey, C. P. 2018. Understanding Fluoroethylene Carbonate and Vinylene Carbonate Based Electrolytes for Si Anodes in Lithium Ion Batteries with NMR Spectroscopy. *J. Am. Chem. Soc.*, *140* (31), 9854–9867.

[163] Leskes, M.; Kim, G.; Liu, T.; Michan, A. L.; Aussenac, F.; Dorffer, P.; Paul, S.; Grey, C. P. 2017. Surface-Sensitive NMR Detection of the Solid Electrolyte Interphase Layer on Reduced Graphene Oxide. *J. Phys. Chem. Lett.*, *8* (5), 1078–1085.

[164] Wolf, T.; Kumar, S.; Singh, H.; Chakrabarty, T.; Aussenac, F.; Frenkel, A. I.; Major, D. T.; Leskes, M. 2019. Endogenous Dynamic Nuclear Polarization for Natural Abundance ^{17}O and Lithium NMR in the Bulk of Inorganic Solids. *J. Am. Chem. Soc.*, *141* (1), 451–462.

[165] Haber, S.; Rosy; Saha, A.; Brontvein, O.; Carmieli, R.; Zohar, A.; Noked, M.; Leskes, M. 2021. Structure and Functionality of an Alkylated $Li_xSi_yO_z$ Interphase for High-Energy Cathodes from DNP-SSNMR Spectroscopy. *J. Am. Chem. Soc.*, *143* (12), 4694–4704.

[166] Lorie Lopez, J. L.; Grandinetti, P. J.; Co, A. C. 2018. Enhancing the Real-Time Detection of Phase Changes in Lithium–Graphite Intercalated Compounds through Derivative Operando (DOp) NMR Cyclic Voltammetry. *J. Mater. Chem. A*, *6* (1), 231–243.

9 Differential Electrochemical Mass Spectrometry for Lithium-Ion Batteries*

Zhiwei Zhao, Long Pang, and Zhi Yang
Dalian Institute of Chemical Physics,
Chinese Academy of Sciences

Yelong Zhang and Zhangquan Peng
Dalian Institute of Chemical Physics,
Chinese Academy of Sciences
Wuyi University

Limin Guo
Dalian Institute of Chemical Physics,
Chinese Academy of Sciences
Dalian University

CONTENTS

* Zhiwei Zhao, Yelong Zhang, Long Pang and Zhi Yang contribute equally to this work.

DOI: 10.1201/9781003299295-9

Lithium-ion batteries (LIBs) have been extensively adopted as power sources for portable electronic devices, electric vehicles and large-scale electrical storage systems, and have been thought to be an indispensable solution for the realization of the grand carbon neutrality target in the forthcoming decades. However, the mass utilization of LIBs is yet encountered with performance degradation and disturbing safety issues. The parasitic reactions, which often occur at the electrode/electrolyte interface and are frequently accompanied by gas evolution, represent a major problem, because some of the evolved gases are flammable and could result in fires or even explosions during battery operation.[1] Moreover, to meet the ever-increasing energy storage demands of contemporary and even future society, LIBs have been developed toward high energy densities by using high-capacity and high-voltage cathode materials. As such, the LIBs often operate beyond the thermodynamic limit of the electrolyte and thus, gas evolution from electrolyte decomposition remains a challenge to tackle. To this end, understanding and controlling parasitic reactions, particularly those accompanied by gas evolution, are essential for building better LIBs.

Differential electrochemical mass spectrometry (DEMS) is a unique technique that can in situ monitor and quantify electrochemically generated volatile products and provide larruping rich information for gas evolution reaction during LIBs operation.[2] In this chapter, we review the historical development of DEMS, present a pedagogical introduction to the basics of DEMS with emphasis placed on experimental setup, recapitulate current understanding of gas evolution of LIBs examined by DEMS, and discuss major current challenges and future opportunities in this rapidly evolving and also very exciting field. This chapter also aims to encourage wider community to broaden the application scope of DEMS by combining with other advanced characterization techniques.

9.1 A BRIEF HISTORY

In this section, we introduce the historical development of DEMS and its penetration into the LIBs field. The evolution of DEMS follows how to interface an electrochemical cell (or a battery) online with a mass spectrometer (MS) for efficiently ionizing and detecting the electrochemically generated volatile species, with required temporal resolution and detection sensitivity. Thus, we focus on inlet systems (e.g., membrane inlet, carrier gas inlet and leak valve inlet) of DEMS in following sections, with more details put on the cell design of the popular carrier gas inlet used in LIBs research.

9.1.1 Membrane Inlet

In 1963, Hoch and Kok proposed a membrane inlet prototype, where a MS was interfaced with a semipermeable membrane, and was used to study the gases dissolved in liquid phases.[3] Soon after, the first coupled electrochemistry (EC)-MS device was reported by Bruckenstein and Gadde[4] in 1971. In their setup, a semipermeable polymer film was used, which could enable the volatile products to penetrate through without the interference of the electrolyte. However, the volatile analyte has a high resistance to pass through this membrane, and thus poor temporal resolution (~minutes level) and only the integral signal of a certain species have been realized. In 1984, Wolter and Heitbaum[5] used a differential pumping system (backing pump and turbomolecular pump) and a porous polytetrafluoroethylene (PTFE) membrane to increase the flux of volatile analyte across the membrane (Figure 9.1a and b), laying the foundation of modern DEMS. The differential pumping emphasizes that the mass signal intensity (i.e., ion current or partial pressure) of species collected by the MS is a differential signal that reflects (more exactly, is proportional to) the flow rate of the volatile analyte through the PTFE membrane and the achieved time resolution can be less than 1 second. Since then, DEMS has been developed extensively as a popular technique in the research fields of electrocatalysis and fuel cells.[6] Over the past three decades, much of the development for the membrane inlet of DEMS focused on the design of electrochemical cells to accommodate various electrodes and catalysts. In the original electrochemical cell of DMES (Figure 9.1b) termed as classical

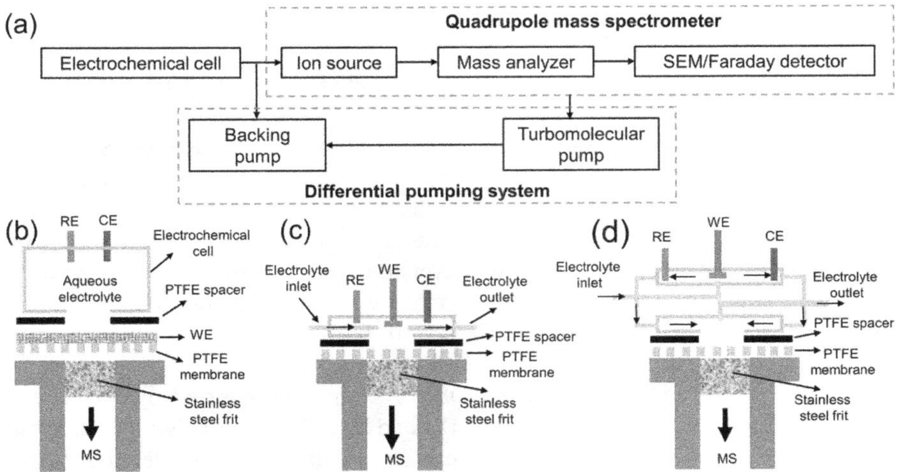

FIGURE 9.1 Schematic representation of (a) DEMS instrumentation, (b) classical membrane inlet electrochemical cell, (c) single thin-layer flow cell and (d) dual thin-layer cell.

DEMS cell, electrode materials are sputtered onto the PTFE membrane as the working electrode that may complicate the construction of the electrochemical interface and increase the difficulty of the DEMS experiment. The advent of a single thin-layer cell[7] (Figure 9.1c) allows to apply DEMS on the flat and single-crystal electrode and increases the versatility of DEMS, but sometimes a poor collection efficiency of volatile analyte could reduce the sensitivity of DMES in a flowing electrolyte environment. Later, a dual thin-layer flow cell[8] (Figure 9.1d) is developed on the basis of the single thin-layer cell, in which the working electrode and vacuum inlet are in separate compartments to improve the collection efficiency of the volatile analyte. This dual thin-layer flow cell also provides the opportunity to combine DEMS with other analytical techniques, such as electronic quartz crystal microbalance[8] and attenuated total reflection-infrared spectroscopy,[9] and has been developed into a commonly used aqueous electrolyte-based DEMS cell. More recently, Trimarco and his co-workers[10] presented a novel membrane inlet method by replacing the PTFE membrane with a silicon microchip, realizing a sub-second time resolution and 100% collection efficiency for quantitative analysis with unprecedented sensitivity.

With the great success of membrane inlet DEMS in aqueous systems (e.g., electrocatalysis and fuel cells), researchers began to apply DEMS in aprotic electrolyte-based LIBs. In 1998, Imhof and Novák[11] used the membrane inlet DEMS to investigate the electrochemical reduction decomposition of carbonate electrolytes on graphite anode of LIBs. Similar DEMS setup was also used to investigate the gas evolution upon the electrolyte oxidative degradation on various cathode materials in LIBs in 1999.[12] However, the DEMS design deviates from the normal operating conditions of LIBs, and the volatile carbonate solvents could easily enter the MS vacuum chamber to increase complexity and difficulty in DEMS measurements. Moreover, the membrane inlet has limitations on the quantification of many gaseous analytes (see Section 9.2.5).

9.1.2 Carrier Gas Inlet

In order to address the inherent insufficiencies of membrane inlet DEMS, Holzapfel et al.[13] reported a new headspace-based gas analysis technique, in which a carrier gas was used to transport gaseous analytes from a 2.0 mL headspace of the electrochemical cell to a quadrupole MS (Figure 9.2). This DEMS cell design is more in line with practical LIBs, reduces the amount of electrolyte used, increases the amount of active materials, enhances the collection efficiency of gaseous analytes, and therefore improves detection sensitivity. However, the quantitative method or protocol was not proposed and the raw ion current of MS was used to conduct qualitative and semiquantitative research. Moreover, the 2.0 mL headspace obviously increases the response time of the DEMS (see Section 9.2.1).

In 2011, a quantitative intermittent DEMS setup was reported by McCloskey et al.,[14] which is suitable for the study of LIBs and Li-O_2 batteries (Figure 9.3a). Their DEMS setup consists of five parts: (i) a carrier gas source, (ii) an electrochemical cell, (iii) carrier gas inlet system containing an 8-port 2-way high-pressure gas chromatography (HPGC) valve, standard volume loop and pressure transducer, (iv) differential pumping vacuum system and (v) MS. There are two positions of the HPGC

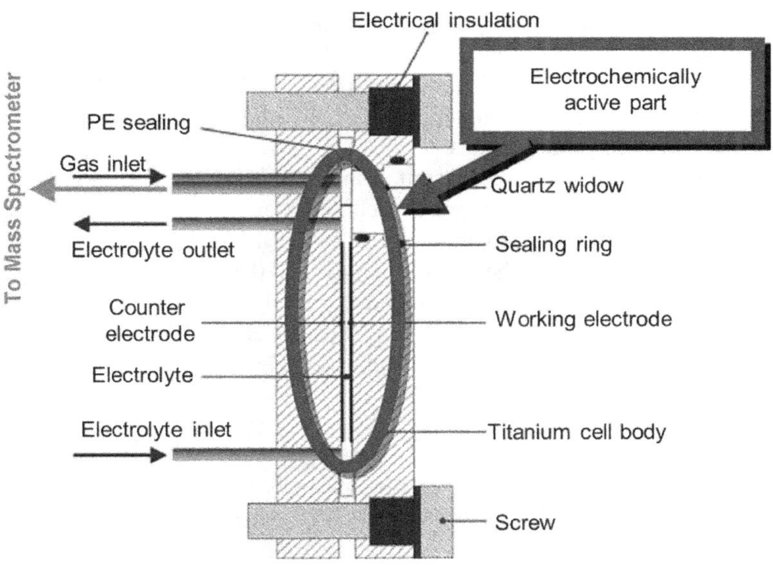

FIGURE 9.2 Schematic drawing of the first DEMS cell based the carrier gas inlet.[13]

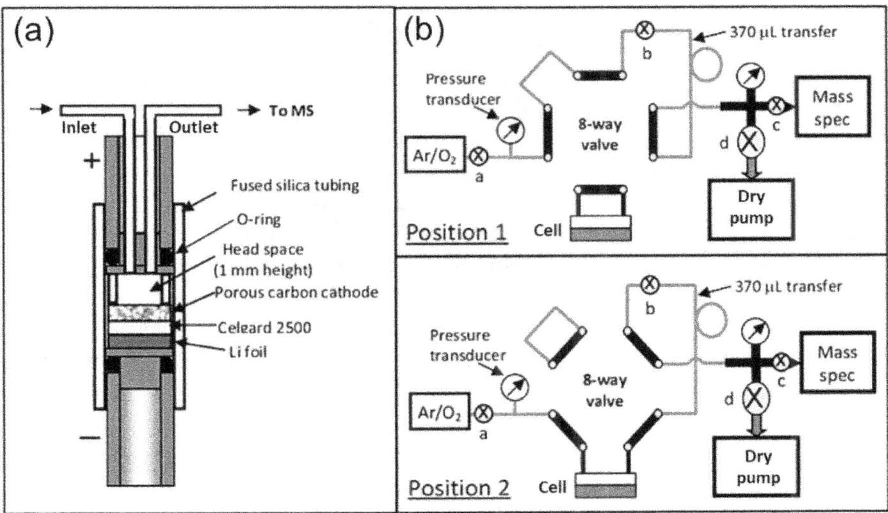

FIGURE 9.3 (a) The schematics of a DEMS electrochemical cell with intermittent inlet to analyze the gas stored in the headspace and (b) the two operation modes of the DEMS, proposed by McCloskey et al. in 2011.[14]

valve (Figure 9.3b). The cell can accumulate gaseous analytes in the headspace at position 1. The electrochemical cell is connected by a capillary tube to the carrier gas inlet and by another to the DEMS sample transfer line at position 2. When the cell is switched from position 1 to position 2, the gaseous analytes in the headspace can be transported to the MS by carrier gas.

To quantify the amount of gas consumed and released in Li-O_2 batteries, they first calibrated the volume of the inlet system including transfer line and cell's headspace by attaching different standard volume loops to the inlet system. According to the ideal gas law ($P_1V_1 = P_2V_2$), the headspace volume and the transfer line volume can be determined, respectively. The O_2 consumed during discharge is directly monitored by measuring the pressure drop of the cell's headspace, based on state equation of gas (note that, no gas is released during discharge and only O_2 is consumed for Li-O_2 batteries). For gas evolution (e.g., O_2, CO_2, H_2) during charge of the Li-O_2 cell, Ar/O_2/CO_2/H_2 mixed gas with known ratio is used to calibrate the DEMS to obtain their respective correction coefficient (see Section 9.2.5). As such, researchers can know the volume ratio (or partial pressure) of gaseous analytes entering the MS. Based on the ideal gas law ($PV = nRT$), the amount (mol) of generated gas could be obtained by the known volume of the carrier gas, pressure and temperature. The details of the valve on and off can be found in ref. [14]. However, the temporal resolution of intermittent inlet DEMS is relatively low (~15 minutes), and it is difficult to obtain potential-dependent gas consumption/evolution information as realized with membrane inlet DEMS.

In 2012, Peng et al.[15] reported a DEMS setup with a continuous carrier gas inlet. In their setup (see Section 9.2.2), a digital mass flow controller was used to accurately control the flow rate of carrier gas and a cold trap in between the DEMS cell and the MS was used to condense volatilized solvent molecules. This setup enabled a simplified quantitative protocol (see Section 9.2.5) and realized an excellent temporal resolution. By optimizing the carrier gas inlet system and the DEMS cell headspace, they achieved for the first time the quantification of gas consumption and evolution of Li-O_2 cells, simultaneously.

However, the continuous carrier gas purging can easily lead to dry out of the electrolyte in the DEMS cell. In 2015, a solution was proposed by Berkes et al.[16] to achieve the long-time monitor of gas evolution of LIBs, where a gas bubbler containing the volatile solvent-based electrolyte was put in front of the DEMS cell, while a cold trap was placed behind the cell (Figure 9.4). During the DEMS experiment, the He gas passed through the gas bubbler (humidifier) first, preventing the volatile components of the electrolyte in the DEMS cell from evaporating into the carrier gas. The gas from the cell passed the cold trap before reaching MS. The cold trap could protect the MS from the interference of the solvent molecules in the carrier gas. Based on this design, the gas evolution of LIBs has been steadily detected for more than 100 hours.

In 2015, Metzger et al.[17] developed a two-compartment cell, where the anode and the cathode compartments were separated by a solid Li^+ ion conductive separator. The species of interest, such as electrolyte additives, were selectively added to either of the compartments to identify their effects on the anode or cathode performance. By using a half cell with a small Li metal wire as counter/reference electrode, the species under study were protected from the strongly reducing environment of the

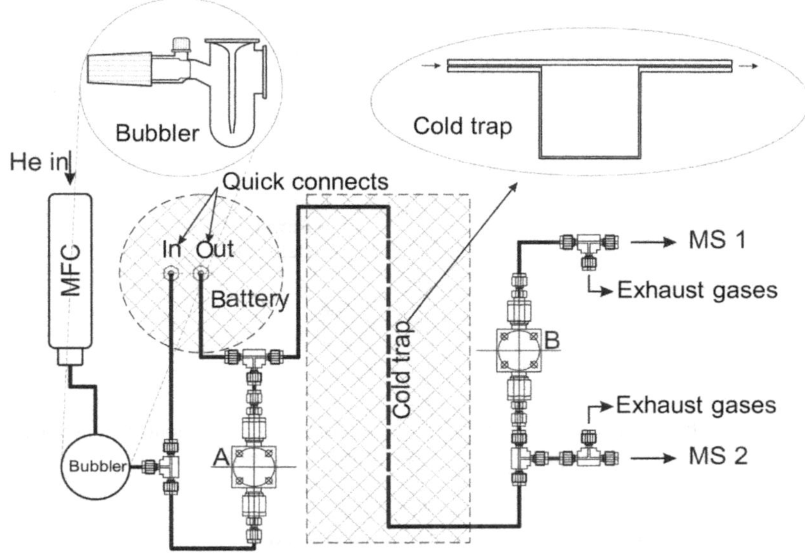

FIGURE 9.4 Schematic diagram of the DEMS used for long-time detection.[16]

Li counter electrode. Therefore, this cell was suitable for fundamental studies of single electrode/electrolyte interface without shuttle effect.

9.1.3 Leak Valve Inlet

The carrier gas inlet of DEMS, even equipped with a solvent-bubbler,[16] still has limitations in the study of LIBs containing volatile solvent-based electrolytes, particularly when longer time (e.g., weeks) monitor of gas/pressure evolution in LIBs is required. A feasible solution to this issue was suggested by Tsiouvaras et al. in 2013,[18] where a calibrated and crimped-capillary leak (~1 μL/min) was used to permit a continuous headspace (9 mL) gas sampling for a long time (~10 hours) without any carrier gas (Figure 9.5). The obtained signal reflects the change of gas concentration in the headspace of the DEMS cell, which was an integral information of gas evolution. Note that the solution may also have its limitation that the working pressure of the battery could deviate from the normal value, because the internal pressure of cell would drop continuously especially for long time tests.

In brief, there are three primary aspects that need to consider for the application of DEMS in LIBs: temporal resolution, continuous working time and quantitative simplicity and accuracy. The inlet methods described above have their own pros and cons. The membrane inlet could easily achieve high temporal resolution but the DEMS cell is slightly complex and limited in aprotic electrolyte-based LIBs. The carrier gas inlet system has a wider applicability for different electrochemical systems with simple and precise quantification but time resolution capability is limited under certain conditions. Although its time resolution capability (~second

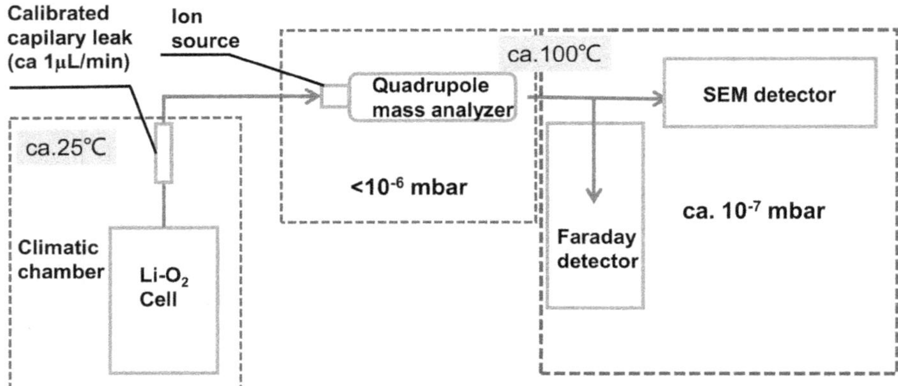

FIGURE 9.5 The schematic diagram of leak valve inlet system without the carrier gas, in which a calibrated crimped-capillary leak (~1.0 μL/min) was used.

level) is relatively low, it is sufficient for LIBs test needs due to their long operating time (~hours level). The leak valve inlet mode simplifies the inlet system but may be limited by the real working pressure decay of the batteries.

9.2 BASIC KNOWLEDGE AND EXPERIMENTAL SETUP

The following sections will describe basic knowledge and experimental setup of DEMS, including electrochemical cell, carrier gas inlet system, MS, electrochemical method and data analysis. The readers can refer to ref. [6] to obtain more details of DEMS with emphasis placed on membrane inlet system.

9.2.1 Electrochemical Cell

Some DEMS cells are commercially available (e.g., EL-CELL GmbH) and their design are based on the work of Peng et al.[15] The DEMS cell typically contains polished stainless steel current collectors and double PTFE/PE/PEEK ferrules to ensure air tightness.[19] The purge gas inlet and outlet of the cell are realized with two glued PEEK capillary tubes, and are usually integrated on the cathode side (design details can also be found in Figures 9.3 and 9.6). The assembly of the DEMS cell is essentially the same as that of traditional LIBs, which is often completed in an Ar-filled glove box with considerable flexibility. The working electrode is often coated on a mesh/or porous current collector for smooth gas flow, and a partially delithiated $LiFePO_4$ is suggested as counter/reference electrode. Sometimes, solid-state electrolyte membrane/separator capable of transporting only Li^+ is preferred because it can avoid any interference from the anode/electrolyte interface. The time constant t_1 of cell could be simply estimated by the ratio between the headspace v_1 (or dead volume) and the flow rate of carrier gas s_1 (i.e., $t_1 = v_1/s_1$). When a 100 μL headspace and 1.0 mL/min flow rate of carrier gas are used, the time constant t_1 of cell is estimated to be 6 s. Note that, the estimated time is less than actual value because

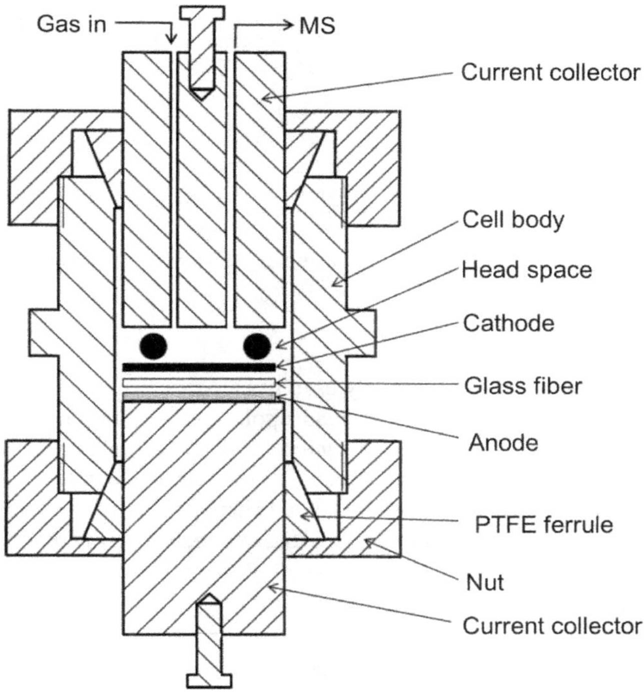

FIGURE 9.6 Schematics of the typical two-compartment DEMS cell.[19]

of vortex phenomena in the carrier gas purging process. Thus, the electrochemical cell is expected to have as high a carrier gas flow rate as possible (its selection will be discussed in Section 9.2.2) and as small dead volume as possible for a high time resolution of DEMS.

9.2.2 Carrier Gas Inlet System

The carrier gas inlet system (Figure 9.7) typically consists of a gas cylinder (source of carrier gas), a digital mass flow controller (precise flow control), a PEEK or stainless-steel capillary tubing, a four-way valve that allows the cell to be transferred from the glove box to the inlet system without air leakage, a cold trap (cooling volatile organic solvents to avoid interference of hydrocarbons) and a filter (filtering some possible solid impurities to prevent clogging the MS inlet). The time constant t_2 of the carrier gas inlet system could also be simply estimated by the ratio between the internal volume of capillary tubing v_2 and the flow rate of carrier gas s_1 (i.e., $t_2 = v_2/s_1$). As a result, we expect to shorten the length of the capillary tubing and reduce its inner diameter (but gas transport cannot be hindered) to obtain better time resolution. Note that the flow rate of the carrier gas needs to choose an appropriate value. If it is too low, the time resolution will be low. Otherwise, if it is too high, the electrolyte will volatilize quickly and the cell will dry out.

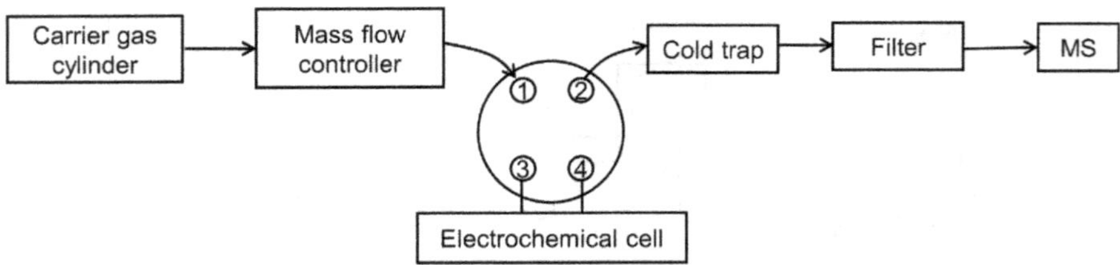

FIGURE 9.7 Schematic diagram of DEMS using the carrier gas inlet method.

9.2.3 Mass Spectrometer

A commercial quadrupole MS is usually used in DEMS system and consists of three parts: (i) electron impact ionization, (ii) quadrupole mass separation and (iii) Faraday and/or secondary electron multiplier detector. Their operation is carried out under high vacuum conditions ($P_{max} < 10^{-6}$mbar) to maintain a good linear relationship between MS signal and amount of the analyte. Therefore, there is an upper limit on the mass flow into the vacuum chamber of MS that depends on the pumping speed of turbomolecular pump. When a pumping speed of $s_2 = 50$ L/s is used, the maximum flux n_{max} into the vacuum chamber is:

$$n_{max} = P_{max} s_2 / RT \approx 2 \text{ nmol/s}, \tag{9.1}$$

where P is the pressure of vacuum chamber of MS ($P_{max} = 10^{-6}$mbar), R is the gas constant and T is the absolute temperature.

In comparison, the flow rate of carrier gas s_1 is usually set to be 0.1–2.0 mL/min under typical laboratory conditions according to experimental system, which means that the flow rate of the carrier gas containing the analyte into the MS is at least greater than 0.1 mL/min ($\approx$75 nmol/s $> n_{max}$). That is, the flow rate of carrier gas containing the analyte is much greater than the allowable influx (n_{max}) of vacuum chamber of MS. Consequently, the front end of the MS is usually set with a backing pump to lower the vacuum collection efficiency, which means that most of the carrier gas containing the analyte (>90%) is vented. When the production rate of gaseous analytes from the electrochemical cell is low, it may not be detected by MS because low vacuum collection efficiency limits the total influx to vacuum chamber of MS.

The time constant t_3 of MS could also be simply estimated by the ratio between the volume of the chamber v_3 and the pumping speed s_2 (i.e., $t_3 = v_3/s_2$). When a 1 L chamber and pumping speed of $s_2 = 50$ L/s are used, the time constant t_3 is estimated to be 20 ms. Therefore, the temporal resolution (delay time) of DEMS relies on the design of carrier gas inlet system and the electrochemical cell rather than the MS itself.

9.2.4 Electrochemical Method

In the LIBs field, galvanostatic and cyclic voltammetric methods are commonly used. However, how to choose the electrochemical method and its parameters (e.g., current density and potential scan rate) may be a blind spot at the beginning of DEMS measurement, which have not been reported in previous studies. As a result, it is necessary to narrate our experience in using DEMS and assist beginners to get their measurements started smoothly.

In general, obtaining a high-quality DEMS result depends on a suitable current density. If the current density is too small, the amount of gas evolution will be below the detection limit of DEMS and cannot be identified. Instead, if the current density is too high, LIB systems cannot sustain due to severe polarization occurring at various locations within the cell. Therefore, a suitable gas evolution rate that matches the DEMS detection capability is the clew to determine the current density.

Based on the considerations above, cyclic voltammetry (empirically with a scan rate 0.1–0.5 mV/s) should be first used to determine the relationship between current density and gas evolution rate. According to the relationship, the appropriate current density range is readily identified. Then, we could apply the identified current density to galvanostatic test of the LIBs and obtain the data of the accompanied gas evolution.

In short, both galvanostatic and cyclic voltammetry techniques can be coupled with DEMS to study the gas evolution of LIBs. However, the choice of test parameters (e.g., the current density of charge-discharge cycle) often relies on experience or the results of the preliminary test, due to the difference in experimental condition, equipment system, battery configuration and so on. In addition, other electrochemical techniques (e.g., potential step) can also be utilized when DEMS test is conducted, depending on requirements of the research.

9.2.5 Data Analysis

The qualitative principle of MS has been widely known, and this subsection mainly describes how to quantitatively analyze and process DEMS data.

The basic quantification principle of DEMS is that the Faradaic current I_{Fi} of electrochemical reactions is proportional to the ion current I_{MSi} (or partial pressure) of a species i detected by MS, and we can simply state:

$$I_{MSi} = K_i \times I_{Fi}. \tag{9.2}$$

The constant K_i includes multiple impact factors, including Faraday current efficiency, collection efficiency, ionization efficiency and detection efficiency, which is dependent on the kind of the electrochemical system, the inlet system and MS. Based on Equation 9.2, the quantitative methods of DEMS can be divided into external standard method (used by membrane inlet) and internal standard method (used by carrier gas inlet). In external standard method, a standard curve between I_{MSi} and I_{Fi} needs to be established using known electrochemical reactions (e.g., oxidation reaction of CO and hydrogen evolution reaction on Pt electrode surface). Therefore, external standard method of DEMS quantification has certain limitations for gases without standard electrode reactions. This may be the reason why the carrier-gas-based inlet mode is widely used in LIBs.

The production rate f_i (mol/s) and Faradaic current I_{Fi} of species i are linearly dependent. For carrier gas inlet system of DEMS, the constant K_i of species i could thus be obtained by purging standard gas to MS and correlating the standard gas composition with their partial pressure (or ion current) in MS. For instance, the standard gas of Ar/CO_2 = V_1:V_2 is passed through the carrier gas inlet to MS to obtain the ion current $I_{MS(Ar)}$ and $I_{MS(CO2)}$, and then the constant K_{CO2} can be calculated by using the following equation:

$$K_{CO_2} = I_{MS(Ar)}/I_{MS(CO_2)} \times V_2/V_1. \tag{9.3}$$

Once K_i is known, the production rate f_i (mol/s) of species i can be calculated via the following equation:

$$f_i = f_g \times K_i \times I_{MSi}/I_{MSg}, \tag{9.4}$$

where f_g is the flow rate of carrier gas and I_{MSg} is the ion current of carrier gas detected by MS.

Finally, the total production n_i (mol) of species i from electrochemical cell is

$$n_i = \int f_i dt. \tag{9.5}$$

In addition, we emphasize a significant but overlooked point in the calculation process of n_i, which is currently lack of widely accepted protocols for reporting DEMS results. Its calculation can be done by following two methods:

1. The data of f_i is first smoothed and integrated to obtain the n_i.
2. The data of f_i are first integrated to obtain the n_i and then smoothed for clear.

The first method seem to contain more artificial factors than the second one, which will result in the difficulty in comparing and evaluating reported DEMS results. Therefore, we suggest that the latter should be used to help readers understand the credibility of the DEMS results.

9.3 ANODE

The initial Coulombic efficiency (CE) of anode materials is a key parameter that determines the energy density of LIBs. Generally, the low CE of anode materials, caused by the interfacial parasitic reactions during the formation and operation of LIBs, will result in a loss of the active lithium and need to be compensated by an additional loading of cathode materials. In this section, we present several rudimentary examples on anode/electrolyte interface accompanied by gassing phenomenon-related parasitic reactions, especially focus on graphite, silicon (Si) and lithium anode.

9.3.1 Graphite

Graphite, with a theoretical capacity of 372 mAh/g, has been one of the most important and widely used anode materials for LIBs, due to its cheap price, abundance and environmental friendliness.[20] However, the gas evolution at graphite anodes can lead to degradation and severe safety issues of LIBs.[21]

DEMS is a powerful technique for in situ analyzing the gas evolution on graphite anode during LIBs operation. One typical example is the operation of a $Li_{1+x}Ni_{0.5}Mn_{0.3}Co_{0.2}O_2$ (NMC 532)/graphite cell.[16] Figure 9.8 illustrated the voltage–time curves and corresponding MS signals over 20 charge/discharge cycles. During the initial charge process, hydrogen ($m/z = 2$), ethylene ($m/z = 26$), carbon monoxide ($m/z = 28$) and carbon dioxide ($m/z = 44$) signals were detected. The second peak of CO_2 at 4.17V was related to the electrochemical oxidative decomposition of the solvent molecules (Figure 9.8a). H_2 and CO_2 evolution was detected during long-term cycling of NMC 532/graphite cell, while C_2H_4 could only be found in the initial cycle (Figure 9.8). Notably, this is the first study to focus on gas evolution in LIBs by DEMS for 20 charge/discharge cycles. The authors claimed that gas generation behavior induced by parasitic reactions would be an invisible and potential threat to NCMs/graphite battery safety.

The repeated destruction and re-formation of SEI film on graphite anode during charge/discharge process is often accompanied by gas evolution. Due to the trace amount of water in the electrolyte, the H_2 signal appeared in the initial charge process (Figure 9.9a).[22] The H_2 peak maximum was very close to the cut-off potential of 0.01V. However, during the subsequent cycles, almost no H_2 evolution was observed in the potential range of 0.01V–0.9V, indicating the formation of a stable SEI layer. However, charging to 1.2V resulted in the reappearance of an obvious H_2 signal in the DEMS profile. Overall, these results can be well explained by the evidence of potential-dependent formation and destruction of the SEI (Figure 9.9b). This observation could be attributed to the highly reactive H^+ from either residual water or parasitic reaction products and the rich reactive sites distributed on the graphite surface.

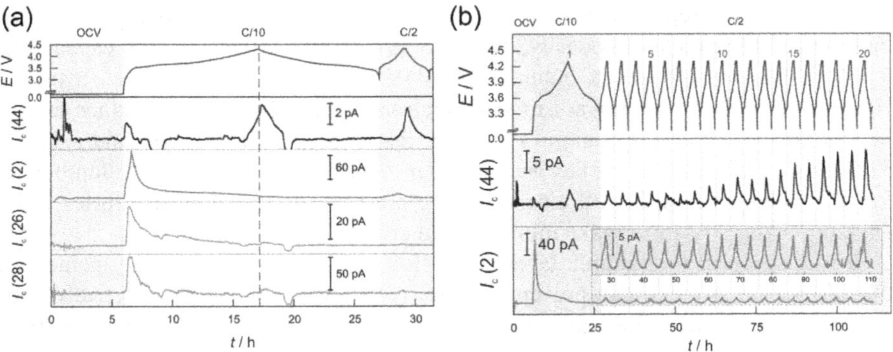

FIGURE 9.8 Mass spectrometric signals of certain m/z values over 20 charge/discharge cycles for a cell with graphite anode and NMC 532 cathode for (a) first two cycles and (b) complete cycling sequence.[16]

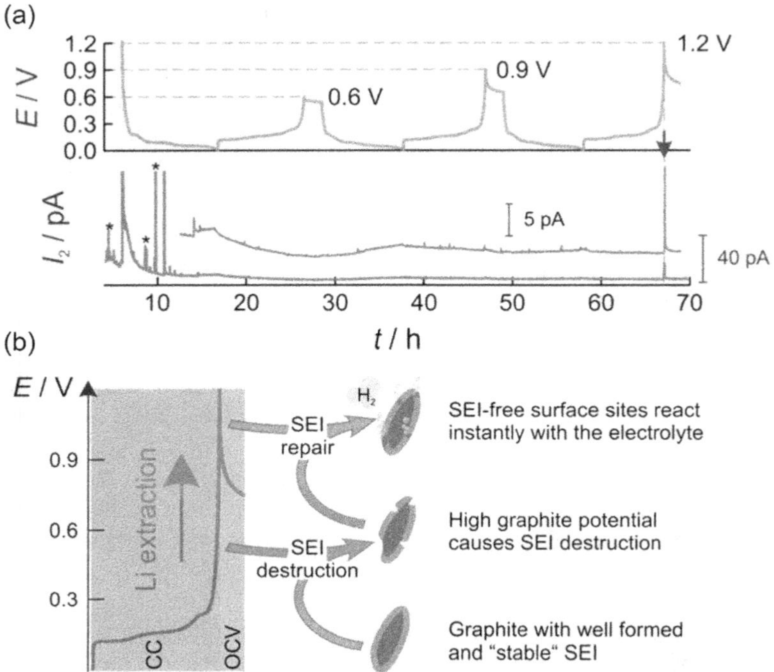

FIGURE 9.9 (a) Voltage profiles and the corresponding H_2 signal from DEMS for a graphite/Li cell discharged to different potentials at 0.1 C. (b) Schematic of the potential-driven SEI destruction on graphite during discharge and the associated H_2 evolution due to reductive electrolyte decomposition. CC and OCV stand for constant current and open circuit voltage, respectively.[22]

9.3.2 Lithium

Li metal is highly regarded as an attractive anode material for rechargeable Li-based batteries with high energy density, owing to its high theoretical specific capacity of 3,860 mAh/g and low redox potential (–3.04 V vs. standard hydrogen electrode).[23] However, the severe gassing at Li metal electrode/liquid electrolyte interface, which occurs during battery operation or storage, leads to poor battery life and safety.[24] Though significant progress has been made in recent years, the gas evolution behavior of Li metal anode is still elusive. There are relative abundant territories to be explored using DEMS to make more reliable and safer Li metal batteries.

It was reported that atomic layer deposition (ALD) coatings directly formed on the surface of Li metal can prevent the electrolyte decomposition, effectively alleviating capacity fading and dendrite formation.[25] DEMS is an efficient tool to study both chemical and electrochemical stability of the protected Li metal anodes against non-aqueous electrolytes under cell operating conditions. For example, gassing behavior on 100 ALD cycles of Al_2O_3-coated Li metal foil was investigated using

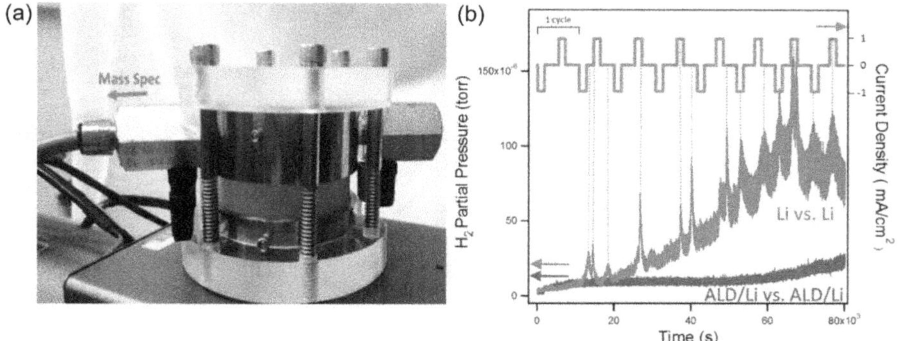

FIGURE 9.10 (a) The assembled symmetric cells (Li vs. Li and ALD/Li vs. ALD/Li). (b) DEMS on symmetric cells cycled at current density 1 mA/cm^2 showing the H_2 evolution for bare Li and for 50 ALD cycles of Al_2O_3 protected Li.[26]

DEMS by Lin et al. (Figure 9.10a).[26] Home-made symmetric cells (Li vs Li and ALD/Li vs ALD/Li) were discharged/charged at 1 mA/cm^2 for half an hour with one hour relaxation time for each cycle (Figure 9.10b). No observable H_2 peak was found for protected Li during the cycling process, indicating enhanced stability of the Li anode/electrolyte interface. The DEMS results strongly support that thin ALD Al_2O_3 coatings directly formed on the Li metal surface considerably enhance the Li anode surface and structure stability, eliminate Li metal corrosion and electrolyte decomposition reaction and thus, will undoubtedly mitigate the inherent poor cycling stability problem of Li anode.

9.3.3 Silicon

The relatively low power and energy density of traditional graphite-based LIBs cannot meet the ever-increasing needs of electric vehicles and large-scale electrical energy storage systems. In the past decades, tremendous research effort has been made to explore new anode materials.[27] Among them, Si has attracted much attention as potential next-generation anode material for Li-based rechargeable batteries, due to its high theoretical gravimetric capacity (4,200 mAh/g), natural abundance, low cost and nontoxic features.[28] Nevertheless, the widespread application of Si anode is hindered by a poor cycling stability resulted from the large volume expansion of over 300% upon repeated insertion/extraction of Li^+ ions.[29] Moreover, the volume variation of Si upon operation leads to an unstable SEI, resulting in low CE during cycling.[30]

Schiele et al. reported the gas evolution during operation of Si-based half cells using two types of electrolytes.[31] They compared a LP57 electrolyte [1 M $LiPF_6$ in ethylene carbonate (EC): ethyl methyl carbonate (EMC) with weight ratio 3:7] with another electrolyte containing fluoroethylene carbonate (FEC) instead of EC as cosolvent. DEMS was used to monitor the gas generation behavior during cell operation. To demonstrate the role of FEC, the charge/discharge profiles of cells at the 1st,

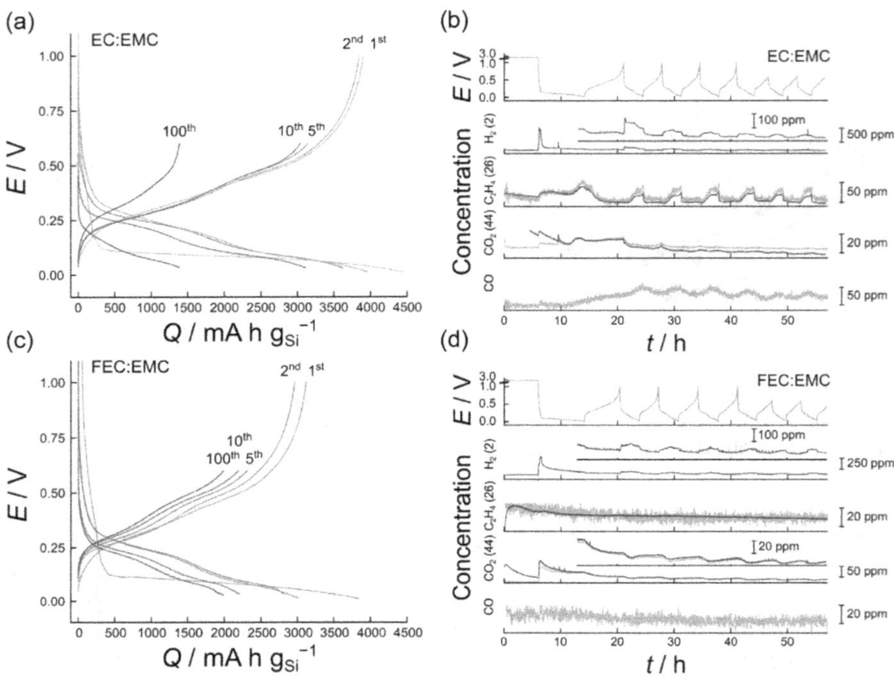

FIGURE 9.11 Charge/discharge profiles and the corresponding mass signals for the Si half-cells using (a and b) 1 M $LiPF_6$ in EC: EMC and (c and d) FEC: EMC.[31]

2nd, 5th, 10th and 100th cycles were compared (Figure 9.11a and c). After 20 cycles, the Si anode with FEC-containing electrolyte showed almost constant capacity, demonstrating the formation of a stable SEI layer in the presence of FEC. However, a rapid decrease in the capacity of the Si anode with EC-containing electrolyte was observed during charge/discharge processes.

Meanwhile, DEMS also provided mechanistic insights into the SEI formation reactions. The gassing behavior of Si anode was strongly electrolyte-dependent. The variation tendencies of H_2 ($m/z = 2$) were similar for the two cells, primarily stemming from electrolyte decomposition and residual water reduction on the Si surface (Figure 9.11b and d). However, the evolution characteristics of C_2H_4 ($m/z = 26$), CO_2 ($m/z = 44$) and CO of the two cells were significant distinctive. For example, C_2H_4 was produced in each cycle for the EC-based electrolytes, which should come from the direct reduction of EC on Si anode. But in the FEC-containing cell, both C_2H_4 and CO evolution were negligible during cycling. In summary, H_2 and CO_2 were mainly produced in the FEC-containing cell, while H_2, C_2H_4 and CO were mainly produced in the EC-based electrolytes. The significant difference in cycle stability and gas evolution of the Si anode in FEC-free and FEC-containing electrolytes demonstrated the critical role of FEC in producing a stable SEI on Si surface.

9.4 CATHODE

The ever-increasing energy demand of modern society continues to drive the development of high-energy LIBs. The cathode materials as one of the determining factors of the capacity and energy density limits of LIBs have been extensively explored in the past few decades. So far, the cathode materials could be loosely grouped into three categories based on their crystal structures: layered oxides ($Li(Ni_xCo_yMn_z)O_2$, $Li(Ni_xCo_yAl_z)O_2$, Li-rich layered oxides), spinel oxides ($Li_xM_2O_4$, M=Co, Ni, Mn, V) and olivine phosphates ($LiFePO_4$ [LFP], $LiFe_xMn_{1-x}PO_4$). However, the introduction of a high-energy density cathode often implies a high activity and/or high operating voltage, which could result in serious parasitic reactions at the cathode/electrolyte interface. To enhance electrochemical performance and life span of LIBs, understanding and stabilizing the cathode/electrolyte interface is therefore of critical importance. In this section, we present an overview of gassing phenomenon at the cathode/electrolyte interface of LIBs, especially on lattice oxygen of cathode, surface impurity and electrolyte chemistry.

9.4.1 Lattice Oxygen

The Li-rich layered oxides, with a chemical formula of $xLi_2MnO_3{\cdot}(1-x)LiMO_2$ (M = Ni, Co or Mn), can provide a high capacity of ~250 mAh/g at an average operating voltage over 3.5 V, and have been considered as alternatives to replace conventional $LiCoO_2$ (~145 mAh/g). To realize the high specific energy, the Li-rich layered oxides often undergo Li^+ ions and oxygen atoms extraction from the Li_2MnO_3 lattice during the initial charge (4.5V–4.6V). However, the extracted oxygen will partly participate in interfacial parasitic reactions (i.e., electrolyte oxidation), resulting in capacity loss.

Novak et al. employed DEMS to follow gas (O_2 and CO_2) evolution to understand the interfacial reaction of 0.5 $Li_2MnO_3{\cdot}$0.5 $LiMO_2$ (HE-NCM).[32] Figure 9.12a presented the potential and gas evolution profiles during the first two cycles using galvanostatic mode and CV. During the first charge, independent of the cell operation modes, the CO_2 started to evolve between 4.2V and 4.7V, and then the CO_2 evolution rate increased significantly once O_2 was produced at 4.7V. In the second charge, negligible O_2 was observed and the CO_2 evolution rate was markedly reduced, indicating Li_2MnO_3 lattice was activated during the first charge. The CO_2 detection onset showed a maximum of 850 mV shift (e.g., 4.55V for C/5 and 3.7V for CV mode), indicating an increase in overpotential of carbonates electrolyte oxidation at high cycling rates. Regardless of the CV or galvanostatic experiments at different rates, however, no potential shift was observed between the O_2 detection onset potential and CO_2 rate increasing onset potential at 4.7V ± 0.05V. The strong correlation between evolution of O_2 and CO_2 indicated that oxygen extraction from the Li_2MnO_3 lattice results in common reactivity characteristics of electrolyte decomposition.

Figure 9.12b depicted the gas evolution mechanism as the function of the applied potential during the first charge, inferred from DEMS analysis. At open circuit voltage (OCV), the EC with high polarizability was more easily enriched at electrode surface compared to DMC/EMC cosolvents and thus EC was considered as the main

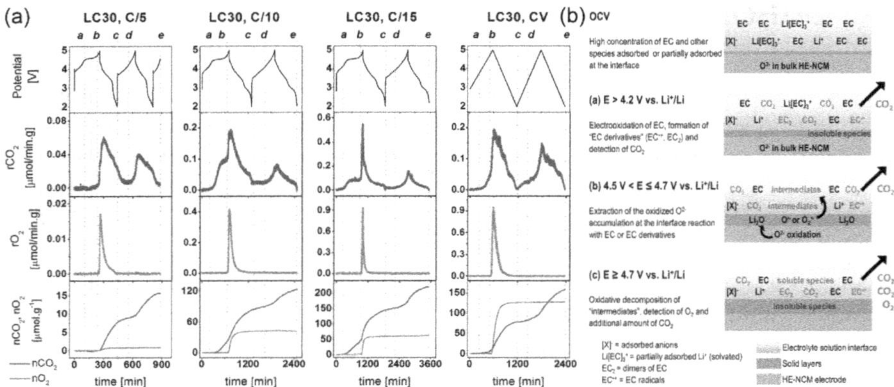

FIGURE 9.12 (a) The voltage profile (top panel) and gas evolution profile (bottom panel) for HE-NCM electrode in the 1 M $LiClO_4$ in a 1:1 mixture (w/w) of EC and DMC. (b) Schematic diagram of electrode/electrolyte interface evolution as a function of the potential.[32]

reactant at electrochemical interface. When the potential was shifted to 4.2V, the EC decomposed to form CO_2 and $C_2H_4O^+$ ethylene oxide radical isomers that could attack other EC molecules to produce additional "EC derivatives" (lower oxidation potential compared to isolated EC) at the interface. At 4.5–4.6 V, the oxygen started to extract along with the activation process of Li_2MnO_3, resulting in Li_2O accumulation and active oxygen species evolution (e.g., superoxide and singlet oxygen), which further reacts with EC and/or EC derivatives to form oxygen-rich intermediates. After the potential reached 4.7 V, these oxygen-rich intermediates further decomposed to evolve O_2 and CO_2.

A question arises naturally: which factor controls the reversibility anionic (lattice oxygen) redox process rather than irreversible O_2 evolution to enhance the cycling stability of Li-rich cathodes. Yu et al. adopted DEMS to monitor the O_2 evolution and further quantify the reversibility of anionic redox of $Li_2Ru_xM_{1-x}O_3$ (model Li_2RuO_3 as the host lattice with different transition metal doping elements), to systematically study the role of M–O covalency in reversible anionic redox.[33] Figure 9.13 showed the charge/discharge profile and gas (O_2 and CO_2) evolution profile of the Li_2RuO_3, $Li_2Ru_{0.75}Ti_{0.25}O_3$, $Li_2Ru_{0.75}Mn_{0.25}O_3$ and $Li_2Ru_{0.75}Fe_{0.25}O_3$ cathodes. Similar amount and onset potential of CO_2 evolution were observed, while the O_2 evolution profiles showed large discrepancy across different cathodes for $Li_2Ru_xM_{1-x}O_3$ (M = Ti, Mn, Fe) with relatively more covalent metal–oxygen interactions, indicating that the difference in oxygen evolution may depend on the electronic structures of the oxides. Coupled with density functional theory calculation and X-ray absorption spectroscopy, they further demonstrated that more ionic substituents and metal–oxygen covalency resulted in irreversible oxygen redox and interfacial parasitic reaction. Clearly, these real-time observations of gas evolution by DEMS are helpful in engineering better Li-rich cathodes for LIBs.

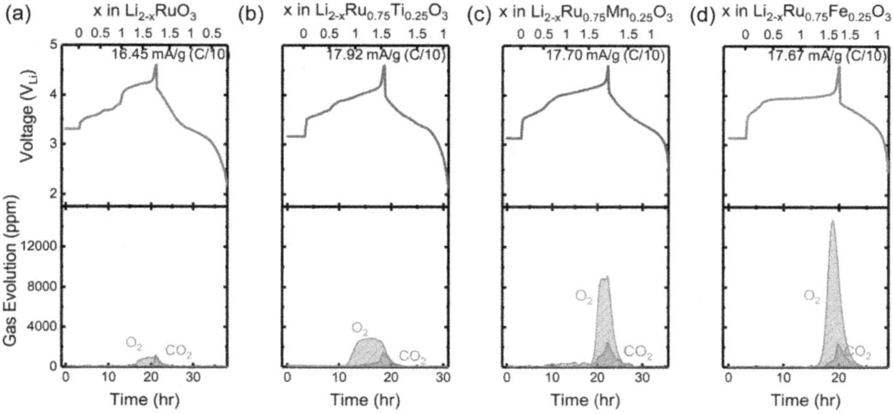

FIGURE 9.13 The charge and discharge profile (top panel) and gas evolution (O_2 and CO_2) profile (bottom panel) of (a) Li_2RuO_3, (b) $Li_2Ru_{0.75}Ti_{0.25}O_3$, (c) $Li_2Ru_{0.75}Mn_{0.25}O_3$ and (d) $Li_2Ru_{0.75}Fe_{0.25}O_3$ cathodes. The anode and electrolyte are Li metal foil and 1 M $LiPF_6$ in a 3:7 mixture by volume of EC: EMC, respectively.[33]

9.4.2 Surface Impurity

It is well recognized that the capacity fading of cathode materials of LIBs is accompanied by O_2 and CO_2 evolution, as demonstrated in the previous subsection. In terms of CO_2 evolution, it may be from multiple sources: carbonate electrolyte, conductive carbon, binder and surface residual Li_2CO_3. It is important to identify these sources for improving battery performance. Many studies have shown that the surface residual Li_2CO_3 of Li-rich layered oxide cathode materials contributed significantly to CO_2 evolution, compared to its own decomposition during the first charge. McCloskey et al. used isotopically labeled DEMS to reveal how residual Li_2CO_3 decomposition affected interfacial parasitic chemistry.[34] In order to demonstrate the direct electrochemical oxidation of Li_2CO_3 in carbonate-based electrolytes, they first synthesized two Li_2CO_3 cathodes with 7% ^{18}O and 99% ^{13}C content, respectively. As illustrated in Figure 9.14a, the CO_2 evolution profile of ^{18}O-labeled Li_2CO_3 followed approximately $2e^-/CO_2$ with 15:1 of C $^{16}O_2$/C $^{18}O_2$ during constant current and potential charge (OCV to 4.8V and then holding at 4.8 V), indicating that degradation of Li_2CO_3 mainly contributes to the evolution of CO_2. Also, the linear sweep voltammetry (LSV) of ^{13}C-labeled Li_2CO_3 cathode from 3.2 to 5.2 V (Figure 9.14b) also presented a CO_2 evolution profile with approximately $2e^-/^{13}CO_2$ starting around 4.0 V, but the $^{12}CO_2$ evolved at 4.6 V, suggesting that the electrolyte decomposition potential was significantly higher than that of Li_2CO_3 cathode. Then, they employed additional control experiments to rule out the possibility that the CO_2 evolution came from a chemical decomposition of Li_2CO_3 via protons attacking formed by electrolyte degradation/residual H_2O. Figure 9.14c showed $^{12}CO_2$ evolution at 4.8V, and no $^{13}CO_2$ was observed on a porous carbon electrode (without Li_2CO_3) by LSV at 10 mV/min. When ^{13}C-labeled Li_2CO_3 was sandwiched between two separators to disallow electronic contact between the Li_2CO_3 and cathode, $^{12}CO_2$ evolved at 4.8V

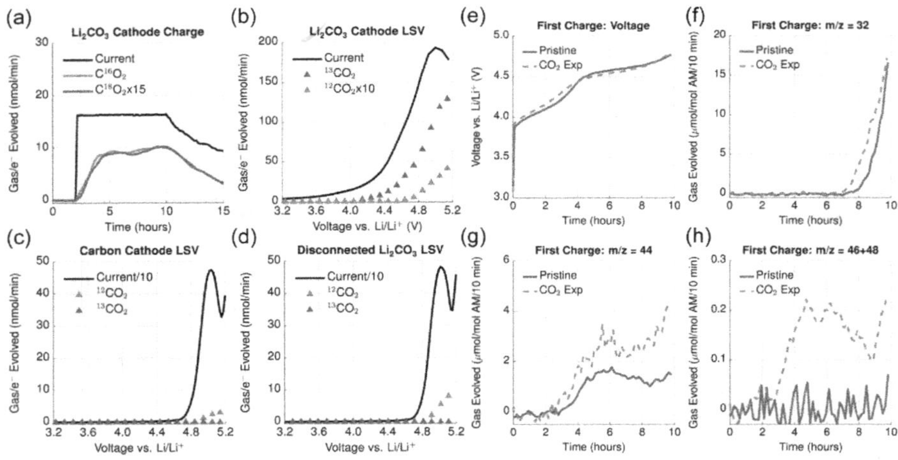

FIGURE 9.14 CO_2 evolution for (a) a Li_2CO_3 cathode with 7% ^{18}O content using a 0.1 C constant current charge, followed by a voltage hold at 4.8V, (b) a 99% ^{13}C-labeled Li_2CO_3 cathode using LSV from 3.2 to 5.2V, (c) a porous carbon cathode using LSV from 3.2 to 5.2 V and (d) porous carbon cathode with 99% ^{13}C-labeled Li_2CO_3 that it was electronically isolated using LSV from 3.2 to 5.2 V. (e–h) Gas evolution during charging for pristine (blue) and C $^{18}O_2$ exposed $Li_{1.17}(Ni_{0.2}Mn_{0.6}Co_{0.2})_{0.83}O_2$ samples. The anode and electrolyte used were Li metal foil and 1.0 M $LiPF_6$ in a 1:1 mixture by weight of EC and DEC, respectively.[34]

and no $^{13}CO_2$ was observed (Figure 9.14d). These results indicated that most CO_2 evolution originated from direct electrochemical decomposition of Li_2CO_3 when the cathode was polarized below ~4.8 V.

Subsequently, they synthesized C $^{18}O_2$ (95% ^{18}O) gas-exposed $Li_{1.17}(Ni_{0.2}Mn_{0.6}Co_{0.2})_{0.83}O_2$ (NMC) cathode, named Li_2C $^{18}O_3$@NMC, to track decomposition of Li_2CO_3 and eliminate the interference of carbonate electrolyte or conductive carbon. As shown in Figure 9.14e, the electrochemical profiles of Li_2C $^{18}O_3$@NMC did not change significantly compared to pristine NMC. However, the gas evolution presented obvious discrepancy between pristine NMC and Li_2C $^{18}O_3$@NMC (Figure 9.14f–h): (i) the NMC lattice had a slightly earlier onset of O_2 evolution for Li_2C $^{18}O_3$@NMC; (ii) the Li_2C $^{18}O_3$@NMC sample showed significantly higher amount of CO_2 evolution originating from decomposition of electrolyte ($m/z = 44$) and Li_2C $^{18}O_3$ ($m/z = 44 + 46$). The above experiment results clearly indicated that direct electrochemical of Li_2CO_3 could trigger instability of NMC lattice (earlier onset of O_2 evolution) to form more reactive oxygen species, resulting in enhanced decomposition of electrolyte. They further concluded that reactive oxygen species from oxidation of Li_2CO_3 catalyzed the destabilization process of NMC surface and proposed that removal of surface impurities of cathode materials could be beneficial for cyclability of LIBs.

9.4.3 Electrolyte Chemistry

The important prerequisites of electrolytes used in LIBs are the Li^+ ion transport properties and electrochemical stability/compatibility that determine most performance metrics of batteries. The organic carbonate-based electrolytes have become a wide choice by virtue of preferred electrochemical window and excellent compatibility with various cathodes and anodes in LIBs. However, gas evolution originated from decomposition of electrolyte represents a major safety concern that can result in ignition and explosion of batteries. In particular, most high potential cathodes, such as $LiCoO_2$, $LiNi_{0.5}Mn_{1.5}O_4$ and $LiMn_2O_4$, operate beyond the decomposition potential of carbonate-based electrolytes (>4.3 V). Therefore, understanding electrolyte interfacial chemistry is highly important to build safe and efficient LIBs.

Abruña and his collaborators used DEMS to study oxidative stability of three carbonates-based electrolytes (i.e., 1 M $LiFP_6$ in EC/DEC, EC/DMC and PC) at three popular cathode materials (i.e., $LiCoO_2$, $LiNi_{0.5}Mn_{1.5}O_4$ and $LiMn_2O_4$).[35] Figure 9.15 presented CVs and corresponding gas evolution profile for three carbonate-based electrolytes and three types of cathodes. During the positive-going scan process from OCV to 5V, two oxidation peaks occurred at around 4.1–4.2 V and 4.7–4.8 V in the three electrolytes, corresponding to partial and complete delithiation processes of $LiCoO_2$ (Figure 9.15a). Although the electrochemical behaviors were similar, the gas evolution behaviors exhibited large differences. In the PC and EC/DMC electrolytes, only CO_2 evolution was observed at the potential region of 4.7–5 V, and no potential-dependent O_2 was detected. Instead, the onset of O_2 evolution occurred at 4.8V and the amount of CO_2 evolution was also the largest for EC/DEC electrolytes. For

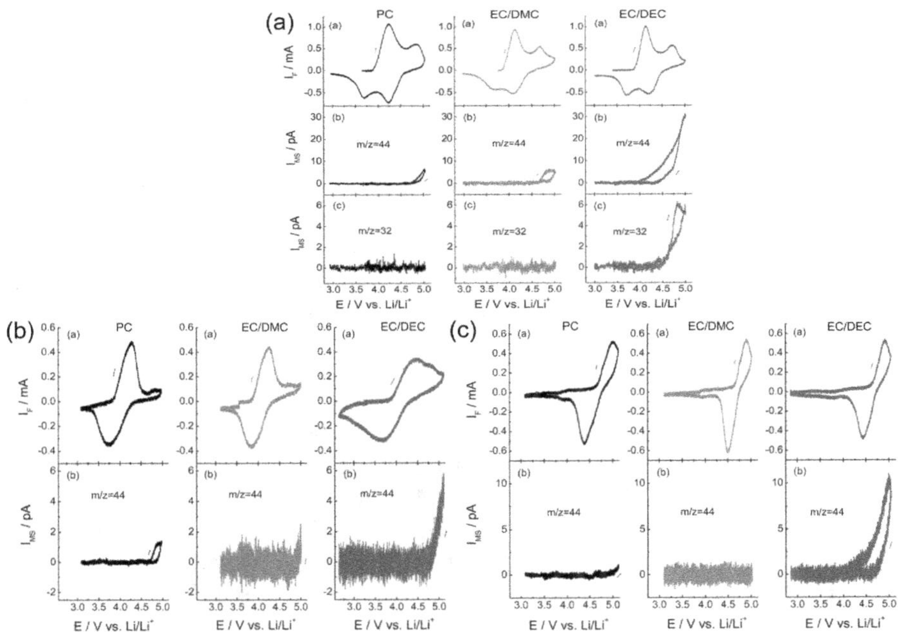

FIGURE 9.15 Cyclic voltammetry profile (top panel) and gas evolution (O_2 and CO_2) profile (middle and bottom panel) for (a) $LiCoO_2$ electrode, (b) $LiMn_2O_4$ and (c) $LiNi_{0.5}Mn_{1.5}O_4$ in 1 M $LiPF_6$+ PC (left panel), 1 M $LiPF_6$+EC/DMC (1:1 wt %) (middle panel) and 1 M $LiPF_6$+EC/DEC (1:1 wt %) (right panel), respectively, at a scan rate of 0.2 mV/s[1].[35]

$LiMn_2O_4$ cathode, a main oxidation (4.3–4.4 V) and a reduction peak (3.6V–3.7V) were observed, corresponding to Li^+ ions extraction and insertion (Figure 9.15b). In EC/DMC electrolytes, the onset of CO_2 evolution occurred at 4.9 V that was higher than 4.7 V for PC and EC/DEC electrolyte. More amount of CO_2 was evolved in the EC/DEC electrolyte than in PC and EC/DMC electrolytes. For $LiNi_{0.5}Mn_{1.5}O_4$ electrode, the EC/DEC electrolyte presented more CO_2 evolution compared to $LiMn_2O_4$ cathode, demonstrating that EC/DEC solvent could be catalyzed by the Mn^{3+}/Mn^{4+} and Ni^{2+}/Ni^{4+} redox couple of $LiNi_{0.5}Mn_{1.5}O_4$ (Figure 9.15c). Conversely, no CO_2 evolution was observed for PC and EC/DMC electrolyte below 5.0V. According to above DEMS data, the EC/DMC mixture electrolyte was the most stable for the three cathode materials, while EC/DEC was the least stable one. Further studies are required to determine how electrode materials (and/or surface properties) to change electrolyte chemistry, including elementary reaction pathways and catalytic active sites.

9.5 CROSS-TALK OF GAS IN FULL BATTERY

The gas evolution shown in the previous sections enables researchers to understand either cathode or anode interfacial chemistry via model half cells, but it is equally important to consider a complete system with both electrodes. This is because the evolved gas from one electrode may migrate to the other, which adsorbs and even participates in interfacial reactions. The CO_2 consumption is recognized as the most typical example. As discussed previously, electrolyte oxidation could generate large amounts of CO_2 especially for high-voltage cathodes, which could be reduced at graphite (especially for lithiated graphite) and Li anode surface to form Li_2CO_3- or $Li_2C_2O_4$-containing SEI, further improving the properties of anode/electrolyte interface. An experimental evidence by Ellis et al. showed the CO_2 gas injected into pouch batteries was almost completely depleted after 100 hours storage.[36] A few works have demonstrated that combinations of cathodes and anodes have significant impacts on the amounts of gas evolution due to the cross-talk effect of gas that changed the interfacial properties.[37]

A question naturally arises: how to quantitatively evaluate this cross-talk effect? Michalak and his co-workers[38] employed DEMS to distinguish process occurring at either cathode or anode (Figure 9.16a). First, they investigated the gas evolution of LNMO/graphite full cell. The charge profile of LNMO/graphite cell showed two plateaus at 4.57 and 4.64 V, corresponding to the Ni^{2+}/Ni^{3+} and Ni^{3+}/Ni^{4+} redox couples (Figure 9.16b). The CO_2 evolution showed a local maximum at the beginning of each charge, then decreased and finally increased again along with Ni^{3+} to Ni^{4+} redox couple. The evolution of H_2 followed a similar trend to that of CO_2. However, C_2H_4 evolution was only observed at the initial charge of cells. Then, the authors substituted graphite with delithiated LFP to establish LFP/LNMO cells, where LFP did not partake in gas evolution and consumption reactions because LFP operated within the stability window of the electrolyte. In this case, H_2, C_2H_4 and CO_2 evolution were observed at initial charge process (Figure 9.16c). The gases from reductive decomposition of the electrolyte on graphite surface were mainly H_2 and C_2H_4, and a small amount of CO_2. When graphite was replaced by LFP, CO_2 evolution as the main gas product with negligible H_2 and C_2H_4 was observed (Figure 9.16d). Thus, CO_2 was

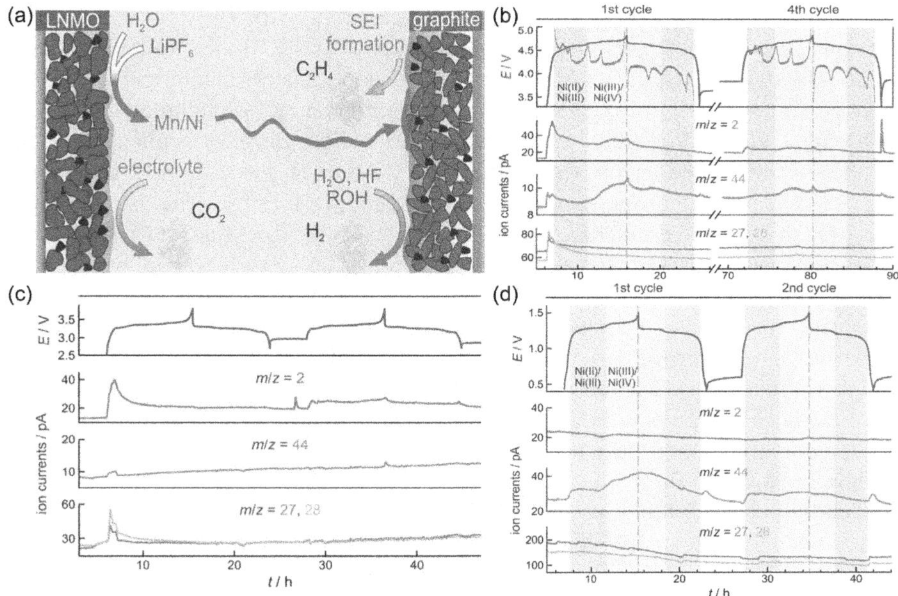

FIGURE 9.16 (a) Schematic diagram of cross-talk between anode and cathode for LIBs. The voltage profiles and corresponding derivative curves (top panel) and gas evolution profiles (bottom panel) for (b) LNMO/graphite, (c) LFP/LNMO and (d) LNMO/LFP cells.[38]

primarily produced at the cathode as a result of oxidative decomposition of electrolyte catalyzed by Ni^{3+} to Ni^{4+} redox couple, and H_2 and C_2H_4 was evolved from the graphite anode. The sum of gas evolution for LNMO/LFP cell and the LFP/LNMO battery was not matching with that of the LNMO/graphite cell perfectly, indicating a cross-talk effect of gas in full cells. The use of DEMS in combination with LFP electrodes to distinguish gas production sources and cross-talk effects could provide more conclusive evidence for gas evolution symptoms of LIBs.

9.6 SUMMARY

To meet the ever-growing demands of portable electronic devices, electric vehicles and large-scale energy storage systems, LIBs need to be further ameliorated in terms of available energy, lifetime and safety. Unfortunately, drastic gas evolution caused by uncontrollable parasitic reactions inside the LIBs directly affects the cell performance and safety. As a result, an in-depth understanding of gas evolution behaviors during cell operation is of critical importance. DEMS has proven to be an indispensable technique for providing qualitative and quantitative information of gas evolution in LIBs.

This chapter has presented a critical review DEMS technique in LIBs field, including historical development, experimental setup and current understanding of interfacial reactions. We have witnessed the rapid development of DEMS over the past three decades. However, DEMS techniques are still in the early or puberty stage at present; there is plenty of space for improvement of DEMS. In the future, the membrane inlet method requires a more advanced semipermeable membrane and a reasonably designed cell, to improve the collection efficiency for quantitative analysis with unprecedented sensitivity and to expand to a wider range of applications (e.g., aprotic solvents-based energy storage device). The carrier gas inlet method requires smaller dead volume to increase temporal resolution. More importantly, a suitable inlet system for commercial batteries (e.g., pouch cell, and steel shell cell) is expected to achieve in situ/operando detection of gas evolution, to assist industrial production, safety monitoring and failure analysis.

In situ acquisition of multidimensional electrochemical information (including gas/liquid/solid intermediates and products, structural and morphological evolutions and phase transitions) have been a dream of former generations of electrochemists and is still a big challenge till now. As each technique has both advantages and disadvantages, a combination of these in situ characterization techniques with DEMS should be developed in the future. The combined use of advanced in situ techniques (such as scanning electron microscopy (SEM), transmission electron microscopy (TEM), atomic force microscopy (AFM), X-ray diffraction (XRD), X-ray photoelectron spectroscopy (XPS), Raman spectroscopy) and DEMS have great potential to give relatively complete mechanistic insights during cell operation. For the combination of different in situ characterization techniques, novel in situ cells with advantages of low cost, high efficiency and convenient installation need to be carefully designed, and thereby improve data quality to accommodate in real-world use conditions of batteries. However, very few works tackle such challenges in this emerging field of in situ characterization techniques, so more efforts should be made to address this issue.

Overall, we have reviewed the state-of-the-art DEMS and its applications in LIBs. We hope this review can offer valuable guidance for the further development of the advanced LIBs with excellent electrochemical performance and high safety. With unremitting research efforts, we believe that DEMS, combined with other advanced characterization techniques, will make significant contributions to understand the battery reaction and process and to design better battery systems.

REFERENCES

[1] Aurbach, D.; Markovsky, B.; Salitra, G.; Markevich, E.; Talyossef, Y.; Koltypin, M.; Nazar, L.; Ellis, B.; Kovacheva, D. Review on Electrode–Electrolyte Solution Interactions, Related to Cathode Materials for Li-Ion Batteries. *J. Power Sources* **2007**, *165* (2), 491–499.

[2] Zhao, Z, Peng, Z. Differential Electrochemical Mass Spectroscopy: A Pivotal Technology for Investigating Lithium-Ion Batteries. *Energy Storage Sci. Technol.* **2019**, *8* (1), 1–13.

[3] Hoch, G.; Kok, B. A Mass Spectrometer Inlet System for Sampling Gases Dissolved in Liquid Phases. *Arch. Biochem. Biophys.* **1963**, *101* (1), 160–170.

[4] Bruckenstein, S.; Gadde, R. R. Use of a Porous Electrode for In Situ Mass Spectrometric Determination of Volatile Electrode Reaction Products. *J. Am. Chem. Soc.* **1971**, *93* (3), 793–794.

[5] Wolter, O.; Heitbaum, J. Differential Electrochemical Mass-Spectroscopy (DEMS) - A New Method for the Study of Electrode Processes. *Ber. Bunsen-Ges. Phys. Chem.* **1984**, *88*, 2– 6.

[6] Baltruschat, H. Differential Electrochemical Mass Spectrometry. *J. Am. Soc. Mass Spectrom.* **2004**, *15* (12), 1693–1706.

[7] Hartung, T.; Baltruschat, H. Differential Electrochemical Mass Spectrometry Using Smooth Electrodes: Adsorption and Hydrogen/Deuterium Exchange Reactions of Benzene on Platinum. *Langmuir* **1990**, *6* (5), 953–957.

[8] Jusys, Z.; Massong, H.; Baltruschat, H. A New Approach for Simultaneous DEMS and EQCM: Electro-Oxidation of Adsorbed CO on Pt and Pt-Ru. *J. Electrochem. Soc.* **1999**, *146* (3), 1093.

[9] Belén Molina Concha, M.; Chatenet, M.; Lima, F. H. B.; Ticianelli, E. A. In Situ Fourier Transform Infrared Spectroscopy and On-Line Differential Electrochemical Mass Spectrometry Study of the NH3BH3 Oxidation Reaction on Gold Electrodes. *Electrochimica Acta* **2013**, *89*, 607–615.

[10] Trimarco, D. B.; Scott, S. B.; Thilsted, A. H.; Pan, J. Y.; Pedersen, T.; Hansen, O.; Chorkendorff, I.; Vesborg, P. C. K. Enabling Real-Time Detection of Electrochemical Desorption Phenomena with Sub-Monolayer Sensitivity. *Electrochimica Acta* **2018**, *268*, 520–530.

[11] Imhof, R.; Novák, P. In Situ Investigation of the Electrochemical Reduction of Carbonate Electrolyte Solutions at Graphite Electrodes. *J. Electrochem. Soc.* **1998**, *145* (4), 1081.

[12] Imhof, R.; Novák, P. Oxidative Electrolyte Solvent Degradation in Lithium-Ion Batteries an In Situ Differential Electrochemical Mass Spectrometry Investigation. *J. Electrochem. Soc.* **1999**, *146* (5), 1702.

[13] Holzapfel, M.; Wursig, A.; Scheifele, W.; Vetter, J.; Novak, P. Oxygen, Hydrogen, Ethylene and CO_2 Development in Lithium-Ion Batteries. *J. Power Sources* **2007**, *174* (2), 1156–1160.

[14] McCloskey, B. D.; Bethune, D. S.; Shelby, R. M.; Girishkumar, G.; Luntz, A. C. Solvents' Critical Role in Nonaqueous Lithium-Oxygen Battery Electrochemistry. *J. Phys. Chem. Lett.* **2011**, *2*, 1161–1166.

[15] Peng, Z.; Freunberger, S. A.; Chen, Y.; Bruce, P. G. A Reversible and Higher-Rate Li-O_2 Battery. *Science* **2012**, *337* (6094), 563–566.

[16] Berkes, B. B.; Jozwiuk, A.; Vračar, M.; Sommer, H.; Brezesinski, T.; Janek, J. Online Continuous Flow Differential Electrochemical Mass Spectrometry with a Realistic Battery Setup for High-Precision, Long-Term Cycling Tests. *Anal. Chem.* **2015**, *87* (12), 5878–5883.

[17] Metzger, M.; Marino, C.; Sicklinger, J.; Haering, D.; Gasteiger, H. A. Anodic Oxidation of Conductive Carbon and Ethylene Carbonate in High-Voltage Li-Ion Batteries Quantified by On-Line Electrochemical Mass Spectrometry. *J. Electrochem. Soc.* **2015**, *162* (7), A1123–A1134.

[18] Tsiouvaras, N.; Meini, S.; Buchberger, I.; Gasteiger, H. A. A Novel On-Line Mass Spectrometer Design for the Study of Multiple Charging Cycles of a Li-O_2 Battery. *J. Electrochem. Soc.* **2013**, *160* (3), A471–A477.

[19] Zhou, B.; Guo, L.; Zhang, Y.; Wang, J.; Ma, L.; Zhang, W.-H.; Fu, Z.; Peng, Z. A High-Performance Li-O_2 Battery with a Strongly Solvating Hexamethylphosphoramide Electrolyte and a LiPON-Protected Lithium Anode. *Adv. Mater.* **2017**, *29* (30), 1701568.

[20] Zhang, H.; Yang, Y.; Ren, D.; Wang, L.; He, X. Graphite as Anode Materials: Fundamental Mechanism, Recent Progress and Advances. *Energy Storage Mater.* **2021**, *36*, 147–170.

[21] Rowden, B.; Garcia-Araez, N. A Review of Gas Evolution in Lithium Ion Batteries. *Energy Rep.* **2020**, *6*, 10–18.

[22] Michalak, B.; Berkes, B. B.; Sommer, H.; Brezesinski, T.; Janek, J. Electrochemical Cross-Talk Leading to Gas Evolution and Capacity Fade in $LiNi_{0.5}Mn_{1.5}O_4$/Graphite Full-Cells. *J. Phys. Chem. C* **2017**, *121* (1), 211–216.

[23] Li, J.; Kong, Z.; Liu, X.; Zheng, B.; Fan, Q. H.; Garratt, E.; Schuelke, T.; Wang, K.; Xu, H.; Jin, H. Strategies to Anode Protection in Lithium Metal Battery: A Review. *InfoMat* **2021**, *3* (12), 1333–1363.

[24] Zhao, H.; Wang, J.; Shao, H.; Xu, K.; Deng, Y. Gas Generation Mechanism in Li-Metal Batteries. *Energy Environ. Mater.* **2022**, *5* (1), 327–336.

[25] Kozen, A. C.; Lin, C.-F.; Pearse, A. J.; Schroeder, M. A.; Han, X.; Hu, L.; Lee, S.-B.; Rubloff, G. W.; Noked, M. Next-Generation Lithium Metal Anode Engineering *via* Atomic Layer Deposition. *ACS Nano* **2015**, *9* (6), 5884–5892.

[26] Lin, C.-F.; Kozen, A. C.; Noked, M.; Liu, C.; Rubloff, G. W. ALD Protection of Li-Metal Anode Surfaces - Quantifying and Preventing Chemical and Electrochemical Corrosion in Organic Solvent. *Adv. Mater. Interfaces* **2016**, *3* (21), 1600426.

[27] Zhang, Y.; Mu, Z.; Lai, J.; Chao, Y.; Yang, Y.; Zhou, P.; Li, Y.; Yang, W.; Xia, Z.; Guo, S. MXene/Si@SiO_x@C Layer-by-Layer Superstructure with Autoadjustable Function for Superior Stable Lithium Storage. *ACS Nano* **2019**, *13*, 2167–2175.

[28] Ma, D.; Cao, Z.; Hu, A. Si-Based Anode Materials for Li-Ion Batteries: A Mini Review. *Nano-Micro Lett.* **2014**, *6* (4), 347–358.

[29] Ashuri, M.; He, Q.; Shaw, L. L. Silicon as a Potential Anode Material for Li-Ion Batteries: Where Size, Geometry and Structure Matter. *Nanoscale* **2016**, *8* (1), 74–103.

[30] Hu, Y.; Qiao, Y.; Xie, Z.; Li, L.; Qu, M.; Liu, W.; Peng, G. Water-Soluble Polymer Assists Multisize Three-Dimensional Microspheres as a High-Performance Si Anode for Lithium-Ion Batteries. *ACS Appl. Energy Mater.* **2021**, *4* (9), 9673–9681.

[31] Schiele, A.; Breitung, B.; Hatsukade, T.; Berkes, B. B.; Hartmann, P.; Janek, J.; Brezesinski, T. The Critical Role of Fluoroethylene Carbonate in the Gassing of Silicon Anodes for Lithium-Ion Batteries. *ACS Energy Lett.* **2017**, *2* (10), 2228–2233.

[32] Castel, E.; Berg, E. J.; El Kazzi, M.; Novák, P.; Villevieille, C. Differential Electrochemical Mass Spectrometry Study of the Interface of x $Li_2MnO_3\cdot$ (1− x) $LiMO_2$(M = Ni, Co, and Mn) Material as a Positive Electrode in Li-Ion Batteries. *Chem. Mater.* **2014**, *26* (17), 5051–5057.

[33] Yu, Y.; Karayaylali, P.; Sokaras, D.; Giordano, L.; Kou, R.; Sun, C.-J.; Maglia, F.; Jung, R.; Gittleson, F. S.; Shao-Horn, Y. Towards Controlling the Reversibility of Anionic Redox in Transition Metal Oxides for High-Energy Li-Ion Positive Electrodes. *Energy Environ. Sci.* **2021**, *14* (4), 2322–2334.

[34] Kaufman, L. A.; McCloskey, B. D. Surface Lithium Carbonate Influences Electrolyte Degradation via Reactive Oxygen Attack in Lithium-Excess Cathode Materials. *Chem. Mater.* **2021**, *33* (11), 4170–4176.

[35] Wang, H.; Rus, E.; Sakuraba, T.; Kikuchi, J.; Kiya, Y.; Abruña, H. D. CO_2 and O_2 Evolution at High Voltage Cathode Materials of Li-Ion Batteries: A Differential Electrochemical Mass Spectrometry Study. *Anal. Chem.* **2014**, *86* (13), 6197–6201.

[36] Ellis, L. D.; Allen, J. P.; Thompson, L. M.; Harlow, J. E.; Stone, W. J.; Hill, I. G.; Dahn, J. R. Quantifying, Understanding and Evaluating the Effects of Gas Consumption in Lithium-Ion Cells. *J. Electrochem. Soc.* **2017**, *164* (14), A3518–A3528.

[37] Michalak, B.; Sommer, H.; Mannes, D.; Kaestner, A.; Brezesinski, T.; Janek, J. Gas Evolution in Operating Lithium-Ion Batteries Studied In Situ by Neutron Imaging. *Sci. Rep.* **2015**, *5*, 15627.

[38] Michalak, B.; Berkes, B. Z. B.; Sommer, H.; Bergfeldt, T.; Brezesinski, T.; Janek, J. R. Gas Evolution in $LiNi_{0.5}Mn_{1.5}O_4$/Graphite Cells Studied in Operando by a Combination of Differential Electrochemical Mass Spectrometry, Neutron Imaging, and Pressure Measurements. *Anal. Chem.* **2016**, *88*, 2877–2883.

10 Thermal Analysis of Li-Ion Batteries

Ankur Jain
University of Texas at Arlington

Mohammad Parhizi
UL Research Institutes

CONTENTS

DOI: 10.1201/9781003299295-10

10.1 FUNDAMENTAL PRINCIPLES OF HEAT TRANSFER ANALYSIS

This chapter presents a brief discussion of the fundamental principles of heat transfer analysis, as applicable for understanding heat transfer in Li-ion cells. Heat transfer refers to the transport of thermal energy from one point to another in accordance with the laws of thermodynamics and is usually driven by a temperature gradient [1,2]. Heat transfer is ubiquitous in engineering materials, devices and systems, and often plays a key role in determining efficiency and safety. As a general example, heat generated during the operation of a computer chip must be adequately removed in order to avoid the overheating of the chip [3]. Similarly, the temperature of a Li-ion cell must be maintained within a narrow temperature window in order to maximize energy conversion/storage efficiency as well as to prevent thermal runaway at elevated temperatures [4,5].

The key parameters of interest in heat transfer analysis are temperature and rate of heat flow. Temperature represents the energy content of the body and is usually measured in Kelvin (K) or degrees Celsius (°C) in SI units. Rate of heat flow refers to the rate at which heat transfer occurs, and is measured in units of W (J/s). Heat flow may also be expressed on a per unit area basis as heat flux, W/m^2. Much of heat transfer analysis aims to determine a relationship between temperature and rate of heat flow for a given geometry, materials and other conditions. Often, the goal of the problem is to understand the nature of heat transfer in steady-state conditions. In contrast, time-dependent behavior may also be of interest in transient problems. Much of heat transfer analysis focuses on determining the temperature field, given other properties and parameters, referred to as the direct problem [6]. In contrast, more limited sets of problems focus on determining conditions, such as heat flux needed to obtain a temperature field that is as close as possible to a desired temperature field [6].

10.1.1 Heat Transfer Mechanisms

Three distinct mechanisms of heat transfer are known – conduction, convection and radiation [1,2]. These are introduced in detail in the following sub-sections.

10.1.1.1 Conduction Heat Transfer

Conduction heat transfer refers to the flow of heat across a solid (or a stationary fluid) due to molecular vibrations [7]. In general, the rate of heat transfer is dependent on the temperature gradient. Mathematically, conduction heat transfer is described by Fourier's law, given by

$$\boldsymbol{q}'' = -\boldsymbol{k} \cdot \nabla T, \tag{10.1}$$

where $\boldsymbol{k}$ is the direction-dependent thermal conductivity vector and $\boldsymbol{q}''$ refers to the directional rate of heat flow. The definition of the gradient operator is different in different coordinate systems. In general, thermal conductivity is anisotropic, i.e., components of thermal conductivity vector (e.g., k_x, k_y and k_z) are, in general, unequal to each other. In several practical cases, however, it is reasonable to assume isotropic thermal conduction, i.e., $k_x = k_y = k_z$. In addition to direction-dependence, it is important to recognize that thermal conductivity is also, in general, a function of temperature, although temperature-independence is often assumed for simplified analysis.

By considering the principle of energy conservation along with Fourier law, a general partial differential equation (PDE) governing the temperature field may be derived. For Cartesian coordinate system, for example, the equation may be written as [7]

$$\frac{\partial}{\partial x}\left(k_x \frac{\partial T}{\partial x}\right)+\frac{\partial}{\partial y}\left(k_y \frac{\partial T}{\partial y}\right)+\frac{\partial}{\partial z}\left(k_z \frac{\partial T}{\partial z}\right)+q''' = \rho C_p \frac{\partial T}{\partial t}, \tag{10.2}$$

where q''', ρ and C_p are the internal heat generation rate, density and heat capacity, respectively. Analytical solutions for the energy conservation equations, such as Equation 10.2, are available for a wide variety of scenarios.

10.1.1.2 Convection Heat Transfer

Convection heat transfer refers to the flow of heat between a solid body and a fluid due to a temperature difference and relative motion between the two [8,9]. Such relative motion results in the setting up of momentum and thermal boundary layers starting at the solid–fluid interface, and flow of heat across the boundary layer. The rate of boundary layer growth depends on a number of parameters, including fluid properties, freestream velocity and temperature, as well as the nature of flow (laminar vs. turbulent). Such relative motion may occur on the external surface of the solid, for example, flow past a cylinder or a flat plate, or internally within the solid, for example, flow through a hollow pipe. In external flows, the boundary layer continues to grow indefinitely, whereas in internal flows, the boundary layers, beginning from the inner surface of the pipe all around circumferentially, grow inward and merge at the center. This is referred to as flow development [8], and the initial region in which the boundary layer is still growing is referred to as the entry length. The developing region prior to the entry length is characterized by greater rates of convective heat transfer.

Mathematically, convective heat transfer is governed by Newton's law of cooling, given by [1]

$$q'' = h(T_s - T_\infty), \tag{10.3}$$

where T_s and T_∞ are temperatures of the solid and freestream fluid, respectively. h is the convective heat transfer coefficient, also often expressed non-dimensionally as the Nusselt number. Depending on the nature of the temperature field, h may be a constant number, or a function of space and/or time. In general, h depends on location, such as distance away from the leading edge, flow conditions, such as freestream velocity, and flow properties, such as viscosity and Prandtl number. Such relationships have been derived analytically using energy conservation analysis for simple problems. For other more complicated problems, correlations based on approximate analysis and/or experimental measurements are available.

10.1.1.3 Radiation Heat Transfer

Radiation heat transfer refers to the flow of heat by electromagnetic radiation [10,11]. Heat transfer between the sun and earth is a common example of radiation. Radiation differs from conduction and convective heat transfer by not requiring a physical medium for heat transfer. The amount of radiative flux from a perfect idealized surface, known as a blackbody depends on its temperature, and is emitted over the entire wavelength spectrum, which, according to Planck's law, peaks at a certain wavelength. The peak wavelength is given by Wien's displacement law that shows that the peak wavelength is inversely proportional to temperature. Based on a number of simplifying assumptions, radiative flux from a surface at temperature T is given by

$$q'' = \varepsilon \sigma T^4, \tag{10.4}$$

where ε is the emissivity of the surface and σ is the Stefan-Boltzmann constant. A blackbody has an emissivity of 1 whereas real surfaces have an emissivity between 0 and 1. Therefore, a realistic surface emits lesser than a blackbody at the same temperature. The emissivity is, in general, a function of wavelength, temperature and direction. The extent of radiative heat transfer between surfaces is often carried out using the concept of view factors, and a balance between radiative flow of heat into and out of the surface. While surface-to-surface view factors are used in several problems, surface-to-gas and gas-to-gas view factors may be necessitated for more complicated problems, such as those in which the gas medium between surfaces also participates in radiative heat transfer.

10.1.1.4 Multi-Mode Heat Transfer

While several practical engineering problems are dominated by one of the three modes of heat transfer outlined above, in some other problems, more than one heat transfer models may be present and significant. Such problems are often called coupled, or conjugate heat transfer problems [12]. As a simple example, heat transfer through a glass window in a house involves heat transfer between the surroundings and the window (on both sides) due to convection and radiation, conduction through the glass as well as radiation through the gap in case of a double-pane window. Careful consideration of how these modes of heat transfer interact with each other in the specific problem of interest is important.

Conjugate heat transfer problems are, in general, more complicated to solve than single-mode heat transfer problems. Exact solutions are known for a few conjugate heat transfer problems, whereas numerical simulations may be needed for other, more complicated problems.

10.1.2 Material Properties

The key material properties relevant to heat transfer analysis include thermal conductivity and heat capacity for thermal conduction analysis, and additionally, thermal properties of the fluid, as well as freestream velocity and temperature for convective heat transfer analysis. Typically, thermal conductivity of materials varies over a broad range, and is in the range of hundreds of W/mK for typical metals, and in the 0.1–1.0 W/mK range for poorly conducting materials such as plastics and polymers. Alloys typically exhibit intermediate thermal conductivity. In specific materials such as composites, anisotropy in thermal conductivity must also be accounted for. Heat capacity may also vary widely, with a very high value of 4,184 J/kg K for water and in the range of hundreds of J/kg K for typical metals. Further, theoretical analysis often assumes such properties to be temperature-independent, which is a reasonable assumption when the temperature difference in the problem is relatively small. When the temperature difference is large, variation over the temperature range must be accounted for, and is commonly done by evaluating properties at the mean temperature. In some specific cases, variation in fluid properties with pressure must also be accounted for.

10.1.3 Heat Transfer Analysis Methods

Heat transfer analysis is usually carried out using either analytical/semi-analytical methods, or numerical simulation techniques. These are introduced briefly in the following sub-sections.

10.1.3.1 Analytical Methods

Analytical methods refer to the derivation of exact solutions for differential equations governing the temperature field. Such differential equations are based on the principle of energy conservation, and, depending on the nature of the problem, may be written either as a steady-state or a transient equation [7]. A number of analytical techniques are available for solving such differential equations, with steady-state problems being, in general, easier to solve. The most commonly used methods include separation of variables technique, Green's functions and Laplace transforms. The separation of variables technique seeks to write the temperature distribution as a product of functions that individually depend on only one variable, and by doing so, transforms the general PDE into a set of ordinary differential equations (ODEs) that often represent a Sturm–Liouville problem and are much easier to solve [7]. Derivation of the final solution also utilizes the principle of orthogonality of eigenfunctions of the problem. Green's function technique utilizes the linear nature of the problem by integrating the analytical solutions of fundamental problems, including the influence of point, line and plane sources [13]. Laplace transform technique is particularly useful for transient problems. In this technique, a Laplace transformation of the governing equations is carried out, resulting in a simpler set of equations not involving the time variable in the Laplace domain. Once solved in the Laplace domain, an inverse Laplace transformation is carried out, either analytically or numerically in order to determine the solution for the problem.

Analytically derived solutions are, in general, simple, elegant and exact. However, analytical solutions exist only for relatively simpler problems, and may not be able to account for complicated geometries and other intricacies that may exist in realistic problems. It is often important to determine what kind of simplifying assumptions may be reasonable to transform a given problem to a configuration that is amenable to an analytical solution and to determine the extent of error involved in such approximation.

10.1.3.2 Numerical Methods

In contrast with analytical methods that seek an exact solution of the energy conservation equation, numerical methods are also often used to approximately solve heat transfer problems. Numerical methods are based on the concept of discretization, which refers to splitting the continuous geometry of the problem into discrete cells [14,15]. The energy conservation equation is then solved appropriately in each cell. The three most commonly used methods for this purpose include the finite-difference (FDM), finite-volume (FVM) and finite-element (FEM) methods. Discretization typically results in a large number of linear algebraic equations involving the unknown temperatures at nodes. Such systems of equations are typically sparse, and may be solved using a number of different computational methods. Stability of such numerical techniques must be carefully considered in solving transient problems. Trade-offs between accuracy and computational time must also be considered. For example, choosing to discretize into a very large number of cells results in highly accurate computation of the temperature field, but also incurs large computational cost. The use of a non-uniform grid, with fine discretization in regions of interest and coarse discretization elsewhere, and adaptive time stepping are commonly used to resolve such challenges. While numerical computation of linear problems is quite straightforward, non-linear problems, such as those involving temperature-dependent properties or source terms, are more complicated and require careful consideration of the discretization method to be used. Numerical computation may be carried out either in commercially available software tools, or by writing one's own code. While commercially available software tools may be elegant and easy to use, it is important for the user to correctly interpret results and ensure the validity of the computed temperature field. Specifically, grid independence and time-step independence checks must always be carried out. These checks refer to refining the grid size and time step sufficiently until the computed temperature field is found to not change significantly with further refinement.

A choice between analytical and numerical computation for heat transfer analysis often needs to be made. Some considerations in this process include the mathematical complexity of the problem, degree of accuracy desired, computational resources and numerical tools available, and the trade-off between accuracy and computational cost.

10.1.4 Coupling between Heat Transfer and Other Physical Phenomena

Heat transfer is often intricately coupled with other physical processes that may occur in engineering materials, devices and systems. Such coupling must be appropriately accounted for and modeled. In some cases, the coupling occurs through material properties. For example, in convective heat transfer, the temperature field is strongly influenced by the nature of the flow field, not only due to direct convective heat transfer between the solid and the flow but also through the viscous dissipation term that appears in the energy equation [8]. The temperature field may, in turn, influence the flow field due to temperature-dependent viscosity of the fluid [9]. As another example, temperature rise in Li-ion cells is influenced by heat generation occurring within the cell. In general, the larger the discharge rate, the greater is the heat generation, and thus temperature rise. Note that the discharge rate is often quantified in terms of the C-rate, which is the inverse of the number of hours in which a fully charged cell is discharged to completion. In extreme conditions, such as at high temperature, such heat generation due to exothermic decomposition reactions itself depends on temperature, since the rates of such reactions are usually strongly temperature-dependent, for example, based on Arrhenius principles. Such two-way coupling may bring about a positive feedback loop, wherein, for example, excessive temperature results in increased heat generation rate, which in turn increases the temperature further. This may result in uncontrolled temperature rise, eventually causing fire and explosion [16].

The coupling between heat transfer and other physics is particularly relevant and important to consider in the case of Li-ion cells because of the multiple processes that occur in the cell. These include electrochemical processes, charge and ion transport, as well as fluid flow that may occur external to the cell.

10.2 Li-ION CELL AS A THERMAL SYSTEM

The performance, safety and reliability of a Li-ion cell and battery pack is directly affected by temperature [4,5]. Therefore, it is important to consider the Li-ion cell and battery pack as a thermal system. An understanding of various aspects of heat transfer in a cell and battery pack, such as heat generation, thermal conduction within the cell, heat removal from the cell, thermal properties and the coupling between heat transfer and electrochemistry may be helpful for improving and optimizing overall thermal performance. Material choices also often strongly influence thermal performance [17].

At low temperatures, the performance of a Li-ion cell deteriorates due to reduced ionic conductivity and increased Lithium plating [4,18]. In contrast, high cell temperature may lead to increased self-discharge, capacity fade and thermal runaway. The extent of such undesirable phenomena often depends on the specific cell chemistry. When the temperature of the cell exceeds a certain threshold, typically around 60°C–80°C, thermal runaway due to a series of exothermic decomposition reactions may occur. Thermal runaway is known to be the underlying cause for several catastrophic events such as fire and explosion in automotive and aircraft battery packs, as well as consumer products such as cellphones. While low temperatures may be of relevance for applications such as electric vehicles or grid storage in extremely cold environment, high temperature concerns arise due to temperature rise caused by heat generated in the cell. Although some work has shown that charging characteristics of a Li-ion cell can be improved by charging the cell at somewhat elevated temperatures

[19], in general, it is also very important to keep the cell temperature within bounds in order to reduce the likelihood of catastrophic safety problems.

The overall thermal performance of the cell, including temperature rise as a function of time, is governed by the interactions between a number of thermal processes and phenomena. These processes are briefly considered in the following sub-sections:

10.2.1 Heat Generation

Li-ion batteries generate heat due to various mechanisms. Heat generation rate within a Li-ion cell is broadly divided into reversible and irreversible heat generation terms as follows [4]:

$$\dot{Q} = I^2 \cdot R - I \cdot T \cdot \frac{dU_{OC}}{dT}, \tag{10.5}$$

where I, R, dU_{OC}/dT and T are the current, resistance, entropy coefficient and temperature, respectively. Also, the first and second terms on the right-hand side of Equation 10.5 are the irreversible and the reversible heat source term, respectively. Equation (10.5) often represents a lumped formulation of heat generation, meaning that the heat generation rate within the cell is considered homogenous. More detailed electrochemical-thermal models consider the irreversible heat source term as multiple heat source terms, including Joule heating from electrodes and electrolytes, heat generation due to contact resistance and heat generation due to electrochemical reaction polarization. In addition, the heat generation rate within a Li-ion cell can be a function of the state of charge (SOC), charge and discharge rates (C-rate) and ambient temperature.

In addition to the heat generation during the nominal battery operation, Li-ion batteries can generate excessive heat at elevated temperatures [16,20]. Elevated temperatures trigger a chain of exothermic decomposition reactions that, in turn, increase the battery temperature even more. If the heat generation rate due to decomposition reactions exceeds the heat dissipation rate to the environment, the positive feedback between the heat generation and temperature leads the battery into a catastrophic failure known as thermal runaway. Multiple decomposition reactions associated with the thermal runaway phenomena have been identified in the literature [16]. These reactions include the decomposition of the solid-electrolyte interphase (SEI) layer, reaction of the electrolyte with anode active material, reaction of the electrolyte with cathode active material, electrolyte decomposition, electrolyte combustion and reaction of released oxygen with the binder.

The heat generation rate due to the decomposition reactions is often described as the product of the mass of reactant, m [kg]; heat of reaction, H [J/kg]; and the reaction rate, R [1/s]:

$$\dot{Q}_{gen} = H \cdot m \cdot R \tag{10.6}$$

The reaction rates, on the other hand, are often described by Arrhenius functions as follows:

$$-R_i = \frac{dx}{dt} = -A\exp\left(-\frac{E_a}{RT}\right)x, \tag{10.7}$$

where x is the concentration of reactants, A is the frequency factor, E is the activation energy, R is the gas constant and T is temperature.

10.2.2 Key Modes of Heat Transfer in a Cell

Heat transfer inside the Li-ion cell is dominated by thermal conduction. Convection is relatively unimportant due to the lack of significant bulk fluid motion within. Due to the heterogeneous nature of the Li-ion cell, thermal conduction through the various constituent materials must be considered. In addition, thermal conduction across material interfaces is also important [21]. The key materials within the Li-ion cells include current collectors (usually made of Al/Cu foils), anode (usually made of carbon or similar material), cathode (usually made of a lithium compound such as $LiCoO_2$ or $LiMn_2O_4$ in a slurry containing other materials such as binders as well), separator (usually made of porous polymer such as polypropylene) and electrolyte (usually ionic salts such as $LiPF_6$ in an organic solvent such as ethylene carbonate). Measurements of thermal properties of these components are available in the literature [22]. Amongst these materials, current collectors usually have the largest thermal conductivity, while separator is usually the least thermally conducting.

Thin layers of these materials (thicknesses in the range of a few tens of microns) are usually pressed together to form a composite sheet, which is then rolled/folded to form a cylindrical/prismatic cell. Due to the nature of such assembly, anisotropy in thermal conduction through the composite may be expected. For example, for a cylindrical Li-ion cell, the sheets are rolled around a thin mandrel, due to which, thermal conduction along the axial direction is expected to occur quite rapidly along the high-thermal-conductivity metal sheets. On the other hand, thermal conduction in the radial direction may be severely impeded because of having to diffuse through low-thermal-conductivity materials such as the separator. In addition, thermal contact resistance may be encountered at multiple interfaces in the radial direction. Similar difference in the nature of thermal conduction between in-plane and out-of-plane directions in a prismatic cell may be expected. Experimental measurements have confirmed such anisotropy in thermal conductivity of Li-ion cells [23]. Further, measurements have shown that the overall radial thermal conductivity of a cylindrical Li-ion cell is lower than thermal conductivity of even the least thermally conducting component, which indicates the strong dependence of overall thermal conduction on interfacial thermal transport [21]. Such dominance of interfacial thermal transport is not entirely surprising, as it occurs commonly in other microsystems as well.

The importance of interfacial thermal transport as the rate-limiting thermal process that determines the overall thermal conductivity of the cell needs to be sufficiently recognized in thermal models of the Li-ion cell. Both measurements [21] and simulations [24] have confirmed the rate-limiting nature of interfacial thermal transport. Calculations based on acoustic mismatch model indicate that interfacial thermal resistance is likely to be much higher than at the cathode–separator interface than at the anode–separator interface [21]. Surface chemistry-based approaches to enhance interfacial thermal transport through surface functionalization have been described [21]. For example, it has been shown that introduction of appropriate bridging chemical species at the interface improves adhesion between the cathode and separator, leading to significant reduction in thermal contact resistance and improvement in overall thermal conductivity of the cell [21].

Despite the heterogeneous nature of materials within the Li-ion cell, it is quite common to carry out thermal analysis by treating the Li-ion cell as a homogeneous material, particularly for thermal simulations of the battery pack containing multiple cells. In such a case, the thermal properties of the representative homogeneous material must be derived based on properties of the constituent materials as well as the composition and nature of interfacial transport between materials. Further, in case the cell is modeled as a homogeneous material, it may be helpful to account for thermal contact resistance between the homogeneous material and the cell casing. Finally, anisotropy in thermal conductivity of the homogeneous material must be modeled appropriately.

10.2.3 Boundary Conditions

In addition to thermal conduction within the cell, thermal boundary conditions at the outer surfaces of the cell are also important in determining the overall thermal performance of the cell. In general, three types of boundary conditions are usually specified in thermal analysis – constant temperature (isothermal), constant heat flux (including the adiabatic, zero-flux condition) and convective heat transfer conditions. Amongst these, the isothermal condition represents the best-possible case for heat removal, in which the surface temperature remains the same regardless of the amount of heat removed, i.e., the medium surrounding the cell acts as an infinite thermal sink. At the other extreme, an adiabatic boundary condition represents the worst possible heat removal scenario, in which the surface is perfectly insulated and no heat removal is possible. In general, the convective heat transfer boundary condition lies somewhere in a spectrum between these two extremes, and is characterized by a convective heat transfer coefficient, h [1]. The value of h generally depends on flow conditions and the nature of the fluid around the cell. For natural convection conditions, in which no forced air is present and flow occurs only due to buoyancy and temperature-dependent density, the value of h is quite small. On the other hand, the greater the extent of forced fluid flow around the cell, the larger is the value of h. Note that the extent of heat removal from the cell does not keep rising rapidly with increase in h. Instead, there is a saturation effect, wherein once h is sufficiently large, further increase in h, for example, by increased fluid flow leads to only small improvement in heat removal, because the boundary is already very close to being isothermal.

In the case of a cylindrical Li-ion cell, the two boundaries available for heat transfer include the radial curved surface and the surfaces at the top and bottom ends of the cell. While the radial curved surface usually has greater surface area, it may not always be more favorable for heat removal due to the lower thermal conductivity in the radial direction than in the axial direction [23]. Even if heat removal from the axial surfaces may be desirable, however, that may not always be possible, since the axial end surfaces of the cell are usually taken up for electrical interconnection. Such interconnection materials are usually made of high-thermal-conductivity metals, and, therefore, they may play an important heat removal role particularly in the case of the thick metal bus bars, and during periods in which no electrical current passes through. A limited amount of studies of thermal conduction through interconnection materials is available, indicating that the amount of heat thus removed may be represented by a reasonably large convective heat transfer coefficient [25].

10.2.4 Thermal Properties

The key thermal properties of relevance for modeling and understanding thermal transport in a Li-ion cell include its thermal conductivity and heat capacity [4,22]. Measurements of these properties for a wide variety of Li-ion cells are available in the literature [22]. Parametric modeling of thermal conduction in larger systems, such as a battery pack comprising a large number of individual cells, is often carried out. In such a case, thermal behavior of the cells is represented by a Resistance–Capacitance network, and the values of various resistances and capacitances are determined based on underlying thermal transport properties of the cells and other materials in the battery pack.

10.2.5 Thermal Management

Thermal management of a Li-ion cell or battery pack refers to the provisioning of appropriate mechanisms to sufficiently remove heat during operation in order to maintain the cell temperature below the safety threshold. In addition, thermal management also includes appropriate heating mechanisms to rapidly warm up the cell in the case of a cold environment, such as in starting of an electric vehicle in a cold climate. Both aspects of thermal management are very important, and integrated thermal management approaches are of particular interest. In addition to maintaining cell temperature within a desirable window, another key aspect of thermal management of a Li-ion cell is to maintain thermal uniformity within the cell. A thermally imbalanced cell with a large temperature gradient may result in faster aging and other undesirable electrochemical effects [4].

Heat removal from a Li-ion cell can be carried out in a number of different ways. The simplest mechanism of heat removal is via free convection. As the cell surface heats up, it causes the air next to the cell to heat up and rise due to lower density at greater temperature. As colder air moves in to replace the rising warm air, heat is removed from the cell by these convective currents. Such free convection-based thermal management is primarily driven by gravity, and, therefore, depends critically on the orientation of the surface. Regardless of orientation, however, free convection is usually a weak effect and may not remove sufficient amount of heat, especially at high discharge rates. In contrast, forced convection driven by an external fluid flow, such as air or a liquid, results in much greater heat removal [26]. Immersion cooling, in which the cell is submerged in a circulating liquid coolant, has attracted much recent attention for Li-ion cell thermal management. While immersion cooling results in excellent cooling, the liquid employed must be dielectric, so as to not interfere with the electrical function of the cell. Additional considerations such as the impact of immersion cooling on cell aging must also be accounted for.

Other more advanced thermal management techniques include two-phase cooling, in which heat removed from the cell is used to drive phase change, either solid to liquid (melting) or liquid to vapor (evaporation). In both cases, the large latent heat of phase change makes it possible to remove significant amount of heat from the cell. However, solid-to-liquid phase change may be self-limiting due to low thermal conductivity of typical phase-change materials [27]. Moreover, this may lead to effective cooling around the surface and not the core of the cell. In the case of liquid-to-vapor phase change, additional challenges such as reduced heat flux due to dry out and the large volumetric expansion due to vapor formation must be addressed.

Heat pipes are devices that comprise a working fluid that evaporates upon heat addition, transports heat to a different location within a closed system and condenses back to continue removing heat in a loop. Heat pipes have been used to a limited extent for thermal management of Li-ion cells. Heat pipes have been placed between cells in a battery pack, and also inserted into a Li-ion cell for heat removal [28]. Heat pipes can remove significant heat, particularly when inserted directly in a cell. A key advantage of a heat pipe is the passive nature of thermal management.

Thermoelectric devices have also been used for thermal management of Li-ion cells. Depending on the direction of the electric current, a thermoelectric device may be used to produce either cooling or heating. As a result, despite the relatively large power consumption, thermoelectric devices may be appropriate for dual-purpose thermal management [29].

Heating of a Li-ion cell in a cold ambient may be carried out through external heat sources. However, such an approach may result in severe temperature gradients. For example, due to poor thermal conductivity of the cell, the outer surface of the cell may warm up rapidly, but the core of the cell may remain cold. In order to prevent such thermal imbalance and to accelerate the heating process, embedded thin-film metal heaters have been used. These heaters may utilize a part of the electrical energy stored within the cell for rapid self-heating. Passing a small current through a resistive sheet embedded within the cell has been shown to produce rapid heating.

10.3 MODELING FRAMEWORKS AND GOVERNING EQUATIONS

Modeling and simulation of heat transfer phenomena in Li-ion batteries vary from simplified zero-dimensional (0D), lumped capacitance framework to more sophisticated, spatially resolved one- (1D), two- (2D) and three-dimensional (3D) models. In all these frameworks, the governing equations describing heat transfer in Li-ion batteries are derived from the principle of conservation of energy. The governing equations appear as ODEs for lumped framework and PDEs for 1D, 2D and 3-D. The resulting ODEs or PDEs are solved using analytical and numerical heat transfer tools [30]. This section briefly describes lumped and specially resolved modeling frameworks and introduces the governing equations in these frameworks. The next chapter presents the analytical and numerical tools for solving heat transfer problems in Li-ion cells and battery packs.

10.3.1 0D Lumped Capacitance Models

0D, lumped capacitance models consider the Li-ion cell as a lumped mass, in which the change in the temperature is only a function of time. In other words, the spatial temperature gradient within the Li-ion cell is neglected. Lumped capacitance framework is valid when the internal thermal resistance due to conduction is much lower than the heat transfer resistance to the environment through convection and radiation. This ratio is defined as the Biot number $Bi = hL/k$ [1], where h is the total heat transfer coefficient, L is the characteristic length and k is the thermal conductivity. Thus, the lumped capacitance models are valid only for small values of Bi number, $Bi < 0.1$. Lumped models are simple and computationally inexpensive, making them an attractive framework for parametric, sensitivity and optimization analyses. Specifically, lumped models may be much faster than spatially resolved models in the existence of non-linearity. However, Li-ion batteries often possess very low thermal conductivities [22,23], resulting in a large Biot number. In addition, the thermal conductivity of Li-ion cells is known to be orthotropic [23], resulting in different values for different directions. Therefore, lumped models may not be a useful framework for these scenarios. Furthermore, capturing temperature gradient within the cells becomes crucial for specific analyses such as thermal management, degradation, and aging studies, where spatially resolved frameworks are necessary.

The conservation of energy for a Li-ion cell in a 0D, lumped capacitance framework is governed by an ODE with an initial condition as follows:

$$mc_p \frac{dT}{dt} = \Sigma \dot{Q}_{\text{ext}} + \Sigma \dot{Q}_{\text{gen}}, \tag{10.8}$$

$$T(t = 0) = T_{\text{ini}}, \tag{10.9}$$

where m is the mass, c_p is the specific heat capacity and T is the temperature of the Li-ion cell. Further, $\dot{Q}_{\text{ext}}$ is the rate of external heat transfer between the Li-ion cell and the environment through convection and radiation, and $\dot{Q}_{\text{gen}}$ is the internal heat generation rate within the Li-ion cell.

10.3.2 Spatially Resolved Framework

For scenarios where the lumped modeling framework is not valid or useful, the governing energy equation may be resolved spatially to account for heat transfer in different directions. For example, spatially resolved modeling frameworks are required to capture the temperature gradient within the Li-ion cell. Depending on the geometry, configurations, boundary conditions and physics involved in the problem, 1-, 2- and 3D modeling frameworks may be required to solve the problem accurately. Reducing the model dimension based on reasonable assumptions is typical for accelerating heat transfer modeling and simulation.

Unlike the lumped capacitance framework, the temperature is a function of both space and time in the spatially resolved modeling framework. For example, in a Li-ion cell generating heat under nominal conditions and being cooled from the outside, the temperature within the cell is a function of both time and space [31]. Specifically, the regions closer to the cell core have higher temperatures due to heat generation, while the regions closer to the cell surface have lower temperatures due to the external heat transfer. Due to the dependency of temperature on both time and space, conservation of energy governing heat conduction within the cell is defined by a more complicated equation, i.e., a PDE with initial and boundary conditions. Li-ion cells come in different form factors, such as pouch, prismatic and cylindrical formats. The PDE governing heat transfer within these cells can be written in a Cartesian coordinate system for prismatic and pouch cells and a cylindrical coordinate system for cylindrical Li-ion cells. The general governing energy equation in a 3D Cartesian and cylindrical coordinate systems framework can be written as follows:

$$\rho c_p \frac{\partial T}{\partial t} = \frac{\partial}{\partial x}\left(k_x \frac{\partial T}{\partial x}\right) + \frac{\partial}{\partial y}\left(k_y \frac{\partial T}{\partial y}\right) + \frac{\partial}{\partial z}\left(k_z \frac{\partial T}{\partial z}\right) + Q''', \tag{10.10}$$

$$\rho c_p \frac{\partial T}{\partial t} = \frac{1}{r}\frac{\partial}{\partial r}\left(k_r r \frac{\partial T}{\partial r}\right) + \frac{1}{r^2}\frac{\partial}{\partial \varphi}\left(k_\varphi \frac{\partial T}{\partial \varphi}\right) + \frac{\partial}{\partial z}\left(k_z \frac{\partial T}{\partial z}\right) + Q''', \tag{10.11}$$

where Q''' is the volumetric heat generation within the cell. In general, Equations 10.10 and 10.11 require six boundary conditions and one initial condition. The boundary conditions may be constant or time-dependent prescribed temperature and heat flux or convection boundary conditions. The mathematical expression of the boundary conditions mentioned above is summarized below:

$$T\big|_{\text{boundary}} = f(t), \tag{10.12}$$

$$-k\left(\frac{\partial T}{\partial n}\right)\Big|_{\text{boundary}} = q''(t), \tag{10.13}$$

$$-k\left(\frac{\partial T}{\partial n}\right)\Big|_{\text{boundary}} = h\left(T\big|_{\text{boundary}} - T_\infty(t)\right), \tag{10.14}$$

where n is the outward normal and q'' is the heat flux. Under appropriate assumptions based on the conditions of the system under study, the Equations 10.10 and 10.11 may be reduced to 2D and 1D frameworks. Reducing the model to 2D and 1D decreases the number of required boundary conditions to 4 and 2, respectively. Furthermore, steady-state solutions are of interest for specific applications, resulting in the elimination of the left-hand side term.

10.4 SOLUTION METHODS FOR GOVERNING ENERGY EQUATIONS

The differential equations described above, i.e., ODEs and PDEs for lumped and spatially resolved frameworks, can be solved using various mathematical techniques. As mentioned previously, in the heat transfer analysis methods, these mathematical techniques can be broadly categorized into analytical/semi-analytical [7] and numerical solutions [14,15]. The analytical and semi-analytical tools provide explicit expressions for temperature as functions of time and space in a continuous domain. Once the expression is known, the temperature can be evaluated at any time and position. On the other hand, numerical techniques convert the continuous domain into a finite number of discrete nodes, elements, and volumes. Thus, the temperature can only be determined at specific points or as average values for each element or volume. A brief overview of the analytical and numerical heat transfer tools and their applications for Li-ion batteries is discussed in the following sub-sections.

10.4.1 Analytical Heat Transfer Tools for Li-Ion Cells

Analytical heat transfer tools such as exact solutions and iterative semi-analytical methods of varying complexities have been used to predict the thermal behavior of Li-ion batteries under various operating conditions for various applications. Analytical heat transfer analysis can range from 0D, lumped capacitance framework to 1D, 2D and 3D models. Due to the complex transport phenomena involved in the operation of Li-ion batteries, exact solutions may only exist for a limited number of problems under simplifying assumptions. Furthermore, exact closed-form solutions to the governing energy equations for lumped and spatially resolved frameworks may only exist if the governing equations and boundary conditions are linear. However, the non-linearity may generally appear in heat generation, external heat transfer and boundary conditions. For instance, in the conservation of energy for the lumped capacitance framework, $\dot{Q}_{\text{ext}}$ in Equation 10.8 may include heat transfer due to radiation defined as $\varepsilon\sigma\left(T_{\text{cell}}^4 - T_{\text{amb}}^4\right)$, resulting in a non-linear ODE that cannot be solved exactly. Moreover, the heat generation, Q''', term in Equations 10.10 and 10.11 can be a non-linear function of temperature, such as Arrhenius heat generations during Li-ion battery abuse conditions known as thermal runaway. Certain assumptions, such as neglecting radiation heat transfer at low operating temperatures or representing Li-ion cell's internal heat generation during nominal conditions as a constant or time- and space-dependent volumetric source term help linearize the governing equations and solve them analytically. Below are some examples of analytical solution techniques used for Li-ion battery applications.

Example I: Solution to the Lumped Model for Li-Ion Cells

Consider a cylindrical, pouch or prismatic Li-ion cell initially at $T = T_0$ and generating heat during a nominal discharge condition and dissipating heat through convection with air at $T = T_\infty$. We are interested in finding the Li-ion cell temperature as a function of time. The lumped formulation of the conservation of energy can be written as follows:

$$mc_p \frac{dT}{dt} = -hA\left(T(t) - T_\infty\right) + q(t) \tag{10.15}$$

$$T(t=0) = T_0 \tag{10.16}$$

Equation above can be written compactly as follows:

$$\frac{d\theta}{dt} + P\theta = Q(t), \tag{10.17}$$

$$\theta(t=0) = \theta_0 = T_0 - T_\infty \tag{10.18}$$

where $\theta = T - T_\infty$, $P = hA/mc_p$ and $Q = q/mc_p$. Equation (10.17) is a first-order linear differential equation with the following solution:

$$\theta(t) = \frac{1}{e^{\int Pdt}}\left(\int e^{\int pdt} \cdot Q(t)dt + c\right) \tag{10.19}$$

For polynomial and sinusoidal Q, the above integral can be determined exactly. Furthermore, for the special case of constant Q with an initial value of zero, the solution given in Equation (10.19) can be simplified as follows:

$$\theta = \frac{Q}{P}(1 - e^{-Pt}) + \theta_0 e^{-Pt} \tag{10.20}$$

where the constant in the exponential term is the inverse of thermal time constant, which is the ratio of heat capacitance to the convection heat transfer thermal resistance.

Example II: Solution to the 1D Heat Equation for Li-Ion Batteries

Analytical heat transfer tools can be used to solve governing energy equations of higher dimensions under specific conditions. For example, under certain assumptions, heat transfer within a cylindrical Li-ion cell may be considered a 1D radial heat transfer problem with internal heat generation and convective boundary conditions. Under this assumption, the governing equation and boundary conditions may be written as:

$$\rho c_p \frac{\partial T}{\partial t} = \frac{1}{r}\frac{\partial}{\partial r}\left(k_r r \frac{\partial T}{\partial r}\right) + Q''', \tag{10.21}$$

$$T \rightarrow \text{finite as } r \rightarrow 0, \tag{10.22}$$

$$-k\frac{\partial T}{\partial r} = h(T - T_\infty), \tag{10.23}$$

$$T(r, t=0) = T_0, \tag{10.24}$$

where Q''' is the heat generation that can be any general function of space and time. The solution to the set of equations described above can be written as an infinite series comprising Bessel functions of space and an exponential function of time. The infinite series solution to the equations above is often derived from the separation of variables techniques. For prismatic and pouch cells, the governing equation will be written in Cartesian coordinate systems, and the solution involves an infinite series comprising sinusoidal functions of space and an exponential function of time. In addition to the separation of variables, Laplace transforms and Green's function are among the methods used to solve the equation above. Green's function approach is a powerful mathematical tool that can be used to solve linear PDEs with generalized heat generation and boundary conditions. The solution procedure for the 1D energy equation in different coordinate systems and subject to various boundary conditions are available in heat transfer books. These methods can be extended to 2D and 3D problems as long as the governing equations and boundary conditions remain linear.

10.4.2 Iterative Semi-Analytical Solutions

Investigating the thermal behavior of Li-ion batteries often requires solving multi-physics coupled problems. For instance, Li-ion battery packs are often accompanied by a thermal management system. Different types of thermal management systems such as fins and cold plates, air/liquid cooling or phase change cooling may be used. These thermal management systems may result in a coupled heat transfer problem. For instance, air/liquid cooling introduces a conjugate heat transfer problem by introducing additional physics, i.e., fluid flow and convection. Similarly, phase-change thermal management systems introduce a coupled problem comprising heat conduction in the cell and phase-change heat transfer in the phase change material. Solving such problems may not be possible using exact analytical models. Iterative semi-analytical methods are often used to solve coupled and conjugate problems, for which an analytical or semi-analytical solution exists for the individual physics. The problem is then decoupled at the physical or mathematical interface and solved individually. Finally, an iterative approach will be used to satisfy the continuity condition at the interface. For example, an iterative method may be used to solve conjugate heat transfer problems. In such problems, the heat transfer within the solid is governed by the conservation of energy, while the conservation of mass, momentum and energy governs the fluid domain. Under certain assumptions, an exact or semi-analytical solution may exist for the individual solid and fluid problem. The application of the iterative approach for Li-ion batteries is presented in the following examples.

Example I: Air/Liquid Cooling Thermal Management [32]

One of the examples of the application of iterative semi-analytical methods for solving problems involving Li-ion cells is air/liquid battery thermal management systems. This problem is a conjugate heat transfer problem involving conduction in the solid and convective heat transfer to the coolant. The two physics are coupled at the interface where heat transfer occurs between the cell and the coolant. By utilizing the continuity of the temperature and conservation of energy at the solid–liquid interface, the conduction and convection problems can be uncoupled and solved separately. Under specific simplifying assumptions, each problem can be solved using analytical methods. The solution of each problem is used as a boundary condition to the other problem in an iterative manner. This iterative process is continued until the interface condition is satisfied and the solution converges within the defined tolerance.

Example II: Phase Change Based Thermal Management [33]

Iterative semi-analytical solutions have been used to investigate the performance of phase change cooling of Li-ion battery packs. Phase-change thermal management systems may offer greater heat removal rate due to their large latent heat during phase change process. The heat generated during the operation of Li-ion cells is absorbed by PCM, resulting in a phase change process. Heat transfer problems involving melting and solidification are non-linear in general. A coupled heat conduction within the cell and phase-change heat transfer within the PCM results in an even more challenging problem to solve. Fortunately, iterative techniques can be used to uncouple the two physics from each other. Once uncoupled, the conduction heat transfer problem within the cell can be solved using standard analytical tools such as separation of variables or Green's function. The phase-change heat transfer problem can be solved using approximate analytical methods such as perturbation method, heat integral method and eigenfunction-based solutions.

10.4.3 Numerical Analysis Tools

Analytical heat transfer tools described above contain many simplifying assumptions. For example, in the above analysis, Li-ion cells are considered a homogenous solid. However, in reality, Li-ion cells are made of different components. For example, a cylindrical Li-ion cell comprises a jellyroll enclosed by a metallic casing. The jellyroll contains porous negative and positive electrodes, a separator, current collectors and an electrolyte mixture. All these components have different thermal properties that can affect the thermal behavior of Li-ion cells. Furthermore, heat generation within the cell is not constant and is coupled to the electrochemical processes occurring within the cell. All these components and mechanisms are neglected when analytical methods are used to solve the conduction problem. An alternative approach to analytical techniques for solving complex problems is using numerical methods. In general, numerical methods approximate differential/integral equations in a continuous domain by a large number of algebraic equations defined in a discretized domain. In numerical methods, both spatial and temporal domains are discretized. The time domain is often discretized using the forward difference (explicit scheme), the backward difference (implicit scheme) or a central difference (Crank–Nicolson scheme) method. The differential equations are often converted to algebraic equations using various methods. The following sub-sections will briefly describe each of these methods.

10.4.3.1 Finite Difference Method (FDM)

FDM converts differential equations into a number of algebraic equations by replacing continuous derivatives with finite differences derived from Taylor's series expansion for a number of nodes. The resulting algebraic equations will then be solved using a computational algorithm developed based on matrix algebra. It is important to note that the accuracy of the FDM depends on the number of nodes considered in the discretization. Also, FDM only provides the value of the parameters such as temperature on the discrete nodes. The spatial derivative in the heat equation is a second derivative. Thus, the FDM discretizes the spatial domain using a second-order central difference method as follows:

$$\frac{\partial^2 T}{\partial x^2} = \frac{T_{i-1}^n - 2T_i^n + T_{i+1}^n}{\Delta x^2} \tag{10.25}$$

10.4.3.2 Finite Element Method (FEM)

Another numerical technique used for solving heat transfer problems is the FEM. FEM divides the spatial domain into smaller elements such as triangles or quadrilaterals, resulting in a finite element mesh structure. Then, the solution of the governing equation is approximated using a simple function within each element. The resulting error due to this approximation may be reduced by reducing the element size, and consequently increasing the number of elements. In addition, the continuity of the temperature and the conservation of the heat flux at the boundaries of adjacent elements are ensured in the solution procedure. The global solution over the entire domain is then obtained from the solution of each element.

10.4.3.3 Finite Volume Method (FVM)

The FVM is established based on the conservation of the flux entering and leaving a control volume. In FVM, the computational domain is divided into sufficiently small elements called control volume by constructing a mesh structure. The differential governing equation is then integrated over each control volume, resulting in a heat balance equation. The heat balance equations are then discretized and solved numerically.

10.5 THERMAL RUNAWAY MODELING

10.5.1 Introduction

Thermal runaway is an important phenomenon that profoundly affects the safety of Li-ion cells and battery packs. Fundamentally, thermal runaway in a Li-ion cell or battery pack arises due to positive feedback between heat generation and temperature rise [4,16,34]. This positive feedback occurs when the external heat dissipation rate to the environment is smaller than the internal heat generation rate within the cell. Thermal runaway may occur due to various off-nominal conditions such as high temperatures, mechanical abuse, short circuit, overcharging and over-discharge. Due to its fundamental basis in heat transfer, much work has been carried out on understanding the various aspects of heat transport in thermal runaway.

Several exothermic decomposition reactions occur in a Li-ion cell. The rates of these reactions are extremely low at room temperature and during nominal operation. However, as temperature increases, the reaction rates, and hence heat generation rates go up exponentially, in accordance with Arrhenius kinetics. The nature of these decomposition reactions varies somewhat with the cell chemistry, but is generally well known, including the values of underlying Arrhenius parameters, based on measurements and modeling. Key reactions occurring in this process include decomposition of the SEI layer, reaction of the negative and positive electrodes with the electrolyte, decomposition and combustion of the electrolyte [16]. These reactions occur in a cascading fashion, so that heat generated by one reaction raises the temperature sufficiently in order to trigger the subsequent reaction, and so on, ultimately leading to fire and explosion.

An important problem in thermal runaway pertains to the propagation of thermal runaway from one cell to another in a battery pack [35]. This is of practical importance in preventing catastrophic failure of battery packs. From a scientific perspective, propagation of thermal runaway is governed by the nature of heat transfer between cells. Key processes that govern thermal runaway propagation are thermal conduction through the interstitial material between cells, radiative heat transfer, heat transfer due to high speed flow of hot vent gases, heat generation due to combustion of vent gases, etc. It is of interest to determine limits in which thermal runaway propagation may not occur. For example, the cell-to-cell gap is an important parameter in this problem. The larger the cell-to-cell gap, the lower is the likelihood of thermal runaway propagation due to increased thermal conduction resistance and reduced view factors for radiative heat transfer between cells. However, larger cell-to-cell gap results in a reduced energy density of the battery pack. Thermal conductivity of the interstitial material is also an important parameter. If the thermal conductivity is too high, there is greater rate of heat transfer from the cell undergoing thermal runaway to its neighbors, which contributes to propagation. On the other hand, too low a thermal conductivity results in excessive thermal insulation of cells, which increases the likelihood of minor thermal excursions leading to initiation of thermal runaway. An interstitial material with an intermediate thermal conductivity may therefore be appropriate.

10.5.2 Numerical Heat Transfer Simulation for Thermal Runaway

A number of interesting phenomena occur once the cell enters thermal runaway. Some examples are the venting of hot and flammable gases that may combust when in contact with ambient oxygen, a potential cell rupture, particulate emission, and radiative heat transfer. Mathematical equations governing these processes are often highly coupled and non-linear. Including these phenomena in modeling and simulation frameworks remains an area of active research interest.

A large amount of research literature exists on numerical modeling and simulations of thermal runaway and propagation. Due to the multidisciplinary nature of thermal runaway, a variety of aspects of thermal runaway have been investigated, including chemical kinetics, heat and mass transfer, fluid flow, combustion and mechanical stresses. Within the context of heat transfer, all three modes of heat transfer, namely, thermal conduction, convection and radiation have been investigated. Key challenges in numerical simulations include the highly non-linear nature of Arrhenius heat generation and other phenomena, which require extremely small time steps, especially close to the thermal runaway event, in order to meet convergence requirements. Optimization of numerical simulation techniques is, therefore, of paramount importance. Development of simplified correlations and design rules based on comprehensive simulations is also of practical importance.

10.5.3 Analytical Heat Transfer Modeling of Thermal Runaway

Due to the highly complicated and non-linear nature of thermal runaway, only a limited amount of work on analytical modeling has been carried out. Most such work relies on appropriate simplifying assumptions and linearization in order to make the problem sufficiently amenable to analytical modeling. Despite such simplifications, analytical modeling is of much importance for developing a fundamental understanding of thermal runaway. For example, analytical modeling may help understand the role of various non-dimensional parameters in the thermal runaway process [34]. Further, analytical results may be used for validation of numerical computation algorithms.

A key question of interest in analytical modeling is to determine the limiting conditions in which initiation or propagation of thermal runaway does not occur. For example, for a cylindrical cell based on a single decomposition reaction, linearized stability analysis has shown [34] that initiation of thermal runaway depends on the value of a single non-dimensional parameter called the Thermal Runaway Number (TRN), which combines the effects of heat generation, thermal conduction and convective heat removal from the cell into a single parameter. For more complicated systems, multilayer diffusion-reaction and convection-diffusion-reaction analysis has been carried out, resulting in the derivation of limiting conditions for the existence of imaginary eigenvalues that imply infinite temperature at large temperatures, and, therefore, instability of the system. In general, the stability of such a system depends on a careful balance between a number of phenomena, including diffusion, reaction and heat removal from boundaries. Such analysis is helpful in the design of safe Li-ion cells and battery packs.

REFERENCES

[1] Bergman, T.L., Levine, A.S., Incropera, F.P., and DeWitt, D.P., 2018. *Fundamentals of Heat and Mass Transfer*, 8th Ed., Wiley.

[2] Lienhard IV, J.H., and Lienhard, V., 2020. *A Heat Transfer Textbook*, 5th Ed., available at https://ahtt.mit.edu/, accessed 11/21/2021.

[3] Choobineh, L., and Jain, A., 2015. 'An explicit analytical model for rapid computation of temperature field in a three-dimensional integrated circuit (3D IC),' *Int. J. Therm. Sci.*, **87**, pp. 103–109.

[4] Bandhauer, T.M., Garimella, S., and Fuller, T.F., 2011, 'A critical review of thermal issues in lithium-ion batteries,' *J. Electrochem. Soc.*, **158**(3), p. R1.

[5] Shah, K., Vishwakarma, V., and Jain, A., 2016, 'Measurement of multiscale thermal transport phenomena in li-ion cells: a review,' *J. Electrochem. Energy Convers. Storage*, **13**(3), 1–13. http://dx.doi.org/10.1115/1.4034413

[6] Özişik, M.N., and Orlande, H.R.B., 2021, *Inverse Heat Transfer - Fundamentals and Applications*, 2nd Ed., Taylor & Francis.

[7] Hahn, D.W., and Özişik, M.N., 2012, *Heat Conduction*, Wiley.

[8] Kays, W.M., Crawford, M.E., and Weigand, B., 2005, *Convective Heat and Mass Transfer*, McGraw-Hill Higher Education.

[9] Bejan, A., 2013, *Convection Heat Transfer*, Wiley.

[10] Howell, J.R., Mengüç, M.P., and Siegel, R., 2015, *Thermal Radiation Heat Transfer*, 6th Ed., CRC Press.

[11] Modest, M.F., and Mazumder, S., 2021, *Radiative Heat Transfer*, 4th Ed., Academic Press.
[12] Dorfman, A.S., 2009, *Conjugate Problems in Convective Heat Transfer*, 1st Ed., CRC Press.
[13] Cole, K.D., Beck, J.V., Haji-Sheikh, A., and Litkouhi, B., 2011, *Heat Conduction Using Green's Functions*, 2nd Ed., Taylor & Francis.
[14] Patankar, S.V., 1980, *Numerical Heat Transfer and Fluid Flow*, 1st Ed., CRC Press.
[15] Minkowycz, W.J., Sparrow, E.M., Schneider, G.E., and Pletcher, R.H., 1988, *Handbook of Numerical Heat Transfer*, Wiley.
[16] Spotnitz, R., and Franklin, J., 2003, 'Abuse behavior of high-power, lithium-ion cells,' *J. Power Sources*, **113**(1), pp. 81–100.
[17] Divakaran, A., Hamilton, D., Manjunatha, K., and Minakshi, M., 2020, 'Design, development and thermal analysis of reusable Li-ion battery module for future mobile and stationary applications,' *Energies*, **13**, pp. 1477.
[18] Linden, D., et al., 2004, *Linden's Handbook of Batteries*, McGraw-Hill.
[19] Yang, X.-G., Liu, T., Gao, Y., Ge, S., Leng, Y., Wang, D., and Wang, C.-Y., 2019, 'Asymmetric temperature modulation for extreme fast charging of lithium-ion batteries,' *Joule*, **3**, pp. 3002–3019.
[20] Parhizi, M., Ahmed, M.B., and Jain, A., 2017, 'Determination of the core temperature of a Li-ion cell during thermal runaway,' *J. Power Sources*, **370**, pp. 27–35.
[21] Vishwakarma, V., Waghela, C., Wei, Z., Prasher, R., Nagpure, S.C., Li, J., Liu, F., Daniel, C., and Jain, A., 2015, 'Heat transfer enhancement in a lithium-ion cell through improved material-level thermal transport,' *J. Power Sources*, **300**, pp. 123–131.
[22] Steinhardt, M., Barreras, J.V., Ruan, H., Wu, B., Offer, G.J., and Jossen, A., 2022, 'Meta-analysis of experimental results for heat capacity and thermal conductivity in lithium-ion batteries: A critical review,' *J. Power Sources*, **522**, pp. 1–25.
[23] Drake, S.J., Wetz, D.A., Ostanek, J.K., Miller, S.P., Heinzel, J.M., and Jain, A., 2014, 'Measurement of anisotropic thermophysical properties of cylindrical li-ion cells,' *J. Power Sources*, **252**, pp. 298–304.
[24] Dhakane, A., Varshney, V., Liu, J., Heinz, H., and Jain, A., 2020, 'Molecular dynamics simulations of separator-cathode interfacial thermal transport in a Li-ion cell,' *Surf. Interfaces*, **21**, pp. 100674:1–8.
[25] Chalise, D., Shah, K., Halama, T., Komsiyska, L., and Jain, A., 2017, 'An experimentally validated method for temperature prediction during cyclic operation of a Li-ion cell,' *Int. J. Heat Mass Transfer*, **112**, pp. 89–96.
[26] Chalise, D., Shah, K., Prasher, R., Jain, A., 'Conjugate heat transfer analysis of thermal management of a Li-ion battery pack,' ASME J. Electrochem. Energy Storage Conversion, 15, pp. 011008:1–8, 2018.
[27] Alexiades, V., and Solomon, A.D., 1993, *Mathematical Modeling of Melting and Freezing Processes*, CRC Press.
[28] Anthony, D., Wong, D., Wetz, D., and Jain, A., 2017, 'Improved thermal performance of a Li-ion cell through heat pipe insertion,' *J. Electrochem. Soc.*, **164**, pp. A961–967.
[29] Mostafavi, A., and Jain, A., 2021, 'Dual-purpose thermal management of Li-ion cells using solid-state thermoelectric elements,' *Int. J. Energy Res.*, **45**, pp. 4303–4313.
[30] Parhizi, M., Jain, A., Kilaz, G., and Ostanek, J., 2022, 'Accelerating the numerical solution of thermal runaway in Li-ion batteries,' *J. Power Sources*, **538**, pp. 231531:1–12.
[31] Shah, K., Drake, S.J., Wetz, D.A., Ostanek, J.K., Miller, S.P., Heinzel, J.M., and Jain, A., 2014, 'An experimentally validated transient thermal model for cylindrical Li-ion cells,' *J. Power Sources*, **271**, pp. 262–268.
[32] Chalise, D., Shah, K., Prasher, R., and Jain, A., 2018, 'Conjugate heat transfer analysis of thermal management of a Li-ion battery pack,' *ASME J. Electrochem. Energy Convers. Storage*, **15**, pp. 011008:1–8.

[33] Parhizi, M., and Jain, A., 2019, 'The impact of thermal properties on performance of phase change based energy storage systems,' *Appl. Therm. Eng.*, **162**, pp. 114154:1–10.

[34] Jain, A., Parhizi, M., Zhou, L., Krishnan, G., 'Imaginary eigenvalues in multilayer one-dimensional thermal conduction problem with linear temperature-dependent heat generation,' *Int. J. Heat Mass Transfer*, **170**, pp. 120993:1–10, 2021.
Jain, A., Parhizi, M., Zhou, L., 'Multilayer One-Dimensional Convection-Diffusion-Reaction (CDR) Problem: Analytical Solution and Imaginary Eigenvalue Analysis,' *Int. J. Heat Mass Transfer*, **177**, pp. 121465:1–11, 2021.

[35] Mishra, D., and Jain, A., 2021, 'Multi-mode heat transfer simulations of the onset and propagation of thermal runaway in a pack of cylindrical Li-ion cells,' *J. Electrochem. Soc.*, **168**, pp. 020504:1–11.

11 Electrochemical Impedance Spectroscopy

Jianbo Zhang, Shangshang Wang, and Kei Ono
Tsinghua University

CONTENTS

DOI: 10.1201/9781003299295-11

11.1 INTRODUCTION

Electrochemical impedance spectroscopy (EIS) is a powerful characterization technique routinely applied in a wide range of electrochemical fields, e.g., corrosion, electrochemical energy systems, and electrochemical sensors. Application of EIS to the R&D of commercial lithium-ion battery (LIB) is showing a significant increase.[1] Recent reviews on the modeling,[2–4] measurement,[5–7] characterization and design,[1,8] and onboard management of LIB using EIS,[9] reflect the breadth, depth, and vitality of this marvelous technique. The ever-increasing application of EIS in LIB attests to the power, the affordability, and the easiness of use of this technique. Meanwhile, it also signifies the need to instruct the inexperienced practitioners on the guidelines for good practice of the art and to warn them against the common pitfalls, as non-judicious use not only hampers the realization of its full power, and leads to wrong interpretation, but also discredits the method itself.

The multi-scale, multi-physics nature of LIB calls for the combined utilization of multiple, complementary characterization techniques, in situ or ex situ, electrochemical or non-electrochemical, to explore the complex structures and processes inside the cell so as to enhance our understanding and improve the LIB performance. While the microscopic and microanalysis techniques, e.g., SEM, TEM, EPS, XRD, and NMR, mostly ex situ and post-mortem, can provide valuable information on the structure and composition of either the surface or the bulk, EIS method, as a non-destructive, in situ and in operando frequency-domain electrochemical method, is effective in separating electrochemical processes in both the bulk and the interface. Such a capability is indispensable for exploring the structure of and the processes across the electrode–electrolyte interface (EEI), especially the formation, properties, and evolution of the protective film over the active materials in LIB.

Specifically, EIS method has a unique capability in differentiating the bulk and interfacial processes such as migration, charge transfer, diffusion, and intercalation, from the analysis of the spectrum obtained via perturbing the test object with small-amplitude AC signal spanning wide frequency ranges. From the spectrum, dominant processes can be spotted. From the features identified in the spectrum, kinetic parameters like charge transfer resistance and exchange current density, transport parameters like ion conductivity and diffusivity, and structural parameters like tortuosity, can be estimated from fitting the spectrum with models. From the shape of the spectrum and the activation energy of the kinetic and transport parameters, the reaction mechanism and the nature of the processes can be differentiated.

However, the nature and the limitation of EIS must be fully understood in order to harness its power. Unlike other frequency-domain microanalysis methods like NMR, which can provide a definitive verdict on the test sample via signature detection and pattern recognition, EIS relies on models to infer understanding from the features in the spectrum. Models in EIS are like a double-edged sword. While the use of physics-based models is what conferring the technique, its power of process diagnosis and prognosis, it may also result in ambiguity as multiple models can be constructed and fit equally well with the same spectrum. Therefore, the EIS method is inherently non-definitive in the sense that the interpretation one can hope to obtain is a well-judged guess, the credibility of which can be reinforced but not proved.

Accordingly, extensive tests are needed to reveal the features and to attribute the features to particular components and processes. In addition, evidence from independent techniques is needed to corroborate the interpretation. In other words, EIS study is necessarily a long process with many cycles of iteration to ensure that the artifacts in measurement are identified and eliminated, that the selected model matches the complexity of the test object and catches the physics in it, and that the interpretation of the spectra and the parameters extracted from the model are validated.

Against the ever-increasing popularity of EIS in the study of LIB, this chapter addresses the needs of the new practitioners on the proper use of this powerful in situ technique so that the common pitfalls can be avoided and its full potential can be realized. In Section 11.2, the principle and process of EIS are explained. In Section 11.3, the characterization of LIB using EIS in the sequence of measurement, features, models, interpretation, and validation are elaborated. In Section 11.4, a compact list of exemplary experimental EIS works in major stages of LIB lifetime is compiled, from which messages on the nature of EIS and guidance on the good practices of EIS study are drawn and discussed.

11.2 PRINCIPLE AND PROCESS OF EIS METHOD

This section first introduces the principle underlying the EIS method, including the three-electrode cell setup, the small-amplitude perturbation to and the response of the electrochemical system, the definition of EIS, and its typical representation in Nyquist and Bode plots. It then divides the process of EIS study into four major stages with inter-stage iteration, namely, measurement, interpretation, validation, and application. The key issues involved in the implementation of the first three stages are elaborated.

11.2.1 PRINCIPLE OF EIS

EIS is a frequency-response technique for which the perturbation to and the response from the system are electrical signals. The transfer function of the system from the alternating current to the alternating potential defines the impedance. EIS probes the processes and properties of both the interface and the bulk of an electrochemical system from the impedances over a wide range of frequencies. A general electrochemical system in an experiment utilizes a three-electrode setup,[10,11] as shown in Figure 11.1a, which includes the working electrode (WE), reference electrode (RE), and counter electrode (CE). The WE is the electrode on which the reaction of interest occurs, the RE is the electrode which has a stable electric potential, and the CE is the electrode which closes the circuit. At the interface between the electrode and electrolyte, an electrical double layer (EDL) exists, as shown in Figure 11.1b. When a potential is applied on the WE (negatively charged in the figure), counter ions (cations in the figure) in the electrolyte solution are attracted to the electrode while the co-ions are repelled away to balance the electrode charge. The EDL consists of an inner Helmholtz plane representing the locus of electrical center of specifically adsorbed ions (normally anions) or the first layer of the adsorbed water, an outer Helmholtz plane representing the locus of electrical centers of solvated counter ions (cations in the figure) closest approachable to the electrode, and a diffuse layer representing the region of charge non-neutrality where ions distribute under the competing influence of the thermal motion and the electrical attraction from the electrode.[12]

EIS measurement uses a periodic electric potential or current perturbation to excite the system, and records the periodic current or electric potential response to this perturbation, as shown in Figure 11.1c. Since the governing equation for the charge transfer reaction in an electrochemical system, the Butler–Volmer equation, is nonlinear, particularly at large overpotentials, the perturbation amplitude must be adequately small to ensure a predominantly linear response. Under such condition, the impedance can be defined as the transfer function of the system using Laplace transform with the transform variable s taking an imaginary value, i.e., $s = j\omega$,[13,14] as shown in Eqn. (11.1) and Figure 11.1d:

$$Z(\omega) = \int_0^{+\infty} \left(\frac{E(t) - E_0}{I(t) - I_0} \right) e^{-j\omega t} dt = Z_r(\omega) + jZ_j(\omega), \tag{11.1}$$

where $\omega = 2\pi f$ is the angular frequency in radian, f is the frequency in Hz, $I(t)$ is the current applied to the system, I_0 is the current flowing through the system at the steady state, $E(t)$ is the electric potential response of the system, E_0 is the electric potential of the system at the steady state, and $Z(\omega)$ is a frequency-dependent complex number whose real part $Z_r(\omega)$ is a resistance and imaginary part $Z_j(\omega)$ is a reactance.

The impedance data are commonly represented by Bode and Nyquist plots, as shown in Figure 11.1e and f, respectively.[a] In Nyquist plot, the imaginary part is plotted against the real part of the impedance. Customarily, the y-axis points to the negative direction so that the impedance spectrum in the 1st quadrant has a negative imaginary part. In this way, the feature of the capacitance, which is more prevalent

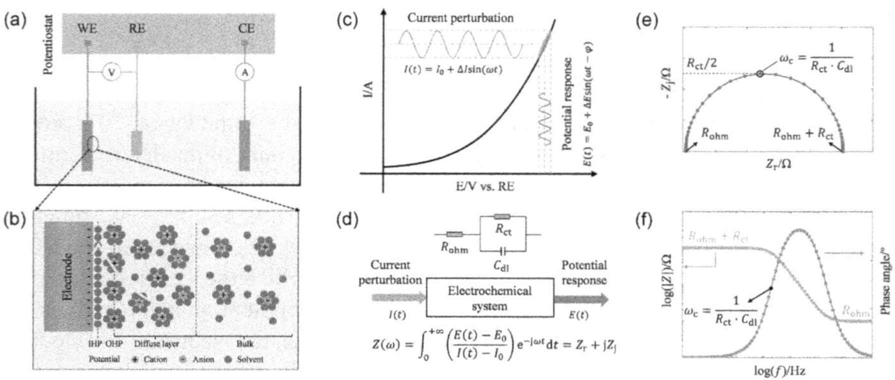

FIGURE 11.1 The principle and typical representations of EIS. (a) The schematic of the three-electrode setup; (b) the schematic of the electrical double layer structure at the electrode/electrolyte interface; (c) the current perturbation and potential response of an electrochemical system[15]; (d) the electrochemical impedance and the equivalent electrical circuit of the system; (e) the Nyquist plot of impedance spectrum; (f) the Bode plot of impedance spectrum.

than inductance in electrochemical systems, falls in the 1st quadrant and hence better visualized. The Nyquist representation is linear in scale for both the real and the imaginary parts; hence, impedance with large magnitude easily dominates the plot. Shifting the origin of the real axis close to the ohmic resistance (crossing point of the spectrum with the real axis at high frequency) can emphasize the impedance from the charge transfer and mass transfer, which are frequently the focus of interest in electrochemical study. In Bode plot, the modulus and phase angle, or the real and imaginary parts, are plotted against the frequency. The Bode representation is logarithmic in scale for the frequency and the impedance magnitude. It contains explicit information on the frequency and phase angle.

The equivalent electrical circuit for a simple electrode with charge transfer and EDL charging/discharging but no adsorption and diffusion are shown in Figure 11.1d. The values of the circuit elements and the characteristic frequency for the $R//C^b$ circuit can be read from the Nyquist and Bode plots, as shown in Figure 11.1e and f.

11.2.2 Stages, Steps, and Iterations in EIS Study

EIS study is a long, iterative process with many intermediate steps. The whole process can be roughly divided into four major stages, as shown in Figure 11.2: (i) Measurement - acquiring a repeatable and valid spectrum; (ii) Interpretation - identifying and understanding the features in the spectrum; (iii) Validation - corroborating the understanding via independent observation; (iv) Application - utilizing the understanding in diagnosis and prognosis, in characterization, design, and control. Iterations within and among these stages are frequently needed to ensure the

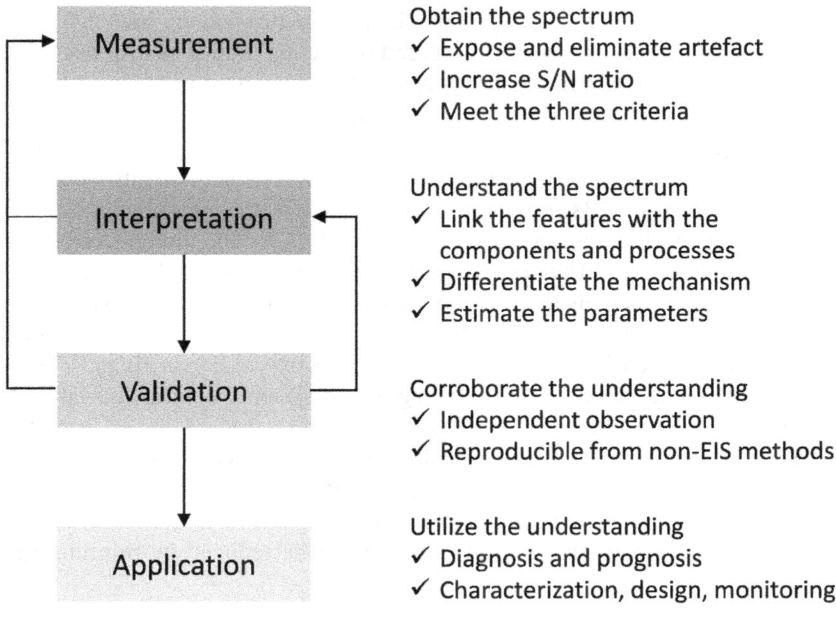
Measurement
Interpretation
Validation
Application
Obtain the spectrum
✓ Expose and eliminate artefact
✓ Increase S/N ratio
✓ Meet the three criteria
Understand the spectrum
✓ Link the features with the components and processes
✓ Differentiate the mechanism
✓ Estimate the parameters
Corroborate the understanding
✓ Independent observation
✓ Reproducible from non-EIS methods
Utilize the understanding
✓ Diagnosis and prognosis
✓ Characterization, design, monitoring

FIGURE 11.2 Four major stages and iterations of EIS study.

reliability of EIS results. The specifics of the first three stages for the study of LIB are elaborated in Figure 11.3, while that of the fourth stage, the application of EIS in different phases of the LIB lifetime, is discussed in Section 11.4. The key elements in the steps and iterations in Figure 11.3 will be explained in detail in the following.

11.2.2.1 Stage 1: EIS Measurement to Acquire Repeatable and Valid Spectrum

The measurement stage encompasses a series of steps, including the design of the experiment, the execution of the design to obtain the impedance spectrum, the check of the linearity, stability, and causality during the measurement, and the confirmation of the Kramers–Kronig (K–K) relation compliance of the obtained spectrum.

11.2.2.1.1 Design of the Experiment

In its early days of development, EIS study is a painstaking effort as researchers have to build their own instrument to measure the spectrum, and have to write their own code to plot and fit the spectrum. Nowadays, many steps in the measurement and data processing have been incorporated into potentiostats as default settings, making EIS an easily accessible, highly automated technique. Accordingly, designing the experiment is becoming more important than its execution to obtain a reliable spectrum. The experiment design includes the instruments selection, the choice of the samples, cell configurations, operation conditions, and cell connections.

To measure EIS accurately, an instrument with adequate accuracy matching the impedance magnitude of the sample should be chosen.[24] A schematic of the accuracy contour plot of an EIS instrument is shown in Figure 11.3a. Line "A" is based on the minimum current resolution, line "B" is based on the wire capacitance, line "C" corresponds to the maximum frequency capability of the instrument, line "D" is based on the wire inductance, and line "E" is based on the maximum measurable current. The target impedance and frequency range of the experiment must lie within these boundaries to guarantee the measurement accuracy.

Cell configuration refers to the number of the electrodes in the electrochemical cell. The two-electrode setup is normally used when it is difficult to introduce an RE, such as in commercial LIBs and fuel cells. In such cases, the measured impedance is the sum of all contributions between the WE and the CE. The three-electrode setup enables the study of the WE independent of the processes taking place at the CE.[25,26] The four-electrode setup, using separate pairs of current-carrying and voltage-sensing electrodes, can eliminate the interference from the contact and probe resistances, hence commonly used for the measurement of electrolyte conductivity.[27] The cell configuration should be selected depending on the system being measured and the purpose of the measurement.

The operation conditions, such as blocking/unblocking and mass transport conditions, determine the state of the cell; hence they affect the nature and quality of the extracted physical parameters. The wires connecting the instrument and cells may have mutual inductance, and this interference can be reduced by minimizing the

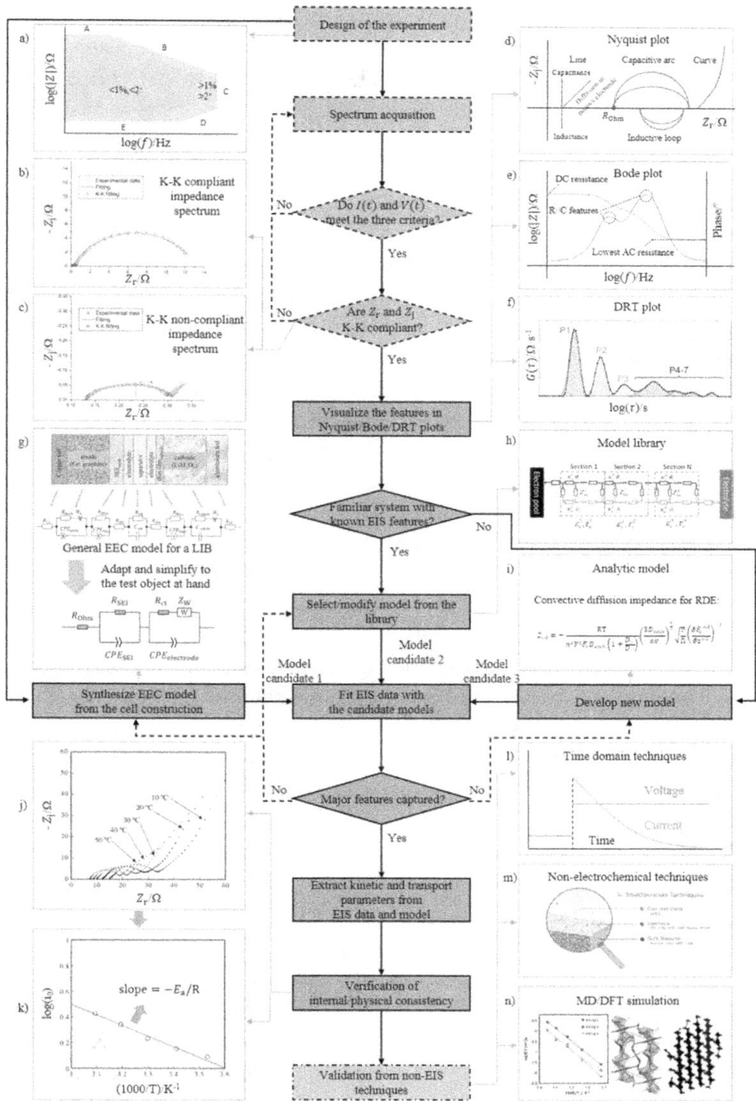

FIGURE 11.3 The steps and iterations in the first three major stages of EIS study. (a) A schematic accuracy contour plot of EIS instruments[15]; (b) the K–K-compliant impedance spectrum[16]; (c) the K–K-non-compliant impedance spectrum[16]; (d) typical EIS features in Nyquist plot; (e) typical EIS features in Bode plot; (f) the time constants in DRT plot[17]; (g) synthesizing equivalent electric circuit model from the cell construction[18]; (h) transmission line model of a porous electrode, an example from the EIS model library[19]; (i) developing new physics-based model, with the analytic model of convective diffusion impedance shown as an example[20]; (j) the EIS of LIB at different temperatures for consistency check[21]; (k) the Arrhenius plot of exchange current density (i_0) obtained from (j)[21] for consistency check; (l) the schematic of time-domain techniques; (m) the schematic of non-electrochemical techniques[22]; (n) the MD/DFT estimation of Li^+ diffusion coefficient in $LiVPO_4F$.[23]

loop area between the two current-carrying wires and the two voltage-sensing wires via a tight-braided twist,[28] and by maximizing the distance between the current-carrying and the voltage-sensing wires.[24] The readers can refer to ref. [15] for more information.

11.2.2.1.2 Spectrum Acquisition

Many steps and settings during both the measurement and data processing have been incorporated as standard procedures or default values into most modern potentiostats, making EIS an easily accessible, highly automated technique. The downside of this convenience is the rather popular blind use of the technique without much understanding of the steps and settings. Actually, the quality of the measured spectrum is highly dependent on these settings, including the selection of potentiostatic or galvanostatic mode, the perturbation amplitude, the frequency range, and the number of cycles to perform the measurement at each frequency.[15]

The proper measurement mode, potentiostatic or galvanostatic, depends on the measured electrochemical systems. In general, the galvanostatic mode is more suitable for non-invasive probing of metal corrosion at the open-circuit potential (zero external current) and for measuring low impedance (low-Z) systems like electrochemical energy devices such as fuel cell and LIB, where the impedance is low and current levels are high. However, for electrochemical systems with high impedance (high-Z), the potentiostatic mode is generally employed.

The suitable amplitude of the perturbing current or potential needs to be determined experimentally. The perturbation amplitude should be small enough to ensure a linear response from the system, while being adequately large to improve the signal-to-noise (S/N) ratio. For the potentiostatic EIS (PEIS) mode, an alternating potential perturbation with a peak-to-peak amplitude of 5–15 mV is commonly used. The frequency range should be set to cover the lowest and the highest characteristic frequencies of the processes in the electrochemical system under study. The typical frequency range for the electrochemical systems, e.g., electrochemical energy conversion and storage devices, corrosion, is 100 kHz–10 mHz. Usually, seven to ten points evenly spaced logarithmically per frequency decade, are needed for measuring an impedance spectrum with sufficient resolution for a detailed data analysis. To minimize the stochastic error, an auto-integration mode can be used to increase the number of cycles for each frequency, for which three or four would suffice for most cases. For some equipments, the user has the options to choose among fast, normal, and low-noise modes so that the instrument will select the number of cycles automatically.

11.2.2.1.3 Check of Linearity, Stability, and Causality

The system should be linear, stable, and causal during the EIS measurement. The constraint of linearity requires that the response to an input perturbation is a linear combination of the input. The linearity can be checked by the shape of the Lissajous plot, which is obtained by tracing the output signal as a function of the input signal at a specific frequency. The Lissajous plot typically shows a tilted oval for a linear response. If there are distortions to the oval shape,[30] the perturbation amplitude needs to be reduced. This check is particularly effective at low frequencies where the influence of nonlinearities is the largest.[29]

The constraint of stability requires that the response to a perturbation shall not grow with time. For example, the current response to a potential perturbation must eventually decay to the initial steady value after the perturbation is removed. The constraint of causality requires that the response to a perturbation shall not precede the perturbation. This constraint ensures that the measured response of the system only results from the applied perturbation, rather than from background noise or self-sustained oscillation which may occur in some systems. The stability and causality can be checked by comparing the plots showing the evolution of the response and the perturbation against time.[30,31]

11.2.2.1.4 Check of Compliance to K–K Relation

An impedance spectrum of a system satisfying the criteria of linearity, stability and causality should be consistent with the K–K relations,[16,32–34]. For such system, the imaginary part of the impedance Z_j, can be calculated from the real part through:

$$Z_j(\omega) = -\frac{2\omega}{\pi}\int_0^\infty \frac{Z_r(x) - Z_r(\omega)}{x^2 - \omega^2}dx. \tag{11.2}$$

If the high frequency limit of the impedance is known, the real part can be calculated from the imaginary part through:

$$Z_r(\omega) = Z_r(\infty) + \frac{2}{\pi}\int_0^\infty \frac{xZ_j(x) - \omega Z_j(\omega)}{x^2 - \omega^2}dx. \tag{11.3}$$

Direct integration of the K–K relations for EIS measurements results has been replaced, in modern use, by regression to K–K-consistent circuit.[35] Figure 11.3b and c shows the K–K compliant and K–K non-compliant impedance spectrum, respectively. The K–K compliance check offers an additional chance to discover any violation of the three criteria in the obtained spectrum.

11.2.2.2 Stage 2: Interpretation to Identify and Understand Features in the Spectrum

After the K–K compliant impedance spectrum is obtained, it needs to be interpreted through the following steps: (i) recognizing the features via visualization in Nyquist, Bode, and distribution of relaxation time (DRT) plots; (ii) constructing, selecting, or developing candidate models for the spectrum; (iii) fitting spectrum with the models and evaluating the models; (iv) extracting kinetic and transport parameters from fitting the EIS data to the best candidate model; and (v) verifying the extracted parameters via evaluating their internal and physical consistency.

11.2.2.2.1 Feature Recognition

Figure 11.3 parts d-f show the use of visualization to recognize the features. To expose and emphasize certain features, it is advisable to try different plots, e.g., Nyquist plot to highlight diffusion and kinetic behavior, Bode plot to highlight the phase angle and frequency-dependent behavior, and DRT plot to separate overlapping processes with closely spaced time constants. The basic circuit elements and their Nyquist and Bode plots are summarized in Table 11.1 for easy access and comparison.

In the Nyquist plot, as shown in Figure 11.3d, the straight line and arc above the real axis usually correspond to capacitive behavior. The ideal capacitance behaves as a straight line with an angle of 90°, and the R//C circuit behaves as a semi-circle. When the capacitance changes into an constant-phase element (CPE), the angle of straight line becomes less than 90° and the semi-circle becomes depressed. The straight line in the high frequency range with an angle of 45° may come from the competing effects of ion migration and EDL charging/discharging in the electrode pore, which can be modeled as a transmission line of resistor and capacitor. The parabolic arc in the low frequency range may relate to ionic diffusion. The loop below the real axis at low frequency, showing inductive behavior, may correspond to adsorption process.

In the Bode plot, as shown in Figure 11.3e, the modulus curve shows the DC resistance at the low frequency limit and the lowest AC resistance at the high frequency limit. The phase angle usually approaches 0° at high frequency owing to the ohmic resistance of the electrolyte. One can eliminate the effect of ohmic resistance by plotting the ohmic resistance-corrected phase angle against frequency, which has an asymptotic value of −90° at high frequency for an ideally polarizable electrode. If there is CPE behavior, the asymptotic value at high frequency approaches $-(90 \cdot n)°$, where the CPE index value n is 1 for a pure capacitor, 0.5 for a Warburg element with infinite diffusion length, 0 for a pure resistance, and −1 for a pure inductor. Values of n other than the above may indicate the non-ideality of the element. In addition, the number of phase angle peaks in the Bode plot indicates the number of R//C processes. It is also worth noting that the process resolution capability of the Bode plot can be enhanced by replacing the variables of the y-axis, i.e., the phase angle or the impedance magnitude, with the differentiation of the phase angle or the logarithm of the impedance magnitude with respect to the logarithm of the frequency.[36]

The DRT method transforms the impedance data in the frequency domain into a function in the time domain. As shown in Figure 11.3f, the resistance per unit relaxation time is plotted against the relaxation time.[17] The number of peaks indicates the number of physical processes with different relaxation time constants, the central position of the peak signifies the characteristic time constant, the height of the peak reveals the polarization strength, the width of the peak manifests the standard deviation, and the area under the peak represents the significance of the polarization.[37,38] For example, for an R//CPE circuit, the height of the peak decreases and the width increases with the decreasing CPE index value n.[39] The integral of the whole DRT curve with the relaxation time in the horizontal axis is the total resistance (DC resistance) of the system.[c] The interested readers are referred to ref. [40] for an overview and ref. [41] for an in-depth discussion of the DRT technique.

11.2.2.2.2 Construction, Selection, and Development of Candidate Models

EIS method needs models to interpret the measured spectrum. Candidate models need to be generated and evaluated.

A rather common practice is to construct or select models from a model library, based on the features recognized via from Nyquist, Bode, and DRT plots. The models

TABLE 11.1
The Basic Circuit Elements and Their Nyquist and Bode Plots

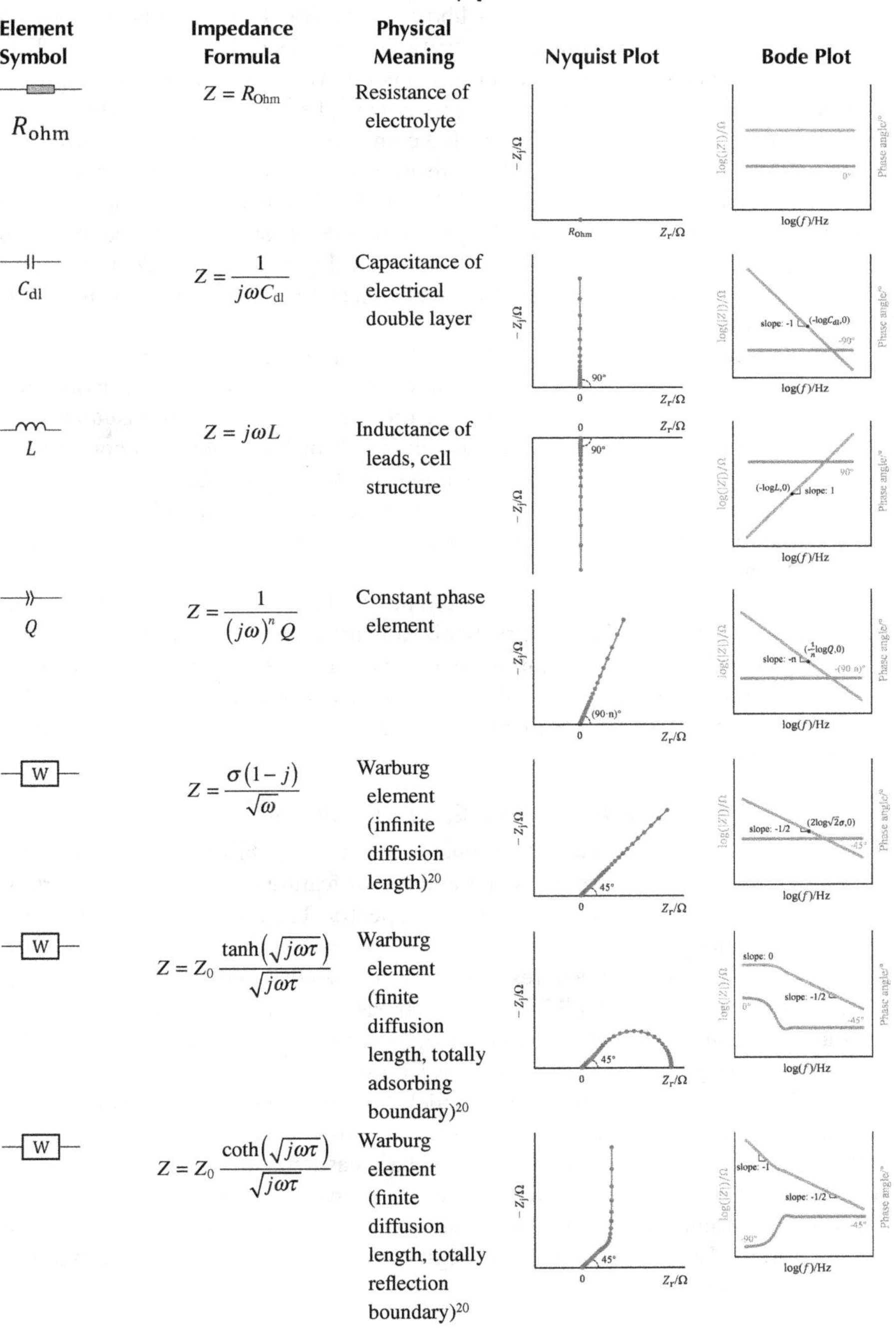

Element Symbol	Impedance Formula	Physical Meaning	Nyquist Plot	Bode Plot
R_{ohm}	$Z = R_{\text{Ohm}}$	Resistance of electrolyte		
C_{dl}	$Z = \frac{1}{j\omega C_{\text{dl}}}$	Capacitance of electrical double layer		
L	$Z = j\omega L$	Inductance of leads, cell structure		
Q	$Z = \frac{1}{(j\omega)^n Q}$	Constant phase element		
W	$Z = \frac{\sigma(1-j)}{\sqrt{\omega}}$	Warburg element (infinite diffusion length)[20]		
W	$Z = Z_0 \frac{\tanh\left(\sqrt{j\omega\tau}\right)}{\sqrt{j\omega\tau}}$	Warburg element (finite diffusion length, totally adsorbing boundary)[20]		
W	$Z = Z_0 \frac{\coth\left(\sqrt{j\omega\tau}\right)}{\sqrt{j\omega\tau}}$	Warburg element (finite diffusion length, totally reflection boundary)[20]		

can be either equivalent electric circuit models (EEC models, EECMs, or ECMs)[d] or analytic models. As an example, the transmission line model (TLM) for a porous electrode[19] is shown in Figure 11.3i.

For easy access and comparison, a library of typical EEC models is collected together with their corresponding characteristic frequencies in Table 11.2. For the R-C circuit, the characteristic frequency is the frequency at which the real part of the impedance equals to the imaginary part, $Z_r(\omega_c) = -Z_j(\omega_c)$; for the R//C circuit, the characteristic frequency is the frequency at which the imaginary part assumes the extremum or equivalently, the real part of the ohmic resistance-corrected impedance equals the imaginary part, $Z_r(\omega_c) - R_{\text{Ohm}} = -Z_j(\omega_c)$; for the TLM of de Levie, the characteristic frequency is the frequency at which the penetration depth of the perturbation equals the pore length.[42] In order to account for the non-ideality of the electrode and achieve a reasonable fitting, it is a common practice to replace the pure capacitor with a CPE in these EEC models.

Another practice in candidate model construction, which we endorse, is to synthesize the EEC model for the tested system based on the cell construction and the expected electrochemical processes[18] before the EIS features are visualized, as shown in Figure 11.3g. This practice has a less risk of losing features conceivable from the cell constructions yet not prominent in the Nyquist/Bode plots. In case large discrepancy shows up in comparison with the models obtained from the first practice, the researchers may be reminded to recheck their design and execution of the experiment.

In case the measured system contains unfamiliar features, or the candidate models constructed based on either the recognized features or the cell construction fail to capture the major and minor features in the spectrum, new models, either EEC or analytic, need to be developed based on the physics governing the test object. The readers are referred to refs. [13,19,20,43] for an in-depth discussion of developing physics-based model.

11.2.2.2.3 Fitting and Evaluating the Candidate Models

All the candidate models need to be evaluated for their capability to fit the EIS data, e.g., how well do they capture the major and minor features in the whole spectrum, and how well they perform against a series of spectra. The most suitable one is chosen for further analysis.

Complex nonlinear least-squares regression is usually used to fit mathematical models to the impedance data.[45] This approach has the advantage that a common set of parameters can be obtained by simultaneous regression of the model to both the real and imaginary parts of the impedance. Commercial programs such as ZView[46] or ZSimpWin[47] are available to fit EEC models to impedance data. Non-commercial programs are also available, such as a recent Python-based program that allows both the regression of arbitrary functions and a measurement model analysis of the data.[34] A combined strategy sequentially using simplex and Levenberg–Marquardt regression is advantageous[48,49] in that the simplex regression can provide good initial guesses, and the Levenberg–Marquardt regression can provide confidence intervals for regressed parameters.

TABLE 11.2
The Library of the Equivalent Electrical Circuit Models with the Characteristic Frequency Annotated

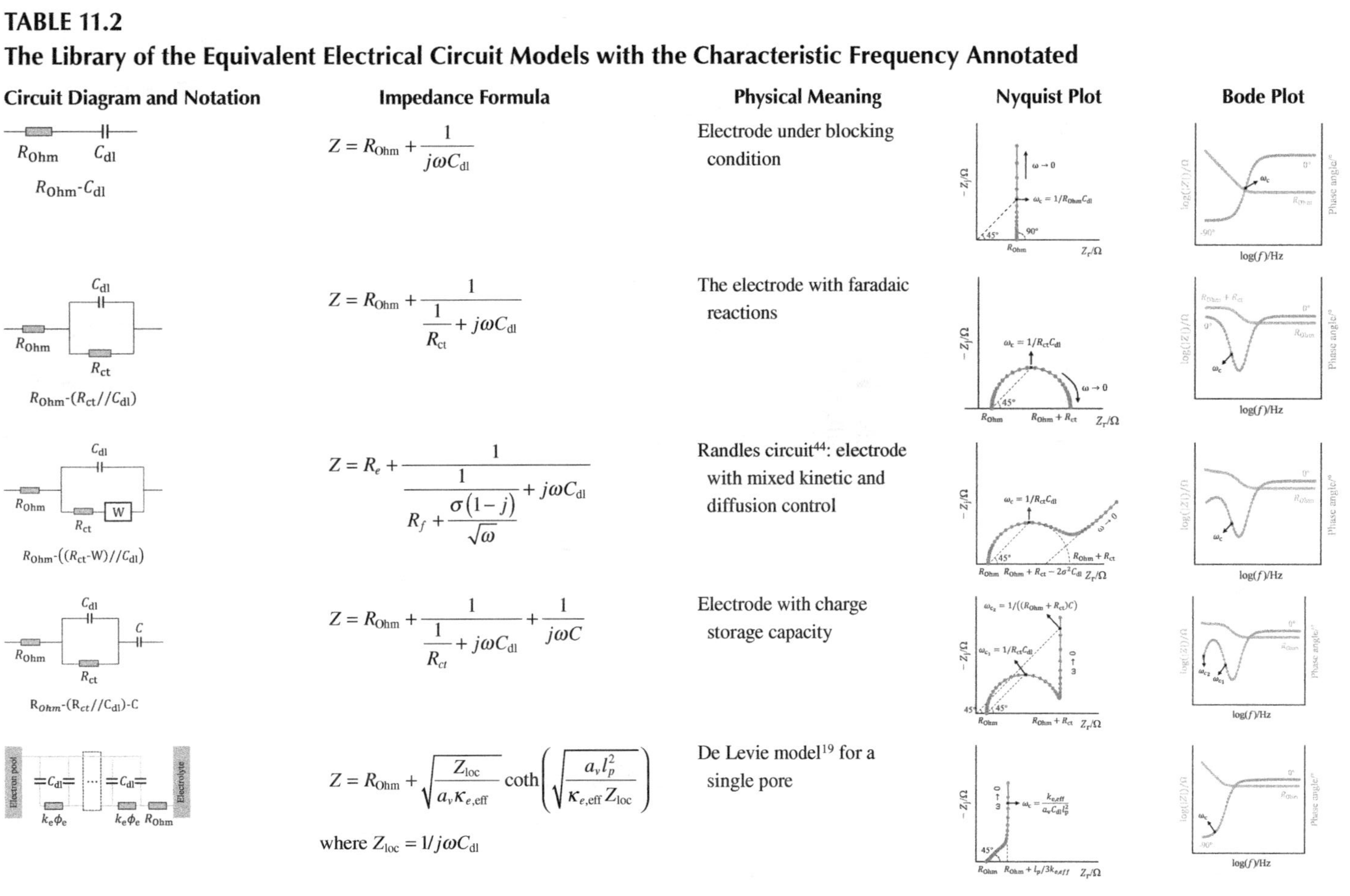

Circuit Diagram and Notation	Impedance Formula	Physical Meaning	Nyquist Plot	Bode Plot
R_{Ohm}-C_{dl}	$Z = R_{Ohm} + \frac{1}{j\omega C_{dl}}$	Electrode under blocking condition	$\omega_c = 1/R_{Ohm}C_{dl}$	ω_c
R_{Ohm}-$(R_{ct}//C_{dl})$	$Z = R_{Ohm} + \frac{1}{\frac{1}{R_{ct}} + j\omega C_{dl}}$	The electrode with faradaic reactions	$\omega_c = 1/R_{ct}C_{dl}$	ω_c
R_{Ohm}-$((R_{ct}$-$W)//C_{dl})$	$Z = R_e + \frac{1}{\frac{1}{R_f + \frac{\sigma(1-j)}{\sqrt{\omega}}} + j\omega C_{dl}}$	Randles circuit[44]: electrode with mixed kinetic and diffusion control	$\omega_c = 1/R_{ct}C_{dl}$	ω_c
R_{Ohm}-$(R_{ct}//C_{dl})$-C	$Z = R_{Ohm} + \frac{1}{\frac{1}{R_{ct}} + j\omega C_{dl}} + \frac{1}{j\omega C}$	Electrode with charge storage capacity	$\omega_{c_2} = 1/((R_{Ohm}+R_{ct})C)$; $\omega_{c_1} = 1/R_{ct}C_{dl}$	ω_{c_2}, ω_{c_1}
Electron pool – C_{dl}, $k_e\phi_e$ … C_{dl}, $k_e\phi_e$, R_{Ohm} – Electrolyte	$Z = R_{Ohm} + \sqrt{\frac{Z_{loc}}{a_v\kappa_{e,eff}}}\coth\left(\sqrt{\frac{a_v l_p^2}{\kappa_{e,eff} Z_{loc}}}\right)$ where $Z_{loc} = 1/j\omega C_{dl}$	De Levie model[19] for a single pore	$\omega_c = \frac{k_{e,eff}}{a_v C_{dl} l_p^2}$	ω_c

The quality of the fit may be evaluated using a graphical comparison of the model fit and the impedance data, or more rigorously, using the weighted χ^2 statistic if the experimental error structure is known.[32] The quality of the candidate models may be evaluated against the criteria like small residual error, narrow confidence intervals, physical consistence of the estimated parameters, and wide applicability to varied cell construction and operation conditions. Physical consistence means that the magnitude of the estimated parameter should be within the well-established range of that parameter, e.g., the double-layer capacitance should be around a range of 10–50 μF/cm^2. Sometimes, multiple candidate models can fit the EIS data equally well. In such case, the model with the simplest structure is preferred, according to the rule of Occam's razor.

11.2.2.3 Stage 3: Validation from Non-EIS Techniques

After understanding of the EIS results is obtained from the interpretation stage, non-EIS techniques, e.g., time-domain and non-electrochemical methods, need to be applied to corroborate that understanding.

Compared with the microscopic and microanalysis techniques, the power of EIS lies in its capability to discern both bulk and interfacial processes in electrochemical systems, as well as to estimate their kinetic and transport parameters, via the analysis of linear response to periodic perturbation across a wide range of frequencies. However, such capability heavily depends on the use of models to interpret the obtained spectrum, and it is well known that multiple models can fit the spectrum equally well. Hence, the EIS method is inherently in-definitive. Consequently, the insight gained from EIS needs to be validated from non-EIS techniques, including the time-domain measurement, the non-electrochemical techniques, and the molecular dynamics (MD)[50] and density functional theory (DFT) simulation.[51] EIS interpretation corroborated from independent observations will carry more weight and give us more confidence in utilizing the understanding for diagnosis and prognosis, material characterization, cell design, and system control.

11.3 CHARACTERIZING LIB USING EIS

EIS study is a long iterative process involving concerted efforts of testing, modeling, and analysis. For simplicity, we de-convolute the intricately coupled process into four major stages (Figures 11.2 and 11.3), the first being measurement, i.e., obtaining a K–K compliant spectrum; the second being interpretation, i.e., understanding the spectrum within the realm of EIS; the third being validation, i.e., corroborating the understanding from independent techniques other than EIS; and the fourth being application, i.e., utilizing the method and insight to the R&D of the electrochemical materials, devices, and systems. This section addresses the major issues in the first three stages in EIS study of LIB, while next section discusses exemplary EIS application in major phases of LIB lifetime. As an electrochemical energy storage and conversion device, LIB is a low-impedance (low-Z) system having its specialties in EIS measurement, features, modeling, interpretation, validation, and application.

11.3.1 EIS Measurement of LIB

Correct measurement of EIS is the foundation of the subsequent feature recognition, modeling, and interpretation. To obtain a reliable spectrum worthy of further analysis, care needs to be exercised with regards to the measurement setup, instrument settings, and test procedures.

11.3.1.1 Measurement Setup

The setup of the EIS measurement includes the instrument, the cell, the test fixture, and the inter connections. Commercial LIB has low impedance with a magnitude on the order of 1 mΩ. Such low impedance may lie out of the dynamic range of the potentiostat with EIS capability. For large format battery, a booster with EIS-compatible frequency response characteristics may be needed.[52] The accuracy contour plot of the instrument should be checked via a preliminary test probing the impedance range of the test object. The instrument should be calibrated to ensure traceability,[1] or at least checked against a standard impedance of the same range, e.g., standard resistor of 1 mΩ[53] for a commercial LIB, to spot any instrument deterioration, deficiency or mismatch between the instrument and the test object. For such low-impedance device, the contact resistance between the cell and the cell holder can become an issue. Up to 10% difference in the cell ohmic resistance was observed over ten EIS measurements using different cell holders and/or reinsertion of the cell into the same holder.[52] A technical note on the cell connections, battery fixture, correction with a surrogate, EIS mode selection, etc., in the measurement of sub-Ω low-Z devices a can be found on the website of Gamry.[54]

The cells of the EIS test for LIB have a large number of variations. Based on the number of electrodes, the cells can be classified as two-, three-, and four-electrode configurations, each with its strength and limitation, hence having to be selected according to the purpose of the study. Commercial cells in various capacities and shapes, such as cylindrical, prismatic, and laminate, are generally measured in two-electrode configuration to obtain the impedance of the whole cell. To differentiate the contribution from the anode and cathode, a half-cell or symmetric cell is commonly employed. The half-cell, in the format of a coin cell or pouch cell, uses Li metal to substitute one electrode of the original full cell. This can help to a certain extent to identify the features from the other electrode. Yet the Li metal itself may have non-negligible, even dominant, complex and time-varying EIS behavior,[55] and may even cause new dynamic behavior in the other electrode due to the less stable solid electrolyte interphase (SEI) on Li metal than that on the graphite.[56] Symmetric cell[57–63] eliminates the interference from the Li metal electrode by using two nearly identical electrodes sampled from the same electrode of the dissembled full cells. However, to access the EIS behavior at different states of charge (SOCs), multiple full cells have to be prepared, their SOC adjusted, and dissembled to fabricate multiple symmetric cells, which can be tedious. Figure 11.4 depicts the processes in preparing anode and cathode symmetric cells.[64]

The three-electrode cell configuration is capable of efficiently isolating the EIS contribution from either positive or negative electrode at different SOCs by introducing a RE inside the cell.[7,65] However, compared with DC measurement, EIS

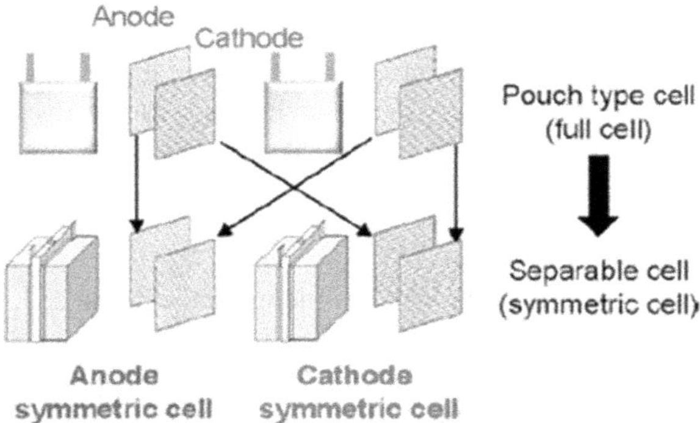

FIGURE 11.4 Schematic of the procedure for preparing anode and cathode symmetric cells from full cells.[64]

measurement with a three-electrode configuration is prone to artifacts if the position,[66] size[67] of RE, or the symmetry of the negative/positive electrodes[68] is inappropriate. It is found that the RE, with a sufficiently long uncoated tip, helps to reduce the artifacts at high frequency, and that averaging the two EIS results with reversed connections can eliminate the high-frequency artifact altogether.[67] The four-electrode cell is used to measure the ionic conductivity and the solvation/de-solvation properties of the electrolyte.[69]

Connections between the cell and the instrument have a larger impact on the EIS measurement than on the time-domain measurement due to the electromagnetic interference among the wires and cables, especially at high frequency and large current. As some cells are fabricated and measured in the glove box, due to the sensitivity of LIB to moisture, long cables are needed, which is susceptible to electromagnetic interference. Measures like coaxial cables, twisted cables, and Faraday cages, etc. may be needed to suppress the interference and to improve the S/N ratio of the measurement.[54]

To increase the sharpness of the features in the measured spectrum and facilitate their subsequent attribution, well-defined model cells using thin-film electrode,[70,71] single particle electrode,[72] thin porous composite electrode,[73] etc., are effective. For porous composite electrode, measuring and comparing the EIS of the electrode subjected to blocking/non-blocking conditions, is an effective way to differentiate EIS features at high/intermediate/low frequency ranges into different components and processes.[65,74]

11.3.1.2 Instrument Settings

Based on the type and size of the perturbation, EIS can be classified into different modes and types. If the current is perturbed and the electric potential response is measured, it is called galvanostatic EIS (GEIS). In contrast, if electric potential is perturbed and current response is measured, it is called potentiostatic EIS (PEIS). In principle, impedance results from PEIS and GEIS measurements should be identical.[75] However, due to the inherent limitation in hardware, e.g., it is easier to measure than to generate high-precision voltage signal in the sub-mV range, GEIS may give more accurate results for low-Z devices.[69] On the other hand, it is found that the galvanostatic measurement takes longer to reach a sinusoidal steady state at a given frequency.[76] If the spectrum needs to extend in the low frequency range, as is the case frequently for LIB, PEIS may be more efficient.

The amplitude of the perturbation should be small enough to ensure a linear relationship between the perturbation and the response, yet large enough to outweigh the interference from the noise. Some groups are exploring the possibility of extracting additional dynamic information from nonlinear (higher harmonic) responses to large-magnitude perturbation.[77,78] Even though such trials still belong to the frequency-domain electrochemical methods and bear many similarities with EIS, they violate the criterion of linearity, hence is not EIS in its strict sense. Aside from the concern with the nonlinearity, another restriction on large-amplitude perturbation is the possible rise of the temperature of the test object, especially for the test under low temperatures. As the impedance depends sensitively on the temperature, the perturbation amplitude has to be restricted, and the temperature near or at the WE needs to be monitored and controlled, using an environmental chamber or thermostat.

For LIB, the GEIS can be further divided into stationary EIS, if the base current is zero, or dynamic GEIS (DGEIS, or commonly termed DEIS), if the base current is non-zero. The purpose of DEIS is to track the temporal evolution of charge transfer and diffusion parameters in intercalation materials, such as lithium manganese oxide ($LiMn_2O_4$) and nickel hexacyanoferrate,[79,80] or to explore the EIS behavior of LIB in operando, i.e., when the cell is under charging or discharging. It is found that the charge transfer resistance from DEIS is somewhat lower than that obtained from stationary EIS, as can be expected from the Butler–Volmer equation, and there is also a slight difference between the charging and discharging, as may be caused by the difference between the activation energy of solvation or de-solvation.[81] However, these effects are very limited in magnitude, possibly because the attainable charge/discharge rate is very low due to the use of a thick porous composite electrode rather than a single particle electrode,[82] or because the resistance due to the SEI film on the active particle is dominant over that of the solvation or de-solvation step in the whole charging/discharging processes.[83] Since the SOC of LIB changes during DEIS measurement, the deviation from stationarity may cause artifacts to appear as a loop at the low frequency range.[84,85] Huang formulated a stationarity condition governing the relationship between the lowest frequency of the perturbation and the base current based on the requirement that the change of the surface concentration of Li ion due to the base current is negligible compared with that from the AC perturbation.[85]

11.3.1.3 Procedure and Sequence

To facilitate the enhancement as well as the attribution of the features in the spectrum, it is effective to perform extensive EIS measurements where certain control variables like temperature,[59] SOC,[86] compression pressure,[87] time during SEI formation,[55] time in storage or cycling number,[63] etc., are systematically varied. The most conspicuous changes in the spectra can be attributed to the component or process that is most sensitive to the varied control variable.

However, the downside of this approach is that the test matrix can easily become bulky and overwhelming. In organizing the test sequence, care and judgment need to be exercised to strike a balance among certainty, accuracy, and efficiency. Take a common test matrix involving a systematic variation of both temperature and SOC as an example. It may be efficient, hence tempting, to set the temperature first and then change the SOC. Yet a more accurate test should be adjusting SOC at the predefined temperature first, and then test the EIS at other temperatures, because the capacity, and hence the SOC of LIB, depends sensitively on temperature.

To check for stationarity, the appropriate rest time after the temperature and SOC adjustment needs to be explored in a trial test. To check for the existence of hysteresis, it is advisable to sweep the control viable in both directions.

11.3.2 EIS Features of LIB

EIS of LIBs share many common features despite the large difference in capacity and chemistry. Some of these features are due to the structure of the porous electrode widely used in electrochemical energy devices, while others are due to the intercalation materials which are specific to LIB. In what follows, the most evident features frequently observed in the spectrum of LIB will be introduced first, then more detailed features exposed in well-defined systems, e.g., three-electrode cells, will be introduced.

11.3.2.1 Major EIS Features of LIB

Figure 11.5a shows a schematic of the prominent features commonly observed in the Nyquist plot of the EIS of a LIB. The spectrum can be divided into three ranges according to the frequency. In the high frequency range, a curve in the 4th quadrant with a large curvature radius, particularly evident for the cylindrical cell, reflects the inductance behavior either from the connecting wires or the spiral cell construction. The intersection of the spectrum with the real axis at high frequency is commonly taken as the ohmic resistance which consists of the ionic resistance of the electrolyte as well as the electronic resistance of the electrode, current collector, and the contact resistances. Actually, the resistance at the intersection overestimates the ohmic resistance.[36] For a more accurate estimation, the interference from the inductance needs to be corrected, which can be easily achieved through first estimating the value of inductance via the fitting of the curve in the 4th quadrant to an R//L circuit, and then subtracting the contribution of the inductance from the spectrum in the 1st quadrant. An alternative approach, less sensitive to errors in fitting the R//L circuit, is to use a measurement model to determine data influenced by the inductance.[88]

In the intermediate frequency range, two depressed semi-circles can usually be found. The first one at higher frequency is smaller in size and is usually attributed to the SEI on the graphite anode. The second one may result from the R//C circuits corresponding to the EDL and charge transfer on the interface between the active materials and the electrolyte. In a two-electrode configuration, the responses from the anode and the cathode may overlap, while in a three-electrode configuration, the inhomogeneity of porous composite electrode may cause frequency dispersion (seen as a depressed semi-circle).

In the low frequency range, a slanted straight line with an inclination of approximately 45° is usually visible, which is commonly ascribed to the diffusion of Li ion in the active materials.

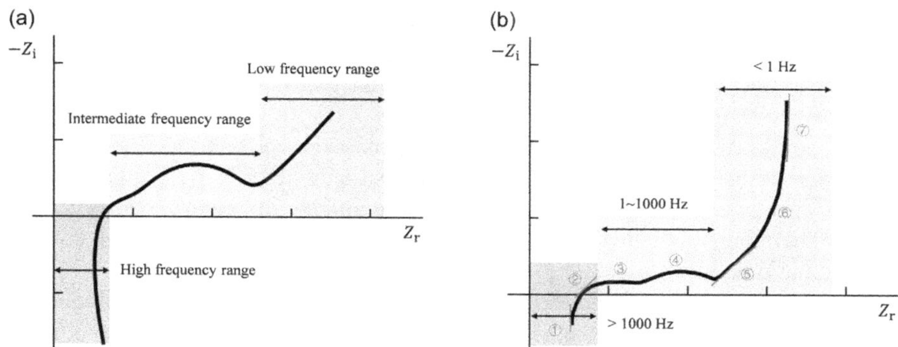

FIGURE 11.5 (a) Major EIS features for LIB in high, intermediate, and low frequency ranges; (b) major and minor features for one electrode: ① line with nearly 90° slope in the 4th quadrant due to inductance of wires or cell construction, ② line segment with 45° slope in the 1st quadrant near the intersection of the spectrum with the real axis due to the competition of ion migration and EDL charging/discharging in electrode pores, ③ first depressed semi-circle in the intermediate frequency range due to SEI or CEI or electric contact at current collector, ④ second depressed semi-circle in the intermediate frequency range due to charge transfer at the surface of the active material, ⑤ line segment with 45° slope in the low frequency range due to the diffusion of ions in the active particle, ⑥ concave curve due to the poly-dispersion of the active particles (shown in the figure) or convex curve due to ion diffusion in the electrolyte of the pores in the low frequency range, ⑦ line with nearly 90° slope in the low frequency range due to charging/discharging of the intercalation material.

11.3.2.2 Minor EIS Features of One Electrode

Using a three-electrode cell configuration, the impedance of one electrode can be isolated and subtle features may be revealed. A more detailed picture of the spectrum of a single porous electrode is depicted in Figure 11.5b.

In the high frequency range, a short line segment with a slope of 45° near the intersection of the spectrum with the real axis is usually visible resulting from the distributed nature of the porous electrode (see de Levie's model in Table 11.2). Depending on the electrode or the compression force, the first semi-circle following the 45° line segment in the intermediate frequency range can be attributed to the SEI film on the graphite,[89,90] or the contact impedance between the electrode and the Al current collector.[87,91] Cathode electrolyte interphase (CEI) can also form on the surface of cathode active material in LIB,[92,93] yet it needs to dominate over the contact impedance or its characteristic frequency needs to fall within the measured frequency range for it to be visible on the spectrum. In the low frequency range, two additional features may be present. At the lowest frequency, a line with a slope approaching 90° reflects the charging/discharging capability of the active materials. Between this 90° line and the 45° line segment following the 2nd semi-circle in the intermediate frequency range, a transition region may appear in different shapes: either a concave curve (as shown) which may be caused by the particle size distribution,[94,95] or a convex arc which may result from the diffusion of Li ion in the electrolyte in the electrode pores.[20,73]

Occasionally, an inductive loop may be observed in the low frequency range, which is attributed to the unstable nature of SEI,[96] the adsorption of the anions in the multi-stage intercalation mechanism,[97,98] or discredited as an artifact from non-stationarity in DEIS.[84,85]

It is noteworthy that the inclination of the line segment and the degree of the depression of the semi-circles described above can only be properly evaluated from visual inspection when the aspect ratio of the real and the imaginary axes is chosen to be 1. Therefore, it is a good practice to always present the Nyquist plot with the two axes having the same scale.

The frequency limits annotated for the high, intermediate, and low frequency ranges in Figure 11.5b are a rough indication of the typical values encountered in the EIS of LIB. It is intended to give the new practitioners a general idea of the frequency ranges corresponding to different processes. The exact values can vary depending on the electrode. Nevertheless, it is worth noting that such rough limits apply amazingly well for cells with more than 10 orders of magnitude difference in capacity and impedance, i.e., from a single active particle (~nAh and ~MΩ),[72] to a coin cell (~mAh and ~Ω),[99] and to a large format commercial cell (~10 Ah and ~mΩ).[100] Specific examples can be found in Table 11.3 of Section 11.4.

11.3.3 EIS Models of LIB

EIS method needs models to link the features in the spectrum to the components and processes in the cell, so as to extract kinetic and transport parameters. Of the many models proposed and utilized in the fitting and interpretation of EIS results, the majority falls under the rubric of EEC models. Using linear electric circuits as the models for the complex, nonlinear electrochemical processes is justified by the fact that EIS is a linear method in which there is a linear dependence between the perturbation and response, due to the application of small amplitude perturbation of current or potential. In other words, electrochemical processes like ion migration and diffusion through the bulk electrolyte, charging/discharging of the EDL, and charge transfer reaction across the EEI under small-amplitude perturbation have exactly the same impedance behavior as the circuit elements, e.g., resistor and capacitor, which are originally developed for the linear electrical processes. Actually, using EEC model to analyze the measured spectrum is a hallmark of EIS compared with other spectroscopic methods. EEC models are easy to handle and intuitively analogous to the electrochemical processes inside the cell. It could be one of the factors behind the popular, nearly ubiquitous usage of EIS in the study of systems involving electrochemical processes.

Given the importance of the model in the EIS study, some authors in the literature claim that the full power of the EIS method can only be realized if physics-based models, rather than EEC models, are used in the spectrum analysis.[4,101] However, the authors of this chapter would argue that such a claim reflects confusion on the term EEC models and over-confidence on the power of the physics-based models. EEC models can have solid physical background, and more sophisticated models can be useless if the test object is not well defined and the simplifying assumptions are not justified. Models of different complexity need to be selected or developed depending on the degree of uncertainty of the test object and the purpose of the study. The problem is not the use of the EEC model, but the non-judicious addition of R//C circuits, arbitrary placement of elements in the circuit, expedient substitution of the capacitor with a CPE in the EEC models etc., just to achieve a good fit. Actually, model selection or development, important as it is, is just one link in the long chains with iterative loops in the EIS adventure. The full power of the EIS method can only be realized when the nature of the method is well understood so that the long processes are followed through with adequate iteration for verification and validation. Specifically, the impedance spectra satisfying the three criteria need to be measured correctly, the model matching the test object and the test objective needs to be selected or derived, and the features identified and parameters estimated need to be reasonable in physics and corroborated from independent techniques, as expressed in Figures 11.1 and 11.2.

Instead of labeling the models as empirical or physics-based, it is more appropriate to classify the models into EEC models when the models have corresponding electric circuits, and analytic models when a closed form of analytic solution accounting detailed processes in the test object is available, yet not lending itself to an easy representation using standard circuit elements. When the complexity in geometry, composition, distribution, etc. of the electrode prevents the derivation of an analytic expression, numerical simulation of partial differential equations has to be resorted to. Such effort can be referred to as numerical modeling as here we don't have a compact circuit or expression to be associated with a model. EEC model with a number of R//C circuits in series (so-called Voigt model) can be used to check the compliance of the spectrum to the K–K relationship, to explore the DRT, and to emulate the electric response when controlling the power to charge/discharge LIB. In such applications, the EEC model is used more like an analog than a model, as no strict correspondence with the the internal components and processes are considered or intended.

11.3.3.1 EEC Models

EEC models, most frequently used in the EIS study, include lumped circuits for simple electrodes and distributed one for porous electrodes. In the following, we first introduce the major circuits for the components and processes in LIB, and then discuss ways to synthesize EEC models for complicated systems.

11.3.3.1.1 Randles Circuit for the Intercalating Materials

For the materials with intercalation reaction, Bruce et al.[102,103] were the first to propose an adion model. The prevailing model nowadays utilizes the Randles circuit in series to an R//C circuit which accounts for the impedance of SEI or CEI film on the surface of the anode or cathode active materials, respectively.

Ho et al. were the first to derive an analytic impedance model in the form of A Randles circuit for intercalating material with a shape of flat slab (Figure 11.6a).[70] To account for the charging/discharging capability and the finite diffusion length of the active materials, the classical Randles circuit, describing the mixed control of kinetic and diffusion for the metal electrode in aqueous solution, is extended with a new analytic expression for the Warburg impedance. Amazingly, with such a simple

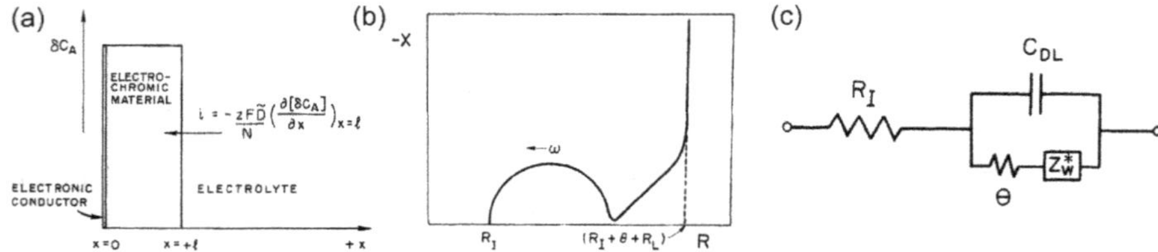

FIGURE 11.6 Ho et al.'s[70] derivation of Randles circuit model for intercalation material. (a) The coordinate system for the solution of diffusion equations for the intercalating species in a flat slab; (b) Nyquist plot of the impedance of thin intercalating slab; (c) Randles circuit for mixed kinetic and diffusion control.

geometry, the spectrum as shown in Figure 11.6b already exhibits the major features of a LIB electrode shown in Figure 11.5b. The 45° line segment in the high frequency range is absent and the semi-circle in the intermediate frequency range is not depressed because the EEI of the flat slab electrode is well defined and not distributed as that in the porous electrode. The first semi-circle in the intermediate frequency range is absent because SEI is not accounted for in the model. The well-defined 45° line segment in the low frequency range following the semi-circle is due to the diffusion inside the active material. The straight line with 90° of inclination is due to the charge storage capacity of the active material. The absence of the transition curve or arc between the 45° line segment and the 90° line is also due to the flat nature of the electrode. The analytic model of Ho can be nicely accommodated in the traditional Randles circuit model with a new expression for the Warburg impedance (Figure 11.6c).

For other shapes of the intercalation materials and other boundary conditions, the same Randles circuit is still applicable; only the Warburg impedance needs to have a different expression.[20,94]

Recently, Bandarenka's group proposed a multi-stage intercalation mechanism to account for the low-frequency inductive loop observed at a number of insertion electrodes in aqueous solution[104] and later found its good applicability also to the intercalating materials, e.g., graphite, lithium iron phosphate (LFP), and lithium titanate (LTO) in aprotic electrolyte.[97,98,105]

11.3.3.1.2 Surface Layer Model for the Protective Film

At the surface of the active materials, protective films are formed due to the reduction and oxidation of solvents, salt, and additives under low and high potentials in the anode and cathode of LIB, respectively. The films have complex compositions and multi-layered structures, resulting in complicated dependence of impedance on the SOC.[92,106] However, when the focus is on the impedance of the whole electrode, the impedance of the films can be simplified as a surface layer model consisting of a resistor representing the de-solvation of Li ion and the subsequent migration across the interphase before charge transfer, and a capacitor representing the dielectric nature of the interphase. There are two variations for the position of the resistor and the capacitor in the EEC models in the literature. The prevailing one adds the R_{SEI}//

C_{SEI} circuit in series to the Randles circuit, while purists argue that physically sound arrangement should be hierarchical in which the resistor R_{SEI} is placed in series to the Randles circuit and the capacitor C_{SEI} in parallel. However, mathematically, the two variations are hardly distinguishable in data fitting in most cases.[73] Such an insensitivity of the EIS method in fitting data to different models reminds us again the inherent indefiniteness of the method. Put differently, a more physics-based arrangement of the EEC model does not guarantee a better fitting, and a better fitting does not prove the physical soundness of the model. Extensive EIS tests substantiated with independent techniques, rather than just good fitting, are needed to gain confidence in the models and the interpretations.

11.3.3.1.3 Transmission Line Models for the Porous Electrode

To account for the distributed nature of the porous electrode, TLMs are widely used with many variations. Two parallel branches of discrete resistors represent the electronic resistance of the network of active materials and the conductive carbon particles, and the ionic resistance in the pores of the electrode, respectively. Distributed interfacial impedance cross-linking the two branches represents the interface impedance between the electrode and electrolyte. For the electrode in an electrochemical capacitor, the interfacial impedance element is just a capacitor, reflecting the charging/discharging of the EDL. Such a condition with no electrochemical reaction at the interface is termed a blocking condition. For the electrode in a fuel cell with electrochemical reactions, the interfacial impedance has an additional element of charge transfer in parallel to the capacitor. For the electrode in a LIB, the interfacial impedance is represented with a Randles circuit containing a Warburg element reflecting the diffusion of Li ion in the active materials. The condition with electrochemical reaction across the interface is termed a non-blocking condition.

With homogeneous elements of the resistors and interfacial impedance, there are analytic expressions derived for the TLM.[19] For the inhomogeneous distribution of the elements, numerical simulation can easily predict the impedance spectrum.[107]

Moskon et al.[3,108] extended the two-branch TLM into four branches to account for the interaction among the active particles and the multiple species in the electrolyte. The four-branch model is transformed to a three-branch and further simplified into a 2D network with more refined structure than the conventional TLM (Figure 11.7). The diffusion of the Li ions in the electrolyte in the electrode pores can be neatly accounted for. The impedance of this refined TLM is verified against the analytic solution from Huang,[19] giving essentially the same results.

11.3.3.1.4 Terminating Circuit for Contact Impedance

On the two sides of the TLM for the porous electrode, terminating circuits may be added to account for impedance from additional processes. Contact impedance between the positive electrode and the current collector may be significant for some batteries and can be represented as an R//C circuit.[108] Diffusion of Li ions in the separator may be added in series as another TLM. However, only under very extreme conditions can the effect of the Li ion diffusion in the separator be discernible in the impedance spectrum.[108]

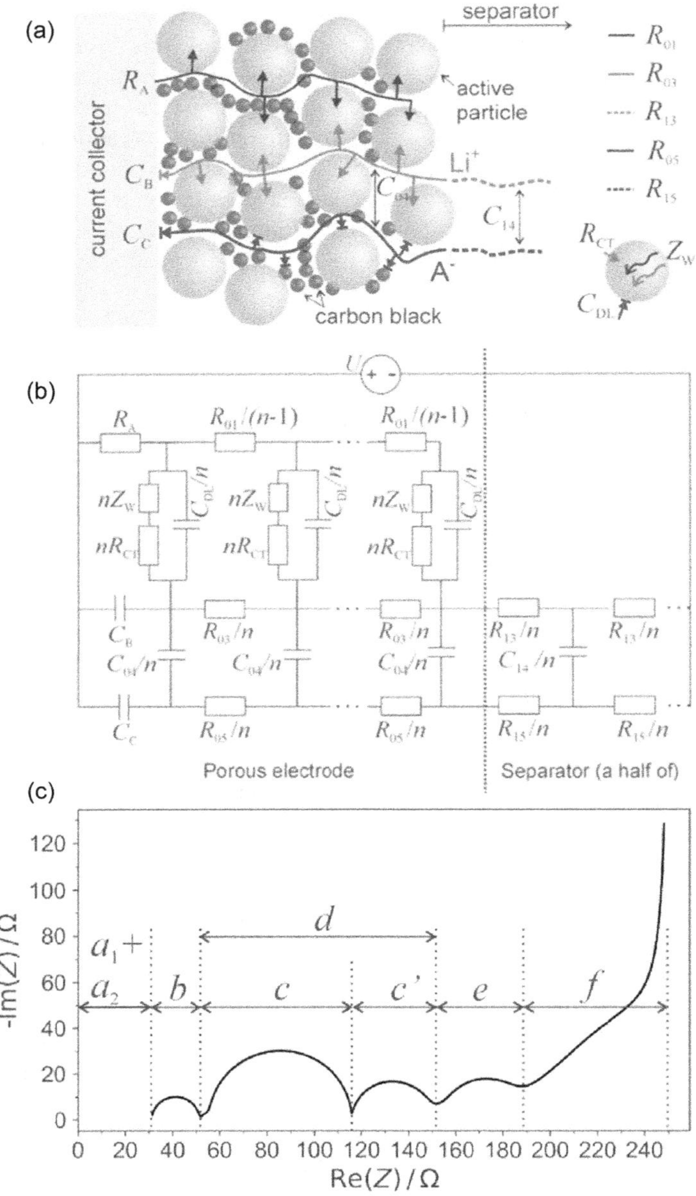

FIGURE 11.7 Refined 2D network TLM accounting for the interaction among the active particles and the multiple species in the electrolyte. (a) Schematic of the movement of charges in a porous lithium insertion electrode. In the legend the meaning of resistors in graph (b) is indicated[3]; (b) TLM for the porous insertion electrode[3]; (c) the Nyquist plot of the impedance spectrum of the porous insertion electrode. Expressions for the segments from a–f can be found in ref. [3].

11.3.3.1.5 Model Synthesis for the Whole Electrode or the Full Cell

One of the beauties of EEC models is that complex models can be easily synthesized from the simple building blocks connected in a way following the Kirchhoff's law. Specifically, the basic circuits introduced above for the components and processes in LIB, as well as that in Tables 11.1 and 11.2, can be selected and combined to formulate EEC model for the full cell based on the knowledge of the construction and composition of the electrodes and the cell. Actually, a general LIB EEC model is constructed in this way,[8,18] as shown in Figure 11.3g. In practice, such a model has to be further reduced, as some circuit units may have a negligible contribution, while other circuit elements, e.g., resistors in series, may be indistinguishable from each other. A more sophisticated EEC model for the whole cell, consisting of TLM for the LFP cathode and TLM for the graphite anode, is constructed and verified via EIS test.[109]

Another way to construct an EEC model starts from the obtained spectrum, that is, to allocate elements and circuits based on the features exposed from spectrum visualization. The features can be identified by visually comparing the Nyquist and Bode plots of the obtained spectrum with the patterns in a library like Table 11.2 or features in Figure 11.5. Such an approach is only effective for the most prominent and distinct features with characteristic frequency differing by two orders of magnitudes. Sophisticated data processing can help to reveal less distinct features. For example, a graphical method involving successively subtracting salient features helps to expose the minor ones.[36] In addition, differentiating the phase angle or the logarithm of impedance magnitude with respect to the logarithm of the frequency and plotting them against the frequency help to reveal processes that may be hard to discern in the normal Bode plot.[36] DRT technique, transforming the impedance into the time domain, demonstrates remarkable capability in identifying processes with a narrow difference in the time constants. Based on the number of time constants exposed from the DRT analysis,[16,110] a Voigt-type model with the same number of R//C circuits was constructed. Such an approach is highly automatic and easy to achieve a good fit. Yet the constructed models, or more appropriately the analog, are purely empirical in nature, and hence cannot be used to characterize the physical processes and electrochemical properties of the system.

It is recommended to synthesize models based on the cell structure and the envisioned electrochemical processes before the visual inspection of the spectrum. Whether the signature of a certain element or process manifests itself in the measured spectrum depends on the quality and resolution of the measurement, the frequency range, and whether there are overlapping features at similar characteristic frequencies. Constructing an EEC model beforehand and comparing it with the measured spectrum give the experimenter an additional chance to evaluate the design and execution of the measurement, and to facilitate the attribution of the features in the spectrum. In contrast, constructing EEC models after the visual inspection or DRT analysis of the spectrum has a higher risk of failing to account for minor or overlapping features, or having little or no physical relevance.

Some variations exist on the positioning of the Warburg impedance (Z_w), the SEI impedance, and the electrode contact impedance in the circuit. The canonical position of the Z_w is in series with the charge transfer resistance. However, it is not uncommon to see Z_w placed outside and in series to the Randles circuit. Similarly, the physically sound position of the resistor and capacitor for SEI and interfacial contact should be a hierarchical structure.[108] However, a simple R//C circuit in series to other circuit components is very popular, possibly due to its amenability to DRT analysis, as well as the insensitivity in fitting performance.

Due to the inherent non-ideality of the porous electrode, using pure capacitors in EEC models is difficult to achieve a reasonable fit. Therefore, it is a rather common practice to use a CPE in place of a capacitor. It is argued that a pure capacitor here is not necessarily physical, as it does not reflect the non-uniform, non-ideal nature of the porous electrode. On the other hand, just because a CPE is capable of obtaining a better even perfect fit, does not mean the whole EEC model is validated. It has to be noted that a CPE with the index value n different from −1, 0, 1/2, 1, is essentially an empirical element equivalent to a distributed R//C circuit, hence capable of fitting to a wide range of spectra irrespective of their physics. The use of CPE elements for complex systems may be unavoidable, but must be kept to a minimum. A good fit using CPE with an index value n significantly different from −1, 0, 1/2, 1, can be taken as a sign of inadequacy of the model. Either the model needs to be improved, or the test system needs to be simplified.

11.3.3.2 Analytic Models

To account for the discrete nature of the intercalation particles, new forms of diffusion impedance for differently shaped active particles are derived.[94] In an elegant work of Huang,[20] which stands between an original paper and a review, a unified framework for the diffusion impedance for active particles of three different shapes (slab, cylindrical, spherical), and two different boundary conditions (reflecting and absorbing), was developed.

For the impedance of porous electrodes composed of spherical intercalating particles, Meyers et al. developed an analytic expression.[111] Using the analytic impedance model for a single spherical particle and the porous electrode, the effects of different inhomogeneity involved in a porous electrode are examined by Levi and Aurbach.[73] To account for the diffusion of Li ions in the pores of the electrode, Sikha et al. derived analytic expressions for the impedance of a half-cell[112] and a full cell.[113]

Huang, through deriving new analytic expressions and developing general frameworks, made a number of significant contributions to the theory of the porous electrode impedance. In ref. [19], Huang relaxed the restrictive assumptions in de Levie's single cylindrical pore impedance model, and developed a general framework based on concentrated solution theory for the porous electrode in the electrochemical capacitor, FC, and LIB. The full model is further simplified into four limiting cases, recovering the de Levie model as the third case when the electronic conductivity is infinite, and reducing to lumped EEC model as the fourth case when the ionic conductivity is sufficiently high. In Figure 11.8, the Nyquist plots of LIB impedance for the full problem and the second limiting case where electrolyte diffusion in the pore of the electrode is negligible are shown, with four characteristic frequencies marked in the plot of the full problem and their formulas annotated. In ref. [114], Huang developed an impedance model for agglomerate (secondary particles) in LIB, taking into account the reaction on the surface of the primary particles inside the agglomerate. In ref.

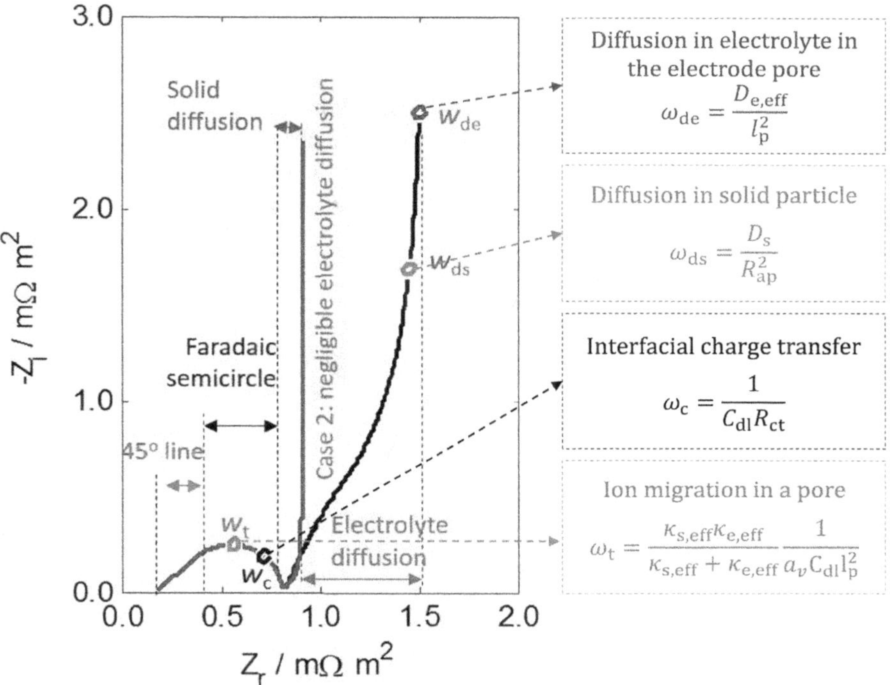

FIGURE 11.8 Nyquist plots of the impedance of a homogeneous porous electrode with insertion reactions for the full problem (black), and the second limiting case where electrolyte diffusion in the pore of the electrode is negligible (blue). Four characteristic frequencies are marked as circles in the plot of the full problem and their formulas are annotated.[19]

[115], Huang developed a three-scale analytic model, accounting for the structure of the primary particle, the agglomerate, and the porous electrode. Using the analytic models, a parametric study is conducted to find the optimal content of active materials in the agglomerate.[114,115] In ref. [2], Huang et al. gave an in-depth expounding on the framework and development of impedance models for the porous electrode, and their variants in the LIB and polymer electrolyte fuel cell (PEFC). What makes this review article special is that it adopted a "hands-on" approach by providing substantial details in derivation. Such an approach not only helps the readers understand the assumptions better, but also acquaints them with the mathematical techniques of impedance modeling that can be used in developing their own physics-based models.

Most of the analytic expressions of the impedance are rather complex, defying a simple representation as an EEC model. However, Moskon et al.[3] designed a sophisticated, physics-based 2D network TLM capable of predicting essentially the same spectrum as that from Huang's analytic model.

11.3.3.3 Numerical Modeling

The analytic expressions of impedance for porous electrodes are derived for rather simplified scenarios, such as homogeneous structure and uniform interfacial and transport parameters. For more realistic situations, numerical modeling is needed to obtain the impedance spectrum. At the simplest level, TLM with preset distributed parameters can be utilized.[107] In ref. [19], Huang proposed to account for the in-plane inhomogeneity via porosity distribution using the volume averaging method, an extension to the conventional treatment assuming pore size distribution. To account for the through-plane inhomogeneity, the transfer matrix method is proposed. To account for the 2D inhomogeneity, multi-dimensional network of effective medium is proposed. The effects of these inhomogeneities on the spectra of the electrochemical capacitor, PEFC, and LIB are compared with that of the homogeneous porous electrode.

At a more advanced level, EIS of porous electrodes with current distribution (hence other distributions such as charge transfer resistance, and concentration) in the thickness direction can be calculated using the well-developed time-domain models of the porous electrode. Time-varying boundary conditions simulating the AC perturbation can be set as input, and the impedance of the system can be calculated from the time-varying response[111] via numerical Fourier transform. Utilizing numerical EIS, Huang et al.[19] found that the charge transfer resistance during charging (numerical DEIS) is smaller than that under the steady state, in line with the experimental results.

Similarly, 2D time-domain models accounting for the current distribution along the surface of the electrodes can be utilized to calculate the impedance of the WE. A number of groups used this approach to examine the proper size and positioning of the RE.[66,116–118]

In principle, it is not a difficult task to obtain the EIS of a complex system via numerical modeling. The concern is that the coupling of too many processes and the blurring due to distribution tend to make the features in the spectrum hard to differentiate, interpret, and validate. After all, obtaining the EIS, via either measurement or simulation, is just one step in the long process of multi-stage, iterative execution of EIS method. The power of EIS hinges on a concerted and balanced efforts in every stage, a message shown in Figures 11.2 and 11.3 and repetitively stressed in this chapter.

11.3.4 Interpretation and Validation of EIS for LIB

After the K–K relation-compliant spectrum is obtained, it needs to be interpreted and validated. Here the interpretation refers to the processes of feature recognition/ attribution and parameter estimation, as well as verification using extensive EIS tests and comparison with well-established knowledge of the test object, while the validation refers to the effort of corroboration using methods beyond EIS.

11.3.4.1 Interpretation

Interpretation of EIS results includes the recognition of the features in the spectra, the attribution of the features to the components and processes inside the test object, and the estimation of the kinetic/transport parameters from the features using EEC or analytic models.

Features in the spectrum refer to the local characteristic shapes of the spectrum, i.e., lines, curves, arcs, and loops, in different frequency ranges. Presenting the spectrum in Nyquist plot with both axes in the same scale facilitates the intuitive spotting of line segments and their inclinations, semi-circles or depressed arcs from visual inspection. Features can be enhanced experimentally by using simplified, well-defined, well-controlled model systems, like half-cell, symmetric cell, and three-electrode cell. Cells using small WE, like a single particle,[72] thin film,[70,71,73,97] or low-mass loading of active materials on the current collector, also help to generate more distinct features.

Identification of the number of processes can be assisted with graphical method presenting transformed data, e.g., transformation of the impedance into time domain via DRT analysis,[119] or differentiation of the phase angle or logarithm of the impedance magnitude with respect to the logarithm of the frequency.[36] While visual inspection of the Nyquist plot can discern two processes with time constants differing by more than two orders of magnitude, DRT is able to differentiate processes with time constants differing by less than one order of magnitude. The graphical method involving successively subtracting prominent features also helps to expose hidden minor features.[36]

The attribution of the features is to assign them to the components and processes in the test object. It is achieved via comparing the features with the pattern library of well-understood systems, i.e., the Nyquist and Bode plots of the circuit elements (Table 11.1) and the typical circuits (Table 11.2), and with the signatures reported in the literature for the porous electrode of similar composition and construction (Figure 11.5). It has to be noted that the apparent absence of certain features doesn't mean their total irrelevance. It may be just because they are not dominant in the present cell construction and test conditions. On the other hand, the presence of a feature may have several possible origins and several equally well-fitting models. Unlike fingerprints, where pattern matching can definitely identify the culprit or exonerate the innocent, the features in the measured spectrum need models to be linked to the processes in the system. The high possibility of different models to fit the same measured spectrum equally well attests to the non-definitive nature of the EIS method. Therefore, the conducting of the EIS study is an inherently iterative process with multi-cycles of verification and validation. The artifacts from measurement need to be exposed via varying and optimizing the test setup and settings. The three prerequisites of being EIS need to be tested against the K–K compliance. The goodness of fitting and the adequacy of the model need to be evaluated with F-test.[120] The attribution of the features needs to be verified against physical relevance and internal consistency from extensive EIS measurements.

With the advance of technologies like big data and artificial intelligence, some steps in the EIS method, like features recognition, model generation, and spectrum fitting, may become highly automated. However, the tentative and iterative nature of EIS study will never change. The judgment of the EIS practitioner will continue to play the central role in obtaining, interpreting, and validating the EIS results.

Measuring and comparing EIS results from systematically changed cell construction or operation conditions help to expose the features and reveal their origins. For example, the nature of the first semi-circle in the intermediate frequency range, which could be attributed to the contact impedance or the SEI, can be differentiated and ascertained by changing the compression force and the contact material,[87] or the temperature.[57] Comparing the EIS of the electrode under blocking/non-blocking conditions is an effective way to separate the processes of ion migration and EDL charging/discharging, dominant in the high frequency range, from the charge transfer reaction and diffusion, dominant in the intermediate and low frequency ranges. The blocking condition in LIB can be realized by changing the electrolyte into non-intercalating one, or adjusting the SOC of the active materials to the fully charged/discharged state.[65,74]

After a feature is attributed to a particular process, the EIS model derived for that process can be used to estimate the kinetic and transport parameters. For a reliable estimation, the test object has to be well defined so that the features are well separated and adequately developed. It is found that the responses of inhomogeneous porous electrodes cannot be analyzed quantitatively without the knowledge of the particle size distribution.[94]

The magnitude of the estimated parameters and their dependence on the potential, temperature, etc. can be used to evaluate the physical consistency of the model. For example, the magnitude of double-layer capacitance on the surface of the current collector or the active particles generally falls in the range of tens of $\mu F/cm^2$, while the intercalation capacitance is much higher and can be checked against the value that can be estimated from OCV (SOC) curves. If the interfacial capacitance is much larger than tens of $\mu F/cm^2$, adsorption processes may be suspected. Of course, the effective surface area of the poly-dispersed active material particles has to be estimated properly.

The dependence of resistance on the potential (SOC) can be used as a criterion differentiating the nature of the underlying processes. It is expected that the resistance of the surface layer (the first semi-circle) has a much weaker dependence on SOC than the charge transfer resistance (the second semi-circle). In the simplest case, there is a concave dependence of charge transfer resistance against SOC if the kinetic step is governed by the Butler–Volmer equation. A deviation from such a dependence may suggest a kinetic step limited by a proceeding or ensuing chemical step.[83] Such speculation can be further verified by changing the concentration or the type of electrolyte, e.g., the aqueous or aprotic.

The magnitude of the activation energy obtained from EIS measurement at different temperatures can be used to differentiate between reaction mechanisms,[69,121] and between electronic (contact) or ionic processes (charge transfer).[57]

11.3.4.2 Validation

Validation of the EIS interpretation means the comparison with and corroboration from independent methods, including time-domain electrochemical methods and non-electrochemical methods such as microscopic characterization or MD and DFT simulation.

Diffusion coefficients can be estimated from a number of electrochemical techniques, e.g., potentiostatic intermittent titration technique (PITT), galvanostatic intermittent titration technique (GITT), EIS, as well as from MD/DFT simulations. The estimated results are often compared with each other to elucidate the strength and limitations of each technique.[6]

In developing the multi-stage intercalation mechanism, the stationarity of the EIS test is validated with quartz crystal microbalance,[105] the three time constants in the EEC model are substantiated with laser-induced current transients, the general applicability of the mechanism is tested in both aqueous and organic electrolyte, and for different active materials.[97] Even with all these corroborations, the proposed mechanism still needs further validation, as a number of concerns remain to be addressed. Specifically, the assumption of the counter ion adsorption before the intercalation/de-intercalation of Li ions entails the permeability of SEI/CEI to counter ions, which needs to be substantiated with independent microscopic characterization. In addition, a rather thick porous electrode is used in the study,[98] which is not a suitable vehicle for exploring sophisticated mechanisms. It is recommended to use an electrode with low loading of the active materials so as to suppress other possible confounding processes and make the test results more amenable to differentiating competing mechanisms. Finally, an analytic model based on the proposed multi-stage intercalation mechanism can be derived in addition to the present EEC model.

In examining the structure and properties of SEI/CEI with EIS, complementary microanalysis techniques like SEM, XRD, and EPS, for the characterization of the surface morphology and composition, are commonly employed.[93] The power of the microanalysis techniques and the EIS method needs to be combined to explore the structure, composition, and properties of such a vital and subtle component in LIB.

In characterizing the tortuosity of separators and porous electrodes, the predictive power of EIS is validated with a model system of densely packed stainless spheres with well-defined diameter.[122] The estimation with EIS is found to be more accurate than the seemingly more advanced X-ray tomography, as the binder is poorly resolved in the tomography.[123]

Kinetic and transport parameters obtained at several temperatures enable the confirmation of Arrhenius law and the estimation of the activation energy, which can be compared and corroborated with the values from other measurements or calculations. The magnitude of the activation energy also helps to reveal the nature of the underlying processes and mechanism. Abe et al.[69] corroborated the activation energy for de-solvation estimated from EIS model with the values predicted from DFT.

In EIS diagnosis of degradation, various postmortem analysis techniques are heavily used to complement each other.[5]

11.4 EXEMPLARY EIS APPLICATION ACROSS THE LIFETIME OF LIB

As a non-invasive technique, EIS is versatile in that it has been employed in nearly every phase of the whole lifetime of LIB, from the characterization of components, diagnosis of cell performance and degradation, to quality insurance in manufacturing, on-board state estimation management, and prognosis for second life. Table 11.3 selects a compact list of high-quality experimental EIS works in different phases of the study and usage of LIB. For more comprehensive discussion of the application of EIS in LIB, the readers are referred to the reviews in refs. [1,8,9].

A number of messages on the nature and features of EIS, as well as guidance on its good practices, can be drawn from comparing the entries in Table 11.3 and from evaluating the measurement, interpretation, validation, and application stages in these exemplary works.

Firstly, using readily affordable instruments and similar settings, EIS is a highly scalable technique applicable to test objects with more than 10 orders of magnitude

TABLE 11.3
Exemplary Experimental EIS Works at Different Phases of LIB Lifetime

Phases	Component/Process/Property		Test Object	Measurement Setup	EIS Mode	Frequency Range	EIS Model	Complementary Techniques	Literature
Material characterization	Separator	Tortuosity	Separators	Model cell, varying the types and concentrations of salts and solvents	PEIS, 5 mV	200–1 kHz	Ionic resistor in series with CPE	SEM, X-ray tomography, model cell with steel spheres in glass tube	122,123
	Conductive binder for cathode	Specific electrical conductivity	Casted thin film of PEDOT:PSS	Swagelok three-electrode cell with Li metal as CE and RE at 10, 20, 30, 40°C	PEIS	100 kHz–0.5 Hz	TML model accounting for electronic and ionic conductivities	CV, rheology, DSC, TGA	124
	Active material	True and apparent diffusion coefficients	Carbon coated $Li_3V_2(PO_4)_3$	Coin cell (CR2032) with Li metal as the anode	N.A.	N.A.	Warburg impedance	GITT, CV, XRD, SEM	125
	Interface	Charge transfer resistance	$LiCoO_2$, single particle (15 μm)	Two-electrode, Li foil as CE and RE (up to MΩ orders)	PEIS, 10 mV (rms)	11 kHz–11 mHz	Randles circuit in series to a intercalation capacitor	CV, potential step chronoamperometry	72
		De-solvation	Interface between ceramic and liquid electrolyte	Four-electrode glass cell, varying salts, solvents, ionic liquid, solid electrolyte, concentration, and T	PEIS, 30 mV	100 kHz–10 mHz	EEC model accounting for grain boundary resistance and interfacial Li ion transfer resistance	Reaction enthalpy from DFT	69

(*Continued*)

TABLE 11.3 (*Continued*)
Exemplary Experimental EIS Works at Different Phases of LIB Lifetime

Phases	Component/Process/ Property		Test Object	Measurement Setup	EIS Mode	Frequency Range	EIS Model	Complementary Techniques	Literature
	Interphase	SEI over Li metal	Li/LMO	Two-electrode pouch cell, three-electrode cell with Li RE, 1.5 by 1.5 cm.	PEIS, 5 mV	1 MHz–5 mHz	EEC model: Two R//C circuits for both electrodes, and one R//C circuit for the film in series with a Z_w	Charging/discharge	55
		SEI over graphite	Li/graphite	Half-cell (button cell) varying salts, solvents, additive	PEIS, 10 mV	100 kHz–10 mHz	EEC model: R//C circuit for the interphase in series to Randles circuit of the interface		106
		CEI over $LiMn_2O_4$	Li/$LiMn_2O_4$	Half-cell (button cell)	PEIS, 5 mV	100 kHz–10 mHz	EEC:R//C circuit for the interphase in series to Randles circuit of the interface	CV, CC cycling, dQ/dE vs V	92
Cell design	Porous electrode	Through-plane tortuosity	Electrodes of different active materials	Pouch bag cell, symmetric electrodes, blocking condition	PEIS, 10 mV	200 kHz–0.5 Hz	TML model for blocking conditions	SEM	122

(*Continued*)

TABLE 11.3 (*Continued*)
Exemplary Experimental EIS Works at Different Phases of LIB Lifetime

Phases	Component/Process/ Property	Test Object	Measurement Setup	EIS Mode	Frequency Range	EIS Model	Complementary Techniques	Literature
	In-plane tortuosity	Electrode	Symmetric coin cell, varying cell size, electrode thickness, electrolyte conductivity, and alignment	PEIS, 10 mV	200 kHz–5 mHz	Newly derived analytic model	SEM	126
	Thickness effects on energy and power density	$LiNiO_2$-based cathode/ graphite	Cylindrical cell, with electrode thickness ranging from 20 to 75 μm. Symmetrical cell. −10°C–60°C	PEIS, 5 mV	100 kHz–0.1 Hz	TML model for blocking/ non-blocking conditions	I–V resistance, SEM	60
	Contact resistance with current collector	$LiFePO_4$/ graphite	Three-electrode coffee bag with Li as CE and RE, varying compression, contact materials. Symmetric coin cell	PEIS	1 MHz–20 Hz, 100 kHz–1 mHz	TML model with terminating impedance		87
	NMC532/AG							

(*Continued*)

TABLE 11.3 (*Continued*)
Exemplary Experimental EIS Works at Different Phases of LIB Lifetime

Phases	Component/Process/ Property	Test Object	Measurement Setup	EIS Mode	Frequency Range	EIS Model	Complementary Techniques	Literature
		Pouch cell (210 mAh), coin cell, full and half-cells. Spectra were collected at −10°C–40°C	PEIS, 10 mV	100 kHz–10 mHz	EEC model	SEM	57	
	T dependence of four types of internal resistances	NCA/graphite	Pouch cell, symmetric cell, −30°C–60°C	PEIS, 5 mV	100 kHz–0.1 mHz	Analytic TLM model for blocking/ non-blocking conditions	SEM	59
	Low T performance	$LiNiO_2$-based cathode/ graphite	Button full and half-cells	PEIS, 10 mV	100 kHz–10 mHz	EEC model: surface layer model in series to Randles circuit	DC-pulse	99
	Aqueous electrolyte	$LiMn_2O_4$ thin-film electrode/ aqueous, organic solution	Three-electrode cell. CE: Pt mesh, RE: NaCl-saturated Ag/AgCl. RT ~ −60°C	PEIS, 5 mV (rms)	100 kHz–10 mHz	EEC model	CV, XRD, ICP-AES	71

(*Continued*)

TABLE 11.3 (*Continued*)
Exemplary Experimental EIS Works at Different Phases of LIB Lifetime

Phases	Component/Process/ Property	Test Object	Measurement Setup	EIS Mode	Frequency Range	EIS Model	Complementary Techniques	Literature
Degradation mechanism	Super-concentrated electrolyte	LMO in dilute and concentrated aqueous and carbonate LiTFSI solutions	3-electrode glass cell, with Li foil and 3 M KCl AgCl/Cl^- as RE	PEIS, 5 mV	10 kHz–10 mHz	EEC model	CV, PITT, XPS	86
	Tab position	$LiFePO_4$/ graphite	18650, coin cell with stainless steel mesh as RE. $T=25°C$	GEIS	1 MHz–10 mHz	Planar TML model with network of R, C, L	Tomography	127
	Charging protocols: CC, CCCV	NCM-NCA (42/58 wt.%)/ graphite	18650, CR2032 type full and half coin cells. $T=0$, 25°C	PEIS, 5 mV	1 MHz–1 mHz	EEC model: surface layer model in series to Randles circuit	dVdQ, SEM, neutron powder diffraction, correlation	128
	Overcharge (effect on diffusivity)	$LiFePO_4$/ graphite	18650 with RE	PEIS, 5 mV	1 MHz–10 mHz	EEC model: surface layer model in series to Randles circuit	Temperature	129
	Cycling with higher cutoff potentials (effect on diffusivity)	NMC111/ graphite	T-cells with gold wire micro RE, $T=25°C$	PEIS, 5 mV	100 kHz–0.1 mHz	EEC model accounting for finite-length spherical diffusion		63

(*Continued*)

TABLE 11.3 (*Continued*)
Exemplary Experimental EIS Works at Different Phases of LIB Lifetime

Phases	Component/Process/ Property	Test Object	Measurement Setup	EIS Mode	Frequency Range	EIS Model	Complementary Techniques	Literature
Manufacturing/assembly	Film drying rate (effect of binder gradient on tortuosity)	Graphite electrode with binder gradient (dried at 50°C, 75°C, 100°C, 125°C)	Symmetric T-cell using a non-intercalating electrolyte	PEIS, 20 mV	200 kHz–10 mHz	COMSOL, based on Newman model	Optical microscope, SEM, EDS	107
	Initial break-in (non-chemical crosstalk)	LCO/graphite	Pouch cell, 210 mAh	N.A.	100 kHz–10 mHz	EEC model	Ultrasonic	130
	Quality assurance	$LiFePO_4$-graphite cell, two batches, 1,100 cells	3 Ah 26,650, 25°C, 50%SOC	GEIS, 50 mA	10 kHz–10 mHz	Four points from Nyquist plot	CCCV, T, weight, inter-correlations among multiple parameters	131

(*Continued*)

TABLE 11.3 (*Continued*)
Exemplary Experimental EIS Works at Different Phases of LIB Lifetime

Phases	Component/Process/ Property	Test Object	Measurement Setup	EIS Mode	Frequency Range	EIS Model	Complementary Techniques	Literature
Field usage	SOC estimation	LFP/LTO	26,650, 2.1 Ah	PEIS, 5 mV	N.A.	Maximum imaginary part of the first semi-circle	Coulomb counting	132
	Internal T estimation	LFP/graphite	26,650, 2.3 Ah, $T=45°C{\sim}{-20}°C$	GEIS, 100 mA	2 kHz–1 Hz	Model accounting for T distribution derived	Surface and internal T measurement	133
	Li-plating-free preheating	NCA/graphite	Pouch cell, RE inserted. −20°C to 5°C. 18650, 2.8 Ah, −20°C to 30°C	PEIS, 5 mV	10 kHz–0.1 Hz	EEC model	Optical microscope, NMR	134
Second life	Nonlinear aging	NMC/graphite	18650, 1.95 Ah	GEIS	10 kHz–10 mHz	Two points from Nyquist plot	SEM, CCCV/CC charging	135

differences in capacity and impedance. The test object can be a single microparticle of the active material with a capacity of ~nAh and impedance of ~MΩ, or a commercial cell with a capacity of ~Ah and impedance of ~mΩ. The electrochemical processes in this wide variety of test objects can be covered nicely in the frequency range between 100 kHz and 10 mHz, with the process of the same nature falling in the same high, intermediate, and low frequency ranges, as schematically shown in Figure 11.5. In other words, EIS results are more sensitive to the nature of the processes than to the size of the test objects. In a sense, the versatility of EIS in characterizing cells of different chemistry, size, shape, etc., stems from this relative insensitivity of the spectrum to the size, i.e., the capacity or impedance, of the cells.

The reason for this insensitivity is that the characteristic frequencies of the key processes in an electrochemical cell are mostly determined by the relative importance of two competing processes, whose property values may be size-dependent individually, yet in a different and cancelling way so as to give size-insensitive time constants when combined (Table 11.2). In particular, the charge transfer resistance and EDL capacitance have an inverse and direct dependence on the surface area of the active material, respectively. Therefore, the characteristic frequency corresponding to the interfacial charge transfer in the intermediate frequency range, determined by the product of the charge transfer resistance and EDL capacitance, is independent of size in principle. In practice, however, the features in the spectrum of a small-sized cell tend to be more distinct and sharper than that of a large-sized cell due to higher degree of homogeneity.

Exploiting the insensitivity of the characteristic frequencies of the interfacial processes to the cell size and the sensitivity of EIS feature distinctness to the cell size, the EISs of cells with different capacities and impedance are employed complementarily. Specifically, commercial cells with all the components and processes being present are used to diagnose the performance and durability, while well-defined, well-controlled model cells, where fewer components, higher homogeneity and hence more salient features are present, are used to characterize the kinetic and transport parameters.

Secondly, to facilitate the separation of the multiple processes, special cell setup and measurement conditions are needed. As a rule of thumb, two orders of magnitude of difference in the characteristic frequencies are needed to differentiate the features of two electrochemical processes in the Nyquist plot. The frequency ranges listed in the Table 11.3 span from 100 kHz to 10 mHz, or 7 orders of magnitude, meaning roughly three to four processes at most with well-separated characteristic frequencies can be clearly differentiated. Therefore, special cell setups like half-cell, symmetric cell, or the three-electrode configuration, though tedious, are indispensable to separate the overlapping processes from the cathode and anode. Additional efforts like subjecting the electrode to blocking and non-blocking conditions during EIS measurement may be helpful in differentiating the feature of the ion migration in the porous electrode from that of the charge transfer and diffusion.[65,74]

Thirdly, EEC models based on the Randles circuit constitute the backbone of the models in the EIS study, and there are many variations to it. It is clear that no model can suit all the purposes and no model is born to be superior to others in every situation. No amount of physics-based modeling can rescue a poorly designed, slovenly executed measurement, in which the three criteria of EIS are violated, or a couple of processes are overlapping in a narrow frequency range. The selection of models must match with the purpose of the study and the nature and the complexity of the test objects. For monitoring and quality control, features alone on the spectra or even single frequency impedance may suffice. To extract kinetic and transport parameters, small-sized cells coupled with analytic and/or EEC models are needed. For exploring mechanisms and developing new analytic models, well-defined and well-controlled test objects, e.g., single particle, thin film, or much reduced loading of active materials, coupled with an extensive test matrix, are needed to make the unknown process dominant and distinct.

Fourthly, the EIS method, powerful as it is in separating multiple processes with different time constants, is mostly used in concert with other complementary techniques which can provide information like structure, composition, and morphology. Such information is needed to justify the selection of models, to validate the assumptions in developing new EIS models, or to corroborate the interpretation of EIS results. The need for the combined use of techniques other than EIS stems from the multi-scale, multi-physics nature of the LIB on the one hand, and the inherent indefiniteness of the EIS method itself on the other hand. Such need is particularly strong if the test object exhibits new, unaccounted features, which may result from unknown composition, unfamiliar structure, and un-identified mechanism.

Fifthly, guidance on good practices in EIS study can be learned from these exemplary works. Regarding measurement:

1. Conduct trial tests and check the impedance magnitude of the test object against the accuracy contour plot of the instrument.
2. Control the temperature tightly during the measurement, using for example thermostat or environment chamber. Cycle the temperature in both directions to check for stationarity.
3. Make multiple tests to ensure that the error bars are sufficiently small before moving to data processing.
4. Make extensive EIS measurements rather than a single shot. Vary cell construction and operation conditions systematically to reveal the trend and the origin of the features. Check the effect of rest time after adjusting the control variables. Strike a balance among certainty, accuracy, and efficiency in arranging the test sequence traversing the test matrix.

Regarding data processing:

1. Check the K–K compliance to ensure that the three criteria for the EIS are met before in-depth data analysis.
2. In presenting the spectrum in Nyquist plots, use the same scale for both axes, add annotation of the decade and characteristic frequencies, normalize the impedance with the electrode area or active material loading, and include error bars.
3. Synthesize EEC models based on the construction and expected processes in the test object beforehand rather than after the visual inspection of the

measured spectrum. Keep the use of empirical implements like CPE and Voigt circuit to a minimum if the purpose is to understand the physics inside the test object.

4. Check and report the confidence interval of the fitted parameters. Use F-test to evaluate the effectiveness of the models.

11.5 CONCLUSION

Against the ever-increasing popularity of EIS in the study of LIB, this chapter addresses the needs of the new practitioners on the proper use so that the common pitfalls can be avoided and the full potential of this in situ, scalable and versatile technique may be realized. In introducing the principle and process of EIS, it is stressed that the EIS study is a long multi-stage iterative process involving multiple cycles of spectrum acquisition, verification and validation. For easy access and comparison, the basic lumped elements and the typical EEC models with their symbols, impedance expressions, representations in Nyquist and Bode plots, and the characteristic frequencies are summarized in tables. The characterization of LIB using EIS in the sequence of measurement, feature recognition, model construction, interpretation, and validation is elaborated, further reinforcing the idea of EIS study being an iterative, concerted, and balanced efforts. Finally, a compact list of exemplary experimental EIS works applied to the major phases in LIB lifetime is provided, from which messages on the nature of EIS as well as guidance on the good practices are drawn and discussed.

ACKNOWLEDGMENT

Jianbo Zhang would like to express his thankfulness to the financial support from the National Natural Science Foundation of China (22179069) and National Key R&D Program of China (2021YFB3800400).

ABBREVIATIONS

CE	Counter electrode
CEI	Cathode electrolyte interphase
CPE	Constant-phase element
DEIS	Dynamic GEIS
DFT	Density functional theory
DRT	Distribution of relaxation time
EDL	Electrical double layer
EEC	Equivalent electric circuit
EEI	Electrode–electrolyte interface
EIS	Electrochemical impedance spectroscopy
GEIS	Galvanostatic EIS
GITT	Galvanostatic intermittent titration technique
IHP	Inner Helmholtz plane
K–K	Kramers–Kronig

LIB	Lithium-ion battery
LICT	Laser-induced current transients
LFP	Lithium iron phosphate
$LiMn_2O_4$	Lithium manganese oxide
LTO	Lithium titanate
MD	Molecular dynamics
OHP	Outer Helmholtz plane
PEFC	Polymer electrolyte fuel cell
PEIS	Potentiostatic EIS
PITT	Potentiostatic intermittent titration technique
QCM	Quartz crystal microbalance
RE	Reference electrode
SEI	Solid electrolyte interphase
S/N	Signal-to-noise
SOC	State of charge
TLM	Transmission line model
WE	Working electrode

NOTES

a Cole-Cole plot is also a common graphical method for visualizing impedance spectra in the complex plane. In a narrow sense, it is a term used for visualizing the permittivity or dielectric relaxation. When the focus of the study is capacitance, complex capacitance plots can be used to expose the relevant features and to extract the values of the properties.

b R//C: resistor and capacitor in parallel; R-C: resistor and capacitor in series

c The DRT plot shows the distribution of relaxation time. The unit of the vertical axis depends on the horizontal axis. It is Ω/s when the horizontal axis is relaxation time, and it is Ω·s when the horizontal axis is frequency.

d EEC model, EECM, and ECM are all used in literatures to refer to the equivalent electric circuit model. In this chapter, we choose to use "EEC model".

REFERENCES

[1] Meddings, Nina, Marco Heinrich, Frédéric Overney, et al. 2020. Application of electrochemical impedance spectroscopy to commercial Li-ion cells: A review. *Journal of Power Sources* 480: 228742.

[2] Huang, Jun, Yu Gao, Jin Luo, et al. 2020. Editors' choice—review—impedance response of porous electrodes: theoretical framework, physical models and applications. *Journal of the Electrochemical Society* 167 (16): 166503.

[3] Moškon, Jože, and Miran Gaberšček. 2021. Transmission line models for evaluation of impedance response of insertion battery electrodes and cells. *Journal of Power Sources Advances* 7: 100047.

[4] Gaddam, Rohit Ranganathan, Leon Katzenmeier, Xaver Lamprecht, and Aliaksandr S. Bandarenka. 2021. Review on physical impedance models in modern battery research. *Physical Chemistry Chemical Physics* 23 (23): 12926–12944.

[5] Waldmann, Thomas, Amaia Iturrondobeitia, Michael Kasper, et al. 2016. Post-mortem analysis of aged lithium-ion batteries: Disassembly methodology and physico-chemical analysis techniques. *Journal of the Electrochemical Society* 163 (10):A2149.

[6] Santos-Mendoza, I.O., J. Vázquez-Arenas, I. González, G. Ramos-Sánchez, and C.O. Castillo-Araiza. 2019. Revisiting electrochemical techniques to characterize the solid-state diffusion mechanism in lithium-ion batteries. *International Journal of Chemical Reactor Engineering* 17 (6).

[7] Raccichini, Rinaldo, Marco Amores, and Gareth Hinds. 2019. Critical review of the use of reference electrodes in Li-ion batteries: A diagnostic perspective. *Batteries* 5 (1): 12.

[8] Choi, Woosung, Heon-Cheol Shin, Ji Man Kim, Jae-Young Choi, and Won-Sub Yoon. 2020. Modeling and applications of electrochemical impedance spectroscopy (EIS) for lithium-ion batteries. *Journal of Electrochemical Science Technology* 11 (1): 1–13.

[9] Wang, Xueyuan, Xuezhe Wei, Jiangong Zhu, et al. 2021. A review of modeling, acquisition, and application of lithium-ion battery impedance for onboard battery management. *eTransportation* 7: 100093.

[10] Bard Allen, J., and R. Faulkner Larry. 2001. *Electrochemical methods: fundamentals and applications*. Wiley.

[11] Bard, Allen J., Fritz Scholz, and Gyorgy Inzelt. 2008. Electrochemical Dictionary.

[12] *Double layer (surface science)*. Available from https://en.wikipedia.org/wiki/Double_layer_(surface_science).

[13] Orazem, Mark E., and Bernard Tribollet. 2017. *Electrochemical impedance spectroscopy*, 2nd edn. John Wiley & Sons.

[14] Lasia, Andrzej. 2002. Electrochemical impedance spectroscopy and its applications. *Modern Aspects of Electrochemistry* 32: 143–248.

[15] Wang, Shangshang, Jianbo Zhang, Oumaïma Gharbi, Vincent Vivier, Ming Gao, and Mark E. Orazem. 2021. Electrochemical impedance spectroscopy. *Nature Reviews Methods Primers* 1 (1): 1–21.

[16] Qu, Huainan, Janak Kafle, Joshua Harris, et al. 2019. Application of ac impedance as diagnostic tool–Low temperature electrolyte for a Li-ion battery. *Electrochimica Acta* 322: 134755.

[17] Weiß, Alexandra, Stefan Schindler, Samuele Galbiati, Michael A. Danzer, and Roswitha Zeis. 2017. Distribution of relaxation times analysis of high-temperature PEM fuel cell impedance spectra. *Electrochimica Acta* 230: 391–398.

[18] Westerhoff, Uwe, Kerstin Kurbach, Frank Lienesch, and Michael Kurrat. 2016. Analysis of lithium-ion battery models based on electrochemical impedance spectroscopy. *Energy Technology* 4 (12): 1620–1630.

[19] Huang, Jun, and Jianbo Zhang. 2016. Theory of impedance response of porous electrodes: simplifications, inhomogeneities, non-stationarities and applications. *Journal of the Electrochemical Society* 163 (9):A1983.

[20] Huang, Jun. 2018. Diffusion impedance of electroactive materials, electrolytic solutions and porous electrodes: Warburg impedance and beyond. *Electrochimica Acta* 281: 170–188.

[21] Takahashi, Masaya, Shin-ichi Tobishima, Koji Takei, and Yoji Sakurai. 2002. Reaction behavior of $LiFePO_4$ as a cathode material for rechargeable lithium batteries. *Solid State Ionics* 148 (3–4): 283–289.

[22] Li, Haoyu, Shaohua Guo, and Haoshen Zhou. 2021. In-situ/operando characterization techniques in lithium-ion batteries and beyond. *Journal of Energy Chemistry* 59: 191–211.

[23] Adams, Stefan, and R. Prasada Rao. 2011. High power lithium ion battery materials by computational design. *Physica Status Solidi* 208 (8): 1746–1753.

[24] *Accuracy Contour Plots – Measurement and Discussion.* Available from https://www.gamry.com/application-notes/EIS/accuracy-contour-plots-measurement-and-discussion/.

[25] Song, J.Y., H.H. Lee, Y.Y. Wang, and C.C. Wan. 2002. Two-and three-electrode impedance spectroscopy of lithium-ion batteries. *Journal of Power Sources* 111 (2): 255–267.

[26] Engebretsen, Erik, Gareth Hinds, Quentin Meyer, et al. 2018. Localised electrochemical impedance measurements of a polymer electrolyte fuel cell using a reference electrode array to give cathode-specific measurements and examine membrane hydration dynamics. *Journal of Power Sources* 382: 38–44.

[27] Sone, Yoshitsugu, Per Ekdunge, and Daniel Simonsson. 1996. Proton conductivity of Nafion 117 as measured by a four-electrode AC impedance method. *Journal of the Electrochemical Society* 143 (4): 1254.

[28] Zhong, Guoan, C.-K; Koh, and Kaushik Roy. 2000. A twisted-bundle layout structure for minimizing inductive coupling noise. Paper read at IEEE/ACM International Conference on Computer Aided Design. ICCAD-2000. IEEE/ACM Digest of Technical Papers (Cat. No. 00CH37140).

[29] Giner-Sanz, Juan José, EM Ortega, and Valentín Pérez-Herranz. 2015. Total harmonic distortion based method for linearity assessment in electrochemical systems in the context of EIS. *Electrochimica Acta* 186: 598–612.

[30] Spinner, Neil. *EIS Data Accuracy and Validity 2019.* Available from https://pineresearch.com/shop/kb/theory/eis-theory/data-accuracy-validity/.

[31] Szekeres, Krisztina J., Soma Vesztergom, Maria Ujvári, and Győző G. Láng. 2021. Methods for the determination of valid impedance spectra in non-stationary electrochemical systems: concepts and techniques of practical importance. *ChemElectroChem* 8 (7): 1233–1250.

[32] Agarwal, Pankaj, Oscar D. Crisalle, Mark E. Orazem, and Luis H. Garcia-Rubio. 1995. Application of measurement models to impedance spectroscopy: II. Determination of the stochastic contribution to the error structure. *Journal of the Electrochemical Society* 142 (12): 4149.

[33] Agarwal, Pankaj, Mark E. Orazem, and Luis H. Garcia-Rubio. 1995. Application of measurement models to impedance spectroscopy: III. Evaluation of consistency with the Kramers-Kronig relations. *Journal of the Electrochemical Society* 142 (12): 4159.

[34] Watson, William, and Mark E. Orazem. *EIS: measurement model program 2020.* Available from https://ecsarxiv.org/kze9x/.

[35] Boukamp, Bernard A. 1995. A linear Kronig-Kramers transform test for immittance data validation. *Journal of the Electrochemical Society* 142 (6): 1885.

[36] Huang, Jun, Zhe Li, Bor Yann Liaw, and Jianbo Zhang. 2016. Graphical analysis of electrochemical impedance spectroscopy data in Bode and Nyquist representations. *Journal of Power Sources* 309: 82–98.

[37] Malkow, K.T. 2019. A theory of distribution functions of relaxation times for the deconvolution of immittance data. *Journal of Electroanalytical Chemistry* 838: 221–231.

[38] Paul, T., P.W. Chi, Phillip M. Wu, and M.K. Wu. 2021. Computation of distribution of relaxation times by Tikhonov regularization for Li ion batteries: usage of L-curve method. *Scientific Reports* 11 (1): 1–9.

[39] Dierickx, Sebastian, André Weber, and Ellen Ivers-Tiffée. 2020. How the distribution of relaxation times enhances complex equivalent circuit models for fuel cells. *Electrochimica Acta* 355: 136764.

[40] Boukamp, Bernard A. 2020. Distribution (function) of relaxation times, successor to complex nonlinear least squares analysis of electrochemical impedance spectroscopy? *Journal of Physics: Energy* 2 (4): 042001.

[41] Danzer, Michael A. 2019. Generalized distribution of relaxation times analysis for the characterization of impedance spectra. *Batteries* 5 (3): 53.

[42] De Levie, R. 1967. *Advances in electrochemistry and electrochemical engineering.* Interscience.
[43] Vivier, Vincent, and Mark E. Orazem. 2022. Impedance analysis of electrochemical systems. *Chemical Reviews* 122 (12): 11131–11168.
[44] Ershler, B. 1948. Discussions Faraday Soc. 1, 269 (1947). *Journal of Physical Chemistry. U. SS* 22: 683.
[45] Macdonald, J. Ross, J. Schoonman, and A.P. Lehnen. 1982. Applicability and power of complex nonlinear least squares for the analysis of impedance and admittance data. *Journal of Electroanalytical Chemistry Interfacial electrochemistry* 131: 77–95.
[46] *ZPlot® For Windows.* Available from https://www.scribner.com/software/146-zplot-and-zview-for-windows/.
[47] *ZSimpWin is a EIS Data Analysis program that does not require user-input on initial values.* Available from https://www.ameteksi.com/products/software/zsimpwin.
[48] Box, George E.P., and Norman R. Draper. 1987. *Empirical model-building and response surfaces.* John Wiley & Sons.
[49] Draper, Norman R., and Harry Smith. 1998. *Applied regression analysis.* Vol. 326. John Wiley & Sons.
[50] Piskulich, Zeke A., Oluwaseun O. Mesele, and Ward H. Thompson. 2019. Activation energies and beyond. *The Journal of Physical Chemistry A* 123 (33): 7185–7194.
[51] Zhang, Tianran, Daixin Li, Zhanliang Tao, and Jun Chen. 2013. Understanding electrode materials of rechargeable lithium batteries via DFT calculations. *Progress in Natural Science: Materials International* 23 (3): 256–272.
[52] Muenzel, Valentin, Anthony F. Hollenkamp, Anand I. Bhatt, et al. 2015. A comparative testing study of commercial 18650-format lithium-ion battery cells. *Journal of the Electrochemical Society* 162 (8):A1592.
[53] *Verification of Low-Impedance EIS Using a 1 mΩ Resistor.* Available from https://www.gamry.com/application-notes/EIS/low-impedance-eis-using-a-1-mohm-resistor/.
[54] *EIS Measurement of a Very Low Impedance Lithium Ion Battery.* Available from https://www.gamry.com/application-notes/EIS/eis-measurement-of-a-very-low-impedance-lithium-ion-battery/.
[55] Orsini, François, Mickaël Dollé, and Jean-Marie Tarascon. 2000. Impedance study of the Li/electrolyte interface upon cycling. *Solid State Ionics* 135 (1–4): 213–221.
[56] Ohashi, Toshiyuki, Ken-ichi Okazaki, Toshiharu Fukunaga, Zempachi Ogumi, and Takeshi Abe. 2020. Lithium-ion transfer at cathode-electrolyte interface in diluted electrolytes using electrochemical impedance spectroscopy. *ChemElectroChem* 7 (7): 1644–1651.
[57] Keefe, A.S., Samuel Buteau, I.G. Hill, and J.R. Dahn. 2019. Temperature dependent EIS studies separating charge transfer impedance from contact impedance in lithium-ion symmetric cells. *Journal of the Electrochemical Society* 166 (14): A3272.
[58] Nara, Hiroki, Keisuke Morita, Tokihiko Yokoshima, Daikichi Mukoyama, Toshiyuki Momma, and Tetsuya Osaka. 2016. Electrochemical impedance spectroscopy analysis with a symmetric cell for LiCoO2 cathode degradation correlated with Co dissolution. *AIMS Materials Science* 3 (2): 448–459.
[59] Ogihara, Nobuhiro, Shigehiro Kawauchi, Chikaaki Okuda, Yuichi Itou, Yoji Takeuchi, and Yoshio Ukyo. 2012. Theoretical and experimental analysis of porous electrodes for lithium-ion batteries by electrochemical impedance spectroscopy using a symmetric cell. *Journal of The Electrochemical Society* 159 (7):A1034.
[60] Ogihara, Nobuhiro, Yuichi Itou, Tsuyoshi Sasaki, and Yoji Takeuchi. 2015. Impedance spectroscopy characterization of porous electrodes under different electrode thickness using a symmetric cell for high-performance lithium-ion batteries. *The Journal of Physical Chemistry C* 119 (9): 4612–4619.

[61] Itou, Yuichi, Nobuhiro Ogihara, and Shigehiro Kawauchi. 2020. Role of conductive carbon in porous Li-ion battery electrodes revealed by electrochemical impedance spectroscopy using a symmetric cell. *The Journal of Physical Chemistry C* 124 (10): 5559–5564.
[62] Kisu, Kazuaki, Shintaro Aoyagi, Haruka Nagatomo, et al. 2018. Internal resistance mapping preparation to optimize electrode thickness and density using symmetric cell for high-performance lithium-ion batteries and capacitors. *Journal of Power Sources* 396: 207–212.
[63] Charbonneau, Valérie, Andrzej Lasia, and Gessie Brisard. 2020. Impedance studies of Li+ diffusion in nickel manganese cobalt oxide (NMC) during charge/discharge cycles. *Journal of Electroanalytical Chemistry* 875: 113944.
[64] Momma, Toshiyuki, Tokihiko Yokoshima, Hiroki Nara, Yuhei Gima, and Tetsuya Osaka. 2014. Distinction of impedance responses of Li-ion batteries for individual electrodes using symmetric cells. *Electrochimica Acta* 131: 195–201.
[65] Landesfeind, Johannes, Daniel Pritzl, and Hubert A. Gasteiger. 2017. An analysis protocol for three-electrode Li-ion battery impedance spectra: Part I. Analysis of a high-voltage positive electrode. *Journal of The Electrochemical Society* 164 (7): A1773.
[66] Hoshi, Yoshinao, Yuki Narita, Keiichiro Honda, Tomomi Ohtaki, Isao Shitanda, and Masayuki Itagaki. 2015. Optimization of reference electrode position in a three-electrode cell for impedance measurements in lithium-ion rechargeable battery by finite element method. *Journal of Power Sources* 288: 168–175.
[67] Raijmakers, L.H.J., M.J.G. Lammers, and P.H.L. Notten. 2018. A new method to compensate impedance artefacts for Li-ion batteries with integrated micro-reference electrodes. *Electrochimica Acta* 259: 517–533.
[68] Levi, Mikhael D., Vadim Dargel, Yuliya Shilina, Doron Aurbach, and Ion C. Halalay. 2014. Impedance spectra of energy-storage electrodes obtained with commercial three-electrode cells: some sources of measurement artefacts. *Electrochimica Acta* 149: 126–135.
[69] Abe, Takeshi, Fumihiro Sagane, Masahiro Ohtsuka, Yasutoshi Iriyama, and Zempachi Ogumi. 2005. Lithium-ion transfer at the interface between lithium-ion conductive ceramic electrolyte and liquid electrolyte-a key to enhancing the rate capability of lithium-ion batteries. *Journal of the Electrochemical Society* 152 (11): A2151.
[70] Ho, C., I.D. Raistrick, and R.A. Huggins. 1980. Application of A-C techniques to the study of lithium diffusion in tungsten trioxide thin films. *Journal of the Electrochemical Society* 127 (2): 343.
[71] Nakayama, N., T. Nozawa, Y. Iriyama, T. Abe, Z. Ogumi, and K. Kikuchi. 2007. Interfacial lithium-ion transfer at the LiMn2O4 thin film electrode/aqueous solution interface. *Journal of Power Sources* 174 (2): 695–700.
[72] Dokko, K., M. Mohamedi, Y. Fujita, et al. 2001. Kinetic characterization of single particles of LiCoO2 by AC impedance and potential step methods. *Journal of the Electrochemical Society* 148 (5): A422.
[73] Levi, M.D., and D. Aurbach. 2004. Impedance of a single intercalation particle and of non-homogeneous, multilayered porous composite electrodes for Li-ion batteries. *The Journal of Physical Chemistry B* 108 (31): 11693–11703.
[74] Pritzl, Daniel, Johannes Landesfeind, Sophie Solchenbach, and Hubert A. Gasteiger. 2018. An analysis protocol for three-electrode Li-ion battery impedance spectra: Part II. Analysis of a graphite anode Cycled vs. LNMO. *Journal of the Electrochemical Society* 165 (10): A2145.
[75] Solchenbach, Sophie, Daniel Pritzl, Edmund Jia Yi Kong, Johannes Landesfeind, and Hubert A. Gasteiger. 2016. A gold micro-reference electrode for impedance and potential measurements in lithium ion batteries. *Journal of the Electrochemical Society* 163 (10): A2265.

[76] Katırcı, Gökberk, Mohammed Ahmed Zabara, and Burak Ülgüt. 2022. Methods—Unexpected effects in galvanostatic EIS of Randles' cells: Initial transients and harmonics generated. *Journal of the Electrochemical Society* 169 (3): 030527.

[77] Murbach, Matthew D., Victor W. Hu, and Daniel T. Schwartz. 2018. Nonlinear electrochemical impedance spectroscopy of lithium-ion batteries: Experimental approach, analysis, and initial findings. *Journal of the Electrochemical Society* 165 (11): A2758.

[78] Harting, Nina, Nicolas Wolff, Fridolin Röder, and Ulrike Krewer. 2019. State-of-health diagnosis of lithium-ion batteries using nonlinear frequency response analysis. *Journal of the Electrochemical Society* 166 (2): A277.

[79] Erinmwingbovo, Collins, V. Siller, Marc Nuñez, et al. 2020. Dynamic impedance spectroscopy of $LiMn_2O_4$ thin films made by multi-layer pulsed laser deposition. *Electrochimica Acta* 331: 135385.

[80] Erinmwingbovo, Collins, Dominique Koster, Doriano Brogioli, and Fabio La Mantia. 2019. Dynamic impedance spectroscopy of nickel hexacyanoferrate thin films. *ChemElectroChem* 6 (21): 5387–5395.

[81] Huang, Jun, Zhe Li, and Jianbo Zhang. 2015. Dynamic electrochemical impedance spectroscopy reconstructed from continuous impedance measurement of single frequency during charging/discharging. *Journal of Power Sources* 273: 1098–1102.

[82] Dokko, Kaoru, Natsuko Nakata, Yushi Suzuki, and Kiyoshi Kanamura. 2010. High-rate lithium deintercalation from lithiated graphite single-particle electrode. *The Journal of Physical Chemistry C* 114 (18): 8646–8650.

[83] Vassiliev, Sergey Yu, Vyacheslav V. Sentyurin, Eduard E. Levin, and Victoria A. Nikitina. 2019. Diagnostics of lithium-ion intercalation rate-determining step: Distinguishing between slow desolvation and slow charge transfer. *Electrochimica Acta* 302: 316–326.

[84] Itagaki, Masayuki, Nao Kobari, Sachiko Yotsuda, Kunihiro Watanabe, Shinichi Kinoshita, and Makoto Ue. 2005. $LiCoO_2$ electrode/electrolyte interface of Li-ion rechargeable batteries investigated by in situ electrochemical impedance spectroscopy. *Journal of Power Sources* 148: 78–84.

[85] Huang, Jun, Hao Ge, Zhe Li, and Jianbo Zhang. 2015. Dynamic electrochemical impedance spectroscopy of a three-electrode lithium-ion battery during pulse charge and discharge. *Electrochimica Acta* 176: 311–320.

[86] Nikitina, Victoria A., Maxim V. Zakharkin, Sergey Yu Vassiliev, Lada V. Yashina, Evgeny V. Antipov, and Keith J. Stevenson. 2017. Lithium ion coupled electron-transfer rates in superconcentrated electrolytes: exploring the bottlenecks for fast charge-transfer rates with LiMn2O4 cathode materials. *Langmuir* 33 (37): 9378–9389.

[87] Gaberscek, Miran, Joze Moskon, Bostjan Erjavec, Robert Dominko, and Janez Jamnik. 2008. The importance of interphase contacts in Li ion electrodes: the meaning of the high-frequency impedance arc. *Electrochemical Solid-State Letters* 11 (10):A170.

[88] Erol, Salim, Mark E. Orazem, and Richard P. Muller. 2014. Influence of overcharge and over-discharge on the impedance response of $LiCoO_2$| C batteries. *Journal of Power Sources* 270: 92–100.

[89] Thomas, M.G.S.R., Peter G. Bruce, and John B. Goodenough. 1985. AC impedance analysis of polycrystalline insertion electrodes: Application to Li1− x CoO_2. *Journal of the Electrochemical Society* 132 (7): 1521.

[90] Levi, M.D., and D. Aurbach. 1997. Simultaneous measurements and modeling of the electrochemical impedance and the cyclic voltammetric characteristics of graphite electrodes doped with lithium. *The Journal of Physical Chemistry B* 101 (23): 4630–4640.

[91] Atebamba, Jean-Marcel, Joze Moskon, Stane Pejovnik, and Miran Gaberscek. 2010. On the interpretation of measured impedance spectra of insertion cathodes for lithium-ion batteries. *Journal of the Electrochemical Society* 157 (11):A1218.

[92] Zhang, Sheng Shui, K. Xu, and T.R. Jow. 2002. Understanding formation of solid electrolyte interface film on $LiMn_2O_4$ electrode. *Journal of the Electrochemical Society* 149 (12):A1521.

[93] Liao, Xing-Qun, Feng Li, Chang-Ming Zhang, Zhou-Lan Yin, Guo-Cong Liu, and Jin-Gang Yu. 2021. Improving the stability of high-voltage lithium cobalt oxide with a multifunctional electrolyte additive: Interfacial analyses. *Nanomaterials* 11 (3): 609.

[94] Song, Juhyun, and Martin Z. Bazant. 2012. Effects of nanoparticle geometry and size distribution on diffusion impedance of battery electrodes. *Journal of the Electrochemical Society* 160 (1):A15.

[95] Diard, J.-P., Bernard Le Gorrec, and Claude Montella. 2001. Influence of particle size distribution on insertion processes in composite electrodes. Potential step and EIS theory: Part I. Linear diffusion. *Journal of Electroanalytical Chemistry* 499 (1): 67–77.

[96] Gnanaraj, J.S., Robert W. Thompson, S.N. Iaconatti, J.F. DiCarlo, and K.M. Abraham. 2005. Formation and growth of surface films on graphitic anode materials for Li-ion batteries. *Electrochemical Solid-State Letters* 8 (2): A128.

[97] Scieszka, Daniel, Jeongsik Yun, and Aliaksandr S. Bandarenka. 2017. What do laser-induced transient techniques reveal for batteries? Na-and K-Intercalation from aqueous electrolytes as an example. *ACS Applied Materials Interfaces* 9 (23): 20213–20222.

[98] Dinkelacker, Franz, Philipp Marzak, Jeongsik Yun, Yunchang Liang, and Aliaksandr S. Bandarenka. 2018. Multistage mechanism of lithium intercalation into graphite anodes in the presence of the solid electrolyte interface. *ACS Applied Materials Interfaces* 10 (16): 14063–14069.

[99] Zhang, S.S., K. Xu, and T.R. Jow. 2004. Electrochemical impedance study on the low temperature of Li-ion batteries. *Electrochimica Acta* 49 (7): 1057–1061.

[100] Wünsch, Martin, Jörg Kaufman, and Dirk Uwe Sauer. 2019. Investigation of the influence of different bracing of automotive pouch cells on cyclic lifetime and impedance spectra. *Journal of Energy Storage* 21: 149–155.

[101] Macdonald, Digby D. 2009. Why electrochemical impedance spectroscopy is the ultimate tool in mechanistic analysis. *ECS Transactions* 19 (20): 55.

[102] Bruce, Peter G., and M.Y. Saidi. 1992. The mechanism of electrointercalation. *Journal of Electroanalytical Chemistry* 322 (1–2): 93–105.

[103] Bruce, Peter G., and M.Y. Saidi. 1992. A two-step model of intercalation. *Solid State Ionics* 51 (3–4): 187–190.

[104] Ventosa, Edgar, Bianca Paulitsch, Philipp Marzak, et al. 2016. The mechanism of the interfacial charge and mass transfer during intercalation of alkali metal cations. *Advanced Science* 3 (12): 1600211.

[105] Yun, J., J. Pfisterer, and A.S. Bandarenka. 2016. Sodium-ion battery. *Energy & Environmental Science* 9: 955.

[106] Zhang, Sheng S., Kang Xu, and T.R. Jow. 2006. EIS study on the formation of solid electrolyte interface in Li-ion battery. *Electrochimica Acta* 51 (8–9): 1636–1640.

[107] Morasch, Robert, Johannes Landesfeind, Bharatkumar Suthar, and Hubert A. Gasteiger. 2018. Detection of binder gradients using impedance spectroscopy and their influence on the tortuosity of Li-ion battery graphite electrodes. *Journal of the Electrochemical Society* 165 (14): A3459.

[108] Moškon, Jože, Jan Žuntar, Sara Drvarič Talian, Robert Dominko, and Miran Gaberšček. 2020. A powerful transmission line model for analysis of impedance of insertion battery cells: A case study on the NMC-Li system. *Journal of the Electrochemical Society* 167 (14): 140539.

[109] Scipioni, Roberto, Peter S. Jørgensen, Christopher Graves, Johan Hjelm, and Søren H Jensen. 2017. A physically-based equivalent circuit model for the impedance of a LiFePO4/graphite 26650 cylindrical cell. *Journal of the Electrochemical Society* 164 (9): A2017.

[110] Heins, Tom Patrick, Ruben Leithoff, Nicolas Schlüter, Uwe Schröder, and Klaus Dröder. 2020. Impedance spectroscopic investigation of the impact of erroneous cell assembly on the aging of lithium-ion batteries. *Energy Technology* 8 (2): 1900288.
[111] Meyers, Jeremy P., Marc Doyle, Robert M. Darling, and John Newman. 2000. The impedance response of a porous electrode composed of intercalation particles. *Journal of the Electrochemical Society* 147 (8): 2930.
[112] Sikha, Godfrey, and Ralph E. White. 2006. Analytical expression for the impedance response of an insertion electrode cell. *Journal of the Electrochemical Society* 154 (1): A43.
[113] Sikha, Godfrey, and Ralph E. White. 2008. Analytical expression for the impedance response for a lithium-ion cell. *Journal of the Electrochemical Society* 155 (12): A893.
[114] Huang, Jun, Hao Ge, Zhe Li, and Jianbo Zhang. 2014. An agglomerate model for the impedance of secondary particle in lithium-ion battery electrode. *Journal of the Electrochemical Society* 161 (8): E3202.
[115] Huang, Jun, Zhe Li, Jianbo Zhang, Shaoling Song, Zhongliang Lou, and Ningning Wu. 2015. An analytical three-scale impedance model for porous electrode with agglomerates in lithium-ion batteries. *Journal of the Electrochemical Society* 162 (4): A585.
[116] Simon, Fabian J., Leonard Blume, Matthias Hanauer, Ulrich Sauter, and Jürgen Janek. 2018. Development of a wire reference electrode for lithium all-solid-state batteries with polymer electrolyte: FEM simulation and experiment. *Journal of the Electrochemical Society* 165 (7): A1363.
[117] Ender, M., J. Illig, and E. Ivers-Tiffée. 2016. Three-electrode setups for lithium-ion batteries. *Journal of the Electrochemical Society* 164 (2): A71.
[118] Baker, Daniel R., Mark W. Verbrugge, and Xu Xian Hou. 2017. A simple formula describing impedance artifacts due to the size and surface resistance of a reference-electrode wire in a thin-film cell. *Journal of the Electrochemical Society* 164 (2): A407.
[119] Steinhauer, Miriam, Sebastian Risse, Norbert Wagner, and K. Andreas Friedrich. 2017. Investigation of the solid electrolyte interphase formation at graphite anodes in lithium-ion batteries with electrochemical impedance spectroscopy. *Electrochimica Acta* 228: 652–658.
[120] Zaban, Arie, Ella Zinigrad, and Doron Aurbach. 1996. Impedance spectroscopy of Li electrodes. 4. A general simple model of the Li– solution interphase in polar aprotic systems. *The Journal of Physical Chemistry* 100 (8): 3089–3101.
[121] Abe, Takeshi, Hideo Fukuda, Yasutoshi Iriyama, and Zempachi Ogumi. 2004. Solvated Li-ion transfer at interface between graphite and electrolyte. *Journal of the Electrochemical Society* 151 (8): A1120.
[122] Landesfeind, Johannes, Johannes Hattendorff, Andreas Ehrl, Wolfgang A. Wall, and Hubert A. Gasteiger. 2016. Tortuosity determination of battery electrodes and separators by impedance spectroscopy. *Journal of the Electrochemical Society* 163 (7): A1373.
[123] Landesfeind, Johannes, Martin Ebner, Askin Eldiven, Vanessa Wood, and Hubert A. Gasteiger. 2018. Tortuosity of battery electrodes: Validation of impedance-derived values and critical comparison with 3D tomography. *Journal of the Electrochemical Society* 165 (3): A469.
[124] Das, Pratik Ranjan, Lidiya Komsiyska, Oliver Osters, and Gunther Wittstock. 2015. Electrochemical stability of PEDOT: PSS as cathodic binder for Li-ion batteries. *ECS Transactions* 68 (2): 45.
[125] Rui, X.H., N. Ding, J. Liu, C. Li, and C.H. Chen. 2010. Analysis of the chemical diffusion coefficient of lithium ions in $Li_3V2\ (PO_4)_3$ cathode material. *Electrochimica Acta* 55 (7): 2384–2390.
[126] Suthar, Bharatkumar, Johannes Landesfeind, Askin Eldiven, and Hubert A. Gasteiger. 2018. Method to determine the in-plane tortuosity of porous electrodes. *Journal of the Electrochemical Society* 165 (10): A2008.

[127] Schindler, Stefan, and Michael A. Danzer. 2017. Influence of cell design on impedance characteristics of cylindrical lithium-ion cells: A model-based assessment from electrode to cell level. *Journal of Energy Storage* 12: 157–166.

[128] Zhu, Jiangong, Mariyam Susana Dewi Darma, Michael Knapp, et al. 2020. Investigation of lithium-ion battery degradation mechanisms by combining differential voltage analysis and alternating current impedance. *Journal of Power Sources* 448: 227575.

[129] Liu, Yadong, and Jian Xie. 2015. Failure study of commercial $LiFePO_4$ cells in overcharge conditions using electrochemical impedance spectroscopy. *Journal of the Electrochemical Society* 162 (10): A2208.

[130] Knehr, Kevin W., Thomas Hodson, Clement Bommier, Greg Davies, Andrew Kim, and Daniel A. Steingart. 2018. Understanding full-cell evolution and non-chemical electrode crosstalk of Li-ion batteries. *Joule* 2 (6): 1146–1159.

[131] Rumpf, Katharina, Maik Naumann, and Andreas Jossen. 2017. Experimental investigation of parametric cell-to-cell variation and correlation based on 1100 commercial lithium-ion cells. *Journal of Energy Storage* 14: 224–243.

[132] La Rue, Aleksei, Peter J. Weddle, Miaomiao Ma, Christopher Hendricks, Robert J. Kee, and Tyrone L. Vincent. 2019. State-of-charge estimation of $LiFePO_4$–Li_4Ti5O_{12} batteries using history-dependent complex-impedance. *Journal of the Electrochemical Society* 166 (16): A4041.

[133] Richardson, Robert R., Peter T. Ireland, and David A. Howey. 2014. Battery internal temperature estimation by combined impedance and surface temperature measurement. *Journal of Power Sources* 265: 254–261.

[134] Ge, Hao, Jun Huang, Jianbo Zhang, and Zhe Li. 2015. Temperature-adaptive alternating current preheating of lithium-ion batteries with lithium deposition prevention. *Journal of the Electrochemical Society* 163 (2): A290.

[135] Schuster, Simon F., Tobias Bach, Elena Fleder, et al. 2015. Nonlinear aging characteristics of lithium-ion cells under different operational conditions. *Journal of Energy Storage* 1: 44–53.

12 Synchrotron X-Ray and Neutron Techniques

Pengfei Yu
Shanghai Institute of Microsystem and Information Technology, Chinese Academy of Sciences

Xiaosong Liu
University of Science and Technology of China

CONTENTS

DOI: 10.1201/9781003299295-12

12.1 BASIC PRINCIPLE OF ADVANCED X-RAY AND NEUTRON TECHNIQUES

To obtain a deep understanding of the materials applied in the fields of physics, chemistry, biology and materials science, powerful measurement techniques are urgently needed for mechanistic studies at multiple time and length scales. Among the various experimental methods, synchrotron X-ray and neutron techniques have stood out as key tools because of their versatile probing from the viewpoints of crystal, local and electronic structures, where the two kinds of techniques are highly complementary. The versatile capabilities of both techniques intrinsically arise from abundant interactions of X-rays and neutrons with matter. In this part, the interactions of X-rays and neutrons are initially introduced in brief, which could be well correlated to the development of various experimental methods. Then, the advantages and disadvantages of the two kinds of techniques are compared.

We must say that the interactions between matter and radiation (X-rays and neutrons) have been fundamental topics with a long history, which already have tremendous implications for many fields of physics. Therefore, this chapter does not give a complete overview but only focuses on the interactions related to the primary experimental methods involved in the studies of rechargeable batteries. More exhaustive discussions can be found in dedicated books on X-ray and neutron techniques.[1–5]

12.1.1 Interactions of X-Rays with Matter

12.1.1.1 What Is an X-Ray?

X-rays are electromagnetic waves with wavelengths (Å to nm) located between UV radiation and gamma rays, which were first discovered in 1895 by Wilhelm Conrad Röentgen at Würzburg University in Germany. From a quantum mechanical perspective, a monochromatic beam can be quantized into photons, the energy of which, given by $\hbar\omega$, can be related to wavelength through

$$\varepsilon = \hbar\omega = \frac{hc}{\lambda}, \tag{12.1}$$

where ε is the energy of a single photon, f is the frequency and c is the speed of light. On account of the unit system of X-rays, Planck's constant $h = 4.14 \times 10^{-15}$ eV·s and thus $hc = 1,240$ eV·nm. If the wavelength boundaries with UV radiation and gamma rays are 10 nm and 0.1 Å, respectively, the energy range of X-rays would lie from 124 eV to 124 keV. However, the boundary between UV radiation and X-rays is not clearly defined and has been set at an energy of 50 or 100 eV with a wavelength of 24.8 or 12.4 nm in previous reports. This ambiguity also exists in regard to the boundary between X-rays and gamma rays. Actually, X-rays and gamma rays are not defined by their energy range but by their source: X-rays are produced by tuning electrons, whereas gamma rays result from atomic nuclear reactions. Therefore, electromagnetic radiation of 150 keV or even higher energy is still an X-ray once the excitation process is related to the electrons. More specifically, X-rays can be classified into soft X-rays and hard X-rays in terms of energy range because of their relationship with the dominant interactions with matter, as discussed below. Soft X-rays cover photon energy starting from the upper energy limit of UV radiation to 3,000 eV, corresponding to a wavelength of ~0.4 nm, above which is the energy range of hard X-rays. However, there is also a lack of a clear definition for the classifications. The intermediate region from 2,000 to ~7,000 eV has been named the "tender X-ray" owing to its exclusive requirement of the hardware configuration. Nevertheless, the X-rays discussed in this chapter are clearly limited to the full range of soft X-rays and partial range of hard X-rays of no more than ~30 keV with negligible interaction of Compton scattering.

X-rays can be produced by two processes, bremsstrahlung radiation and characteristic X-ray emission. The former results from the sudden stopping, breaking or slowing of high-speed electrons under a strong electromagnetic field, where the X-rays have a continuous energy distribution. The latter originates from a fluorescent process with electrons of the outer shell decaying into the hole of the inner shell induced by the incident electrons, where the energy of X-rays is discrete and characteristic. The two processes can exist simultaneously in X-ray tubes with certain metal targets, such as Fe, Cu and W, which are the most common devices used to produce X-rays. However, both processes are inefficient in X-ray tubes, where only ~1% of the electrical energy can be converted into X-rays, while most of the energy is released as waste heat. To remove the excess heat, X-ray tubes must be specially designed. To date, the engineering strategies of heat dissipation have become the bottleneck to achieve X-ray tubes with high power and thus limit the further improvement of X-ray intensity. To produce X-rays with high qualities, a specialized source named synchrotron radiation has been designed, which has been widely used in recent decades. The production of X-rays in this facility basically relies on the bremsstrahlung process of electrons at a speed close to that of light, the properties of which are then tuned along beamlines constructed by a set of optical devices.

12.1.1.2 General Description of the Interactions

When a beam of X-rays passes through matter, the intensity is reduced. This results from the intrinsic interactions of the photons with the components of matter because of the electromagnetic nature of X-rays. The primary interactions are the scattering or absorption of X-rays when they interact with electrons. These interactions can be classified by the evolution of X-ray photons in terms of their energy and momentum. If photons change their propagation direction without a change in energy, that is, momentum changes with energy retained, the interaction process is called **elastic scattering**. In the first-order approximation, which describes the scattering of photons off the unbound electrons, elastic scattering can be simplified as **Thomson scattering**. When the scattering of photons is closely related to the bound states of electrons described by the second-order approximation, the scattering is called **resonant elastic scattering**. If photons totally transfer the energy to the electrons, the interaction is called the **photoelectric effect**. Generally, the bound electrons are likely to be kicked out of the shells and become free if the photon energy is much larger than their binding energy. However, when the photon energy equals the energy of electronic transitions from occupied shells to unoccupied shells, **resonant absorption** within the atoms occurs. With the formation of holes in the deep shell in the photoelectric process, electrons in the shallow shell backfill the holes accompanied by the release of characteristic energy. The energy can be released in a form of light named **fluorescence**. Otherwise, the energy excites electrons of shallow shells out of atoms, which are named **Auger electrons**. With part of the photoelectrons and Auger electrons flying out of the matter directly, the other part of the electrons can eject other electrons of the outer shells out of atoms with partial energy transferred, where the excited electrons are called **secondary electrons** or thermal electrons. This process occurs as a series of interactions until the kinetic energy of electrons becomes too low to create further excitation. The process has been thought to be dominant in the ionization of atoms, which is also the major origin of beam damage. If photons collide with electrons with partial energy transferred, where the scattered photons change both energy and momentum, the interaction is called **inelastic scattering or Compton scattering.** Such a process becomes dominant when the photon energy is high enough (usually tens or hundreds of keV), where the electrons can be treated as free electrons. Historically, this was of considerable importance, as it evidenced the particle property of X-rays in addition to volatility. With a further increase in photon energy, interactions with the electric field of electrons and the nuclear field occur, which are named **pair production** and **triplet production**, respectively. The interaction of X-rays with atomic nuclei, called **photodisintegration**, can even occur once the photon energy reaches a very high value. In addition, there is still a possibility that some photons transmit through matter without any changes in energy and momentum, viz., no interactions. For a polyenergetic beam of X-rays produced by synchrotron sources, all of these interactions might take place simultaneously. However, which interaction process will dominate depends on the energy of photons and the type of elements in matter. Because of the elements contained in battery materials (3d and 4d transition metals [TMs] and typical ligands, including C, N, O, F, Si, P and S), the X-ray techniques used to investigate the scientific issues of rechargeable batteries are mainly based on elastic scattering and photoelectric effects, although a few techniques related to Compton scattering have been applied to bulk investigations. Therefore, elastic scattering and photoelectric effects will be given more qualitative and quantitative descriptions.

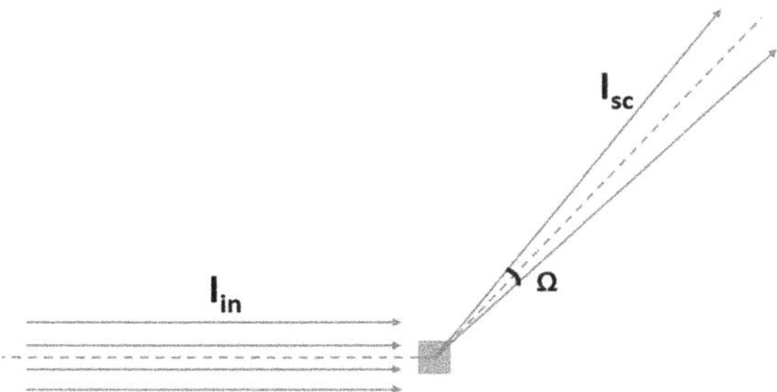

FIGURE 12.1 Schematic layout of a generic scattering experiment used to define the differential cross-section ($d\sigma_s/d\Omega$).

12.1.1.2.1 Elastic Scattering

To quantify the scattering of X-rays with matter, the cross-section σ is defined as I_{sc}/I_{in} according to an experiment (Figure 12.1), where I_{sc} and I_{in} are the intensities of the scattered and incident X-rays, respectively. This quantity is very important since it is the joint point between experiment and theory. Since the scattered light is usually detected in a limited solid angle ΔΩ, the differential scattering cross-section, $d\sigma/d\Omega$, is more commonly used, which is defined as

$$\frac{d\sigma_s}{d\Omega} = \frac{I_{sc}}{\Delta\Omega I_{in}} \tag{12.2}$$

The intensity I_{in} is the number of photons passing through a unit area per second. The number of scattered photons recorded per second in a detector is defined as the intensity I_{sc}, where the detector is positioned at a distance R away from the object and subtends a solid angle ΔΩ.

Assuming the electric field of incident light is E_{in}, the density of photons is proportional to $|E_{in}|^2/\hbar\omega$ since the energy density is proportional to $|E_{in}|^2$. The number of photons I_{in} can be related to $|E_{in}|^2/\hbar\omega c$ by multiplying the speed of light c. Similarly, the number of photons I_{sc} in solid angle ΔΩ is proportional to $R^2\Delta\Omega|E_{sc}|^2/\hbar\omega c$, where E_{sc} stands for the electric field of scattered light. Therefore, the differential cross-section is given by

$$\frac{d\sigma_s}{d\Omega} = \frac{|E_{sc}|^2 R^2}{|E_{in}|^2}. \tag{12.3}$$

Since the scattering of X-rays is intrinsically related to electrons in matter, the differential cross-section of electrons is assessed by the vibration picture of electrons under the electric field of incident light.

The relationship between E_{sc} and E_{in} can be derived by

$$\frac{E_{sc}(R, t)}{E_{in}} = -r_0 \frac{e^{ikR}}{R} |\hat{\varepsilon} \cdot \hat{\varepsilon}'|, \tag{12.4}$$

where E_{in} can be expressed as $E_0 e^{ikR}$ for the incident wave. r_0 is referred to as the Thomson scattering length or the classic radius of an electron, which is defined as

$$r_0 = \frac{e^2}{4\pi\varepsilon_0 mc^2} = 2.82\times10^{-5}\,\text{Å}, \tag{12.5}$$

where ε_0 is the dielectric constant in vacuum, m_e is the mass of electrons and c is the speed of light. $-|\hat{\varepsilon}\cdot\hat{\varepsilon}'|$ reveals the direction relationship of the incident light and scattered light, where $\hat{\varepsilon}$ is the basic vector of the incident field and $\hat{\varepsilon}'$ is that of the radiated field. The factor −1 indicates the existence of a 180° phase shift between the incident and scattered waves.

Accordingly, the differential cross-section of the electron is given by

$$\frac{d\sigma_s}{d\Omega} = r_0^2|\hat{\varepsilon}\cdot\hat{\varepsilon}'|^2, \tag{12.6}$$

which describes the Thomson scattering process of an electromagnetic wave by a free electron.

The factor $|\hat{\varepsilon}\cdot\hat{\varepsilon}'|^2$ is defined as the polarization factor of scattering, P, which depends on the polarization characteristics of the X-ray sources and scattering geometry (Figure 12.2).

$$P = |\hat{\varepsilon}\cdot\hat{\varepsilon}'|^2 = \begin{cases} 1 & \text{Linearly polarization w / vertical scattering} \\ \cos^2\psi & \text{Linearly polarization w / horizontal scattering} \\ \frac{1}{2}\left(1+\cos^2\psi\right) & \text{Unpolarization} \end{cases} \tag{12.7}$$

The total cross-section for Thomson scattering can be obtained by integrating the differential cross-section over the entire spherical surface. Thus, the total cross-section σ_T is equal to $4\pi r_0^2 \times 2/3 = 0.665 \times 10^{-24}\text{cm}^2 = 0.665$ barn. Both the differential and total classic cross-section for the scattering of X-rays by a free electron are constant and independent of energy. This result is particularly accurate for hard X-rays, where the photon energy is high enough that even atomic electrons can be treated as free electrons in a good approximation. However, this approach breaks down in the range of soft X-rays, where the photon energy passes a threshold for the resonant excitation of electrons from deeply bound atomic states. Such a process has been mentioned above as resonant scattering.

The elastic scattering of X-rays by atoms is the superposition of scatterings of electrons from different positions. The superposition is closely related to the phase difference $\Delta\Phi(r)$ from different positions, where a typical schematic is shown in Figure 12.3.

$$\Delta\Phi(r) = (k-k')\cdot r = Q\cdot r, \tag{12.8}$$

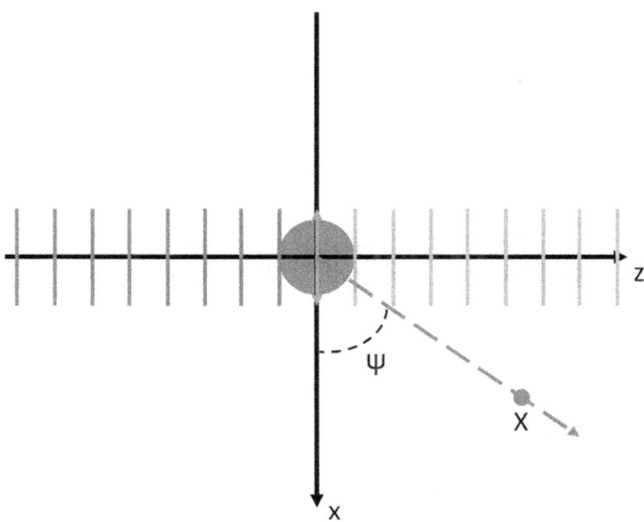

FIGURE 12.2 Schematic of an X-ray scattered by an electron. The incident wave with the polarized electric field along x propagates along the z-axis. The outgoing wave is detected at point X within the x–z plane, where the direction vector deviates from the x-axis at the angle of Ψ.

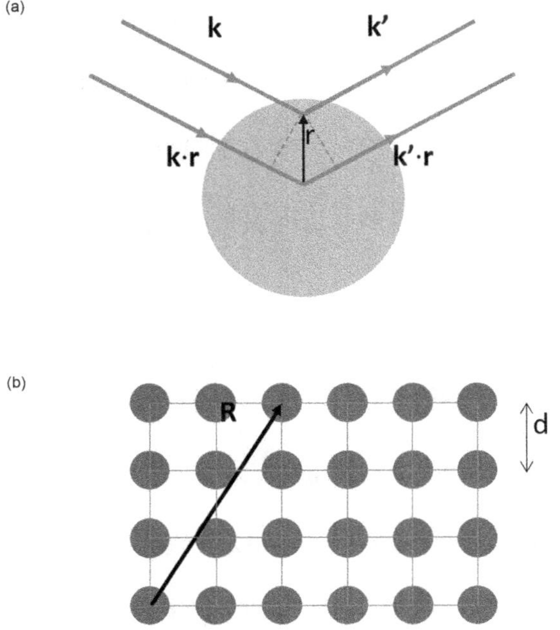

FIGURE 12.3 Schematic of scattering from (a) an atom and (b) a crystal.

where $\boldsymbol{k}$ and $\boldsymbol{k}'$ are the wave vectors of the incident and scattered X-rays, respectively. $\boldsymbol{Q}=\boldsymbol{k}-\boldsymbol{k}'$, is defined as the wave vector transfer or scattering vector. With $|\boldsymbol{k}|=|\boldsymbol{k}'|$, $|\boldsymbol{Q}|=2|\boldsymbol{k}|\sin\theta=4\pi/\lambda\cdot\sin\theta$.

Therefore, the total scattering length of an atom is

$$-r_0\int e^{i\boldsymbol{Q}\cdot\boldsymbol{r}}\rho(\boldsymbol{r})d\boldsymbol{r}=-r_0 f^0(\boldsymbol{Q}), \tag{12.9}$$

where $\rho(\boldsymbol{r})d\boldsymbol{r}$ is the number of electrons at $\boldsymbol{r}$. $f^0(\boldsymbol{Q})$ is known as the atomic form factor. In a long wavelength limit with Q approaching 0, the scattered waves by electrons at different parts in an atom are not coherent. Therefore, $f^0(\boldsymbol{Q})=Z$, which is the number of electrons in an atom.

The above assessment is suitable for the Thomson scattering process, where the photon energy is much larger than the binding energy. However, if the photon energy is much less than the binding energy, the scattering length of an atom should be reduced. In this case, the atomic form factor can be corrected with the term $f'(\hbar\omega)$. Moreover, when energies approach the absorption edge, the electrons respond with a phase delay. Therefore, the atomic form factor should be further corrected with the term $f''(\hbar\omega)$. The final form of the atomic form factor is

$$f(\boldsymbol{Q},\hbar\omega)=f^0(\boldsymbol{Q})+f'(\hbar\omega)+if''(\hbar\omega), \tag{12.10}$$

where f' and f'' are dispersion corrections that are strongly dependent on the photon energy. These corrections are related to the resonant scattering. Soft X-ray resonant scattering has been widely used to study the structure of complex polymer materials. However, resonant scattering is not elaborated in this chapter because of its nascent utilization in rechargeable batteries to date.

A real Thomson scattering experiment of solid materials usually contains more superpositions since materials have lattice arrangements and atom distributions in the unit cell. Similar to the atomic form factor, a crystal has a form factor expressed as

$$F^{\text{crystal}}(\boldsymbol{Q})=\sum_j f_j(\boldsymbol{Q})e^{i\boldsymbol{Q}\cdot\boldsymbol{r}_j}\sum_n e^{i\boldsymbol{Q}\cdot\boldsymbol{R}_n} \tag{12.11}$$

The first term is known as the structure factor of the unit cell, while the second term is called a sum over lattice sites. X-rays scattered by these units could be coherent because of the comparable wavelength of the hard X-rays with the separation distance of atoms. Coherent scattering only occurs when the scattering vector happens to meet

$$\boldsymbol{Q}\cdot\boldsymbol{R}_n=2\pi\times n \tag{12.12}$$

where $\boldsymbol{R}_n$ is the lattice vector and n is a positive integer.

$$\boldsymbol{R}_n=n_1\boldsymbol{a}_1+n_2\boldsymbol{a}_2+n_3\boldsymbol{a}_3 \tag{12.13}$$

where $\boldsymbol{a}_1$, $\boldsymbol{a}_2$ and $\boldsymbol{a}_3$ are the basic vectors of the crystal in three dimensions.

Therefore, the scattering from crystals forms a series of discrete sparkling points, as revealed on two-dimensional (2D) detectors. The intensity at each point is modulated by the absolute square of the unit cell structure factor. Therefore, the intensities of the scattering peaks can be used to deduce the positions of the atoms in the unit cell. This is the basic principle of **X-ray diffraction (XRD)**, which has been in turn applied to identify the crystal structure. An exhaustive description can be found in the dedicated chapter of this book. With the same principle, Thomson scattering can also be used to identify larger periodical structures, where the technique is named **small-angle X-ray scattering (SAXS)**. The only difference from XRD is that SAXS is usually performed with a scattering signal collected in an angle range below 10°. Therefore, XRD can also be called **wide-angle X-ray scattering (WAXS)** because of the signal collection at high angles. Moreover, the elastic coherent scattering signal from low angles up to very high angles, viz., in a large range of Q values, can be collected to reveal the atomic **pair distribution function (PDF)**, which is an especially powerful probe for the local structure of materials with poor crystallinity. For other applications, coherent scattering can also be a strategy used to form **computerized tomography (CT)**, where the phase information can be used to characterize the relative position of different units. Because of the wide applications of coherent characteristics, Thomson scattering is also named coherent scattering.

Apart from the interactions with bulk matter, the interactions of X-rays with interfaces need to be specifically described because of their exclusive expression and utilization. When X-rays hit the surface of matter from a vacuum, a phenomenon called **refraction** occurs. The refractive phenomenon can be described by Snell's law, as shown in Figure 12.4a, which is the same as all the other kinds of electromagnetic waves. Therefore, the refraction index $n_r = \cos\alpha / \cos\alpha'$, which is defined as 1 in vacuum. For X-rays, the refraction index can also be expressed as

$$n_r = 1 - \delta + i\beta, \tag{12.14}$$

Where the real part δ has a relationship with $f^0(\boldsymbol{Q})$ in Thomson scattering and is of order 10^{-5} in solids.

$$\delta = \frac{\lambda^2 \rho r_0}{2\pi} f^0(\boldsymbol{Q}), \tag{12.15}$$

where λ is the wavelength of X-rays and ρ is the electron density.

The imaginary part β is much smaller than δ and can be related to the absorption form factor $f''(\hbar\omega)$.

$$\beta = \frac{\lambda^2 \rho r_0}{2\pi} f''(\hbar\omega) \tag{12.16}$$

Therefore, n is less than 1.

A refraction index less than 1 implies that total external **reflection** can occur for X-rays below a certain incident grazing angle called the critical angle, α_c

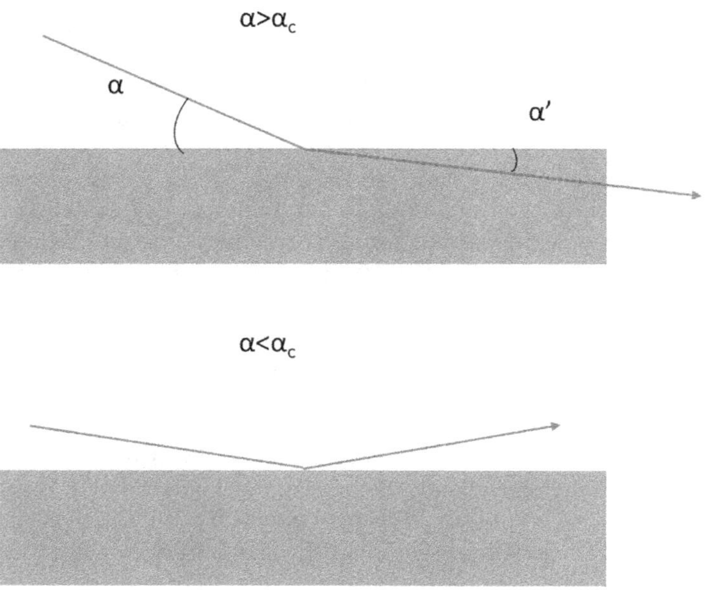

FIGURE 12.4 Refraction of X-rays at a glancing angle larger than the critical angle α_c (at the top). Total external reflection occurring at a glancing angle smaller than the critical angle α_c (at the bottom).

(Figure 12.4b). Using Snell's law with $\alpha=\alpha_c$ and $\alpha'=0$ as well as $n=1-\delta$ while ignoring the β term for simplicity, the critical angle can be related to δ by

$$\alpha_c = \sqrt{2\delta}. \tag{12.17}$$

Therefore, α_c can be evaluated to be of order 10^{-3} radians. The first important application of total external reflection is to form a curved surface that enables focusing optics. The second wide application is based on the increase in surface sensitivity when X-rays enter at an angle of **grazing incidence (GI)** less than α_c. A so-called evanescent wave within the refracting medium propagates parallel to the flat interface with a typical penetration depth of only a few nanometres. This phenomenon enables X-rays to become a valuable tool to study surfaces and interfaces with a GI coherent scattering geometry, such as **GIXRD, GISAXS and X-ray reflectivity (XRR)**.

12.1.1.2.2 Photoelectric Absorption

An incident X-ray photon interacts with an atom and ejects one of the bound electrons from certain shells. The photon transfers all of its energy to the electron. The electron can overcome the binding energy and fly out of the atom with a certain kinetic energy ε_k, leading to the ionization of the atom.

$$\varepsilon_k = \hbar\omega - \varepsilon_b \tag{12.18}$$

where $\hbar\omega$ is the incident photon energy and ε_b is the binding energy of the electron. This process, named **photoelectric absorption**, can only be enabled when $\hbar\omega$ is larger than ε_b, which is the principle of **X-ray photoemission spectroscopy (XPS)**.

The photoelectric interaction can be quantified by the linear absorption coefficient μ. According to Figure 12.5, μ is defined by

$$-\mu dx = \frac{dI}{I(x)}. \tag{12.19}$$

If the photon intensity at $x=0$ is set as I_0, then we have

$$I(x) = I_0 e^{-\mu x}. \tag{12.20}$$

The absorption can also be expressed according to the interaction of quantized photons with electrons at the microscale.

$$dN = N_p \rho_e dx \sigma_a, \tag{12.21}$$

where N_p is the number of photons, which corresponds to the X-ray intensity. ρ_e is the linear density of electrons. σ_a is the absorption cross-section. dN_p is the number of photons absorbed at the distance x, which corresponds to dI. The equation indicates that the absorption number is proportional to the total number of photons, number of electrons $\rho_e dx$ and absorption cross-section σ_a.

Therefore,

$$I(x)\rho_e dx \sigma_a = I(x)\mu dx \tag{12.22}$$

Then, the relationship of the absorption coefficient with σ_a can be derived by

$$\mu = \rho_e \sigma_a = \left(\frac{\rho_m N_A}{M}\right)\sigma_a, \tag{12.23}$$

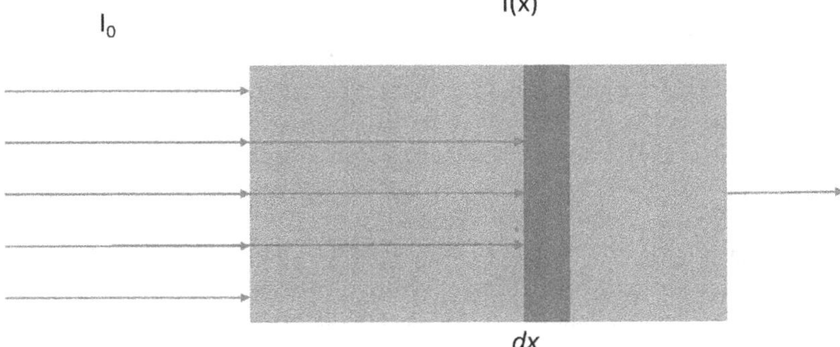

FIGURE 12.5 The penetration of an X-ray beam through a sample accompanied with absorption.

where N_A, ρ_m and M are Avogadro's number, the mass density and molar mass, respectively.

For composite materials with several kinds of elements, the expression can be extended by

$$\mu_{\text{total}} = \sum_j \rho_{e,j}\sigma_{a,j}. \tag{12.24}$$

In practice, a more useful quantitative indicator of photoelectric absorption is not the absorption cross-section or linear absorption coefficient but the mass absorption coefficient, μ/ρ_m. Since the term is constant for each element, the μ/ρ_m of a composite material can be readily calculated by

$$\left(\frac{\mu}{\rho_m}\right)_{\text{total}} = \sum_j w_j\left(\frac{\mu}{\rho_m}\right)_j, \tag{12.25}$$

where w is the weight ratio of the component.

It should be noted that even though photoelectric absorption is dominant in a specific energy range, other processes, including Thomson and Compton scattering, can also contribute to the attenuation of X-rays into matter. Therefore, a more accurate absorption coefficient and cross-section need to be corrected on account of other scattering interactions.

The absorption cross-section can be further related to the imaginary part of the scattering form factor in a classic picture of an electromagnetic wave by

$$f''(\hbar\omega) = -\left(\frac{k}{4\pi r_0}\right)\sigma_a. \tag{12.26}$$

The absorption cross-section is found to be dependent on the photon energy accordingly.

A more specific description of σ_a can be obtained through quantum mechanical calculation, which is complex and beyond the discussion in this chapter. Nevertheless, σ_a can be revealed to be proportional to the fourth order of the atomic number, Z^4, and inversely proportional to the third order of the photon energy, ε^3. The dependencies can be identified by the experimentally determined values, as shown in Figure 12.6, where four elements are selected. Since the cross-sections of the four elements are normalized by ε^3 and Z^4, they approximately appear as a single curve with a value of 0.0215 barn (1 barn $= 10^{-24}$cm^2). Below certain characteristic energies, the scaled cross-sections drop to another value, approximately 0.0022 barn. The discontinuous jumps of the absorption cross-section are called absorption edges, which are element-specific. The characteristics arise from the electron transition from the inner core shell to the outer shells when the photon energy is slightly larger than the threshold values, at approximately the binding energy of certain shells. The probability of absorption increases greatly, leading to a significant increase in the absorption cross-section, viz., the jump. Such a process is named resonant absorption, which is the basis of **X-ray**

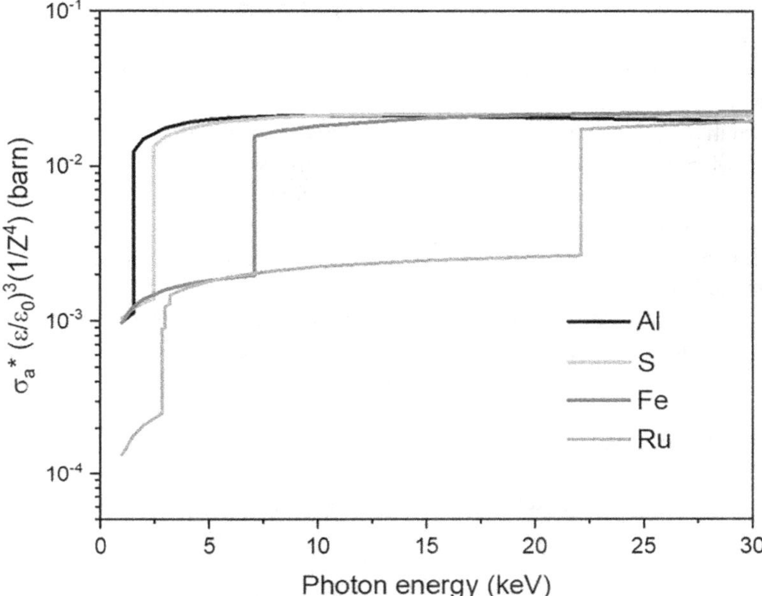

FIGURE 12.6 The normalized absorption cross-section as a function of photon energy for selected elements.

absorption spectroscopy (XAS). For a specific element, there is usually more than one absorption edge. Taking Ru as an example, the absorption edge at 22.12 keV is named the K-edge, which is the largest jump and corresponds to an electron transition from the 1s orbital or K shell. The absorption edges at 2.838, 2.967 and 3.224 keV are named L-edges, which correspond to the transition from the L shell. The three characteristics result from the degeneracy of the electron energy of the L shells. The energy of 2s electrons is lower than that of 2p electrons due to the potential drops with screening of the nuclear charge, leading to the L_I edge. The splitting of the 2p orbital due to spin–orbital coupling results in the L_{II} and L_{III} edges. Because of element selectivity, photoelectric absorption has been used to filter photons in a certain energy range with proper filter materials. On the other hand, this is one of the strategies used to achieve CT, where the absorption contrast can be formed because of the difference in X-ray absorption.

Accompanied by photoelectric absorption, the holes created are subsequently filled by electrons from the outer shells. The transition energy can be simultaneously released by the emission of a photon, the process of which is known as fluorescence. Therefore, fluorescent X-rays are specific for elements. The characteristic energy can be evaluated by empirical laws, which was first discovered by Moseley on the *Kα* line.

$$\varepsilon_{K\alpha}[\text{keV}] \approx 1.017 \times 10^{-2}(Z-1)^2, \tag{12.27}$$

where $\varepsilon_{K\alpha}$ is the energy of the $K\alpha$ line and Z is the atomic number for the element. The element sensitivity of fluorescent radiation can be utilized to establish techniques for the nondestructive analysis of samples such as **X-ray fluorescence spectroscopy (XRF)**.

Alternatively, the energy released can be used to kick another electron in the outer shell out of the atom, where the emitted electrons are called Auger electrons. The ejected electrons travel only a short distance before they lose their kinetic energy because of inelastic scattering by other electrons. This scattering can be quantified by the inelastic mean free path (IMFP), which has been demonstrated to be dependent on the kinetic energy of electrons, as shown in Figure 12.7. The relationship of the IMFP with kinetic energy is derived from the experimental values in several materials, which infers its universal nature. With the increase in kinetic energy from several eV to thousands of eV, the IMFP is shown to first decrease sharply and then increase in a relatively slow slope. The maximum IMFP is no more than 10 nm with very low kinetic energy, while the minimum IMFP is only several Å with kinetic energy in the range of 50–100 eV. The limitation of the IMFP makes the techniques based on the detection of Auger electrons surface-sensitive. The inelastic scattering of Auger electrons results in subsequent excitations. Collision leads to the ejection of new electrons accompanied by the ionization of the affected atom. Alternatively, Auger electrons can be slowed down to produce bremsstrahlung radiation with the emission of low-energy X-rays. Most of the kinetic energy of electrons turns into heat by collisional interactions, resulting in beam damage.

It should be mentioned that fluorescence and Auger processes occur in competition. Overall, the fluorescence yield P_f and Auger yield P_A from all shells constitute a probability of 100%. That is, $P_f + P_A = 1$. The situation is the same when it refers

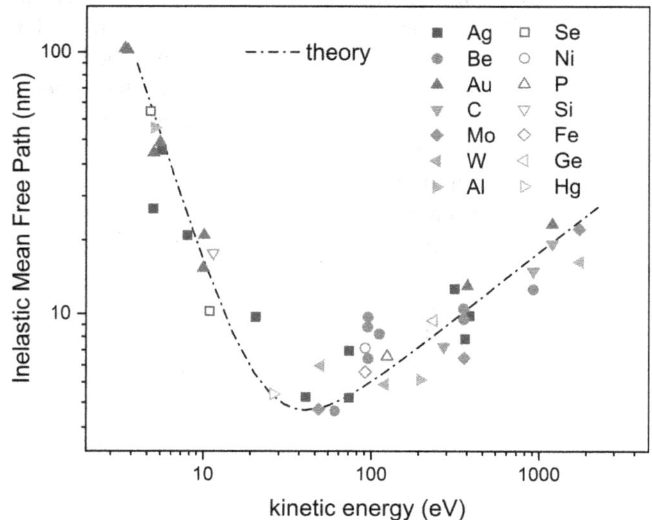

FIGURE 12.7 Inelastic mean free path of photoemission electron as a function of the kinetic energy. (Data taken from ref. [6].)

to the decay process for a specific shell. Taking K-shell emission as an example, the fluorescence yield $P_{f\text{-}K}$ can be empirically given by

$$P_{f-K} = \frac{Z^4}{10^6 + Z^4}. \tag{12.28}$$

Accordingly, the fluorescence yield and electron yield per K vacancy as a function of atomic number can be derived, as shown in Figure 12.8. For low-Z elements, Auger yield is dominant, while fluorescence emission becomes more crucial for high-Z elements.

The simple step-like characteristic in the absorption cross-section discussed above is only expected for isolated atoms. For assemblies of atoms, σ_a could appear in more structures when the photon energy is scanned in the vicinity of an absorption edge, the structures of which result from a variety of physical processes. For energies within approximately ±10 eV of the edge, σ_a usually appears to overshoot the step-like jump. This corresponds to transitions of core electrons to unoccupied bound states. Because of the higher density of such bond states compared with that of unbound states, σ_a turns into a peak called the white line, the name of which originates from the recording method by photographic films in history. For higher photon energies, photoelectrons are excited into unbound states so that the electrons can propagate from centre atoms as spherical waves. The waves may be back-scattered by the neighbouring atoms and then superposed with themselves, causing oscillations of σ_a. With photon energy in the range of 10–50 eV above the absorption edge,

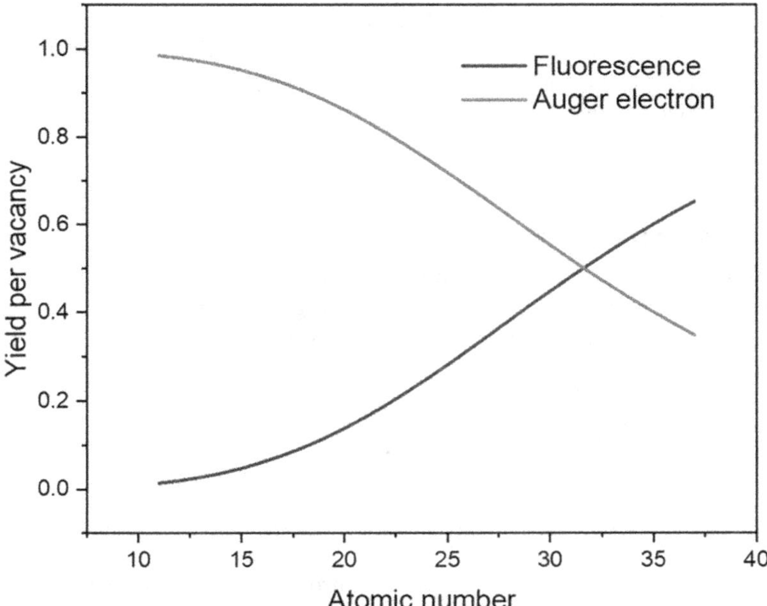

FIGURE 12.8 Fluorescence and Auger electron yield per K vacancy as a function of atomic number.

the electrons ejected with low kinetic energy have large probabilities of experiencing multiple scattering. In a broad sense, the spectroscopic regions of white lines and oscillations resulting from multiple scatterings are known as **X-ray near-edge fine structures (XANESs)**. With photon energy 50–1,000 eV higher than the edge, photoelectrons have sufficient kinetic energy to primarily undergo single scattering. This region is called the **extended X-ray absorption fine structure (EXAFS)**, which has been widely used to determine the local structure of materials.

12.1.1.2.3 Magnetic Interactions

The interactions of X-rays with matter discussed thus far originate from the interactions between the electric fields of X-rays and the charge of electrons. Actually, the magnetic fields of X-rays can also interact with the spin and magnetic moments of electrons. Due to magnetic scattering, magnetic structures can be obtained. However, magnetic scattering is much weaker than charge scattering, which makes the development of magnetic scattering relatively slow. This weakness can be quantified by the amplitude ratio of magnetic to charge scattering for a single electron.

$$\frac{A_{\text{magnetic}}}{A_{\text{charge}}} = \left(\frac{h\omega}{mc^2}\right) \tag{12.29}$$

For X-rays with 5.11 keV, the amplitude ratio is 0.01, so the intensity ratio of diffraction peaks is 10^{-4}. Considering that only relatively few atomic electrons contribute to the magnetic scattering, the intensity ratio is further decreased by an additional factor of approximately 10^{-2}. This magnetic interaction also occurs during the absorption process. For example, the different absorption of left- and right-handed circularly polarized light leads to a technique named **X-ray magnetic circular dichroism (XMCD)**. These rich magnetic interactions and relevant X-ray techniques are beyond the scope of this chapter.

12.1.2 Interactions of Neutrons with Matter

12.1.2.1 What Is a Neutron?

The neutron, first discovered by James Chadwick in 1932, can be generally treated as a classic particle because of its comparably large mass, $m_n = 1.675 \times 10^{-27}$ kg. The energy ε and momentum p can be expressed classically by

$$\varepsilon = \frac{1}{2} m_n v^2, \tag{12.30}$$

$$p = m_n v, \tag{12.31}$$

where v stands for the propagation speed of neutrons. The neutron particle is revealed to be composed of one up quark with a charge of $2/3e$ and two down quarks, each with a charge of $-1/3e$. Therefore, the total charge amount of neutrons is zero, which results in the absence of an electric dipole moment or at least out of detection owing to the limit. Hence, neutrons primarily interact with matter through short-range nuclear forces instead of Coulombic forces. Moreover, the arrangement of quarks related to the charge distribution with a spin of 1/2 gives rise to a magnetic dipole moment of −1.913 μ_N, where the nuclear magneton $\mu_N \approx -0.96624 \times 10^{-26}$ J/T. This enables the magnetic interactions of neutrons with matter, primarily the unpaired electrons in atoms.

The neutron particle can also be represented as a matter wave, which was first proposed by Louis De Broglie. According to quantum theory, the energy ε and the momentum p of a particle can be linked to the frequency f and wavelength λ of the matter wave by the Planck–Einstein equation and the de Broglie equation, respectively.

$$\varepsilon = \hbar\omega, \tag{12.32}$$

$$p = \frac{h}{\lambda} = \hbar k, \tag{12.33}$$

where Planck's constant $h = 6.626 \times 10^{-34}$ J·s; the reduced Planck's constant $\hbar = h/2\pi$ and $k = 2\pi/\lambda$.

Considering the multiple collisions of neutrons with moderators in a nuclear reactor, which is the major source of neutrons, the probability distribution of neutron speed is closely related to the temperature of the moderators. Therefore, the kinetic energy of neutrons for experiments can correspond to a characteristic temperature T, which is given by

$$\varepsilon = k_B T, \tag{12.34}$$

where the Boltzmann constant $k_B = 1.381 \times 10^{-23}$ J/K or 8.617×10^{-5} eV/K. In combination with the equations on energy,

$$\varepsilon = \frac{mv^2}{2} = \frac{h^2}{2m\lambda^2} = \frac{\hbar^2 k^2}{2m} = k_B T. \tag{12.35}$$

According to the above relationships, the other characteristic values of the neutrons can be estimated with one specified characteristic value. For example, a neutron with an energy of 1 MeV corresponds to a wavelength of 4.04×10^{-4} Å, a temperature of 1.16×10^{10} K and a speed of 1.38×10^7 m/s. A neutron with a temperature of 300 K has an energy of 25.8 meV, a wavelength of 1.78 Å and a speed of 2.22×10^3 m/s. The former can cross the sample at a thickness of 15 cm in approximately 11 ns, while the latter will do so in approximately 70 μs. However, both processes require a much shorter time than the mean lifetime (~880 s) of a free neutron, which will be unstable and decay into protons.

Because of the large differences in their characteristics, neutrons can be further classified according to their energy range, as shown in Table 12.1.

TABLE 12.1
Classification of Neutrons

Types		Energy Range (eV)	Wavelength (Å)
Fast neutrons		>1,000	<0.009
Slow neutrons	Epithermal neutrons	1–1,000	0.009 – 0.3
	Thermal neutrons	~0.025	~1.8
	Cold neutrons	$5\times10^{-5} - 0.025$	1.8 – 40.45
	Very cold neutrons	$5\times10^{-7} - 5\times10^{-5}$	40.45 – 404.5
	Ultracold neutrons	$<5\times10^{-7}$	>404.5

Similar to the classification of X-rays, the energy boundaries of neutrons also vary in different references. However, the order of magnitude of the bound is almost consistent, which is sufficient to clarify our discussion. Epithermal, thermal and cold neutrons are often used in the study of condensed matter. In particular, thermal neutrons are very suitable for probing the phase structure and dynamics of lattice vibrations because of the comparability of wavelength and energy with atomic distances and phonon energy, respectively.

Because of the instability of free neutrons, the only abundant source of neutrons is the atomic nucleus where both protons and neutrons exist. The neutrons can be released from the nucleus by various nuclear reactions. Using nuclear reactions within lighter nuclei where the number of neutrons approaches or equals that of protons, such as ^{9}Be, neutrons with a relatively low yield in the range of 10^{8}–10^{11} per second can be produced. Such reactions can be used to build small neutron sources, which can be portable and suitable for certain experiments less desired for high flux, such as diffraction and imaging. However, for most of the experiments calling for high flux (10^{15}–10^{18} per second), the production of neutrons relies primarily on large-scale facilities based on the nuclear reactions of heavy nuclei where the number of neutrons is considerably larger than that of protons, such as W, Hg and ^{235}U.

One of the major sources is the nuclear reactor, which produces neutrons by the fission reaction of the heaviest nuclei, such as ^{235}U. The energy of neutrons up to MeV needs to be lowered by moderators such as water by a multiple scattering process, which eventually results in an energy distribution with a mean energy of approximately 0.025 eV. Neutrons colliding with moderators emerge as a continuous stream with different energies. Therefore, neutrons with a narrow energy band need to be selected by monochromators such as pyrolytic graphite, germanium and copper or a set of choppers. However, this strategy wastes a majority of the neutrons.

An alternative source is based on spallation reactions, where neutrons are produced by the striking of protons with high speed on solid targets made of heavy metals. The neutrons are released in pulses of several times per second (10–60 Hz). Therefore, the energy of neutrons can be differentiated in flight time when the neutrons arrive at detectors. This technique is called time of flight (TOF), where a monochromator is no longer needed.

Neutrons can also be produced by the fusion of deuterium (D) and tritium (T) with the formation of a helium isotope, which is called a fusion reaction. Such a reaction is intriguing because of its notable advantage of low deposition heat. Desktop accelerator-based sources based on DT fusion cannot produce as many neutrons as reactor or spallation sources. Instead, inertial fusion, which can be triggered by the compression of D and T at a significantly high density and temperature utilizing lasers, can be a route to achieve much brighter sources. However, the practical application of such sources is still a long way off.

12.1.2.2 General Description of the Interactions

Because of the zero charge, neutrons primarily interact with nuclei of matter through nuclear forces rather than Coulombic forces with electrons. Since the nuclear forces are in a range of 10^{-15}m, much smaller than the atomic size (~10^{-10}m), neutrons can propagate a long distance among the atomic framework of solid matter without any interactions. For example, the number of neutrons decreases by ~1% per mm compared with 99% per mm or more for X-rays when penetrating aluminium sheets. This high penetration capability means that surface effects can be neglected in most cases when neutrons are applied in measurements. The only exception is **neutron reflection**, which occurs on a flat surface at a very small critical angle. The penetration capability of neutrons is considerably different for each element. This enables **neutron imaging** (**NI**), which is particularly suitable for detecting the macroscopic distribution of structures in the aggregation of light elements. Moreover, neutrons deposit little energy in samples, which eliminates the beam damage to a great extent. In summary, neutron techniques are nondestructive and primarily bulk-sensitive, which is particularly suitable for in situ studies under extreme conditions, such as high pressure or temperature with specific containers or biological environments.

However, the high penetration power of neutrons is a double-edged sword. The interaction probability is extremely small, and the neutron sources inherently provide lower flux compared with X-ray sources. To obtain acceptable signals, most neutron measurements need to be performed with large amounts of samples at large-scale dedicated neutron sources, restricting their widespread applications.

The nuclear interactions primarily include two types, **scattering** and **capture**. When a neutron is scattered by a nucleus, its speed and direction change with the change in the number of neutrons and protons in the nucleus. Scattering interactions can be further divided into elastic and inelastic scattering. For elastic scattering, the total kinetic energy of neutrons and nuclei is unchanged with only the direction changed. This interaction can be used to lower the speed of neutrons with high energy. The average energy loss is $2\varepsilon m_A/(m_A+1)^2$ when neutrons with kinetic energy ε encounter nuclei with atomic weight m_A. To reduce the initial energy ε_0 in the range of MeV to ε_n, usually in the range of meV, the number of collisions $n = \dfrac{\log(\varepsilon_n/\varepsilon_0)}{\log\left[(m_A^2+1)/(m_A+1)^2\right]}$ is needed. According to the expression, nuclei with smaller m_A should be preferred to reduce the speed of neutrons with fewer elastic collisions. This is why water is preferred as a moderator in neutron reactors, where the m_A of hydrogen is 1. For inelastic scattering, both the propagation direction and the total kinetic energy of neutrons and nuclei change, where the change in energy is related to the excitation of nuclei. If no excited states are present in the nuclei, neutrons only interact with the nuclei by elastic scattering.

The scattering interaction above is only a general description. Specifically, for the interaction of neutrons with the bound nuclei in condensed matter, elastic and inelastic scattering can be classified by the energy change of neutrons for simplicity. The elastic scattering in condensed matter can be used to determine the bulk crystal structure and local structure, where the relevant techniques are **neutron diffraction** (**ND**), **small-angle neutron scattering** (**SANS**) and the **neutron pair distribution function** (**nPDF**). Furthermore, the elastic scattering forms the basis of neutron reflection, which results in a technique called **neutron reflectometry** (**NR**). NR is surface-sensitive to exploring thin films with flat multiple layers, which is intuitively anomalous to the high penetration capability of neutrons. All the neutron techniques above are theoretically analogous to X-ray techniques, both of which originate from wavelengths comparable to the periodical dimension. However, neutron techniques are complementary to the corresponding X-ray techniques because of their respective advantages. What makes neutrons unique is their powerful capabilities in exploring the dynamics of condensed matter through techniques based on inelastic scattering, including **inelastic neutron scattering** (**INS**) and **quasi-elastic neutron scattering** (**QENS**).

In addition to being scattered by nuclei, neutrons can also be captured. Neutrons fuse into nuclei, forming excited states. Because of the further decay process, the capture reaction can be divided into two types, fission and absorption. Fission refers to the capture reaction occurring in heavy nuclei ($Z > 80$), where an excited nucleus roughly splits into two equal fragments and a few light particles, nearly always neutrons. Therefore, the fission reaction is the major strategy to produce neutrons. Absorption refers to the capture reaction accompanied by the emission of lighter particles (e.g., protons), electrons and a series of gamma rays. Absorption reactions can be used to detect neutrons, which becomes the basis of neutron techniques. The obstacle of direct detection on electroneutral neutrons can be bypassed through the detection of charged particles (e.g., protons or deuterons) arising from the absorption reaction. ^{3}He, ^{6}Li and ^{10}B are commonly used as absorbers in detectors and monitors because of their large absorption to thermal neutrons. The absorption reaction can also be used for shielding purposes. For example, pieces of ^{113}Cd metal, B_4C or ^{157}Gd-impregnated paint can be used to cover sample appendages to reduce useless scattering from the environment. More importantly, a series of neutron techniques have been established on the basis of absorption reactions. By detecting the gamma ray emitted in 10^{-12} to 10^{-9} s during neutron radiation, **prompt gamma activation analysis** (**PGAA**) can be developed for element analysis. If the gamma ray, which is usually emitted from the residual nucleus with a long lifetime after the release of charged particles, is collected for element analysis, the technique is called **neutron activation analysis** (**NAA**). Furthermore, by resolving the kinetic energy of the charged particles flying out of the sample surface, which is directly related to the thickness penetrated by the particles, the element distribution along with the distance from the surface can be obtained by the **neutron depth profiling** (**NDP**) technique. This technique is usually suitable for the nondestructive probing of light elements, such as ^{3}He, ^{6}Li, ^{10}B, ^{14}N and ^{17}O. However, strong absorption reactions have some problems in some specific applications. Absorption should be limited as much as possible when tuning the speed of produced neutrons. For example, deuterium is preferred as a moderator compared with hydrogen because of the much smaller absorption probability for neutrons, although hydrogen has higher moderating efficiency. Additionally, substantial absorption is averse to scattering experiments owing to the large intensity decrease of neutrons. To suppress the absorption reaction, a good choice is to use isotopes with weak absorption probabilities for substitution.

Owing to the intrinsic magnetic dipole moment, neutrons interact with unpaired electrons in matter through electromagnetic forces. Different from the substantial weakness of magnetic scattering compared with charge scattering for X-rays, **neutron magnetic scattering (NMS)** is of similar intensity compared with nuclear scattering owing to the similar strength of the two interactions. This enables the unique powerful capability of neutron scattering in exploring magnetic structures. For magnetic matter, the peak features from magnetic scattering are always superposed with those features from nuclear scattering. Magnetic scattering could give additional peaks in the nuclear diffraction patterns for antiferromagnetic materials. However, magnetic scattering only gives additional intensity on the nuclear diffraction patterns for ferromagnetic materials due to the equivalence of the magnetic unit cell and crystallographic cell. Fortunately, the anisotropic nature of magnetic scattering resulting from the dipolar nature of the magnetic interaction, which is different from the isotropic nature of nuclear scattering, can be used to separate the contributions from the two kinds of scattering. Specifically, if the electronic moments are aligned by an applied magnetic field, the contribution from magnetic diffraction can be removed when the scattering vector is parallel to the induced magnetization. The scattering and absorption interactions are described quantitatively in the following because of their important applications in studies of rechargeable batteries.

12.1.2.2.1 Scattering

The probability of neutron scattering can also be intuitively quantified by the cross-section in an experiment.

$$\sigma_s = \frac{I_{\text{sc}}}{I_{\text{in}}}, \tag{12.36}$$

where σ_s stands for the total scattering cross-section. I_{in} indicates the intensity of incident neutrons, corresponding to the number of neutrons per second, and I_{sc} stands for that of scattered neutrons.

The intensity of the incident neutron beam can be expressed as

$$I_{\text{in}} = |A|^2 v, \tag{12.37}$$

where A and v are the amplitude and the speed of the monochromatic neutron described as a plane wave, respectively. $|A|^2$ represents the neutron density.

The intensity of the scattered neutron beam can be derived as

$$I_{\text{sc}} = 4\pi r'^2 \times \frac{b^2|A|^2}{r'^2} v = 4\pi b^2 |A|^2 v, \tag{12.38}$$

where r' is the distance from the centre of the nucleus. b represents the amplitude change of the outgoing spherical wave relative to the incident plane wave, which can be used to evaluate the interaction of the neutron with the nucleus and is thus named the **scattering length** on account of its dimension.

Therefore, the total cross-section for a bound nucleus is

$$\sigma_s = 4\pi b^2. \tag{12.39}$$

The scattering length can be regarded as the effective range of the nuclear scattering potential. This can be identified by the typical magnitude of the scattering length in the range of 10^{-15}–10^{-14}m, which is the same order as the interaction distance of the nuclear force. Therefore, the cross-section usually has a magnitude of 10^{-28}m^2. It should be noted that b and σ_s for neutron interactions can only be obtained by experimental measurements because the values cannot be predicted through the properties of the nucleus according to the current theory. Therefore, the scattering length of the nucleus becomes the basic unit in scattering calculations, which theoretically is the product of the atomic form factor and scattering length of neutrons if analogous to X-rays.

The scattering length and cross-section are characteristics of the nucleus, and they vary with the neutron energy and nucleus. Generally, the elastic scattering length and cross-section are independent of neutron energy less than 1 MeV, while the inelastic scattering length cross-section is proportional to the reciprocal of neutron energy, which is also called the $1/v$ law and will be discussed in detail when involving the absorption cross-section. Therefore, the evolution of the total scattering length and cross-section with energy for a specific nucleus depends on which type of scattering is dominant.

The scattering length evolves across the periodic table, as shown in Figure 12.9, where the trend is the same for the cross-section. The values are considerably scattered, although a roughly gradual increase is exhibited. The random evolution, especially for adjacent elements, highlights the specialty of neutron scattering compared with normal X-ray scattering. Since Thomson X-ray scatterings are based on electrons, they have limited resolving power among Fe, Co and Ni or among O^{2-}, Na^+, Mg^{2+}, Al^{3+} and Si^{4+}, which have similar or equal numbers of electrons. The site occupancies and element distributions (not oxidation states) in the structure can be readily resolved by neutron scattering. The strong variations intrinsically arise from the strong dependence on the spin of the nucleus. The spin-dependent nature even makes scattering lengths as well as cross-sections sensitive to the isotopes for a given element. Assuming the spin of the nucleus is I, the total spin of the neutron and nucleus can be either $I + 1/2$ or $I - 1/2$. The scattering lengths of the two spin states are represented by b_+ and b_-. The scattering length of a nucleus is the average of b_+ and b_-, which can be calculated by

$$b = p_+b_+ + p_-b_- = \frac{I+1}{2I+1}b_+ + \frac{I}{2I+1}b_-, \tag{12.40}$$

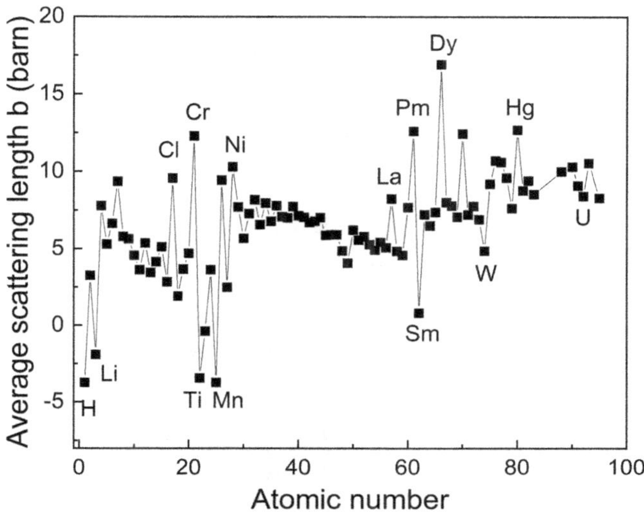

FIGURE 12.9 Average coherent scattering length for natural elements.

where p_+ and p_- are the probabilities of the two spin states, which can be calculated by the number of orientations of the angular momentum. For example, ^{1}H has a triplet state with $J=1$ and the singlet state with $J=0$ because of nuclear spin $I=1/2$. The values of b_+ and b_- are 10.81 fm and −47.42 fm, respectively. The value of b is −3.74 fm. In comparison, the values of b_+ and b_- are 9.53 fm and −0.98 fm, respectively, owing to the nuclear spin of 1 for ^{2}H. The value of b averaged for ^{2}H is 6.67 fm. Similarly, the scattering length of an element is the average value of its different isotopes. For example, hydrogen is composed of three isotopes, ^{1}H (protium), ^{2}H (deuterium) and ^{3}H (tritium), with relative abundances of 99.985%, 0.015% and approaching zero, respectively. The scattering length of the natural hydrogen is −3.7384 fm, which is almost the same as that of ^{1}H.

In an experiment, the scattering in a given direction instead of all directions is measured. Accordingly, the differential scattering cross-section is defined by $d\sigma_s/d\Omega$, where Ω is the solid angle. Since the solid angle of the spherical surface is 4π, $d\sigma_s/d\Omega = b^2$ when Ω approaches zero for the scattering of neutrons by a bound nucleus.

12.1.2.2.2 Coherent and Incoherent Scattering

Analogous to the calculations of the atomic form factor for X-ray scattering, the differential scattering cross-section describing the scattering by a sum of bound nuclei can be derived based on the cross-section of the single bound nucleus.

$$\frac{d\sigma_s}{d\Omega} = \left| \sum_j b_j e^{(iQ\cdot r_j)} \right|^2, \tag{12.41}$$

where $\boldsymbol{Q}$ is the scattering vector and $\boldsymbol{r}$ is the position vector of the nucleus.

This can be further divided into coherent and incoherent differential cross-sections.

$$\left(\frac{d\sigma_s}{d\Omega}\right)_{coh} = \sum_{jk} \left(\bar{b}\right)^2 e^{\left[iQ\cdot(r_k - r_j)\right]} \tag{12.42}$$

$$\left(\frac{d\sigma_s}{d\Omega}\right)_{\text{inc}} = \sum_{j=k} \left(\overline{b^2} - \bar{b}^2\right) \tag{12.43}$$

According to the expression, coherent scattering essentially describes interference between waves by the scattering of a single neutron from all the nuclei in a sample, which is scattering angle-dependent. In structure probing experiments, any variation in the intensity with Q derives from coherent scattering.

Incoherent scattering can be considered the remaining part of the total scattering with the coherent scattering subtracted. It involves the wave correlations at the same nucleus between different times. In most situations, the incoherent scattering intensity is independent of the scattering angle. The effect makes the incoherent scattering only contribute to the structureless background in diffraction experiments.

Incoherent scattering intrinsically originates from the random deviations from the average scattering length of an element because of its isotope and spin-dependent nature. However, there are other sources for the random deviations under some specific conditions. One is the random site occupation by two or more atoms of similar size and ionization state, such as the alloy of Cu and Zn (β-brass), which is called chemical incoherence. Another is the random orientation of atomic magnetic moments in paramagnetic materials. However, the magnetic incoherent scattering is Q-dependent due to the dipole magnetic interaction.

It should be noted that incoherent scattering is unique for neutron techniques compared with X-ray and electron scattering experiments. This is because the latter two are dominated by charge scattering. The charge density usually has no fluctuation on a given site in most situations except the random variations in the element species or ionization states. Some special neutron applications have been developed for the control of incoherent scattering. For example, the incoherent scattering can be suppressed by substituting ^{1}H with ^{2}H because of their large differences in the incoherent scattering cross-section (80.3 barn for ^{1}H vs. 2.05 barn for ^{2}H), which is especially important in applications of neutron scattering to structural biology and polymer science. On the other hand, the incoherent scattering dominated by ^{1}H can be used to determine the concentration of hydrogen in nuclear fuels and the solubility limit of hydrogen in metallic materials. More importantly, incoherent scattering is intrinsically suitable when studying dynamics, such as diffusion, because of its wave correlations of the same nucleus between different times.

Inelastic scattering is always involved when exploring the dynamics. To quantify the energy loss process, **the double-differential cross-section** $\frac{d^2\sigma_s}{d\Omega d\varepsilon}$ has been developed. Inelastic scattering can usually be divided into three types according to the magnitude of energy loss. With the largest energy loss of the three, **neutron**

Compton scattering occurs. It is suitable to examine the dynamic process at short time and length scales by recording the momentum distribution and mean kinetic energies of single particles. At the opposite end, inelastic scattering with a very small energy loss occurs. It is suitable to explore the dynamics at long time and length scales, such as diffusional atomic motion. Since the energy transitions are generally unquantized, unsymmetric broadening features centred on the elastic peak can be observed. Therefore, inelastic scattering is called **quasi-elastic scattering**. The extent of the broadening reveals the jump rate of the atoms and their orientation in the diffusive process. Both inelastic scatterings are incoherent because of the incoherent nature of scattering from single particles. In between the two extremes, inelastic scattering occurs because of the interaction with a variety of different phonon modes, such as bond stretching and bending modes. A series of inelastic peaks appear in a neutron measurement, which is called **neutron vibrational spectroscopy**. In comparison with Raman and infrared vibrational spectroscopy, the most notable merit is that all vibrational modes can be measured because the neutron interaction is not governed by the optical selection rules. Because neutrons interact with phonon modes, which are synergic movements of all atoms, inelastic scattering is coherent.

12.1.2.2.3 Absorption

In addition to scattering, the other most likely process is the absorption of neutrons into nuclei. The nuclei are excited to a higher energy level and then relaxed by the emission of gamma or beta rays, which makes absorption not contribute to diffraction patterns. To quantify the absorption process, the absorption cross-section σ_a is defined as

$$\sigma_a = \frac{I_{\text{abs}}}{I_0}, \tag{12.44}$$

where I_0 and I_{abs} are the intensities of incident neutrons and absorbed neutrons, respectively.

The absorption cross-section follows the $1/v$ law; that is, σ_a is commonly observed to vary inversely with the speed of neutrons below a resonance range. The physical origin of the law is that the absorption probability increases in proportion to the time that the neutron stays in the vicinity of the nuclear potential. Therefore, the values of σ_a can be estimated over an energy range by using the $1/v$ law based on the tabulated value at 25.3 meV. When the kinetic energy of absorbed neutrons is close to that of an excited state of the compound nuclear composed of neutron and initial nuclear, the evolution of the cross-section appears as a peak called resonance absorption. For heavy nuclei, large and narrow resonances appear in the energy range of several eV. The resonances become too close to resolve in the keV region and are sparser and very broad in the MeV region. For light nuclei, only broad and small resonances appear in the MeV region. For intermediate nuclei such as Ni and Fe, resonances appear at intermediate intensities below 1 keV and broaden in comparison to heavy and light nuclei.

Absorption cross-sections can also be related to the isotopic composition. For example, σ_a of ^{11}B is only 0.0055 barn, whereas ^{10}B has a very large σ_a of 3,835 barn.

Therefore, the averaged absorption cross-section of natural B is very large despite only 20% natural abundance of ^{10}B. Neutron scattering experiments could be feasible for ^{11}B-rich matter because of the suppression of strong absorption. A similar case occurs in natural Gd composed of ^{154}Gd ($\sigma_a = 85.1$ barn) and ^{157}Gd ($\sigma_a = 259{,}000$ barn).

The effect of absorption on scattering lies in not only the decrease in the neutron number but also in the change in the scattering cross-section. Usually, the change is small and thus negligible. However, the scattering cross-section can vary appreciably near an absorption resonance, which is similar to the resonant scattering of X-rays. Because of the effect of absorption, the scattering length can be expressed as a complex number, $b = b' - ib''$, which is analogous to the expression of the total atomic form factor for X-rays. b'' can be related to the absorption cross-section σ_a by

$$b'' = \frac{k}{4\pi}\sigma_a. \tag{12.45}$$

12.1.3 The Needs of X-Ray and Neutron Techniques

X-ray radiation interacts with electrons of matter through Coulombic forces because of their electromagnetic nature. However, neutral neutron particles interact with nuclei through nuclear forces as well as electrons through magnetic forces owing to their 1/2 spin. The intrinsic difference in interactions with matter results in separate specialties of X-ray and neutron techniques as follows.

1. The large probabilities of interactions of X-rays with matter place the penetration depth at the μm scale. In most cases, X-ray techniques are thought to be bulk-sensitive. However, the penetration depth cannot represent the bulk information in some special situations, such as the detection of metal workpieces with thicknesses on the cm scale. In contrast, surface-sensitive measurements are easily performed by the GI of X-rays or the collection of surface-sensitive signals such as photoelectrons.

   In comparison, the much smaller probabilities of interactions of neutrons with matter place the penetration depth at the mm–cm scale. Neutron techniques can almost always obtain true bulk information, although few surface-sensitive methods can be fulfilled, such as NR. The penetration makes neutron techniques very suitable for in situ experiments, especially under extreme conditions, such as high temperature and high pressures, usually with numerous surrounding shields.
2. A high flux of X-rays can be easily obtained through multiple sources. For dedicated synchrotron facilities, the flux can be of order from 10^{12} to $10^{18}\,s^{-1}\cdot 0.1\%$ BW. Even for X-ray tubes, the flux can reach 10^{9}–$10^{11}\,s^{-1}\cdot 0.1\%$ BW. A sufficiently high flux allows for wide applications of many kinds of X-ray techniques in the laboratory. Large interaction probabilities as well as high photon flux enable a short experiment time, with the amount of sample

at the mg scale. On the other hand, these properties result in beam damage, although X-ray techniques are often thought to be nondestructive.

In comparison, neutron flux is intrinsically difficult to improve. The brilliance of modern reactor neutron sources is approximately 10^{14} of magnitude lower than that of a third-generation synchrotron source, although this shortcoming has been partly overcome in recent pulsed neutron sources. Therefore, most applications of neutron techniques have been severely restricted in the laboratory owing to the lack of effective compact neutron sources. The extended application to more research fields has been further confined because of the scarcity of dedicated neutron facilities compared with synchrotron facilities. Consequently, neutron techniques are rarely used as a priority when the desired information can be obtained by other techniques. Much smaller interaction probabilities, as well as deficient neutron fluxes, lead to long experiment times while requiring a large sample amounts (at least tens of grams). These two properties, as well as small energy transfers from thermal neutrons, result in much less beam damage compared with X-rays. Nondestructive neutron techniques have been widely applied to study living biological systems.

3. In addition, X-rays can be focused at the nanoscale through numerous optical devices owing to the high flux and strong interaction with matter. This enables high spatial resolution in X-ray experiments, especially in X-ray imaging. In comparison, the lower source brilliance and weak interaction with matter make neutron beams difficult to focus down to spot sizes below 1 mm. The spatial resolution, such as for NI, is often determined by the thickness of the scintillator (tens of μm at least).
4. The difference in scattering characteristics distinguishes between X-ray and neutron scattering techniques.

   The Thomson scattering of X-rays is dependent on the electronic density around the nucleus. Therefore, it is nonsensitive to light elements, such as H and Li, because of the excessively low electronic density. In addition, it cannot distinguish the adjacent elements in structural units such as Fe, Co and Ni and iso-electronic ions such as O^{2-}, F^-, Na^+ and Mg^{2+} because of the approaching or equivalent electronic number. A structure composed of elements with large differences in atomic number is also unsuitable for resolution by X-ray scattering. Although the Z-dependent issue in scattering can be partly overcome using absorption interactions, the wavelength comparable to the length scale of the structure will be changed.

   In comparison, cross-sections are scattered across the periodic table for neutrons. Neutron scattering is sensitive to light elements, such as H and Li. It can distinguish adjacent elements and iso-electronic ions without changes in the length scale of the structure. Neutron scattering specializes in resolving structures composed of both light elements and very heavy elements. However, incoherent scattering and absorption reactions might disturb the structure probing. Fortunately, both interferences can be suppressed by isotopic substitution. In addition, owing to the isotropic nature of the nuclear forces, the scattering intensity is independent of the scattering angle, which gives neutron scattering high Q-sensitivity.

5. Normal X-ray scattering has difficulties probing magnetic structures because of the much larger nuclear weight compared with electron weight. In comparison, neutron techniques are powerful for probing the static and dynamic magnetic properties of matter, such as magnetic ordering, magnetic excitations and spin fluctuations, because the strength of magnetic forces is comparable to that of nuclear forces.
6. X-ray techniques are powerful when exploring the properties related to electronic behaviours. Therefore, the wide applications of absorption techniques are unique for X-rays, which are used to resolve oxidation states or even band structures. Absorption phenomena also occur for neutrons, which results in some neutron absorption techniques with element sensitivity. However, the radioactivity of samples must be tested before they leave the beamlines because of the nuclear reactions.

In comparison, neutron techniques are powerful when exploring the properties related to atomic behaviours. Therefore, inelastic neutron scattering has been widely used in exploring the dynamics of atoms at the atomic scale. However, kinematic restrictions cannot access full energy and momentum transfer, which can be complemented by inelastic X-ray scattering to some extent.

As discussed above, X-ray techniques and neutron techniques are complementary. The former is based on interactions with electrons, specializing in the investigation of electronic behaviour. The latter is primarily based on interactions with nuclei, specializing in the exploration of atomic behaviour. Such complementarity can be noted, particularly in the study of rechargeable batteries, which always experience synergetic changes in electrons and ions.

12.2 ELASTIC SCATTERING TECHNIQUES

The characteristic performances of lithium-ion batteries, including capacity, reversibility and cyclability, are closely related to the structures and evolutions of the battery components, including cathodes, anodes, electrolytes and the resultant interfaces, at multiple length scales. Therefore, a deep understanding of the structures of the equilibrium states and nonequilibrium states at multiple length scales is desirable for battery development. As discussed, elastic scattering interactions of X-rays and neutrons enable a series of powerful techniques for structure exploration. In this section, XRD, SAXS and PDF analysis are introduced with several typical case studies in sequence, which are sensitive to crystal structure, mesoscale structure and local structure, respectively. The corresponding neutron techniques are considered complementary because of the high penetration and the sensitivity to light elements. This chapter clarifies the specialties and applications of various elastic scattering techniques in gaining insight into Li-ion battery electrodes and electrolyte functions.

12.2.1 XRD

XRD is a powerful tool for resolving crystal structures of crystalline or partially crystalline battery materials. The working principle of XRD is that the outgoing X-ray beam by Thomson scattering interfaces constructively or destructively along with scattered angles in space, which depends on the distribution of electrons along with the arrangement of atoms in materials. The scattered angles are related to the lattice parameters, which can be described by the structure factor. The relationship can be expressed with Bragg equation $2d \sin \theta = n\lambda$ in brief. The inference intensity is closely related to the element type of the structure unit. In addition to resolving fine crystal structures, XRD can also be used to obtain phase composition and microstructure information, such as grain size, strain and texture.

In practice, XRD experiments have been widely performed in the laboratory. Even in situ studies can be readily carried out with commercially available in situ cells. However, lab-XRD usually collects signals using the reflection mode due to the limitations of photon flux and penetration depth. The reflection mode intrinsically requires too much time to complete a single measurement because of the movement of detectors. This limits the time-resolved capability of lab-XRD experiments in exploring short reaction processes, such as the mechanism of battery rate performance.

Owing to the high brilliance, high collimation and tuneable energy of X-rays at synchrotron facilities, synchrotron XRD experiments can be performed in transmission mode with a large increase in penetration depth as well as spatial and time resolution. This makes synchrotron XRD particularly suitable for in situ/operando studies in batteries. Using high-energy X-ray beams generally larger than 40 keV, commercial batteries, such as coin cells and 18650 cells, instead of customized cells with windows can be investigated directly to ensure the reality of the reactions. Apart from the time-resolved in situ XRD, more variants of XRD methods have been developed, among which microbeam XRD and energy-dispersive XRD are worth mentioning. The former enables mapping of the phase distribution with a transverse spatial resolution of hundreds of nm. The latter is capable of achieving longitudinal space resolution by collecting signals at a fixed angle with polychromatic incident X-rays.

12.2.1.1 Time-Resolved In Situ XRD

In situ/operando XRD experiments have been widely performed to study the structural changes of battery materials during electrochemical processes. Almost all kinds of electrode materials, such as insertion-type materials, conversion-type materials and alloy materials, have been studied by in situ synchrotron XRD. Here, a discussion of the structural changes of layered oxide cathodes under equilibrium conditions by in situ XRD is presented. The other application of in situ XRD to the structural changes of $LiFePO_4$ under nonequilibrium conditions will be discussed later.

Layered oxide cathodes are the most important kind of cathode in lithium batteries because of their 2D voids for lithium transport. Layered oxide materials usually have a hexagonal structure with R-3m symmetry. The structure can be generally described as close-packed Li-O layers alternatively stacked between two TM-O "slabs" along the c direction. According to the stacking period and coordination structure of Li-O, the stacking in most of the layered oxides can be assigned to the O3-type. With lithium deintercalated from the matrix in lithium-ion batteries, a series of phase transitions usually occur in sequence. In situ XRD can be used to track the quasi-equilibrium phase transition elaborately because of its quick scanning capability. For example, in situ XRD has revealed that $LiCoO_2$ experiences phase transitions from H1 to O1 by H2 and O1a when charged to voltages higher than 4.8 V.[7] For $LiNiO_2$, the H1 and H2 hexagonal phases coexist at the very early state, and the transition to the H3 phase at the end is observed. Regardless of the specific phase transitions, a common trend of lattice parameters in the major newly formed phases is the expansion along the c axis with contraction in the a and b directions. The former is usually thought to originate from an increased repulsion force between the two neighbouring oxygen layers, while the latter results from the shrinking TM–O bonds of the TM–O_6 octahedron because of TM oxidation.

However, an in situ XRD study showed an abnormal evolution of lattice parameters in Li_2MoO_3 (Figure 12.10).[8] In the charge process, a solid solution reaction

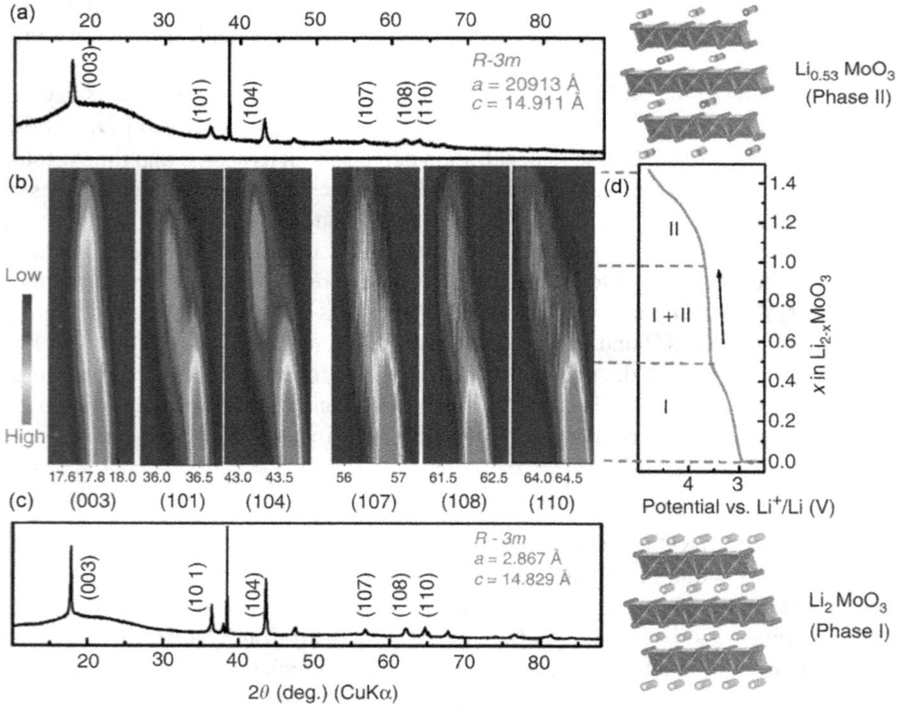

FIGURE 12.10 (a) XRD pattern of the $Li_{2-x}MoO_3$ electrode immediately after charging to 4.8 V. (b) Contour plot of diffraction peak evolution of (003), (101), (104), (107), (108) and (110) during delithiation. (c) XRD pattern of the Li_2MoO_3 electrode before charging. (d) Charge curve at a current density of 10 mAg^{-1} from the open-circuit voltage (OCV) to 4.8 V during XRD data collection. (Reproduced with permission from ref. [8]. Copyright 2014 Nature Publishing Group.)

occurs with the delithiated number in the range of 0–0.5. All the major peaks shift to lower angles, indicating an increase in lattice parameters. Thereafter, a new phase with larger lattice parameters forms a delithiated number in the range of 0.5–1, indicating expansion along the a, b and c directions at the same time. Such abnormal behaviour has been explained by Mo–Mo bonding interactions in the Mo-O layers. The evolution trend of lattice parameters as well as tuning strategies can be used to design cathode materials with better cyclability.

In situ synchrotron XRD is more specialized in resolving the change in the non-equilibrium structure, which helps to obtain a deep understanding of battery kinetic performance. A good example is the investigation of the structural origin of the high kinetic performance in $LiFePO_4$.[9] $LiFePO_4$ has attracted great scientific interest because of its high stability when applied as a cathode in commercial batteries. A quasi-equilibrium two-phase reaction mechanism has been verified by many XRD studies, where both phase components $LiFePO_4$ and $FePO_4$ have very poor electronic conductivity. A poor rate capability can be inferred by these characteristics. However, nanosized $LiFePO_4$ presents a high rate capability. Delmas et al. proposed a domino-cascade model to understand the rate capability by the careful analysis of ex situ XRD data. They found the strict coexistence of $LiFePO_4$ and $FePO_4$ throughout the whole reaction process without any partially lithiated phase. The high rate capability is therefore attributed to the quick movement of the two-phase interfaces during the deintercalation/intercalation in particles.[10]

However, the structural changes of $LiFePO_4$ at high rates are challenged by in situ XRD investigations. The emergence of an intermediate phase with Li composition in the range of 0.6–0.75 is revealed first by in situ XRD in microsized $LiFePO_4$ at a 10 C-rate.[11] This indicates an alternative reaction path under nonequilibrium conditions. However, the reaction mechanism of nanosized $LiFePO_4$ at a high rate needs to be further clarified. In situ XRD measurements on nanosized $LiFePO_4$ in the first five cycles at a 20 C-rate were made by using an AMPIX cell with a conductive glassy carbon window, which could suppress the undesired delay of reactions near the window and improve the cyclability of in situ batteries.[12] The in situ XRD patterns as well as the galvanostatic charge and discharge profiles are shown in Figure 12.11a. All the major peak features can be indexed to either the lithium-rich phase $Li_{1-\alpha}FePO_4$ or the lithium-poor phase $Li_{\beta}FePO_4$. The deviation from the ideal composition of $LiFePO_4$ and $FePO_4$ results from the increase in interfacial energy per volume with the nanocrystallization of particles. In addition, additional features appear in the ranges of 8.15° – 8.4°, 13.95° – 14.1° and 15.15° – 15.4°. Specifically, as shown in Figure 12.11b, the reflections (200) and (301) start to broaden asymmetrically at higher angles in the latter half of the charge process, which is more obvious during the discharge process. In comparison with theoretical simulations, such features have been attributed to the formation of a solid solution phase with lithium composition in the range of 0–1 in contrast to that of the microsized intermediate phase. In situ XRD studies on both microsized and nanosized $LiFePO_4$ confirm the hypothesis that the continuous structure changes occur at a high rate and are responsible for the high rate capability. These studies highlight the crucial role of in situ XRD in exploring the nonequilibrium phase transitions of battery materials.

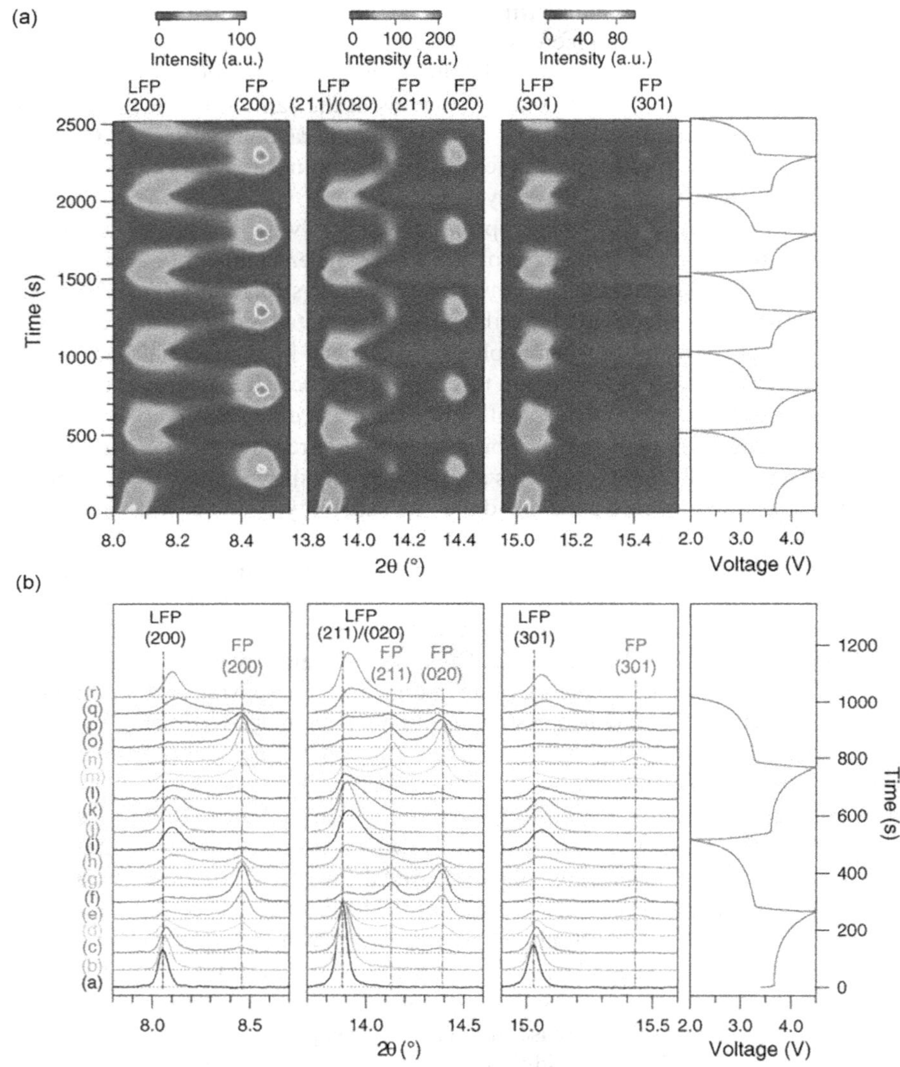

FIGURE 12.11 In situ XRD patterns of a $Li/LiFePO_4$ cell at a rate of 10C. (a) The image plot for (200), (211), (020) and (301) reflections during the first five cycles and (b) selected individual diffraction patterns during the first two cycles are stacked against the voltage profile. (Reprinted with permission from ref. [12]. Copyright 2014 American Association for the Advancement of Science.)

12.2.1.2 Microbeam XRD

Apart from time-solved XRD, the development of space-resolved XRD, called microbeam XRD, can be feasible because available X-rays with small spot sizes can eventually be obtained through a series of highly efficient focusing optics along the beamline owing to the high brilliance and high collimation of synchrotron X-rays. Specifically, for hard X-rays, the spot size can reach below 20 nm through Fresnel zone plates and multilayer Kirkpatrick–Baez mirrors. However, the practical beam size is around 0.2–5 μm because of the challenges from the accuracy of optics, elimination of various vibrations, beam stabilization and temperature control. Therefore, microbeam XRD can be used to resolve the heterogeneous mechanism of battery materials, such as structural changes in single particles and the distribution of phases in certain areas, which sometimes determine the battery performance at the macroscale. Microbeam XRD can be fulfilled with a monochromatic incident beam or polychromatic incident beam. The former is limited to cases when polycrystal materials have grain sizes below the beam size so that statistical Bragg diffraction can be obtained within a multitude of grains. The latter can be used when the above statistical requirement cannot be met. With polychromatic radiation, the Bragg conditions can be satisfied for each grain.

Utilizing operando microbeam XRD, the structural changes within individual particles of $LiFePO_4$ have been comparatively revealed at low and high rates.[13] Because of the comparability of beam size (1.7 um) and particle size (0.14 μm), the statistical diffraction rings can be broken up into separate spots, each representing an individual single crystalline grain. To increase the detection probability of reacted grains with good statistics, the coffee bag cell measured in transmission mode rotates over an angle perpendicular to the X-ray beam. A single diffraction spot on a specific diffraction ring defined by θ and a specific position of the ring defined by η represents an individual $LiFePO_4$ particle (Figure 12.12a). η is related to the orientation

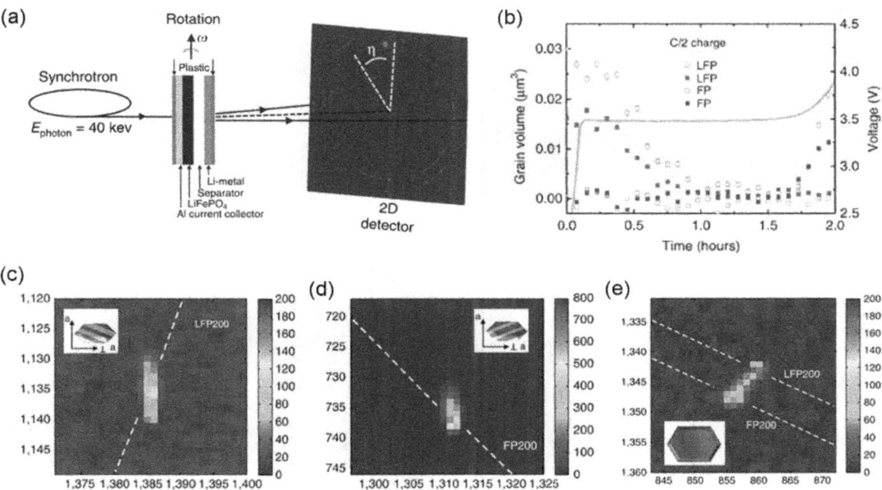

FIGURE 12.12 (a) Schematic representation of the microbeam X-ray diffraction experimental setup. During the exposure, the sample was continuously rotated around the vertical axis. (b) Charging voltage curve (C/5) including the evolution of a 2D (200) LFP and (200) FP peak showing the progressive FP formation and LFP disappearance. Vertical streaks are observed for the (200) reflections of the (c) LFP and (d) formed FP phases, indicating the formation of a platelet-shaped domain. (e) (200) Reflection at a 2C charging rate showing coexistence of the LFP and FP phases within a single grain. The dashed lines indicate the powder rings for the (200) reflections of the LFP and FP phases, indicated as LFP200 and FP200, respectively. (Reprinted with permission from ref. [13]. Copyright 2015 Nature Publishing Group.)

of crystalline particles, and the shape and size of the spot can reveal the shape and dimension of the phase domain. As shown by the 0.2 C-rate case in Figure 12.12b, the average transformation time of individual $LiFePO_4$ is slower than the prediction from the mosaic reaction model. In addition, streak-like diffraction spots can be observed when charged at a low rate of 0.2C (Figure 12.12c and d), indicating the exhibition of platelet-like $LiFePO_4$ and $FePO_4$ domains with well-defined interfaces. With a charge rate up to 2C (Figure 12.12e), the narrowing of the streaks indicates an increase in platelet domain thickness. The appearance of a double-peak feature demonstrates the diffusion of the interface with the formation of unstable intermediate phases within a single grain. This work emphasizes the significance of exploring phase transitions in individual grains to understand battery performance.

Apart from the phase evolution in individual particles, the distribution of the $FePO_4$ phase across the electrode sheet at the quasi-equilibrium state, corresponding to the charge distribution of Fe^{3+}, has been studied by microbeam XRD.[14] The distribution across the cross-section of the electrode sheet, which is charged to 50% SOC (state of charge) in the Swagelok-type cell, was first raster-scanned vertically and horizontally, as shown in Figure 12.13a, with a beam size of 2×5 μm. With a charge rate of 3 A/g (~18 C), the concentration of $FePO_4$ decreases away from the top

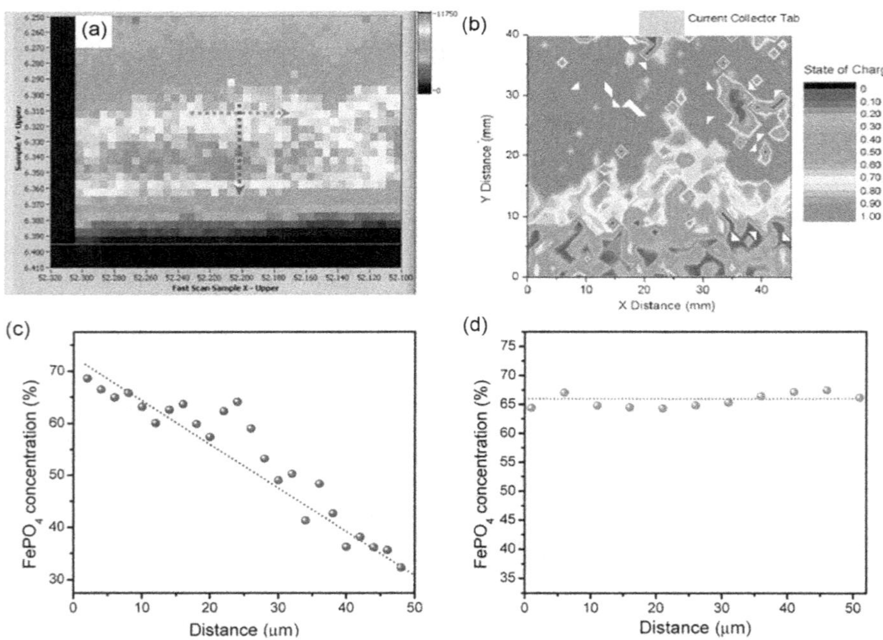

FIGURE 12.13 (a) Fe fluorescence image of the cross-section of a circular $LiFePO_4$ electrode. The approximate locations of the vertical and horizontal scans are superimposed. (b) $FePO_4$ phase concentration profile of the prismatic electrode at 50% SOC (charged at 2 A/g). $FePO_4$ phase concentration versus scan distance along (c) the vertical direction and (d) the horizontal direction of the $LiFePO_4$ electrode at 50% SOC (charged at 3 A/g). (Reprinted with permission from ref. [14]. Copyright 2010 American Chemical Society.)

surface, which nears the current collector (Figure 12.13c), while remaining constant in the plane (Figure 12.13d). In comparison to the even distribution in both directions with a small rate of 20 mA/g (~0.11 C), the gradient distribution is thought to result from the limited electronic conductivity. The limitation from electronic conductivity can be further identified by the uneven distribution in the rectangular sheet, which is charged to 50% SOC at 2 A/g (~12 C) with a pouch cell and scanned with a beam size of 2×25 μm (Figure 12.13b). The concentration of $FePO_4$ decreases away from the tap, which is used to conduct electrons. In addition, some uncharged islands can be found, which might result from isolation by delamination or cracking during battery preparation.

12.2.1.3 EDXRD

Although microbeam XRD enables scanning of the cross-section of electrode materials, it still has difficulty in operando experiments. In comparison, the method called energy-dispersive X-ray diffraction (EDXRD) is more specialized for the operando investigation of the structural changes along the depth direction of composite electrode sheets.[15] In an EDXRD experiment, as shown in Figure 12.14, polychromatic X-rays are used for incident beams with spot sizes on the scale of tens of micrometres. As the penetration of the whole cell in operando experiments, the X-rays need to have an energy range up to 200 keV, which calls for high-energy light sources. The scattered beam is collected by an energy-resolved detector at a fixed angle. The effective detection lies in the gauge area intersected by the incident beam and the scattered beam. As the diffraction process from all the other parts along the penetrating path does not interfere with the signals, the method is particularly suitable for operando experiments using real batteries. The diffraction spectrum can be collected one time owing to the energy dispersion, where Bragg's condition can be modified as

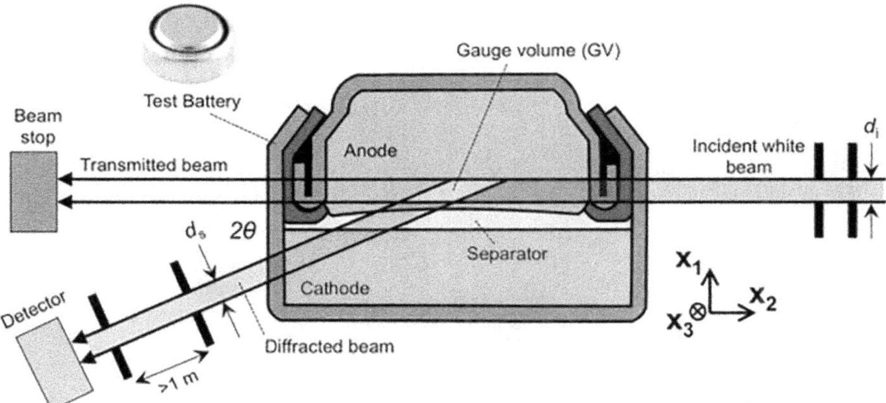

FIGURE 12.14 Schematic showing white-beam EDXRD from a gauge volume placed within a battery. The beam path to the detector, shown in blue, is determined by the collimation slit settings d_i and d_s. By moving the battery, the gauge volume can be placed at many locations. (Reproduced with permission from ref. [15]. Copyright 2020 Royal Society of Chemistry.)

$$d = \frac{hc}{2\varepsilon \sin\theta}, \tag{12.46}$$

where h is Planck's constant and c is the speed of light.

By moving the electrode accompanied with the battery along the x_1 direction, the phase structure at different positions can be distinguished because of the focusing beam size, providing the depth profiling capability. Most of the applications in battery fields have focused on structural changes in this dimension.

The homogeneity across the section in the $LiFePO_4$ prismatic cell has been studied under operando conditions by EDXRD.[16] As one unit of repeating layer structure with cathode and anode switched in prismatic cell is 400 μm, a whole mapping from cathode to anode can be achieved by scanning with 40 μm resolution ten times. A typical depth profile at a specific SOC is shown with contour imaging (Figure 12.15a and b) in the range of energy up to 200 keV. The regions of the cathode and anode can be clearly distinguished according to various XRD features from the electrode and corresponding current collectors. One depth profile contains ten EDXRD spectra in total. The fraction of the $FePO_4$ phase at a certain depth can be extracted from the corresponding EDXRD spectra for the cathode. During the discharge process, phase changes along the depth direction are profiled at 19 SOCs. The phase changes for the cathode on both sides of the current collector are revealed by the fraction of the $FePO_4$ phase, as shown in Figure 12.15c. The vertical spacing between the fraction at the same depth decreases during discharge, indicating a decrease in the local reaction rate. The decreasing trends on the two sides are different. The above two phenomena are thought to originate from heterogeneous reactions. In addition, the largest phase relaxations in the fully discharged state are observed at 20 and 140 μm. This indicates that the nonequilibrium conditions increase away from the current collector. All the inhomogeneous distributions are thought to be closely related to the

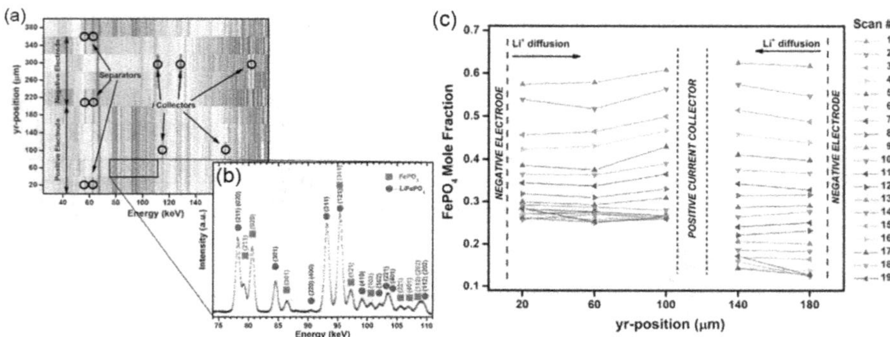

FIGURE 12.15 Internal phase mapping and structural information. (a) Diffraction contour plot showing a one-dimensional cross-section through an electrochemical cell. Colour represents the diffraction intensity, with blue being the lowest and red being the highest. (b) Typical energy-dispersive diffraction spectrum of the lithium iron phosphate positive electrode. The Bragg reflections of both phases are labelled by their hkl coordinates. (c) Time evolution of the iron phosphate mole fraction as a function of position (depth) in the electrode layer. (Reprinted with permission from ref. [16]. Copyright 2015 Elsevier.)

electronic conduction. In contrast to the case of $LiFePO_4$, the in situ EDXRD study of $LiMn_2O_4$ based on the 2,032 coin cell indicates an opposite trend of the spatial phase distribution.[17] Specifically, the region far away from the current collector has been found to react ahead of the region near the current collector. It can be inferred that the inhomogeneous phase changes are related to lithium transport, which has been identified in subsequent in situ EDXRD.[18]

The problem of EDXRD is the low peak resolution, which primarily results from the limited energy resolution of detectors. This makes the method less sensitive to the peak shift during the reaction and thus more suitable for qualitative analysis. To improve the quantitative capability, an alternative method called energy-scanning confocal XRD has been developed. The polychromatic incident beams are achieved by energy scanning through a monochromator with high energy resolution. Accordingly, the limitation of the energy resolution from detectors is removed as energy-resolved detectors are no longer needed. The energy-scanning confocal XRD method improves its spatial and depth resolutions for quantitative analysis at the cost of time-resolved capability.

12.2.2 Neutron Diffraction

ND or neutron powder diffraction (NPD) is a structure probing method that is complementary to XRD. NPD has the same principle as XRD, following Bragg's law. The major characteristics of neutrons making NPD unique are the sensitivity to low-Z elements, such as Li and O, and the high penetration into the sample, where more discussions can be found above. The relatively larger coherent cross-sections of Li, especially ^7Li, than those of TMs make it specialized in resolving the structure information of Li, such as positions and occupancies. The much higher penetration is beneficial for the study of bulk structure and in situ/operando experiments. On the other hand, NPD experiments call for large sample amounts and can interfere with other parts containing low-Z elements in the beam. Interference from H can be particularly problematic in battery studies because of its wide exhibition in separators, electrolytes and binders. Fortunately, commercial batteries, which are typically studied in situ/operando to obtain real insight, can intrinsically solve the two problems. On the one hand, electrodes are usually coated on both sides of current collectors with a large area mass density, and the total quantity is very large. On the other hand, the quantities of electrolyte, conductive agents and binders have been minimized as much as possible.

By combining XRD and NPD measurements, the interphases at certain temperatures are found with structure refinements.[19] Specifically, $Li_{0.5}FePO_4$ and $Li_{0.75}FePO_4$ are found as solid solution phases at 350°C and 370°C, respectively, which have olivine-type structures with Li^+ disorder. When cooling back to room temperature, another metastable phase, $Li_{\sim 0.64}FePO_4$, appears instead. The average Li-O bonding lengths in $Li_{\sim 0.64}FePO_4$ become longer accompanied by the shortening of the Fe-O bonding lengths in comparison to $LiFePO_4$ owing to the exhibition of Fe^{3+}. The discrepant bonding length might be responsible for the metastability of the intermediate phase. The results could help to gain further insight into the mechanism of the high rate capability of $LiFePO_4$.

In addition, lithium diffusion in $LiFePO_4$ has been studied by TOF NPD to reveal the origination of the rate capability, as shown in Figure 12.16.[20] Using Rietveld refinement for the NPD pattern, the anisotropic atomic displacement parameters for lithium can be displayed as green ellipsoids to illustrate thermal displacement. The diffusion path of lithium can be derived from the preferable direction of displacement, which points to the face-shared vacant tetrahedra. This is the first direct experimental observation of a curved one-dimensional chain for lithium motion. In addition, the lithium distribution along the [010] direction at 620 K has been visualized by analysing NPD with the maximum entropy method.

The other major application of NPD in batteries lies in solid-state electrolytes. By combining XRD and NPD, the structures of Li_6PS_5Cl, Li_6PS_5Br and Li_6PS_5I were comparatively studied.[21] Cl and Br are found to occupy two S sites, whereas I stays at an independent site. Only one Li site can be found in Li_6PS_5Cl, while two Li sites can be found in Li_6PS_5Br and Li_6PS_5I. This study highlights the critical role of halide in the distribution of atoms, which should be responsible for the high ionic conductivity

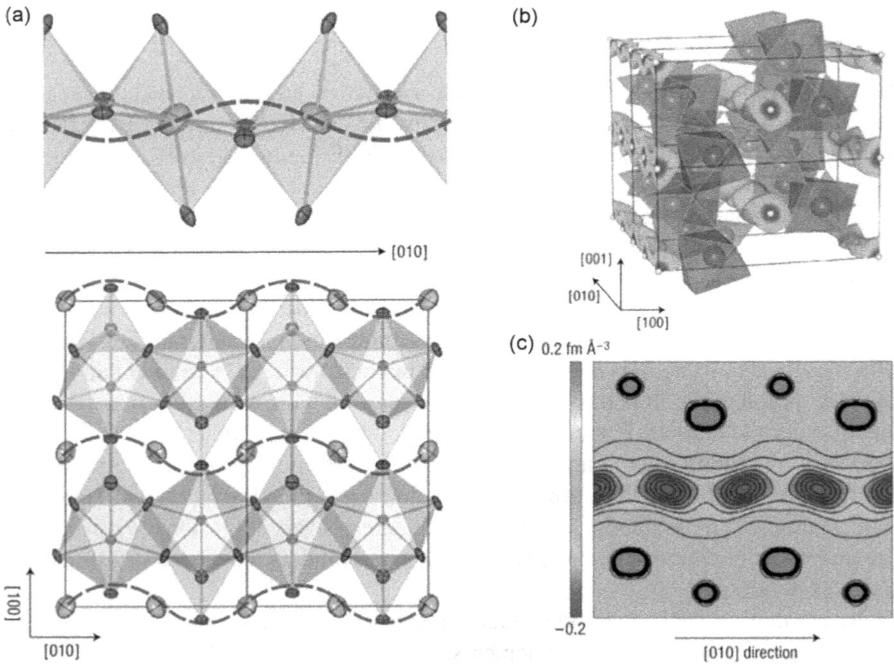

FIGURE 12.16 (a) Anisotropic harmonic lithium vibration in $LiFePO_4$ shown as green thermal ellipsoids and the expected diffusion path. (b) Three-dimensional Li nuclear density data shown as blue contours, which are calculated by the MEM using neutron powder diffraction data measured for $Li_{0.6}FePO_4$ at 620 K. The brown octahedra represent FeO_6, and the purple tetrahedra represent PO_4 units. (c) Two-dimensional contour map sliced on the (001) plane at $z = 0.5$; lithium delocalizes along the curved one-dimensional chain along the [010] direction, whereas Fe, P and O remain near their original positions. (Reproduced with permission from ref. [20]. Copyright 2008 Nature Publishing Group.)

in Li_6PS_5Br. In situ/operando NPD can be easily performed in a pure solid-state electrolyte. For $Li_7La_3Zr_2O_{12}$, in situ NPD indicates that the partial melting of carbonate during the heating process can promote the phase formation of $Li_7La_3Zr_2O_{12}$.[22]

12.2.3 Small-Angle X-Ray Scattering

SAXS is a tool specialized in structures with larger periods. The underlying principle of SAXS is almost the same as that of XRD, which has a very similar expression on the total cross-section. The only difference is that the atomic form factor for XRD is substituted by the form factor of the nanoscale unit for SAXS. Therefore, SAXS is particularly good at resolving ordered structures at the particle level, while XRD is particularly good at the atomic level. Specifically, the former scale is approximately 1–100 nm, while the latter is approximately 0.1–1 nm. In experiments with X-rays from a Cu Kα source ($\lambda = 1.54$ Å), the former scattering features are located in 2θ from 0° to 10°, which lies in a "small angle" range. The latter features appear in the 2θ range from 10° to 100°, which categorizes XRD as WAXS. However, the techniques can be classified better by the Q range because the scattering angle can be varied according to the wavelength of the incident X-ray. The Q range of SAXS is approximately 0.06–6 nm^{-1} in comparison to 6–60 nm^{-1} for XRD. The technique performed in the Q range below 6 nm^{-1} is known as ultrasmall-angle X-ray scattering. It has been mentioned that there is no strict boundary of the classification.

Different from the simplification of the atomic form factor into atomic number Z under the long wave limitation for XRD, the form factor of the nanoscale unit for SAXS is related to the atomic distribution in the unit. Therefore, SAXS has a powerful capability in resolving the structural information of the unit, such as morphology (size and shape) and atomic composition and position, apart from the arrangement of the unit related to the structure factor. To form the scattering signals, the average scattering length density (SLD) of the units should be different from that of the matrix. The unit can be nanocrystals, bio-macromolecules, polymer domains or even defects. The scattering intensity is proportional to the square of the differential value. Therefore, SAXS cannot tell whether the contrast comes from void spaces or solid domains with the same absolute differential value.

Form factor and structure factor analysis make SAXS widely applicable in various forms of materials. SAXS can be performed on one particle, which can reveal information on the form factor, including the size, shape and atomic composition. SAXS can be used to study a group of uncorrelated particles, such as liquid solutions, where the size distribution of the particle can be obtained. For a group of correlated particles, such as nanostructured materials, SAXS can reveal structural information according to both the form factor and structural factor.

SAXS experiments can be performed in the laboratory with X-ray tubes. However, it can take a long measurement time to obtain high-quality data because of the low flux. Similar to synchrotron XRD, in situ/operando SAXS with time-resolved and/or space-resolved capability can be achieved because of the merits of synchrotron X-rays, including high flux, micro size and tuneable energy.

For battery studies, SAXS has been widely used to resolve nanosized structures and their spatial distributions in electrodes and electrolytes. The first example is a comparative study on the conversion reaction of a series of Fe fluorides, oxyfluorides and oxides.[23] At the discharge state, Fe^0 can be formed through the conversion reaction in all Fe-based compounds. The size and distribution of Fe^0 particles have been studied by SAXS simulation (Figure 12.17a and b). Because of the forms of profiles and inversions in the gradient with increasing Q, the SAXS data need to be modelled as monomodal distributions of spheroidal Fe particles with a structure factor based on a hard-sphere model to describe the interparticle interactions. As shown in Figure 12.17c, fluoride-rich electrodes produce larger Fe^0 nanoparticles with broader size distributions than oxygen-rich electrodes. For conversion at high temperature, the SAXS features shift to lower Q values, indicating an increase in Fe particles. However, the particle diameters from both lognormal and Gaussian distributions change little with increasing temperature for FeO_xF_{2-x}. For pure Fe fluorides and oxides, the shape of Fe particles can be well modelled by perfect spherical geometry, with each particle surrounded by an average of three to four neighbouring nanoparticles with a separation of 1–9 Å. In contrast, FeO_xF_{2-x} electrodes produce abnormal Fe nanoparticles with no well-defined neighbour at high temperature. An increase in polydispersity may indicate a much larger size distribution or distortion of particles. The difference in particle size, size distribution and shape of Fe^0 particles can be attributed to the difference in O and F anion chemistry.

Another typical application is the investigation of the solvent shell structure in liquid electrolytes. The microscopic structure of electrolytes composed of EC with classic solvents in lithium-ion batteries, including PC, EMC and DEC, has been studied by SAXS.[24] As shown in Figure 12.18b, the SAXS curves of PC and EC/PC solutions are almost featureless, indicating the exhibition of a similar microscopic structure. In contrast, the spectra of EC-EMC and EC-DEC show features in a Q range from 0.02 to 0.5 $Å^{-1}$. The featureless curve might result from the cyclic carbonate properties of EC and PC. The cyclic EC mixed with linear EMC or DEC could

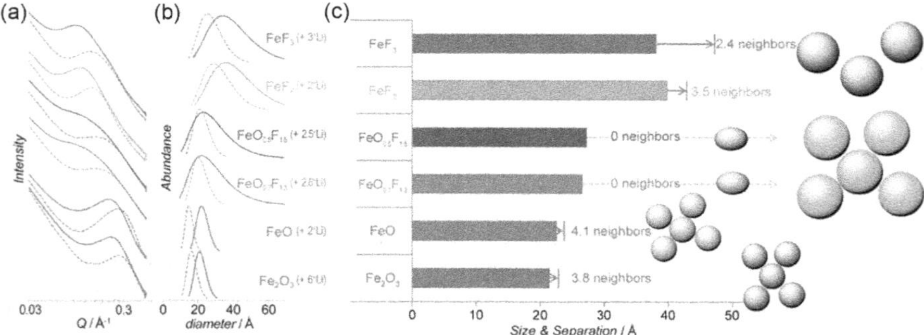

FIGURE 12.17 (a) SAXS data and (b) the corresponding (lognormal) particle size distributions for Fe_0 nanoparticles formed through electrochemical conversion of a series of Fe electrodes at 25°C and 60°C. (c) The Fe_0 nanostructure resulting from lithiation of different Fe electrodes at 60°C. The dominant (mode) primary particle size (bars), the average particle–particle separation (arrows) and the number of well-defined neighbouring particles are indicated. (Reproduced with permission from ref. [23]. Copyright 2014 American Chemical Society.)

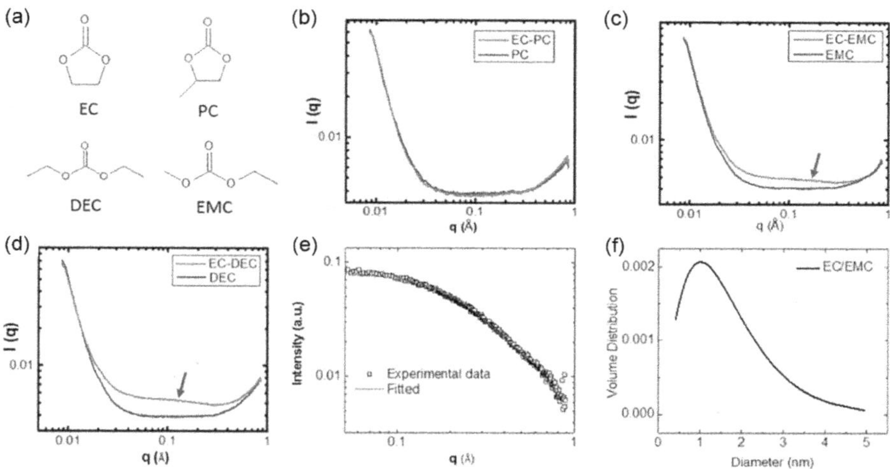

FIGURE 12.18 (a) Chemical structures of EC, PC, DEC and EMC. (b) SAXS data of EC/PC and PC. (c) SAXS data of EC/EMC and EMC. (d) SAXS data of EC/DEC and DEC. (e) Fitted curve of subtracted data of the EC from the EC/EMC solution. (f) Size distribution obtained after fitting the spherical model to the subtraction result of (e). (Reproduced with permission from Ref. [24]. Copyright 2019 the Electrochemical Society.)

cause the rearrangement of molecules. The evolution related to EC can be obtained by subtracting the EMC data from the EC-EMC data (Figure 12.18c). This reveals that EC molecules can form clusters at ~1 nm (Figure 12.18f) in EMC and DEC by fitting SAXS curves with a spherical model in the range of 0.05–0.4 $Å^{-1}$. In addition, an $LiPF_6$ contact ion pair with a size of ~6 Å can be found in the EC-EMC electrolyte, which is absent in EC-PC mixtures. The solvent structure as well as the nonideal behaviour of $LiPF_6$ can be helpful to the design of new electrolytes.

12.2.4 Small-Angle Neutron Scattering

Similar to NPD, SANS has advantages in in situ/operando studies in battery materials because of the high penetration and the sensitivity to low-Z elements, such as H, Li and O, in comparison with SAXS.

One of the major applications is the investigation of solid electrolyte interphase (SEI) evolution. For example, the interaction of a concentrated LTFSI electrolyte with a mesoporous carbon electrode and thus the influence on the dynamics of the electrochemical reaction have been studied by operando SANS (Figure 12.19a).[25] The microstructural information can be obtained by analysing the Porod region at low *q*, the peak feature due to mesopore ordering and the microporous region at high *q*. Specifically, the intensity decay at the low *q* range indicates carbon surface fractality. The peak position and intensity related to mesopores reveal the pore–pore distance and the contrast between pores and carbon frameworks, respectively. The intensity at the high *q* value reflects the micropore scattering intensity (MPSI) because of the constant background. By fitting the SANS spectra, the peak position and intensity

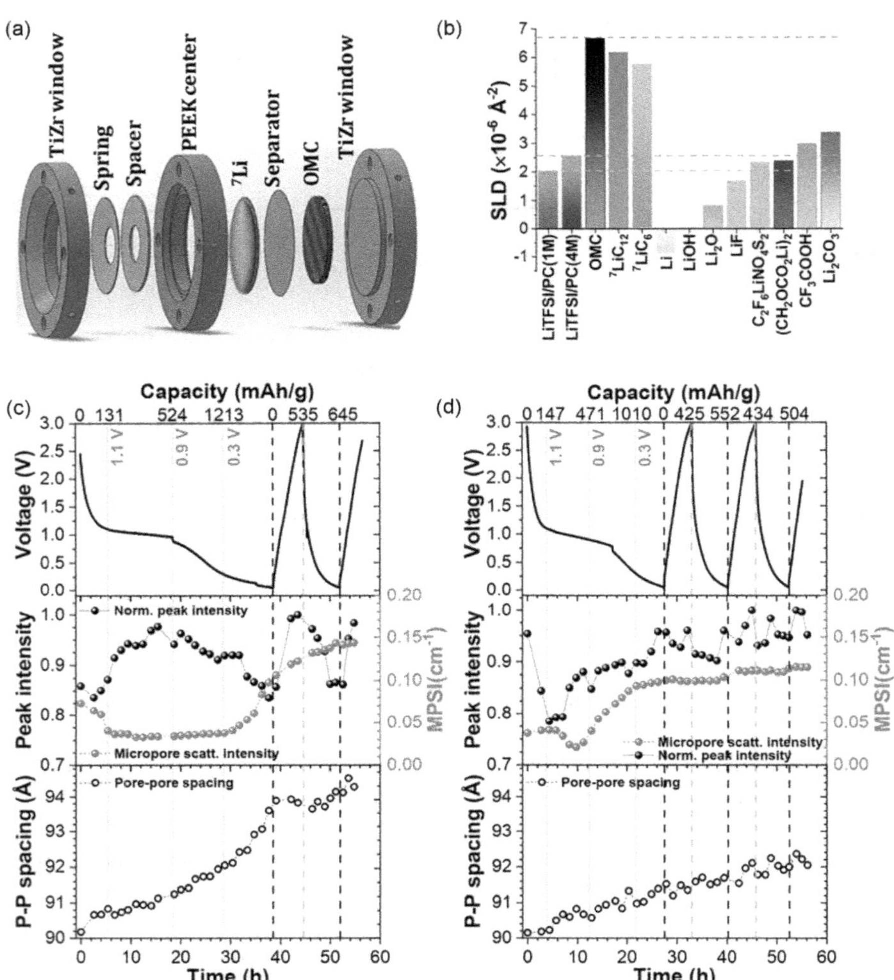

FIGURE 12.19 (a) Schematic view of the operando SANS cell. (b) The scattering length density (SLD) values for dominant components present in the cell at OCV (electrolyte, carbon) and those that may form as a result of cycling (lithiated carbon, lithium) and electrolyte reduction (major SEI components). Operando electrochemical and SANS results for (c) the 1M and (d) the 4M electrolyte systems. Top: electrochemical cycling profile; middle: integrated Gaussian peak intensity for pore ordering and micropore (high q) scattering intensity (MPSI); Bottom: pore–pore spacing determined by the Gaussian peak position. (Reproduced with permission from ref. [25]. Copyright 2019 Royal Society of Chemistry.)

related to mesopores and scattering intensity from micropores are extracted. The evolution of corresponding feature values during the electrochemical reaction using 1M and 4M electrolytes are comparatively presented in Figure 12.19c and d. Generally, the evolution in the first discharge process of the two cases can be divided into four stages. An analysis of the specific evolution can be performed by combining the SLD values from different components, as shown in Figure 12.19b. At the first stage with discharge from the OCV (open circuit voltage) to 1.1 V, the mesopore peak intensity decreases, while the MPSI of the 1M cell decreases similarly. This result indicates that $Li^+(PC)_4$ absorbed on and then filled the micro- and mesopores with the displacement of Ar in the 1M cell. For the 4M cell, the mesopore peak decreases while MPSI remains constant. This indicates that the Li(PC)-TFSI aggregates only absorb on mesopores, which is attributed to the high electrolyte viscosity. The pore–pore spacing increases from ~90.1 to ~90.9 Å for the 1M cell, which indicates expansion in the carbon framework. However, no corresponding change was found for the 4M cells, indicating no expansion in the matrix. The difference in expansion is consistent with the absorption picture.

After discharging from 1.1 to 0.9 V at the second stage, the mesopore peak intensity increases slightly below 1.1 V while MPSI remains constant. This results from the co-intercalation of $Li^+(PC)_4$ into graphitic layers, while the small increase in intensity is attributed to SEI formation with the reduced products of TFSI in mesopores, which have a slightly lower SLD than that of $Li^+(PC)_4$ and thus increase the contrast. For the 4M cell, the micropore peak intensity remains constant and starts to sharply increase from 1.0 to 0.9 V. This result indicates the domination of changes in SEI chemistry on the mesopores after the completion of pore filling with $Li^+(PC)_4$-TFSI. In addition, the change in MPSI, which is constant above 1.0 V and decreases to a minimum at 0.95 V, reveals the adsorption of $Li^+(PC)_4$-TFSI onto micropores below 1.0 V. The higher viscosity of the 4M electrolyte might be responsible for the requirement of adsorption for a stronger driving force. The electrode in the 1M cell shows a small expansion from ~90.8 to 91.2 Å, whereas the electrode in the 4M cell shows an initial jump of approximately 0.7 Å, followed by little change. The former is related to the expansion resulting from co-intercalation, while the latter is due to adsorption into the micropores, which is nearly identical to the occurrence of the 1M cell at higher voltages.

When discharged to the third stage, the mesopore peak intensity of the 1M cell decreases continuously. This can be explained by the formation of SEIs, such as $ROCO_2Li$, with the reduction of PC and Li^+-PC. However, the MPSI increases slightly for the 1M cell. This result might be due to the small increase in the SLD of the carbon matrix caused by the co-intercalation of solvated Li into graphitic layers. For the 4M cell, the mesopore peak intensity increases only slightly, indicating few reactions in the mesopores. In contrast, the MPSI starts to sharply increase from 0.95 to 0.3 V, evidencing a significant change in the chemical composition of the micropores. The pore–pore spacing for the 1M cell increases rapidly from ~91.2 to 92.1 Å, which is likely caused by the co-intercalation of larger solvated Li^+ species. The relatively small changes with $\Delta d = 0.5$ Å for the 4M cell originate from the similar levels of micropore filling compared to the $Li^+(PC)_4$-TFSI.

At the final discharge stage from 0.3 to 0.05 V, the mesopore peak intensity for the 1M cell continues to decrease until the end of the first discharge, which is again attributed to SEI formation. A sudden increase of 15% in the MPSI occurs from ~0.3 V to the end of the discharge. This can be explained by the dramatic change in contrast because of the filling into the micropores with Li, which has a negative SLD. For the 4M cell, the micropore scattering changes little, which can be understood by the saturation of micropores with Li-rich compounds. Instead, the mesopore scattering increases strongly. This can be explained by the domination of micropore scattering by the Li species with a low SLD. The pore–pore spacing for the 1M cell increases from 92.1 to 93.9 Å, consistent with the change in the MPSI and co-intercalation of solvated Li. However, the same increase for the 4M cell cannot be found during Li interaction in the micropores. This study reveals the complicated dynamics of SEI formation and pore filling along with the concentration of electrolytes during the electrochemical process. It shows the powerful capability of operando SANS in SEI-related studies by exploring microstructure information.

12.2.5 Pair Distribution Function

While XRD has difficulty resolving structures with poor crystallinity, structural investigations of materials lacking long-range order, such as nanocrystals, disordered materials, glasses and even liquids, need to make use of the total scattering signals, including Bragg diffraction and diffuse scattering. The atomic PDF indicative of the local structure can be derived from the total scattering signals through a Fourier transformation of the coherent scattering function $S_{\mathrm{coh}}(\boldsymbol{Q})$.

$$G(r) = \frac{2}{\pi}\int_{0}^{Q_{\max}} Q\left[S_{\mathrm{coh}}(Q) - 1\right]\sin(Qr)\,dQ, \tag{12.47}$$

where $Q_{\max}$ is the maximum magnitude of the scattering vector $\boldsymbol{Q}$. $G(r)$ is called the reduced atomic PDF, which is the presented data indicative of the local structure.

$S_{\mathrm{coh}}(\boldsymbol{Q})$ can be related to the measured intensity $I(\boldsymbol{Q})$ by

$$I(\boldsymbol{Q}) = I_{\mathrm{coh}}(\boldsymbol{Q}) + I_{\mathrm{inc}}(\boldsymbol{Q}) + b.g. \propto \sigma_{\mathrm{coh}} S_{\mathrm{coh}}(\boldsymbol{Q}) + \sigma_{\mathrm{inc}} S_{\mathrm{inc}}(\boldsymbol{Q}) + b.g., \tag{12.48}$$

where σ_{coh} and σ_{inc} are coherent scattering cross-section and incoherent scattering cross-section, respectively. With the intensity corrections from the background, detector efficiency, dead time absorption, multiple scattering, etc., $S_{\mathrm{coh}}(\boldsymbol{Q})$ can be calculated by

$$S_{\mathrm{coh}}(\boldsymbol{Q}) = 1 + \frac{I_{\mathrm{coh}}(\boldsymbol{Q}) - \Sigma c_i \left|f_i(\boldsymbol{Q})\right|^2}{\left|\Sigma c_i f_i(\boldsymbol{Q})\right|^2}, \tag{12.49}$$

where c_i and $f_i(\boldsymbol{Q})$ are the atomic concentration and the atomic form factor for the ith type of atom, respectively.

The definition of $G(r)$ can be expressed quantitatively as

$$G(r) = 4\pi r \rho_0 \left[g(r) - 1\right], \tag{12.50}$$

where ρ_0 is the atomic density of the system. $g(r)$ is called the pair correction function and can be further related to the position vector of atoms $\boldsymbol{r}$ by

$$g(r) = \frac{1}{N}\sum_{i}\sum_{j \neq i}\left\langle \delta\left(\boldsymbol{r} + \boldsymbol{r}_i - \boldsymbol{r}_j\right)\right\rangle, \tag{12.51}$$

where *i* and *j* represent the *i*th and *j*th atoms, respectively. *g*(*r*) stands for the probability of finding any atom in the volume unit around position $\boldsymbol{r}$ with an atom at the origin. For amorphous materials, $\boldsymbol{r}$ can be replaced by its amplitude *r* representing a radial distance. The physical meaning of *g*(*r*) can be better understood by introducing the radial distribution function *R*(*r*).

$$R(r) = 4\pi r^2 \rho_0 g(r) \tag{12.52}$$

R(*r*)*dr* means the number of atoms existing within the thickness *dr* at distance *r* from the origin.

Although *g*(*r*) and *R*(*r*) have clearer physical meanings, *G*(*r*) is usually plotted instead because of its straight relationship with the experimental values. According to the term $Q[S(\boldsymbol{Q}) - 1]$ in the *G*(*r*) expression, the data at higher *Q* values are particularly important for the derivation of PDF. High-quality total scattering data should ensure a high signal/noise ratio even for weak diffuse scattering, cover a large reciprocal space (>30 Å^{-1}) and have high q resolution. Therefore, synchrotron X-rays are particularly suitable for PDF experiments because of the high brilliance, tunability of energy up to 45 keV or above, and limited q broadening related to energy monochromaticity and beam divergence. Although some PDF experiments can be performed recently with lab XRD configurations, it is always at the cost of too much test time (tens of hours), large sample amount (hundreds of mg at least) and decreasing feasibility for preparation (for air-sensitive samples).

PDF analysis is very suitable to study the mechanism of the conversion reaction because of the nanosized products. Among various conversion materials, FeS_2 has compelling performance as a cathode in lithium-ion batteries. On account of the ambiguous structural information of the intermediate phases, the local structure of the products at various charged states has been studied by operando PDF.[26]

As shown in Figure 12.20, the first discharge profile displays one plateau feature. PDF data show a change of FeS_2 to Fe and Li_2S-like domains from the viewpoint of local structure. Furthermore, principal component analysis (PCA) identifies the existence of the third component. It forms intermediately and then is consumed. The local structure is found to be different from that observed in later cycles. In addition, the intensity of peaks for the fully discharged state decreases even at 14 Å, indicating poor long-range order.

As shown in Figure 12.21, the first charge profile contains two plateaus and a sloped region in between, corresponding to 2.3 mol of lithium storage. The first plateau provides 1 mol of lithium to the capacity, while the second plateau above 2.5 V contributes 0.7 mol of lithium. The sloped region between them corresponds to 0.6 mol of lithium, which cannot be neglected. Over the first charge, the domains of Fe and Li_2S transform to another local structure, which is different from both the initial FeS_2 and the product at full discharge. Above the lower plateau, the first three peaks lying in the range of 1.5–5 Å shift to lower r, corresponding to the shortening

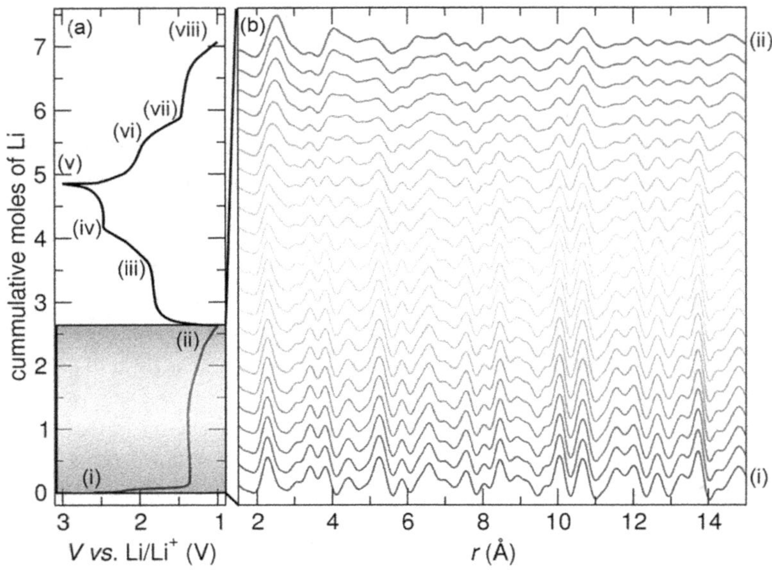

FIGURE 12.20 (a) The first 1.5 cycles of operando electrochemistry at ≈C/17. (b) The corresponding PDFs for the first discharge show the conversion of (i) FeS_2 to (ii) Fe- and Li_2S-like products. (Reproduced with permission from ref. [26]. Copyright 2017 American Chemical Society.)

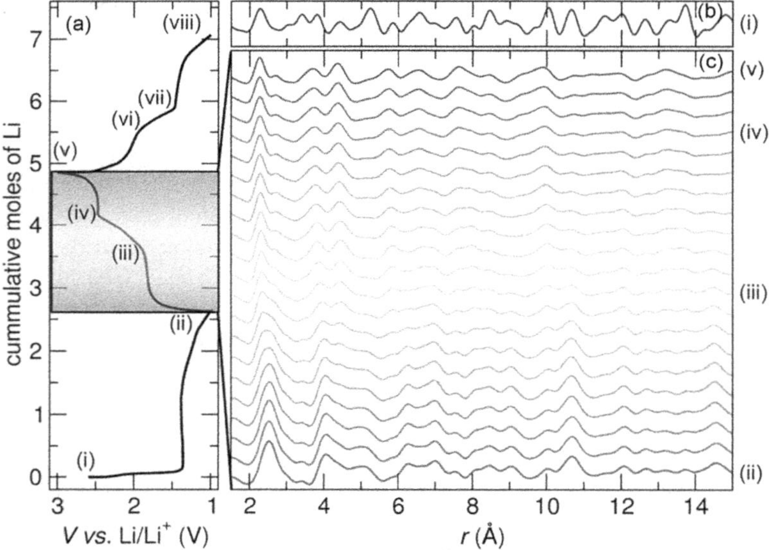

FIGURE 12.21 (a) First 1.5 cycles of operando electrochemistry at ≈C/17. (b) PDF of FeS_2 before cycling. (c) The PDFs for the first charge show conversion of (ii) discharge products to (iii) an intermediate phase and then (iii) to (iv) the shifting of pairwise interactions to lower r over the sloped region and (-v) the high potential plateau with continued charging. (Reproduced with permission from ref. [26]. Copyright 2017 American Chemical Society.)

of the specific pair distance. The gradual shift suggests that the charge product is structurally similar to the intermediate phase. The product has much less long-range order, as indicated by the low intensity at 14 Å, which is different from the initial FeS_2.

The local structure of the intermediate ternary product was studied by fitting the PDF of $Li_{1.5}FeS_2$ collected at 1.74 V based on the calculated structure. As shown in Figure 12.22, the three kinds of motifs used for fitting are all based on FeS_4 tetrahedra since the FeS_6 octahedra can be excluded by preliminary PDF fitting. The "tetrad" motif has both edge- and corner-sharing FeS_4 tetrahedra. The second one has corrugated chains with edge-sharing tetrahedra, while the third one has linear chains of edge-sharing tetrahedra. All the fittings based on the three motifs can capture most of the features well. However, both the fit and the differential curve show that the tetrad motif could fit the PDF data in the range of 1.5–5.5 Å best. However, the overall quality of fitting decreases with the range extending to higher r. This indicates that the local structure is different from the more average structure. In addition, many features can be captured when the tetrad motif is used to fit the data at various

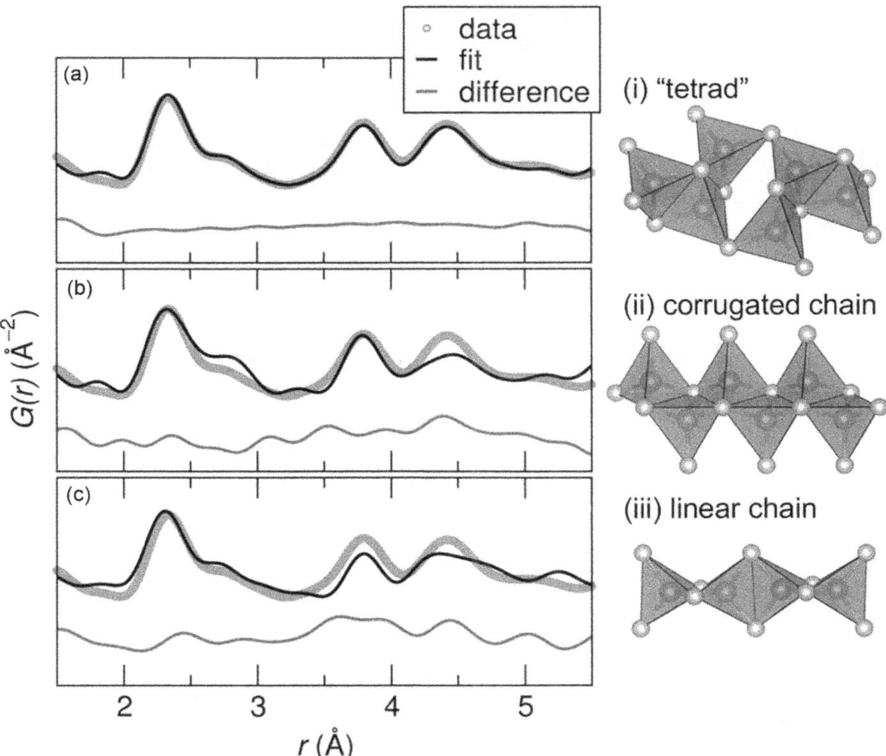

FIGURE 12.22 Fits of three calculated structures with the composition $Li_{1.5}FeS_2$ with tetrahedrally coordinated Fe to data collected at 1.74 V and 1.2 mol of Li during the second discharge and various local structure motifs, including (a) "tetrad", (b) corrugated chains of edge-sharing tetrahedra and (c) linear chains of edge-sharing tetrahedra. (Reproduced with permission from ref. [26]. Copyright 2017 American Chemical Society.)

states charged above the lower plateau, further identifying the similarity between the intermediate state and the charge product.

The hybrid modes of storage in the charge process are revealed from the viewpoint of the local structure, which is beneficial for screening promising candidates. This study shows the powerful capability of operando PDF in studying the mechanism of lithium storage at the nanoscale.

The other interesting use of PDF is located in the structural study of electrolytes. PCA enables the extraction of dissimilar features indicative of cation–solvent interactions from common features resulting from anion–solvent interactions. The application of PDF to solutions would aid in the development of new electrolytes in lithium-ion batteries and beyond lithium-ion batteries.

12.2.6 Neutron Pair Distribution Function

PDF analysis can be performed by neutron total scattering, where the principles are very analogous to those of X-rays. The major merit of nPDF lies in the sensitivity to light elements, such as Li and O. Accordingly, lithium transport in solid-state electrolytes and lattice oxygen redox (OR) in cathodes with high energy density are two important issues in which nPDF are specialized. In the following, the application of nPDF will be introduced by the case study of OR in high-voltage $LiCoO_2$.[27]

Due to the high scattering length of O, as shown in Figure 12.23a, nPDF data are mainly contributed by O-related atom pairs. Among the specific contributions of O-related species to nPDF, O–O pairs make the greatest contribution. In comparison, as shown in Figure 12.23b, the O–O pairs make a much smaller contribution in the X-ray PDF (xPDF), identifying the specialty of the nPDF.

In Figure 12.24b, the evolution of the shortest O–O distances with different charged states is presented. The distance values can be extracted by model-free fitting with a Gaussian profile function and are shown in Figure 12.24c with those derived from XRD and NPD for comparison. The O–O distances provided by nPDF

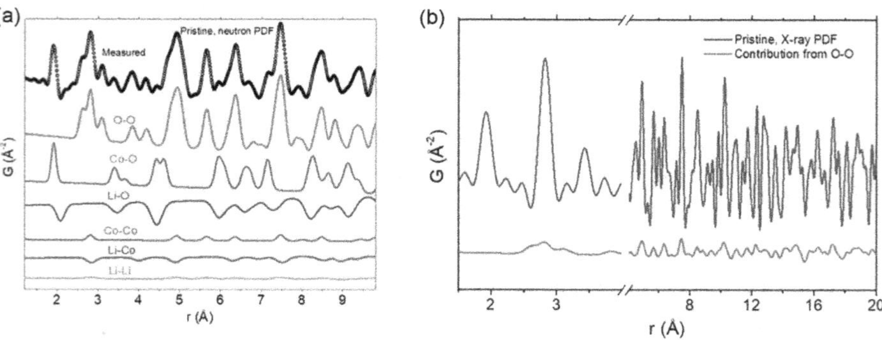

FIGURE 12.23 (a) The measured nPDF pattern for pristine $LiCoO_2$ with contributions from all possible atomic pairs. Specific atomic pair contributions are calculated using the model obtained by fitting the measured NPDF data. (b) The contribution from the O–O pair in the X-ray PDF of the pristine $LiCoO_2$ sample. (Reprinted with permission from ref. [27]. Copyright 2021 American Association for the Advancement of Science.)

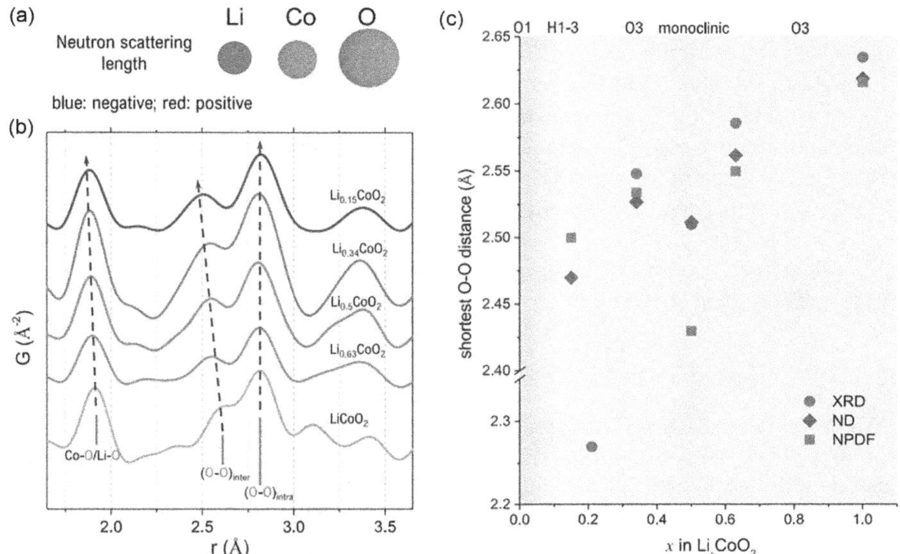

FIGURE 12.24 (a) The relative neutron scattering length of Li, Co and O. Note that the scattering length of Li is negative, while the scattering lengths of Co and O are positive. (b) Ex situ NPDF of $LiCoO_2$ during the charging process. (c) Evolution of interlayer O–O pair distances in $LiCoO_2$ during charging. (Reprinted with permission from ref. [27]. Copyright 2021 American Association for the Advancement of Science.)

should be more accurate than by XRD and NPD (shorter than 2.3 Å) for highly charged $LiCoO_2$ because of the occurrence of high disorder. This result reveals that the O–O distance decreases continuously during delithiation. The decrease below $x = 0.5$ is thought to arise from structural ordering. The dramatic shortening of the interlayer O–O distance even after 4.5 V is thought to result from the reduced repulsion among O^{2-} anions with the oxidation of O. The O redox has been probed by mapping resonant inelastic X-ray scattering (mRIXS), which is also in agreement with the Bader charge calculation. Meanwhile, the shortening of O–O distances can also be observed directly by aberration-corrected bright-field scanning transmission electron microscopy. In addition, the migration of TM Co at a highly charged state can be excluded mostly because of the undesirable PDF fitting using the structural model with the migrated Co ions. This study provides an elaborate description of the OR configuration, which is valuable to the deep understanding of anionic chemistry. The uniqueness of the nPDF in resolving local structure information involving rich light elements has been highlighted.

12.3 X-RAY ABSORPTION AND EMISSION SPECTROSCOPIES

Apart from the long-order changes in structure, electrochemical reactions in batteries always involve the evolution of electronic structure and local structure because of the simultaneous participation of electrons and ions. Therefore, deep insights into the reaction mechanism from the viewpoint of electronic and local structures are desired, guiding the improvement of battery performance. X-ray absorption and emission spectroscopies based on the photoelectric effect have a powerful capability of resolving the electronic structure and local chemical environment. In this section, an overview and applications of a series of techniques, including hard X-ray absorption spectroscopy (hXAS), soft X-ray absorption spectroscopy (sXAS), soft X-ray emission spectroscopy (sXES) and resonant inelastic soft X-ray spectroscopy (sRIXS), are introduced. This section clarifies the specialties of emission and absorption spectroscopies that have relatively wide applications in the battery field. This emphasizes the necessity of understanding electrochemistry from the viewpoint of electronic and local structures.

12.3.1 Overview of XAS, XES and RIXS

XAS, also known as X-ray absorption fine structure (XAFS), is based on the absorption process of X-rays by atoms in matters described quantitatively above. As discussed there, XAS covers absorption features in a certain energy range in the vicinity of the absorption edge and can be further divided into XANES and EXAFS. The abbreviation XANES is usually widely used in the hard X-ray range, while the corresponding part is called near-edge X-ray absorption fine structure (NEXAFS) in the soft X-ray range.

For NEXAFS, the absorption features are generally understood by the excitation of electrons at the core level to the unoccupied bound states. The absorption process described by the absorption coefficient μ follows Fermi's golden rule with the electric dipole approximation.

$$\mu \propto \left|\left\langle \Phi_f \middle| \hat{e} \cdot r \middle| \Phi_i \right\rangle\right|^2 \cdot \delta_{E_f - E_i - hv} \tag{12.53}$$

Here, Φ_i and Φ_f are the initial and final states in transition, respectively. $\left|\Phi_f \middle| \hat{e} \cdot r \middle| \Phi_i\right|$ is called the dipole matrix element, describing the transition rate between the two states. In the one-electron approximation, all other electrons in the initial and final states can be neglected because of their "inactive" and "frozen" status. Φ_f can be treated as the initial state Φ_i with a core hole represented as c and an excited electron e'. The expression can be simplified as

$$\mu \propto \left|\left\langle \Phi_i c e' \middle| \hat{e} \cdot r \middle| \Phi_i \right\rangle\right|^2 \cdot \delta_{E_f - E_i - hv} \approx \left|\left\langle e' \middle| \hat{e} \cdot r \middle| c \right\rangle\right|^2 \cdot \rho, \tag{12.54}$$

where ρ is the density of unoccupied states. According to the electric dipole selection rule, only the transitions with the orbital quantum number of the final state differ by 1 from the initial state ($\Delta l = \pm 1$), and the spin is conserved ($\Delta s = 0$). For example, the electric dipole transitions from 1s to 4p and 2p to 3d are permitted, but the transition from 1s to 3d is forbidden, which is permitted in the electric quadrupole transition. For XANES, the features have been more often analysed based on multiple scattering processes, which might be because p unoccupied states are usually more delocalized than d unoccupied states. Although the two pictures are often competitive, XAS features in this region can provide electronic and structural information localized on the excited sites, such as oxidation state, ligand type and geometric symmetry.

With photon energy increasing further away from the absorption edge, the excited electrons have enough kinetic energy to fly out of the central atoms. The backscattered electron waves by neighbouring atoms can interfere with the initial electron wave, leading to oscillations of absorption features (i.e., EXAFS). The EXAFS process can be further described quantitatively.

According to the description, EXAFS can be defined as the disturbance represented by the correction factor $\chi(E)$ on the isolated atomic absorption coefficient.

$$\mu(E) = \mu_0(E)\left(1+\chi(E)\right) \tag{12.55}$$

Therefore, the oscillations can be extracted from experimental data by

$$\chi(E) = \frac{\mu(E) - \mu_0(E)}{\mu_0(E)}. \tag{12.56}$$

On the other hand, the expression of oscillations can be derived step by step based on the interference picture. The oscillation can be related to the ideal interferences with a scattered spherical wave from one neighbouring atom by

$$\chi(k) = \frac{F(k, \pi)}{kR^2}\cos\left(2\pi\frac{2R}{\lambda}\right), \tag{12.57}$$

where $F(k, \pi)$ is the magnitude of the complex backscattering factor and R is the distance between the absorber and the neighbour. k is the magnitude of the wave vector and is defined as $2\pi/\lambda$, which can be further derived from the kinetic energy of emitted electrons, $\varepsilon_k = \varepsilon_{\text{photon}} - \varepsilon_{\text{ionization}}$, as $k = \frac{1}{h}\sqrt{2m_e\left(\varepsilon_{\text{photon}} - \varepsilon_{\text{ionization}}\right)}$. By considering the multiple absorbers and neighbours, the equation can be expanded as

$$\chi(k) = \sum_i \frac{N_i F_i(k,\pi)}{kR_i^2}\cos(2kR_i). \tag{12.58}$$

However, several effects must be considered in the real interference process. The first effect results from the nonideal "soft boundary" for the neighbouring atoms. Therefore, the backscattering photoelectron wave cannot reverse its direction instantaneously. This means that the scattered wave can have a phase delay, and thus the term $\delta(k)$ is introduced into the EXAFS expression. The second effect is due to the relaxation of electrons in absorbers in response to the hole in the core level. Because the effect is modest but not negligible, the amplitude reduction term S_0^2 is added to describe the resulting destructive interference. Usually, S_0^2 has a typical value between 0.7 and 1.0. The third factor originates from the inelastic scattering process with other electrons. This results in an increase in wavelength, which overall suppresses the EXAFS signals. The further photoelectrons propagate, the more likely inelastic scattering occurs. Therefore, an R-dependent damping term needs to be added. Meanwhile, decay of propagated electrons into the core holes of absorbers also suppresses the EXAFS signal. Therefore, the mean free path of the photoelectron should also be included. In total, the damping term is introduced as $e^{\frac{-2R}{\Lambda(k)}}$. This term reveals the intrinsic cause of EXAFS for local probing up to ~5 Å. The fourth factor is attributed to the disorder of atoms. The disorder can either come from the atomic vibration around their equilibrium position, which is closely related to temperature, or result from the multiple chemical environments related to local structures. Both cases can be modelled by terms called cumulant expansion. In the first approximation, it changes the EXAFS equation with a term dependent on the mean square radial displacement, σ^2, also named the Debye–Waller factor.

The final EXAFS expression can be written as

$$\chi(k) = S_0^2 \sum_i \frac{N_i F_i(k,\pi)}{kR_i^2} e^{-\frac{2R_i}{\Lambda(k)}} e^{-2k^2\sigma_i^2} \cos\left(2kR_i + \delta_i(k)\right). \tag{12.59}$$

Based on the experimental oscillations and their theoretical expression, a simulation can be performed. According to the expression, the local structural information, including the coordination number, interatomic distance and disorder, can be derived quantitatively through the simulation.

It should be noted that the xPDF is also capable of exploring the local structure. The differences between EXAFS and xPDF primarily lie in three aspects. First, xPDF collects the information about all atomic pairs. One single peak can have contributions from several kinds of pairs with similar bond lengths, which are difficult to separate. In contrast, EXAFS is an element-specific method that provides atomic distance information around the X-ray absorber of interest, which is able to deconvolute the total xPDF into the individual, element-specific xPDFs. However, EXAFS still has difficulty resolving two kinds of atoms that stay at a similar distance away from the absorber. In this scene, the contributions might be separated according to their k-dependent response only if the two kinds of atoms have a large difference in electron number (or atomic number). Second, EXAFS can only detect atomic arrangements around absorbers in a very short range (~5 Å) because of rapid damping. However, xPDF is capable of determining longer distances, thus allowing the differentiation of the short- and medium-range order in a given structure. Finally, the number of independent data points in EXAFS usually lies in the range from 15 to 30. It limits the number of degrees of freedom in the fitting, which is of order 10, and causes a lower precision in estimating distances (usually three significant digits). In contrast, xPDF can offer a high precision because of the large number (~10^3) of data points, even if refining a structure with tens of atoms in asymmetric positions, which calls for 10^2 variables. Therefore, EXAFS and xPDF are regarded as complimentary tools in resolving local structures.

As discussed above, XAS has powerful capabilities in resolving electronic structures and local structures. On the other hand, normal XAS experiments can be set up in a simple configuration because only the number of photons or electrons needs to be counted to quantify the absorption coefficient without any requirement of space and energy resolution on detectors. Therefore, XAS has been widely used in the investigation of battery materials. However, XAS is almost an exclusive technique at synchrotron facilities due to the requirement of energy tunability and high photon flux. Although lab-scale XAS instruments have been developed recently, the collection of XAS data has many limitations in the form and thickness of samples, element concentration, measurement time and so on. With the development of light source and beamline optics, variants of XAS techniques have been further achieved. For example, transmission X-ray microscopy has been achieved by using a Fresnel zone plate for space-resolved capability. To improve the time-resolved capability, a fly scanning mode in sXAS and quick XAFS (QXAFS) in hXAS have been achieved by improving the movement of monochromators. Moreover, energy-dispersive XAFS (DXAFS) has been developed for rapid measurement (~ms) in the application of polychromatic X-rays and position-sensitive detectors. In addition, the utilization of high-energy resolution fluorescence detectors improves the energy-resolving power of hXAS and is especially suitable for the investigation of heavy elements.

12.3.1.1 XES

X-ray emission spectroscopy (XES) is based on the decay process with the emission of fluorescence after photoelectric excitation and is thus assigned to the so-called photon-in/photon-out (PIPO) technique. The decay process involving the electron transitions follows the dipole selection rules. For example, decay from 2p to 1s or from 3p to 2d is permitted, while decay from 3s to 1s is forbidden. The energy of fluorescence equals the energy difference between the two electronic levels, independent of incident X-rays. Empirically, the energy of Kα corresponding to the 2p to 1s transition can be estimated by Moseley's law:

$$h\omega \sim (Z - b_s)^2, \tag{12.60}$$

where Z is the atomic number and b_s is the shielding constant. Any emission spectral line is the fingerprint of an element. The energy of fluorescence is resolved by utilizing a spectrometer to provide valuable information with respect to the occupied states, including oxidation state, spin state and even the nature of ligands (valence to core emission), which are complementary to detected information from XAS.

The experimental setup of XES requires a monochromatic X-ray larger than the absorption threshold and an energy-resolved spectrometer. The requirements can be met relatively easily compared with those of XAS in the laboratory. However, XES has a higher requirement on the photon flux because of the energy-resolved number of fluorescence photons. The weakness in the signal can be alleviated by using a new spectrometer, such as the von Hamos configuration, and an alternating excitation source, such as an electronic gun. In this scene, synchrotron XES still has great advantages because of its much higher photon flux.

12.3.1.2 RIXS

In contrast to the case of normal XES, fluorescence can also be emitted during the resonant excitation of an absorption process. Based on this, another PIPO technique, resonant inelastic X-ray scattering (RIXS), has been developed. The RIXS process can be quantitatively expressed by the Kramers–Heisenberg Formula, which is beyond the discussion in this chapter. The RIXS process can be further divided into direct and indirect decay processes. Direct RIXS involves electronic decay from occupied states to the core level with the production of electron–hole excitation. The resulting fluorescence has characteristic energy, which is independent of the energy of incident X-rays. Indirect RIXS refers to the backfilling of core holes by excited electrons disturbed by the strong core–hole potential present in the intermediate state. This causes the loss of energy and momentum of photons to some other excitations at a much lower energy scale. Therefore, the scattered photons have a slightly lower energy than the incident photons, leading to the energy dependence of fluorescence. RIXS is a powerful tool for probing the low energy excitations in condensed physics, including excitons, charge transfer, dd-excitations, lattice vibrations and so on. The significance of RIXS in chemistry has been recently revealed in probing the bulk OR in battery cathodes. However, the utilization of RIXS in battery studies is in its infancy because of the scarcity of experimental instruments and the complexity of theoretical explanations.

12.3.1.3 Comparison of Hard X-Ray and Soft X-Ray Absorption

XAS has become an indispensable tool for studying electrochemical mechanisms in batteries because of its powerful resolving capability and relatively easy accessibility. Most applications in batteries involve either hXAS or sXAS, relying on their research requirements. Therefore, hXAS and sXAS are given a further introduction in comparison. The basic difference between hXAS and sXAS lies in the energy range, where the classification of X-rays can be found in the first section of this chapter.

Because of the larger interaction of soft X-rays with matter, soft X-rays have a penetration distance three orders of magnitude smaller than hard X-rays. The difference in penetration results in two major differences between hXAS and sXAS. First, hXAS can be performed in the atmosphere, while sXAS can only be performed in an ultrahigh vacuum environment. This makes hXAS very suitable for in situ/operando experiments, while sXAS is very difficult for in situ studies. In recent decades, in situ/operando sXAS experiments have made considerable progress because of many pioneering studies in in situ cell design. Second, hXAS can be performed in transmission mode by penetrating samples within tens of micrometres, providing bulk information. However, sXAS is usually performed in reflection mode with a penetration depth of ~100 nm in solids, providing less bulk information. The transmission mode of sXAS can only be achieved with a sample thickness below 100 nm, which is widely used in scanning transmission X-ray microscopy (STXM).

Another critical contrast between hXAS and sXAS lies in the excitation process. hXAS covers the K-edge of 3d TMs, corresponding to the electronic transition from the 1s to 4p orbitals. However, sXAS is suitable for measuring the L-edge of 3d TMs and the K-edge of low-Z elements, such as C, N, O and F, which correspond to the electronic transition from the 2p to 3d orbitals and the transition from the 1s to 2p orbitals, respectively. Therefore, hXAS and sXAS cover most of the elements exhibited in ordinary battery materials, showing their critical role in battery studies. The difference in excitation can be further revealed by the resolving capability of hXAS and sXAS. For the K-edge of 3d TMs, hXAS usually contains XANES and EXAFS. The absorption edge in XANES can be used to determine the oxidation states. However, the edge features often overlap with other broadening shoulder peaks related to the delocalized 4p states, disturbing the analysis of oxidation states. Pre-edge features, corresponding to the dipole forbidden transition from 1s to 3d orbitals, often appear because of the hybridization of 3d and 4p states with distorted coordination. Therefore, pre-edge features with detection on 3d states are more often used to probe the coordination geometry. Moreover, the broadening XANES features above the edge and EXAFS features are specialized in exploring the local structure. As a whole, it can be concluded that the specialty of hXAS lies in the investigation of local structure. In contrast, sXAS is more powerful in resolving electronic structures. On the one hand, the dipole-allowed transition from 2p to 3d enables direct probing of the localized 3d unoccupied states near the Fermi level, which is very sensitive to the oxidation states, spin states and orbital splitting and orientation. On the other hand, the pre-edge features of the ligand K-edge can reveal the hybridization of the 2p states of the ligand and 3d states of TMs, directly probing the bonding interaction of TMs and ligands. In addition, soft X-ray EXAFS (sEXAFS) can hardly be achieved because of the limitation of monochromators. Therefore, the specialty of sXAS lies in the investigation of electronic structure.

The third major difference originates from the decay processes in hXAS and sXAS. As mentioned in Section 12.1, two kinds of decay modes, the fluorescence process and Auger process, are present after the resonant excitation. In the hard X-ray range, the fluorescence yield is dominant, while the electron yield becomes the major part in the soft X-ray range. Therefore, hXAS measurement has been primarily performed through transmission or fluorescence mode. Because of deep penetration, hXAS usually provides bulk information. In contrast, sXAS can be performed by collecting the total fluorescence yield (TFY) and total electron yield (TEY). TEY mode usually has much stronger signals compared with TFY mode because of the dominated Auger process and effective suppression of background noise. The TFY mode enables us to obtain relatively bulk information (~100 nm). In comparison, the TEY mode is sensitive to the surface information (5–10 nm) because of the limited mean free path of excited electrons. The surface sensitivity can be further improved by selectively collecting excited electrons with a certain kinetic energy, resulting in the development of the partial electron yield (PEY) mode and Auger electron yield (AEY) mode.

As discussed, hXAS and sXAS comprise different levels of information, which are complementary, on the TM-L coordinated structure units. The complementarity of the information can be further achieved by hard X-ray Raman scattering (hXRS), which is based on the inelastic scattering process. The local structure of ligand and unoccupied states in the bulk can be revealed by hXRS because of the feasibility of sEXAFS for the light elements and bulk sXAS. Moreover, soft XES and sRIXS can provide information related to the occupied states near the Fermi level, which helps provide a whole description of the electronic structure combined with sXAS for the investigation of the redox mechanism. Analogous to XAS, XES and RIXS can also be performed with hard X-rays, providing information on bound occupied states, which is outside the scope of this chapter.

12.3.2 Applications

12.3.2.1 Redox Reactions through XAS, XES and RIXS

Apart from the evolution of structures in long-range order, which primarily results from lithium storage, redox reactions are another major topic in batteries because of the intrinsic charge transfer process. Because of their sensitivity to oxidation states, hXAS, sXAS, sXES and sRIXS have been demonstrated to be powerful in exploring redox mechanisms with their wide applications in batteries. The specialties of each technique in redox investigation are assessed according to their typical cases.

hXAS is specialized in the in situ/operando investigation of the redox mechanism because of the relatively high penetration of hard X-rays. The different contributions of Ni, Co and Mn to the electrochemical reaction in $Li_{1.2}Ni_{0.15}Co_{0.1}Mn_{0.55}O_2$, especially at high rates, have been studied in situ.[28] As shown in Figure 12.25a, the first charge profiles contain two reaction regions, where the first region lies in the range from OCV to 4.4 V with a capacity of ~100 mAh/g, while the second one is primarily a plateau above a voltage of 4.4 V with a capacity of ~220 mAh/g. The Ni K absorption edge (Figure 12.25b) shifts to higher energy when charged below 4.4 V, indicating the oxidation of Ni in the first region. However, the shift of the Co and Mn K absorption edges cannot be easily clarified. Therefore, the Fourier transform of EXAFS has been used to further track the evolution of Co and Mn, where the first peak at approximately 1.5 Å corresponds to the metal–oxygen interaction in the first coordination sphere, and the second peak at approximately 2.5 Å is related to

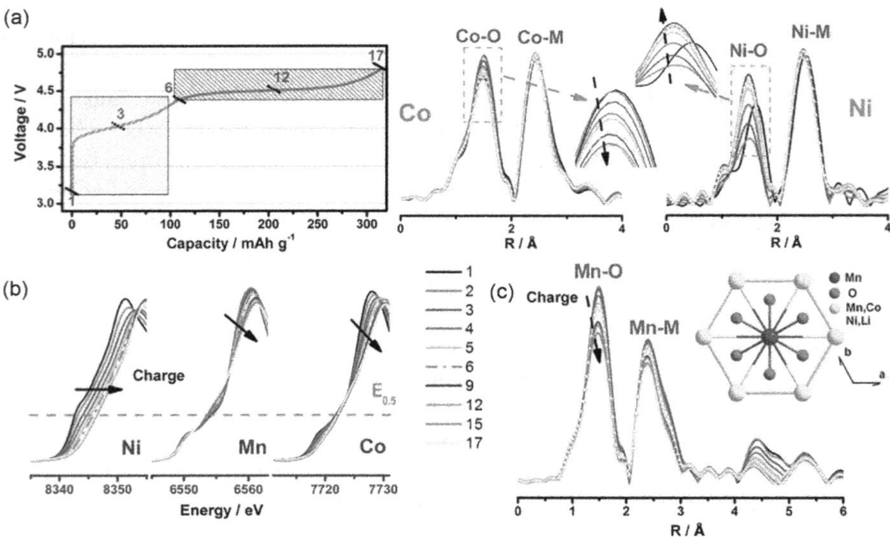

FIGURE 12.25 (a) The first charging curve of $Li_{1.2}Ni_{0.15}Co_{0.1}Mn_{0.55}O_2$ during an in situ XAS experiment (current density: 21 mA/g). Selected XAS scan numbers are marked on the charge curves. (b) Normalized XANES spectra and (c) magnitude of the Fourier-transformed Mn, Co and Ni K-edge spectra of $Li_{1.2}Ni_{0.15}Co_{0.1}Mn_{0.55}O_2$ collected during the initial charge process. (Reprinted with permission from ref. [28]. Copyright 2014 John Wiley and Sons.)

the metal–metal interaction in the second coordination sphere. As shown in Figure 12.25c, the intensity of the Ni-O and Co-O peaks mainly changes below 4.4 V. In contrast, the intensity of the Mn-O peak changes during the whole charge process. This indicates that the first region is mostly related to the oxidation of Ni and Co, while the second region is mainly related to Mn.

The semiquantitative analysis of the capacity contributions of Ni, Mn and Co based on the extracted characteristic parameters is shown in Figure 12.26. The half-height energy position at 4.4 V reveals that the oxidation state of Ni is 4+. The remaining capacity below 4.4 V comes from the oxidation of Co from 3+ to 3.5+. Debye–Waller factors σ of the first coordination shells through simulation, which reveal the local disorder with delithiation, are used to track the contribution of Mn and Co. $\sigma_{Co\text{-}O}$ increases significantly below ~4.4 V, corresponding to the oxidation of Co in this region. In contrast, $\sigma_{Mn\text{-}O}$ increases in the entire range up to 4.8 V. The changes below ~4.4 V can be caused by local structure changes induced by delithiation. Only

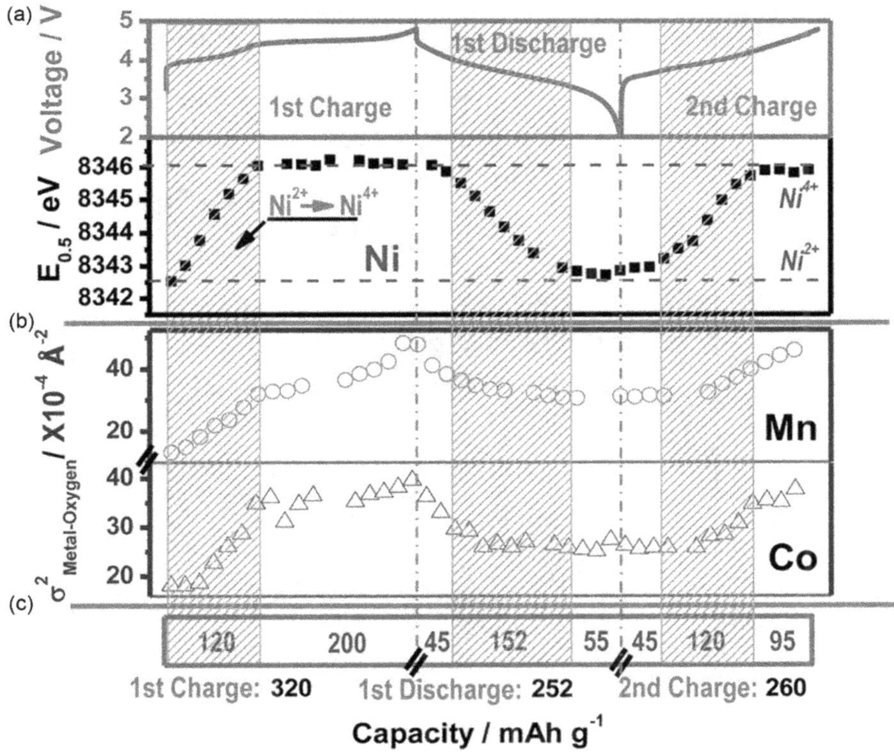

FIGURE 12.26 Quantitative correlation of the capacity with the reaction sites. Evolution of (a) the half-height energy position of the Ni K-edge XANES spectra and (b) Debye–Waller factors of Mn-O and Co-O during in situ XAS measurements. (c) The capacity delivered in a corresponding voltage range. (Reprinted with permission from ref. [28]. Copyright 2014 John Wiley and Sons.)

changes in $\sigma_{Mn\text{-}O}$ in the second region indicate that local structural changes occur mainly around Mn sites.

The relationship of the reaction kinetics with redox sites has been further clarified. As shown in Figure 12.27, the half-height energy position of the Ni K-edge within approximately 160 s evolves during the 5 V constant voltage charge, representing the charge compensation from Ni. The intensity of the Co-O peak mainly decreases within the first 200 s, indicating the oxidation of Co in this time range. However, the intensity of the Mn-O peak decreases over the entire observation process (900 s), indicating the much slower kinetics around Mn sites. This result is consistent with the quantitative analysis based on $\sigma_{Mn\text{-}O}$ at this time scale. This study indicates the key role of the Li_2MnO_3 domain in limiting the rate capability of Li-rich layered materials. In addition, in situ/operando hXAS is shown to be a powerful tool to study redox mechanisms, especially at high rates, by combining XANES and EXAFS analysis.

Although XANES of the TM K-edge can be used to reveal the oxidation states, the evolution of broadening features is sometimes not substantial, as shown in the above case. In comparison, the spectra of the TM L-edge are more suitable for the analysis

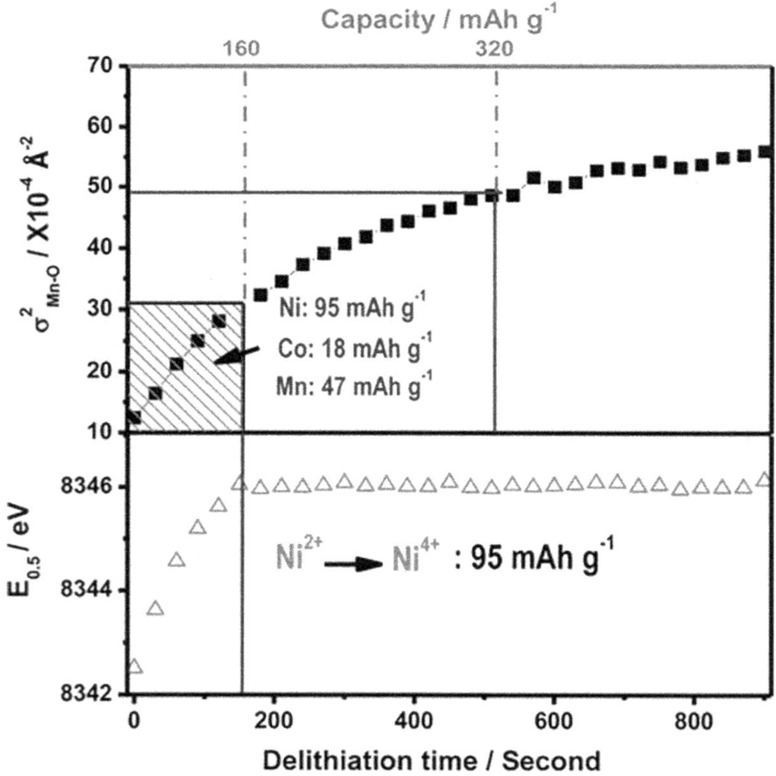

FIGURE 12.27 A quasi-quantitative analysis of the delithiation process of the $Li_{1.2}Ni_{0.15}Co_{0.1}Mn_{0.55}O_2$ electrode during 5 V constant voltage charging. (Reprinted with permission from ref. [28]. Copyright 2014 John Wiley and Sons.)

of oxidation states because of the sharp features of the bound d unoccupied states, which are more related to the number of valence electrons. Take hXAS of the Fe K-edge and sXAS of the Fe L-edge of Li_xFePO_4 (Figure 12.28) as examples, where x equals 0, 0.75 and 1. The edge shift in the XANES spectrum of the Fe K-edge and the intensity change of the pre-edge features are very small with the oxidation from Fe^{2+} to Fe^{3+} accompanied by delithiation. However, the sXAS of the Fe L_3-edge displays complete changes of sharp peak features from Fe^{2+} to Fe^{3+}. However, the higher sensitivity of sXAS to oxidation states often comes at the cost of less flexibility in the sample environment than that offered in hXAS.

Based on the sXAS of the Fe L-edge, an in-depth understanding of the evolution of the electronic structure with (de)lithiation in Li_xFePO_4 nanoparticles and single

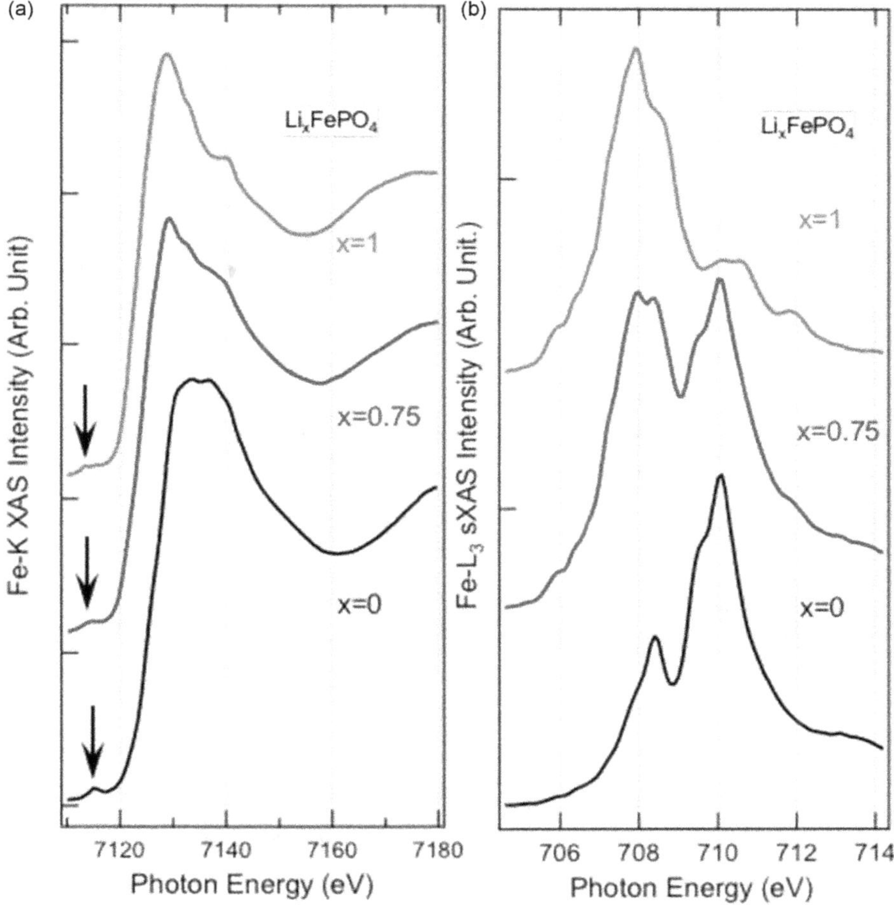

FIGURE 12.28 Comparison of (a) the hard X-ray Fe K-edge XAS and (b) the soft X-ray Fe L_3-edge soft XAS. (Data of (a) reproduced with permission from ref. [29]. Copyright 2017 John Wiley and Sons. Data of (b) reproduced with permission from ref. [30]. Copyright 2012 American Chemical Society.)

crystals has been completed.[30] As shown in Figure 12.29a–c, the Fe L_3-edge sXASs of $LiFePO_4$ with chemical and electrochemical delithiation are fitted with the references of $LiFePO_4$ and $FePO_4$ to quantify the phase evolution. The spectra of chemically delithiated Li_xFePO_4 fit perfectly, indicating the strict transformation between $LiFe(II)PO_4$ and $Fe(III)PO_4$. In contrast, the fitting to the spectra of electrochemically delithiated Li_xFePO_4 always displays subtle deviation, as shown in Figure 12.29d and e. The spectral deviation from fitting is attributed to the formation of an intermediate phase within a very limited region. Moreover, the difference in the lithium distribution on the two opposite sides of the electrode sheet can be revealed based on the fitting (Figure 12.29f). More $LiFe(II)PO_4$ component (29%) can be found on the side facing the current collector than the other side facing the electrolyte. This indicates the limitation of electric conductivity on the reaction progress.

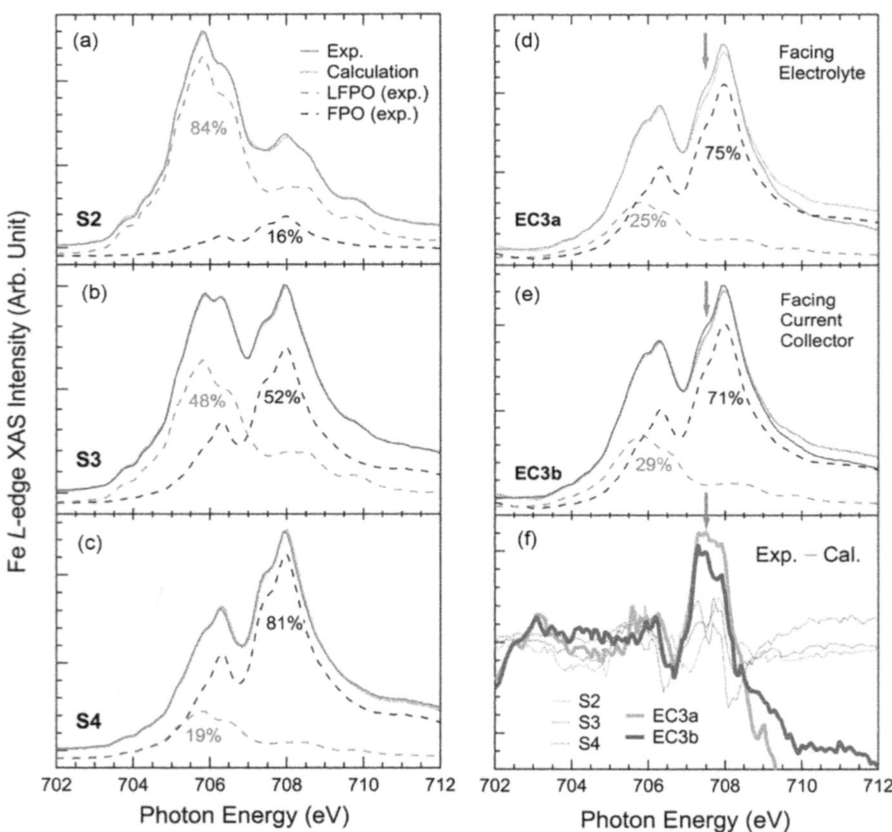

FIGURE 12.29 (a–c) Comparison of the experimental (solid lines) and two-phase fitting spectra (dotted lines) of three partially chemically delithiated Li_xFePO_4: S2–S4. (d and e) Same comparison of partially electrochemically cycled Li_xFePO_4 cathode facing electrolyte (EC3a) and current collector side (EC3b). (f) Difference between experimental and two-phase fitting spectra of the electrochemically (thick lines) and chemically (thin lines) samples. (Reprinted with permission from ref. [30]. Copyright 2012 American Chemical Society.)

The Fe 3d states in the two phase components, $LiFe(II)PO_4$ and $Fe(III)PO_4$, have been further studied. The spectra profiles are simulated by CTM4XAS, where the two ligand parameters Ds and Dt are induced to describe the D4h symmetry. Based on the simulations, the energy distribution of 3d states can be calculated as shown Figure 12.30a and b. The dz^2 state is lowered in $LiFePO_4$, which is related to the elongation along the *z*-axis of the FeO_6 octahedron. The role of the local structure on sXAS can be further displayed (Figure 12.30c and d) by feature simulation with the removal of the multiplet effect resulting from atomic 2p-3d overlap. The dxy orbital is situated above the degenerated *dxz*/*dyz* and dz^2 states, forming a separate peak feature. The direct experimental identification of Fe 3d states is performed

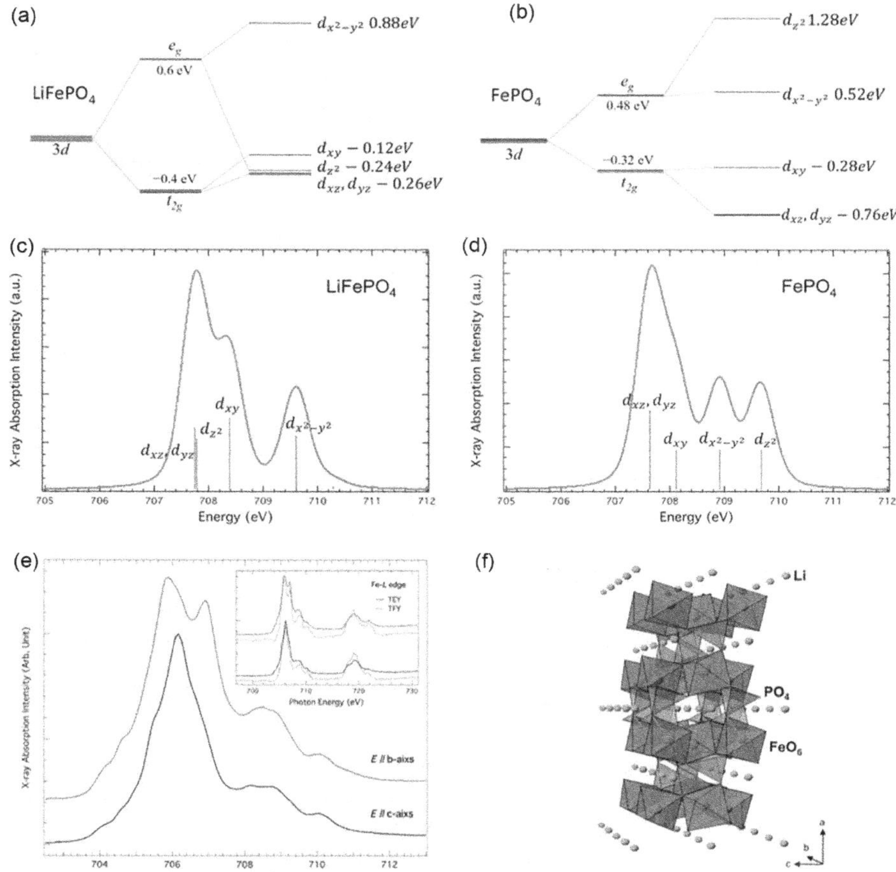

FIGURE 12.30 Crystal field diagram of (a) $LiFePO_4$ and (b) $FePO_4$ derived from the multiplet calculations. Theoretical calculations for (c) $LiFePO_4$ and (d) $FePO_4$ without the multiplet effect from the 2p-3d overlap. (e) Fe L_3 TEY XAS spectra of the LiFePO4 single crystal obtained for the electric field vector along the *b*-axis (red) and *c*-axis (black). (f) Crystal structure of $LiFePO_4$. (Reprinted with permission from ref. [30]. Copyright 2012 American Chemical Society.)

utilizing polarized sXAS detection on single-crystal $LiFePO_4$. As shown in Figure 12.30e, a striking polarization-dependent feature appears at 707 eV when the electric field direction of the incident X-ray is along the lithium diffusion direction, *b* axis. Because the *dxy* orbital is approximately aligned to the lattice *b* axis, the strong enhanced 707 eV peak should correspond to the 3*dxy* orbital according to the dipole approximation for resonant excitation. The experimental assignment is consistent with the feature simulation of the 3d states above, indicating the dominant role of the crystal field in the energy position of the spectral feature of the 3*dxy* state. This study sheds light on understanding the phase transformation and transport mechanism of $LiFePO_4$. In addition, it highlights the powerful capability of sXAS in deeply resolving the interplay between the local structure and the electronic structure.

Complementary to sXAS, which probes the unoccupied 3d states, sXES provides a probe of the occupied 3d states, which is more directly associated with the delithiation process of battery cathodes. By combining sXAS and sXES, the redox mechanism of $LiFePO_4$ in the charge and discharge process has been described completely.[31] The Fe L_3-edge sXES spectra of Li_xFePO_4 ($x=0$, 0.3, 0.7 and 1) are shown in Figure 12.31a. In comparison with sXAS, the features of sXES are much broader, which makes them more difficult to separate. Nevertheless, one major peak at ~706 eV and a hump feature at ~710 eV appear at the Fe L_3-edge of $LiFePO_4$. With delithiation from $LiFePO_4$ to $FePO_4$, the major peak shifts to ~707.5 eV, while the hump monotonically decreases. The features of $LiFePO_4$ and $FePO_4$ can be further revealed by the exhibition of the two isosbestic points shown in Figure 12.31b, indicating the presence of three occupied states. The states can be further visualized through the multipeak fittings shown in Figure 12.31c and d. It should be noted that a small feature close to $FePO_4$ is needed to fit the spectra of $LiFePO_4$ due to the signal interference of the $FePO_4$ on the surface. The major peak of $FePO_4$ represents the integrated configuration of the $3d^5$ states. Because of the high spin state of Fe 3d in Li_xFePO_4, the five electrons in $FePO_4$ half-fill the five 3d orbitals with the majority spin state. With lithiation, one extra electron introduced in $LiFePO_4$ has to take the opposite spin state, triggering a strong on-site Coulombic interaction. This results in the splitting of the $3d^6$ states, as shown by the separated sXES features of $LiFePO_4$. The energy splitting corresponds to the energy of the on-site Coulombic repulsion. $U_{\text{on site}}=3.0$ eV. The Coulombic energy penalty lowers the band of majority spin states, reshuffling the entire valence band. Such a nonrigid band evolution with the electron in the minority spin state close to the Fermi level eliminates the involvement of O 2p states during delithiation, which is critical for the safety of battery cathodes.

The unoccupied states regulating the lithiation process are measured by sXAS and hXRS. Because Fe L-edge spectra are dominated by the multiplet effect, the energy of unoccupied states cannot be directly obtained by the spectral features. Instead, the hybridized features at the O K pre-edge are used to determine the energy of the Fe 3d unoccupied states because of the small influence from the multiplet effect. As shown in Figure 12.32a, the pre-edge feature at 530–533 eV is obvious for $FePO_4$, indicating that the 3d unoccupied states sit right above the 3d valence state but below the broad O 2p states. For both $LiFePO_4$ in nanoparticles and single crystals, very weak features are displayed. However, the bulk hXRS results (Figure 12.32b) indicate no pre-edge feature for $LiFePO_4$ at all, indicating surface interference on sXAS. The

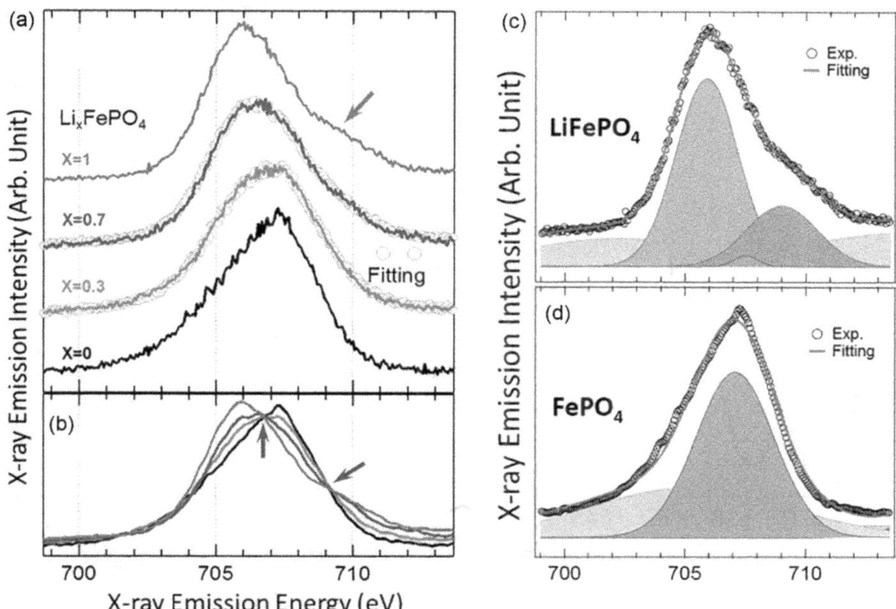

FIGURE 12.31 Soft X-ray emission spectroscopy of the Fe L_3-edges of a series of Li_xFePO_4 nanoparticles. (a) The fittings (open circles) of the intermediate states are performed by linear combinations of the spectra of $x=0$ ($FePO_4$) and 1 ($LiFePO_4$). (b) Two isosbestic points (blue arrows) are obvious when all the sXES data are stacked together. Peak fittings of the experimental Fe-L_3 sXES data of (c) $LiFePO_4$ and (d) $FePO_4$. The sXES features of $LiFePO_4$ include a low-intensity $FePO_4$ feature at 707 eV and two separated features at 706 and 709 eV. (Reprinted with permission from ref. [31]. Copyright 2015 Royal Society of Chemistry.)

absence of a pre-edge feature indicates that Fe^{2+} stays in a highly ionic state. The evolution indicates the critical role of the interaction with lithium in changing the Fe–O covalency.

The schematic diagram of the electronic states studied above is summarized in Figure 12.33. Due to the Coulombic interactions, electrons with minority spins in $LiFePO_4$ are pushed up closely to the Fermi level. The electrons are responsible for the charge process. For the discharge process, intrinsic unoccupied states of Fe 3d close to the Fermi level accept electrons. Therefore, only Fe 3d states have been involved in the entire electrochemical process, leading to safe battery operations. This unique electronic structure results from the interplay of the surrounding crystal field and electron spin state instead of the simple number of 3d electrons. The critical role of sXES in battery studies by directly probing the pDOS has been revealed in this work.

The powerful capability of combining sXAS and sXES can be further shown in the direct verification of the Zaanen–Sawatzky–Allen (ZSA) diagram, which is closely related to the redox type and intrinsic limit for voltage. Olalde-Velasco et al. aligned the energy of sXES of the TM L-edge and sXAS of the F K-edge using the TM-F hybridized feature in various difluorides.[32] A direct experimental probe of

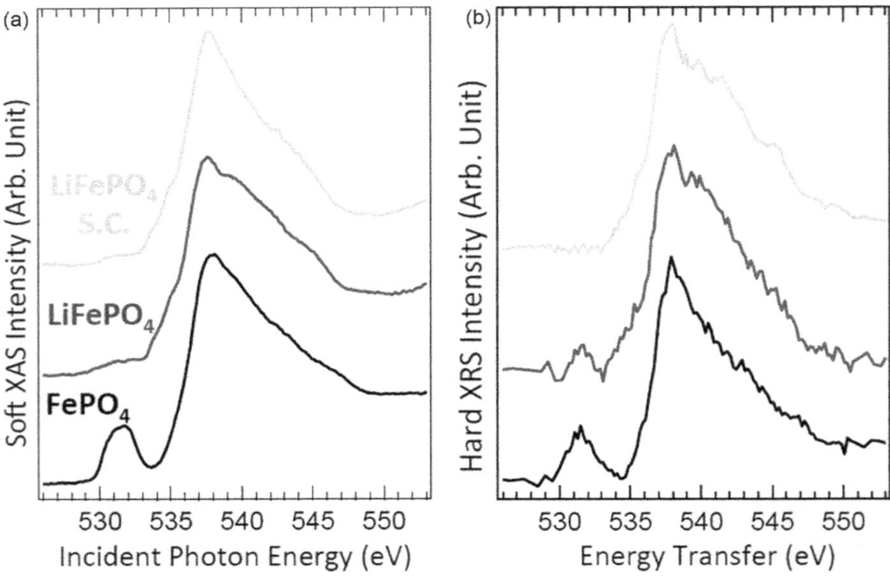

FIGURE 12.32 (a) Soft X-ray absorption spectroscopy and (b) hard X-ray Raman spectroscopy of the O K-edges of $FePO_4$ nanoparticles (black), $LiFePO_4$ nanoparticles (green) and $LiFePO_4$ single crystals (red). (Reprinted with permission from ref. [31]. Copyright 2015 Royal Society of Chemistry.)

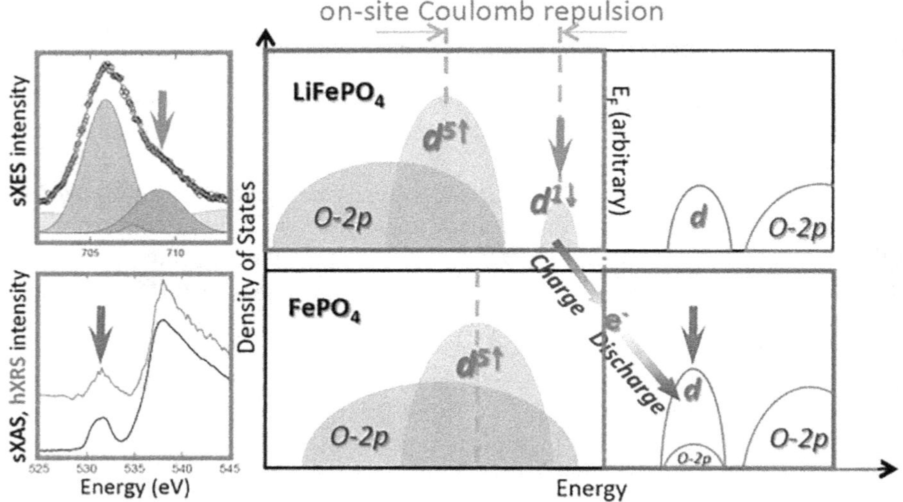

FIGURE 12.33 Summary of the sXES, sXAS and hXRS results and a schematic of the electronic state distributions in $LiFePO_4$ and $FePO_4$. (Reprinted with permission from ref. [31]. Copyright 2015 Royal Society of Chemistry.)

Mott–Hubbard and charge-transfer insulators has been achieved by using the experimental U_{dd} and Δ_{CT} for the first time. Later, Hong et al. extended the method to perovskite oxides.[33] The energetic barriers for electron transfer and surface deprotonation can be estimated according to the aligned pDOS on an absolute energy scale, which are correlated with oxygen evolution reduction (OER) activity.[34]

Compared with normal sXAS and sXES, sRIXS shows higher resolution in resolving oxidation states and chemical environments. The unique capability of sRIXS can be demonstrated by two typical cases. The first is the identification of unusual Mn^+ redox by sRIXS in a Prussian Blue analogue material, $Na_{1.24}Mn[Mn(CN)_6]$. As shown in Figure 12.34a, an emerging peak for the charged state can be found at approximately 643.5 eV in comparison to the sXASs in the discharged state. According to the atomic multiplet calculations, the peak is assigned to the monovalent Mn^+ coordinated with carbon. In contrast to the subtle changes in sXAS, the sRIXS mapping in Figure 12.34b shows a sharp contrast.[35] This feature can be assigned as the direct probe of Mn^+. The low spin $3d^6$ electron states of Mn^+ coordinated with carbon result in a specific configuration with fully occupied t_{2g} states and empty e_g states. Such a configuration naturally enhances the d–d excitation between the t_{2g} and e_g bands because of the forbidden transition within the t_{2g} band compared with Mn^{2+}, leading to the enhanced feature in sRIXS. The excitation energy lies at ~643.4 eV calculated by the subtraction of the energy loss value from the elastic peak position, which matches well with the energy of the sXAS feature of Mn^+. This work has clarified a 90-year speculation on the exhibition of Mn^+, which showcases the powerful capability of sRIXS for revealing novel TM chemistry.

The second case revealing the specialty of sRIXS is related to the direct probe of OR, which triggers great interest in designing cathodes with anionic redox for high energy density. Detection of O redox by sXAS is primarily influenced by the intensity and position of TM-O hybridization features at the pre-edge. sRIXS is capable of distinguishing the OR and hybridization features because of its new dimension

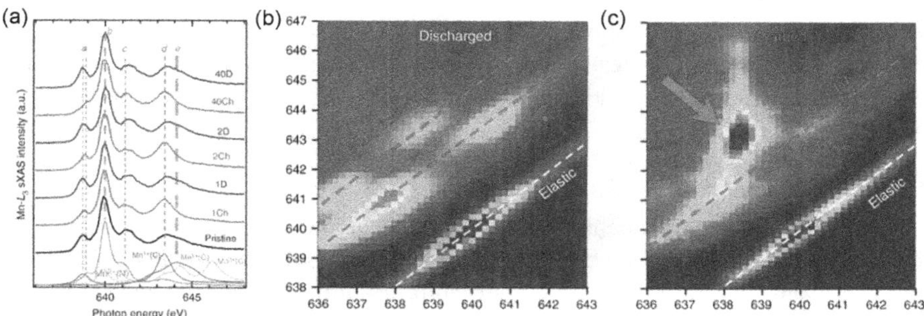

FIGURE 12.34 (a) Mn L_3-edge sXAS spectra collected on a series of MnHCMn electrodes at the charged (reduced), marked with "Ch", and discharged (oxidized), marked with "D", states of the 1st, 2nd and 40th cycles. Calculated spectra of MnII(N), MnIII(C), MnII(C) and MnI(C) are plotted at the bottom. Mn L_3-edge sRIXS maps on MnHCMn electrodes at (b) the discharged (oxidized) and (c) charged (reduced) states. (Reproduced with permission from ref. [35]. Copyright 2018 Nature Publishing Group.)

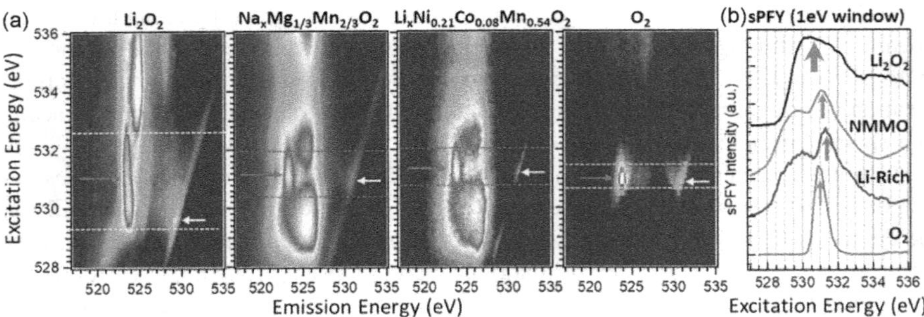

FIGURE 12.35 (a) Direct comparison of the full mRIXS profile of oxidized oxygen states in four systems: Li_2O_2, $Na_{2/3}Mg_{1/3}Mn_{2/3}O_2$, $Li_{1.17}Ni_{0.21}Co_{0.08}Mn_{0.54}O_2$ and O_2. (b) sPFY spectra of the four systems extracted from mRIXS maps by integrating the intensity within 0.5 eV of the 523.7 eV emission energy. (Reproduced with permission from ref. [36]. Copyright 2020 American Chemical Society.)

of emission energy. The resolving capability of O redox can be demonstrated by the comparative study of O_2, CO_2, Li_2O_2 and the oxidized oxygen state in representative Na/Li-ion battery electrodes ($Na_{2/3}Mg_{1/3}Mn_{2/3}O_2$ and $Li_{1.17}Ni_{0.21}Co_{0.08}Mn_{0.54}O_2$), as shown in Figure 12.35a.[36] All the sRIXS spectra of these oxidized oxygen species display a fingerprint at an emission energy of ~523.7 eV. However, the distributions of these features along excitation energies are different despite the roughly same expansion along emission energies. Specifically, Li_2O_2 shows the longest distribution, while O_2 shows the shortest distribution. Another difference lies in the excitation energy position of the features, which can be clearly seen in the partial fluorescence yield spectra (Figure 12.35b) with the signal selectively collected within a ±0.5 eV window centred at the emission energy of 523.7 eV. Additionally, an enhanced feature of low-energy excitation close to the elastic line marked with a white arrow is displayed in all the sRIXS spectra, which is the Raman characteristic corresponding to certain vibration modes. Clearly, the excitation energies of this feature are different among these species. The differences strongly suggest that O redox should be distinguished from the molecular configuration of Li_2O_2 and O_2, where the high TM–O bonding interaction plays a key role.

12.3.2.2 Probe of Local Structure

Owing to the resolving capability of the local structure through EXAFS analysis, another major application of hXAS is the investigation of reaction mechanisms in batteries from the viewpoint of local structure. Figure 12.36 shows an example of the hXAS investigation on the first charge process in $Li_2Ru_{0.75}Mn_{0.25}O_3$ (LRSO).[37] Through PCA on XANES, four principal components can be found in the in situ dataset. By combining the charge process shown in Figure 12.36a, transformations among components I, II and III involve the first voltage step related to cation redox, while the transformation from component III to IV dominates the second voltage step at approximately 4.2 V related to the O redox. EXAFS analyses of the four components, which are reconstructed with the multivariate curve resolution-alternating least-squares (MCR-ALS) method, are shown in Figure 12.36b–e. Component I is

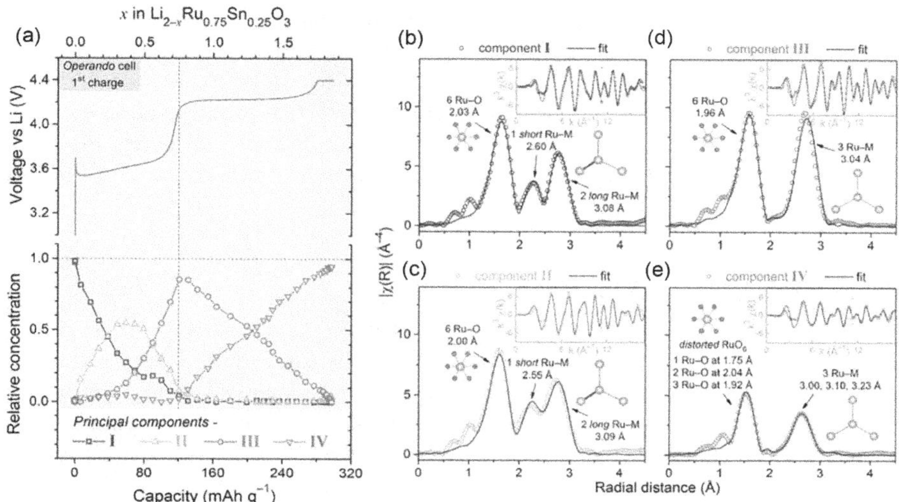

FIGURE 12.36 (a) The staircase-like charge voltage profile (top of figure) measured in the operando cell shows two steps, approximately 3.6 V (cationic redox, green background) and 4.2 V (anionic redox, red background). The bottom of the figure tracks the relative concentrations of the four principal components reconstructed using the MCR-ALS method, explaining nearly 100% of the variance in the experimental data with a lack of fit of 0.039%. Magnitude of Fourier transforms of k^3-weighted EXAFS oscillations for the four components (b–e) along with fitting results. The $|\chi(R)|$ plots are not corrected for phase shifts. Insets show the quality of fits in k-space. (Reproduced with permission from ref. [37]. Copyright 2017 American Chemical Society.)

identical to the pristine LRSO in the presence of three peaks. Similar to that of I, $|\chi(R)|$ of component II has three peak features corresponding to six Ru–O (2.00 Å) distances, one short Ru–M (2.55 Å) distance, and two long Ru–M (3.09 Å) distances. This indicates that the removal of Li cannot alter the symmetry of the local structure with only the shrinkage of Ru–O because of Ru oxidation. With further delithiation, component III forms with the consumption of components I and II. Only two peak features appear in the $|\chi(R)|$ of component III, where the first one represents six Ru–O (1.96 Å) distances and the second peak comes from three Ru–M (3.04 Å) distances. While the shrinkage of the Ru–O distance is attributed to further Ru oxidation, the merging of the two kinds of Ru–M distances indicates the Li-driven rearrangement of the second shell. Furthermore, the shortest O–O distance can be calculated as 2.48 Å in component III. Although the interatomic distances change for components I, II and III, the structural order can be preserved upon Li removal, as demonstrated by their nearly unchanged $|\chi(R)|$ peak intensities and Debye–Waller factors.

For component IV, which is related to anionic oxidation, $|\chi(R)|$ shows two peaks located at positions similar to those of component III with much lower intensity. This indicates that the fully charged phase becomes disordered. The first peak can be best fitted using a distorted RuO_6 coordination with three types of Ru–O distances. The second peak needs to be fitted using three slightly different Ru–M distances.

The distorted RuO_6 coordination has been experimentally verified for the first time by EXAFS modelling, which is consistent with the growth of the XANES pre-edge peak. One significantly shorter (1.75 Å) Ru–O distance can be found than the other five distances, rationalizing the formation of a "peroxo-like" configuration $(O–O)^{n-}$ between two such short Ru–O distances. The O–O distance can be estimated to be less than 2.33 Å, which is significantly shorter than the distance of 2.48 Å in component III.

12.3.2.3 Study of Surface Chemistry

XAS has also been widely used in surface studies, which is crucial to battery performance. Owing to the surface sensitivity of electron yield, sXAS is almost always the prior technique apart from XPS to investigate the surfaces of battery materials, such as electrodes and solid-state electrolytes. The formation of a Li_2CO_3 layer on air-exposed $Li_7La_3Zr_2O_{12}$ (LLZO) has been studied by sXAS, as shown in Figure 12.37.[38] Specifically, the build-up of Li_2CO_3 in air compared with that in Ar is comparatively monitored using sXAS spectra of the O K-edge as well as the C K-edge. The TEY spectrum of pure Li_2CO_3 at the O K-edge shows a peak at 534.1 eV, which corresponds to the electron transition from the O 1s to π* (C=O) orbital. However, only an absorption peak at a lower energy (533.0 eV) can be found in the TEY spectrum of LLZO exposed to Ar, indicating the lattice oxygen of LLZO in the absence of Li_2CO_3. The TEY of LLZO exposed to air shows a very similar feature to that of pure Li_2CO_3, indicating the formation of Li_2CO_3. When the probing depth is increased to ~100 nm under TFY mode, the peak at 533.0 eV related to lattice oxygen appears in both LLZOs. An additional peak at 534.1 eV can be detected for the air-exposed LLZO, identifying the location of Li_2CO_3 on the pellet surface. The comparison of the TEY spectrum at the C K-edge in the three samples can also support the formation of Li_2CO_3 on the air-exposed LLZO. A closer observation can reveal the minor feature at 533.0 eV in the TEY spectrum of air-exposed LLZO, indicating that the thickness of Li_2CO_3 is

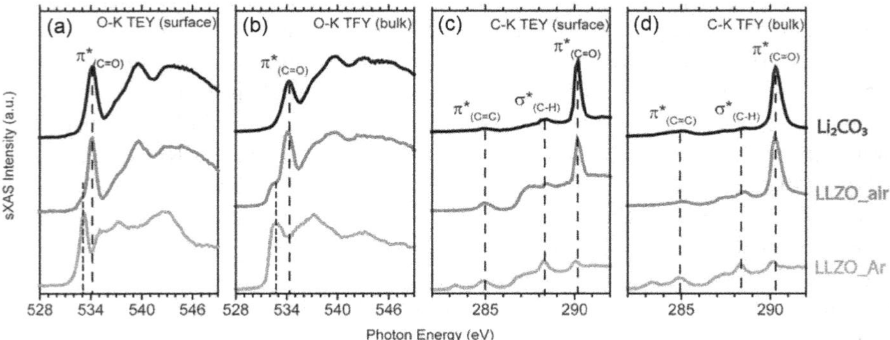

FIGURE 12.37 K-edge XAS spectra of LLZO_Ar, LLZO_air and the Li_2CO_3 reference collected in (a) TEY and (b) TFY modes; C K-edge XAS spectra of LLZO_Ar, LLZO_air and the Li_2CO_3 reference collected in (c) TEY and (d) TFY modes. (Reproduced with permission from ref. [38]. Copyright 2014 Royal Society of Chemistry.)

slightly smaller than the penetration depth of the TEY mode (~10 nm). Therefore, the growth of Li_2CO_3 can also be semiquantitatively estimated through the sXAS study.

Despite the unique capacity of surface detection and rough depth profiling, sXAS requires experimental conditions in an ultrahigh vacuum chamber, leading to difficulty in performing in situ/operando measurements. By combining the high collimation of the synchrotron X-ray beam and grazing incident geometry, a surface-sensitive hXAS called total reflection fluorescence (TFY-XAS) can be established with the restriction of the penetration depth to several nm. TFY-XAS is particularly suitable for in situ study of the electrode surface under operating conditions. The first application of this tool to batteries is the in situ investigation of the surface structural evolution in $LiCoO_2$.[39] The experimental configuration is shown in Figure 12.38c. The relationships of fluorescence intensity and penetration depth with the grazing angle in theoretical calculations and practical experiments are shown in Figure 12.38a and b, respectively. When the incident angle is set to 0.2° and 2.2°, the

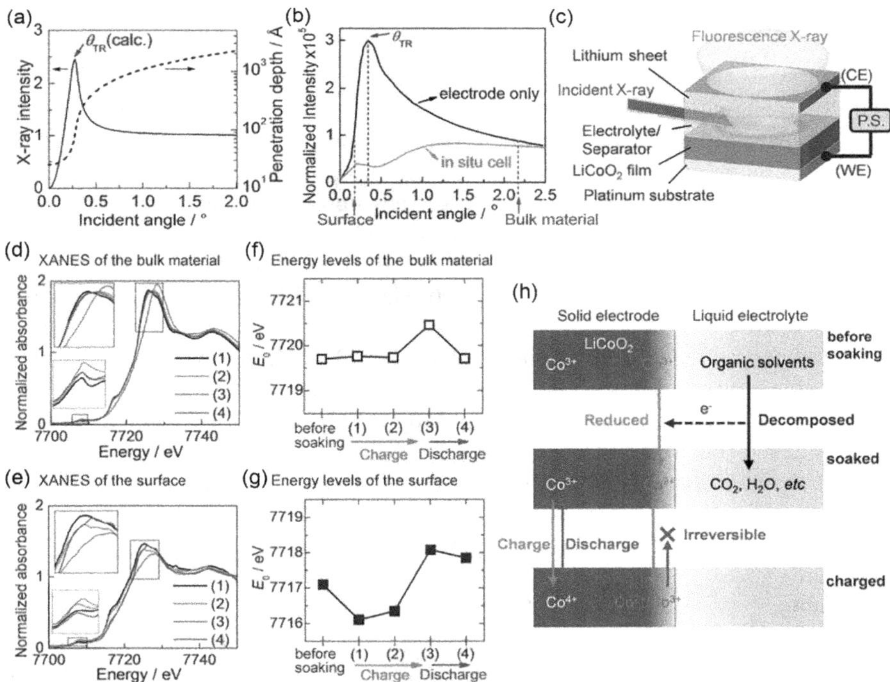

FIGURE 12.38 (a) Calculated X-ray intensity and penetration depth of the 7.7 keV incident X-ray to $LiCoO_2$. (b) Incident X-ray angle dependence of the Co fluorescence from $LiCoO_2$/Pt obtained before electrolyte soaking (black) and after electrolyte soaking (red). (c) The spectro-electrochemical cell used for in situ TRF-XAS measurements. (d–g) Co K-edge XANES spectra of the $LiCoO_2$ film measured at four different points by applying a voltage and measuring the corresponding absorption energy levels of the bulk material (d and f) and the surface (e and g), respectively. (h) The proposed reaction of the initial degradation that occurs at the electrode/electrolyte nanointerface. (Reproduced with permission from ref. [39]. Copyright 2012 John Wiley and Sons.)

corresponding penetration depths can be estimated to be 3 and 100 nm, respectively. As shown in Figure 12.38d–h, the measurements indicate that a gradual degradation at the surface of the $LiCoO_2$ thin-film electrode occurs in the charge and discharge process. Co^{3+} can be reduced immediately after dipping into the organic electrolyte even at an OCV. Such irreversible changes on the surface are quite different from the reversible redox reactions in the bulk.

12.4 NEUTRON ABSORPTION AND INELASTIC SCATTERING

The unique role of neutron techniques in battery research is mainly revealed by the direct probe of lithium compared to X-ray techniques. Here, three kinds of exclusive neutron techniques, including NDP, inelastic neutron scattering (INS) and QENS, are introduced, which can provide concentration and dynamics information on lithium. The introduction focuses on applications, avoiding in-depth discussions on principle and experimental setup.

12.4.1 NDP

NDP is a quantitative analytical method to detect the distribution of light elements (Li, B, N, He, Na, etc.) as a function of depth away from a solid surface based on the absorption process. Specific to lithium, NDP is primarily based on the neutron absorption of ^{6}Li because of its much larger absorption cross-section (940.4 barn) compared with that of ^{7}Li (0.045 barn), which can be expressed as

$$^{6}\text{Li} + n_{\text{thermal or cold}} \rightarrow {}^{4}\text{He}\left(2.06\ \text{MeV}\right) + {}^{3}\text{H}\left(2.73\,\text{MeV}\right). \qquad (12.61)$$

A general experimental setup is shown in Figure 12.39. A thermal or cold neutron beam coming out of a neutron guide hits a sample composed of lithium in a high vacuum chamber, where cold neutrons are usually better for the measurement because of the enhancement of the absorption cross-section. The produced particles (^{4}He and ^{3}H) lose their kinetic energy due to scattering when passing through the sample to

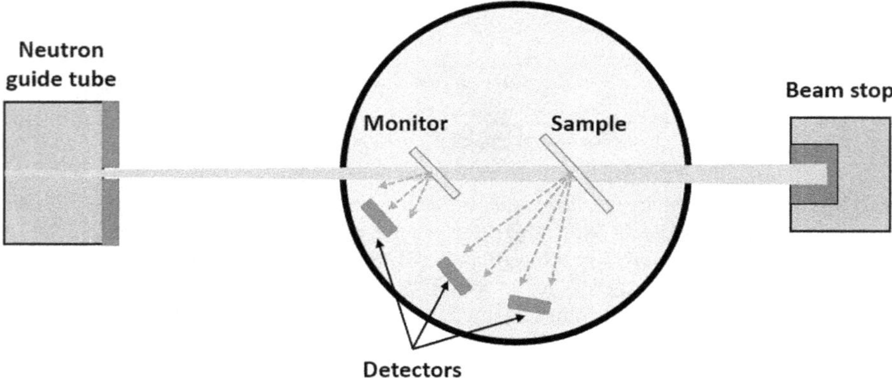

FIGURE 12.39 Schematic illustration of the NDP setup.

the detector, referred to as stopping power, which is directly related to the composition and density of the sample. The energy and the number of detected particles can be used to determine the location of the reaction and the abundance of lithium there, respectively. The depth range of NDP varies from several μm to tens of μm with a spatial resolution of tens of nm, depending on the atomic composition of the samples and the detected particles. Because of the weak absorption, where <0.01% of incoming neutrons are consumed, and the low energy deposition, NDP is considered a nondestructive technique.

NDP is particularly suitable for the in situ investigation of thin-film solid-state batteries on the spatial distribution of lithium and the corresponding temporal evolution, which would provide irreplaceable insights into fundamental issues in practical batteries. For example, lithium plate-stripping processes at the interface of a solid electrolyte and anode have been studied in situ in comparison to symmetric Li/garnet/Li (LGL) cells with asymmetric Li/garnet/carbon (LGC) cells.[40] Figure 12.40a and b depicts the series of NDP spectra of a Li/garnet/CNT asymmetric cell during the plating and stripping processes in one cycle, respectively. The lithium concentration at the surface region of the CNT film can be revealed in the kinetic energy range of detected particles from ~2,150 to 2,500 keV. The intensity in this region increases with the ongoing plating progress, indicating the accumulation of lithium in the CNT film. When turning to the stripping process, the near-surface peaks start to decrease. The total NDP spectra in multiple plating–stripping cycles are shown with the 2D spectra in Figure 12.40c. The integrated counts in the energy range as well as the voltage profiles are plotted with cycles in Figure 12.40d, indicating the evolution of lithium concentration along with electrochemical cycling. More than half of the plated lithium cannot be reversibility stripped back during the first cycle. For the second and third cycles, the accumulation of lithium continues, improving the contact at the interface, as indicated by the decreased polarization. In the following 15 cycles, a similar amount of irreversible Li accumulates in the surface region in each cycle, indicated by the linear increase in intensity along with cycles. The counts vary in one cycle with the sign change of the current, indicating reversible changes in the lithium layer. With the accumulation of lithium to a certain extent, the process is irreversible due to the loss of contact with the electrolyte and insufficient transport through the CNT host during the last few cycles. The NDP curve becomes flat with the occurrence of a short circuit when the current density increases to 120 $\mu A/cm^2$. The evolution of the lithium concentration at different depths is further resolved by integrating the NPD data within several small energy ranges, as shown in Figure 12.40e. Both the count increment and variation are relatively small in the inner layers between 2,014 and 2,230 keV, indicating the stability of the garnet electrolyte. The layers between 2,230 and 2,374 keV contain periodic variations, which result from reversible Li plating–stripping. The outermost layers between 2,374 and 2,446 keV mainly have a linear increment. These results indicate that the formation of a reversible layer near the CNT/garnet interface is crucial to the reversible Li plating–stripping process. The same NDP study was performed in a symmetric Li/garnet/Li cell (not shown here), revealing better reversibility with higher maximum current densities because of the conformal contact between Li and the garnet electrolyte.

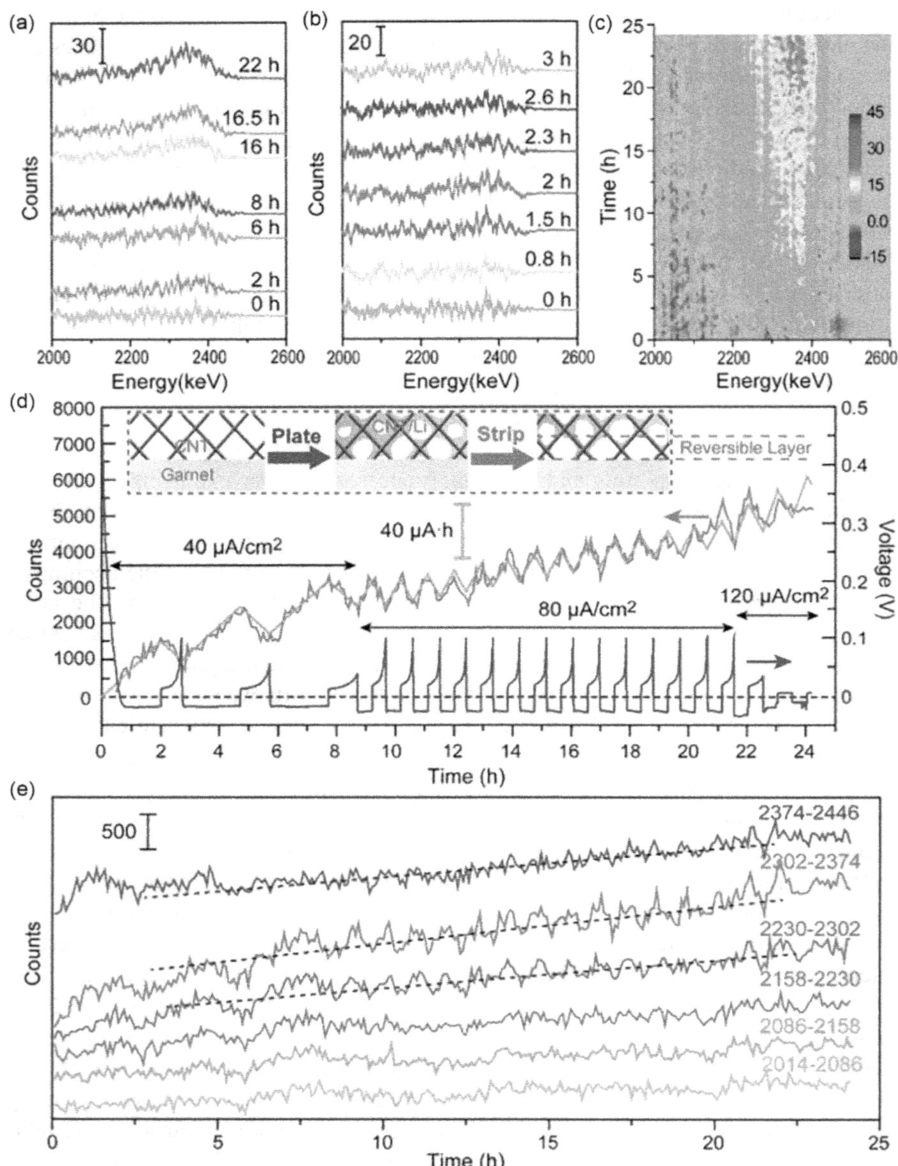

FIGURE 12.40 Typical NDP spectra of the Li/garnet/CNT asymmetric cell, with the background collected before cycling omitted, at different times (a) and during one plating–stripping cycle (b). (c) 2D projection of the NDP spectra collected at 5 minutes intervals during cycling. (d) The corresponding voltage profile (blue), charge curve after multiplying a constant of variation (green) and the integrated NDP count curve (red) of asymmetric cell cycling at different current densities. The inset illustrates the Li plating–stripping behaviour in the reversible surface layer. (e) The integrated NDP data within several small energy (keV) ranges indicate depth-dependent Li deposition during cycling. (Reprinted with permission from ref. [40]. Copyright 2017 American Chemical Society.)

12.4.2 INS and QENS

The kinematic properties of battery materials at the atomic scale play important roles in battery performance, especially the decisive effect of lithium transport on the rate performance. Therefore, investigations on the kinematic properties of battery materials are critical to in-depth understanding of the reaction mechanisms in batteries and thus their performance. Neutrons are particularly suitable for studying atomic kinematics in matter because of the comparability of the wavelength with interatomic distances and the energy with kinematic excitations. Atomic motion can be revealed by elastic scattering with the analysis of anisotropic contribution in the Debye–Waller factor and deviation from harmonicity at elevated temperatures (refer to the case study of NPD). Moreover, the particular strength of exploring kinematics can be demonstrated by inelastic scattering with the accessibility of detecting excitations at multiple time and length scales. The kinematic excitations from a fixed state to another, which have an energy of several meV corresponding to the oscillation frequency on the order of THz, are usually detected by inelastic neutron scattering (INS) with the corresponding inelastic peaks well separated from the elastic peak. The kinematic excitations usually include molecular vibrations and rotations, lattice vibrations and local distortions. In contrast, the diffusion of nearly free atoms in either the liquid or solid phase with the excitation energy distributed from zero to few μeV can be detected by QENS with the shape broadening of the elastic peak. The peak broadening in QENS depends on Q values, which can be used to distinguish the atomic/molecular transport mechanism and calculate the corresponding characteristic quantities.

Both INS and QENS have the same experimental setups shown in Figure 12.41. A monochromatic neutron beam with an energy of several meV is focused onto the sample. The detection on the scattered neutrons can be fulfilled by either direct geometry or indirect geometry. The former resolves the energy of scattered neutrons with the TOF methods, whereas the latter measures the scattered neutrons using an analyser crystal after sampling, commonly referred to as a triple-axis spectrometer. Because the energy of excitations is comparable to the kinetic energy of incident neutrons, the energy changes of the scattered neutrons can be measured with high accuracy.

Generally, the total intensity of scattered neutrons measured in INS and QENS contains the contributions from incoherent and coherent scattering from the sample. For coherent scattering, the spatial correlations can tell the pair–pair correlation function related to the collective motion of atoms in specific structures. For incoherent scattering, the probability of a single particle appearing in a position after a certain time can be derived from the self-correlation function, thus determining the time and length scale of the self-motions. The specific contribution of the two kinds of scattering is element dependent. For lithium, the coherent scattering of INS has often been used to illustrate lithium dynamics in the analysis of lattice waves (phonons) because of the relatively large coherent cross-section and moderate incoherent cross-section. In contrast, the incoherent scattering of QENS has been majorly discussed to reveal the lithium diffusion process due to rare discussions on coherent QENS deciphering collective diffusion.[41] It should be noted that inelastic scattering experiments for lithium are difficult due to the very large absorption cross-section.

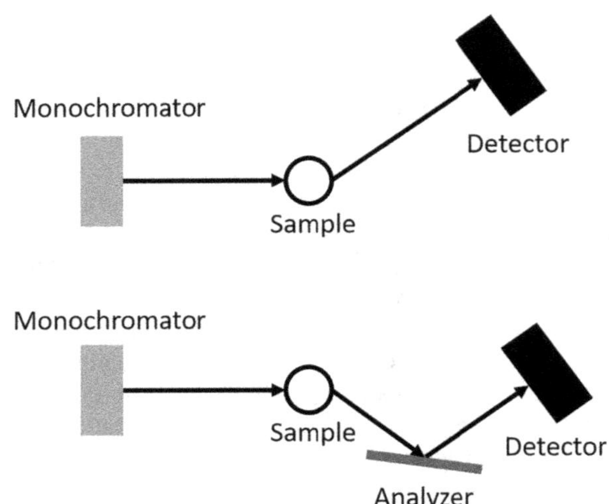

FIGURE 12.41 Geometries of inelastic neutron scattering setups.

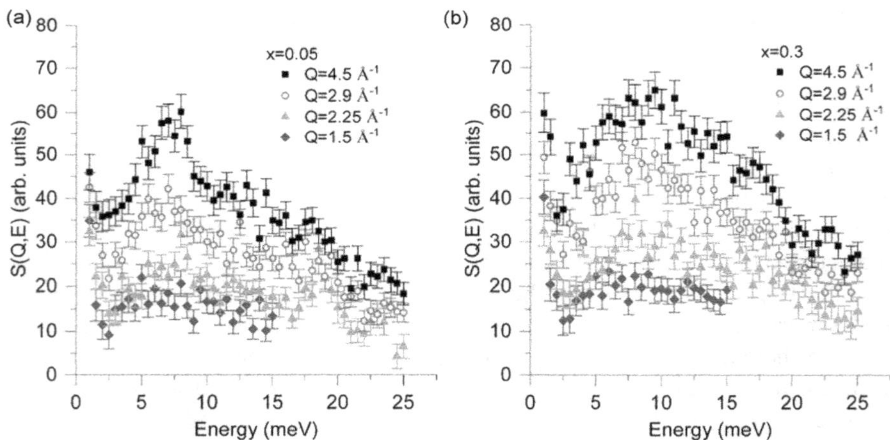

FIGURE 12.42 An excess of intensity is found above the Debye level across a range of measured momentum transfers for both (a) the low Li_2SO_4 content sample ($x=0.05$) and (b) the high Li_2SO_4 content sample ($x=0.3$). This excess is identified as the Boson peak. (Reprinted with permission from ref. [42]. Copyright 2018 Elsevier.)

The evolution of the Boson peak, which is one of the most unique features of disordered systems, in the two members of the family of superionic glasses, $xLi_2SO_4{\cdot}(1-x)LiPO_3$, $x = 0.05$ and 0.3, has been studied utilizing INS.[42] As shown in Figure 12.42, the Boson peak becomes more prominent as O increases to the maximum value of $4.5\,Å^{-1}$. This indicates a change in the lattice dynamics and the underlying structure due to the addition of sulfate ions.

The Q dependence of the dynamic response in constant energy scans has been focused on resolving the relative contribution from coherent and random phase components. As shown in Figure 12.43, for both samples, the inelastic structure factor $S(\boldsymbol{Q}, \mathrm{E})$ at 15 meV exhibits a bump near 2.7 Å^{-1} in comparison to the spectra at (~7 meV) and below (~3 meV) BP peak positions. This is substantially different from the elastic structure factor $S(\boldsymbol{Q}, 0)$, indicating the contribution from random phases. On the other hand, the qualitative similarity between $S(\boldsymbol{Q}, E = 7\text{ meV})$ and $S(\boldsymbol{Q}, 0)Q^2$ indicates the exhibition of a certain number of coherent phases.

The contributions from coherent and random phases can be separated by the quantitative fit of $S(\boldsymbol{Q}, E)$ by

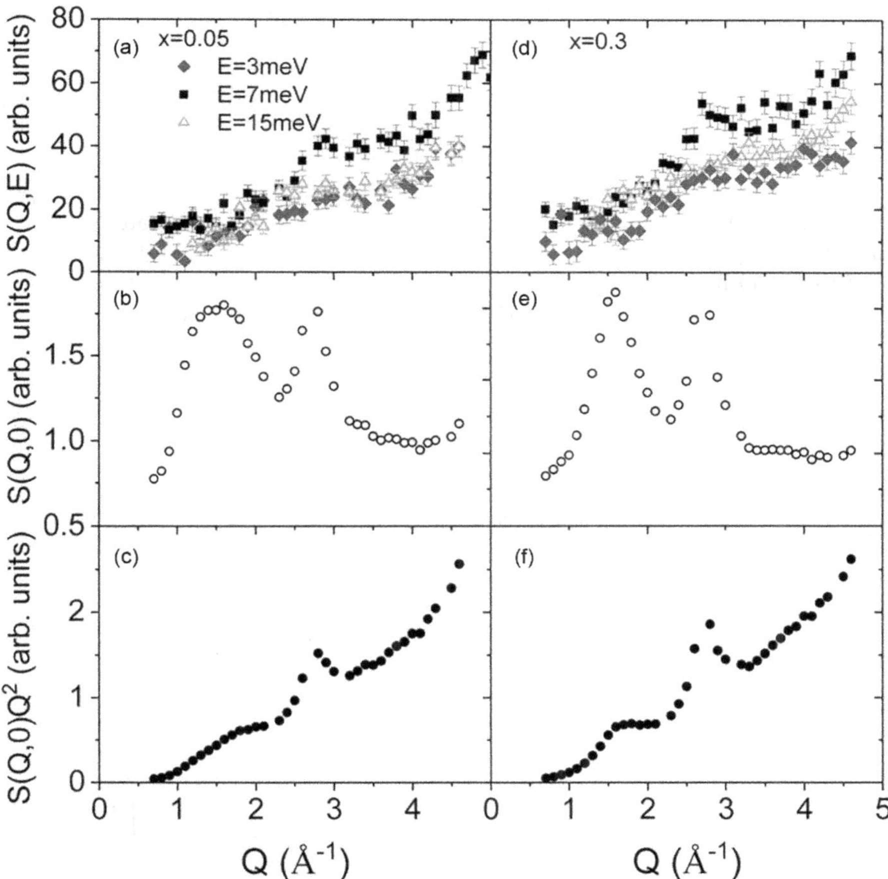

FIGURE 12.43 The constant energy scans are plotted for the $x = 0.05$ (a) and $x = 0.3$ (d) compositions. The feature near 2.7 Å^{-1} coincides with the second sharp diffraction peak in the corresponding liquid-like structure factors, as displayed in panels (b) and (e), respectively. Since phonon-like coherent vibrations are expected to result in a Q dependence of the S (Q, E) that reproduces features of the S (Q, 0) multiplied by Q^2, this quantity is plotted in panels (c) and (f) for a direct comparison. (Reprinted with permission from ref. [42]. Copyright 2018 Elsevier.)

$$S(\boldsymbol{Q}, E) = C_1 S(\boldsymbol{Q}, 0) Q^2 + C_2 Q^2 + \text{constant}, \tag{12.62}$$

where C_1 and C_2 represent the relative weights of the two phases. As shown in Figure 12.44, the relative amount of random phase dynamics increases with increasing energy for $x = 0.3$, totally different from the evolution trend for $x = 0.05$. The enhanced random-phase component is thought to result from the presence of Li_2SO_4. The trend of the coherent phase contribution is less clear despite the downward tendency in the high energy range. The increase in random phase dynamics should be related to the enhancement of the BP feature to the high energy side, which might promote lithium diffusion.

The lithium mobility and anionic stability of lithium-ion conductors with LISICON and olivine structures have been studied from the viewpoint of lattice dynamics by combining phonon DOS measurements using INS with theoretical calculations.[43] The measured and computed phonon DOSs in several LISICON conductors are shown in Figure 12.45. The phonon DOSs of Li_3PO_4 and $Li_{3.4}Ge_{0.4}P_{0.6}O_4$ have higher energies than those of Li_3PS_4 and $Li_{3.25}Ge_{0.25}P_{0.75}S_4$, indicating the significant softening of phonon modes with oxygen substituted by sulfur. The sublattice vibrations, including anion, lithium and nonmobile structural units, can be separated using the calculation of phonon DOS. Lithium-ion vibrations can be found to shift from 40–70 meV to 40–50 meV when replacing oxygen in Li_3PO_4 with sulfur to form Li_3PS_4. In contrast, the features in phonon DOS broaden without significant softening with

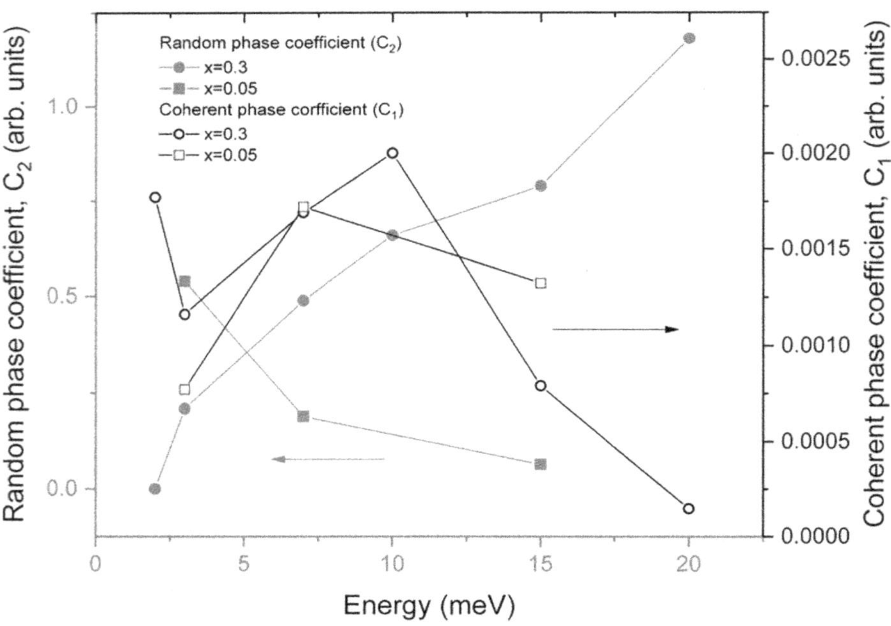

FIGURE 12.44 The coefficient of the random (C_2) and coherent (C_1) phases of the fitting function is plotted as a function of energy for each of the two compositions. (Reprinted with permission from ref. [42]. Copyright 2018 Elsevier.)

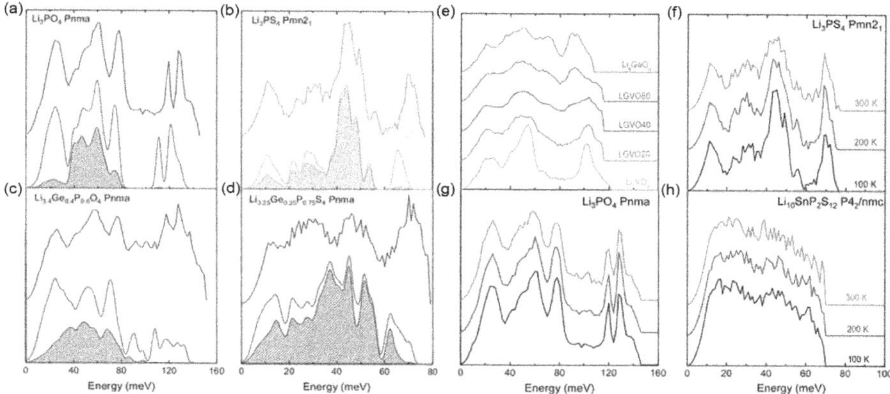

FIGURE 12.45 Experimental phonon DOSs collected at 100 K are shown on top, while computed phonon DOSs at 0 K are shown on the bottom, as well as the computed lithium-projected DOSs (shaded) for (a) Li_3PO_4 (Pnma), (b) Li_3PS_4 (Pmn21), (c) $Li_{3.4}Ge_{0.4}P_{0.6}O_4$ (Pnma) and (d) $Li_{3.25}Ge_{0.25}P_{0.75}S_4$ (Pnma). Measured phonon DOS as a function of aliovalent substitution in (e) $Li_{3+x}Ge_xV_{1-x}O_4$ ($x = 0$, 0.2, 0.4, 0.8 and 1) and temperature at 100, 200 and 300 K for (f) Li_3PS_4, (g) Li_3PO_4 and (h) $Li_{10}SnP_2S_{12}$, respectively. (Reprinted with permission from ref. [43]. Copyright 2018 the Royal Society of Chemistry.)

aliovalent cation substitution, as demonstrated by the broadened features with systematically increasing Ge substitution in $Li_{3+x}Ge_xV_{1-x}O_4$, where $x = 0$, 0.2, 0.4, 0.8 and 1. In comparison to negligible broadening with isovalent cation substitution, the broadening of lithium-ion vibrations should result from increasing disorder with the formation of structural defects.

As shown in Figure 12.46a, the downshifting of the lithium phonon band centre, which changes little with temperature, can be correlated to the decreasing enthalpy of lithium migration and thus the improved lithium mobility. Therefore, lithium transport can be tuned by replacing oxygen with sulfur. Moreover, the downshifting of the anion phonon band centre has been found to reduce the stability against electrochemical oxidation (Figure 12.46b). This indicates that the mobility of the anion sublattice might be the kinetic origin of the electrochemical stability of the solid-state electrolyte. By replacing oxygen with sulfur, the LISICON structure has the largest reduction in the oxidation potential because of the largest downshift of the anion phonon band centre. The evolution trends highlight a trade-off between lithium-ion mobility and oxidation stability for the design of super ion conductors.

The trade-off can be overcome when exploring conductors with olivine structures, as shown in Figure 12.47, which exhibit low lithium band centres but high anion band centres. Unfortunately, conductors in olivine structures have an apparent activation energy much higher than the intrinsic migration enthalpy, which is attributed to the exhibition of anti-site defects, limiting the long-range ion conductivity. This work highlights the critical role of lattice dynamics in developing new conductors with enhanced conductivity and stability. Undoubtedly, INS is one of the right tools to perform the investigation on lattice dynamics.

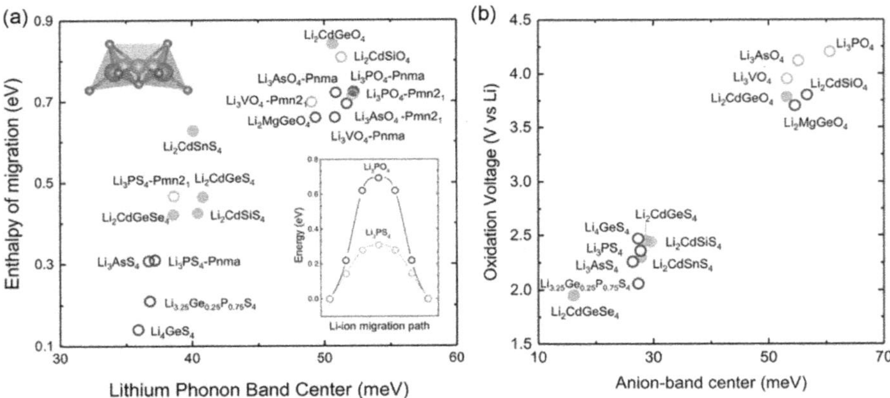

FIGURE 12.46 (a) Correlation between computed enthalpy of migration and oxidation potential with computed phonon band centre. The computed enthalpy of migration of 15 stoichiometric LISICONs known in the ICSD and two computed structures correlated well with the computed lithium band centre at 0 K. (b) Correlation between the computed stability oxidation potential and the computed anion band centre of 13 stoichiometric LISICONs. (Reprinted with permission from ref. [43]. Copyright 2018 the Royal Society of Chemistry.)

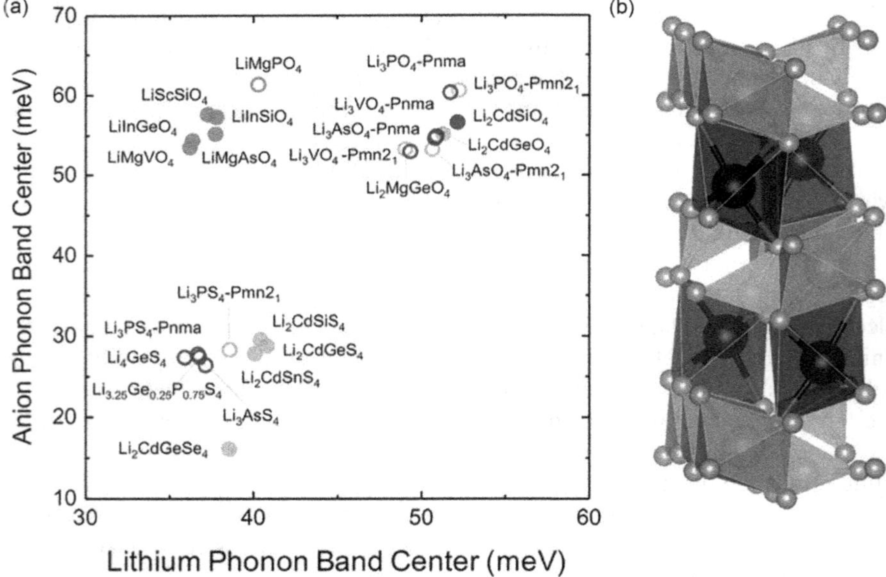

FIGURE 12.47 (a) Anion band centres and Li-band centres of olivine compounds compared to LISICON compounds. (b) Olivine is closely related to the LISICON structure because both have hexagonal anion sublattices, but Li occupies the octahedral interstice in Olivine, unlike the tetrahedral sites in LISICON. (Reprinted with permission from ref. [43]. Copyright 2018 the Royal Society of Chemistry.)

The diffusion dynamics of a series of lithium sulfate-substituted phosphate glass materials $x\text{Li}_2\text{SO}_4\cdot(1-x)\text{LiPO}_3$ have been studied by QENS.[44] The mean square displacement (MSD) for atoms, which is calculated according to the intensity in the elastic window, is plotted as a function of temperature in Figure 12.48. A linear dependence is revealed in the low temperature range, indicating the harmonic vibrations of atoms around their equilibrium position as described by the Debye–Waller factor. Deviations from this linear behaviour appear in the vicinity of room temperature for both samples with $x = 0.3$ and 0.4, indicating the occurrence of anharmonic motions. Therefore, the diffusive dynamics are expected to be measured in the QENS above this temperature.

The QENS was performed at 450 K for the two compositions, where the spectrum for $x = 0.4$ is shown in Figure 12.49. The broadening of the incoherent central peak related to the diffusive motions can be extracted from the QENS spectra with the removal of backgrounds from the sample holder and any fast component contribution from the sample, and the elastic peak convoluted by the resolution measured at 50 K with a delta function. The remaining features involving diffusion dynamics can be fit with a Lorentzian function.

The Lorentzian full width at half maximum Γ as a function of Q^2 is used to quantify the local diffusion process, as shown in Figure 12.50. A linear relationship indicates isotropic translational diffusion. The slope is related to the diffusion constant by $\Gamma = 2DQ^2$. The diffusion for the two compositions is the same within the statistical precision, with a diffusion constant $D \sim (55 \pm 4) \times 10^{-8}\text{cm}^2/\text{s}$. In contrast, no QENS signal can be observed for the two samples with low Li_2SO_4 contents ($x = 0$ and 0.05).

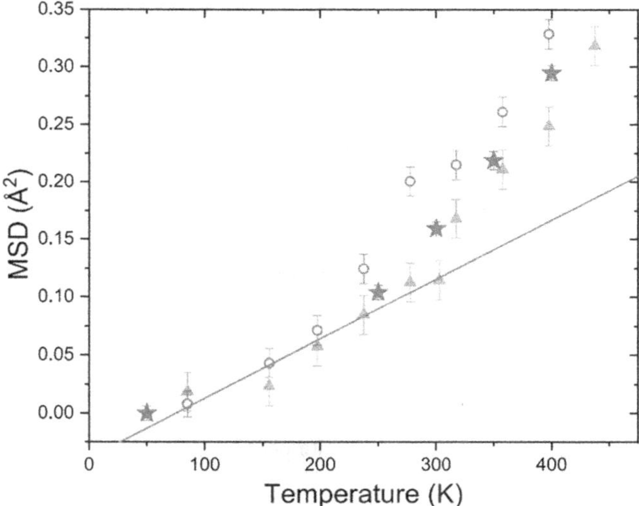

FIGURE 12.48 The elastic component of the backscattering signal is monitored as a function of temperature to obtain the mean square displacement. Open red circles: $x = 0.3$; green triangles and magenta stars: $x = 0.4$. (Reprinted with permission from ref. [44]. Copyright 2019 Elsevier.)

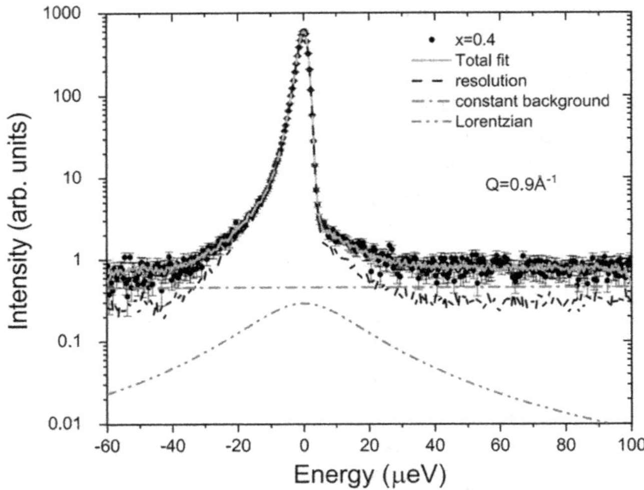

FIGURE 12.49 QENS signal and Lorentzian fitting for $x = 0.4$ at 450 K. (Reprinted with permission from ref. [44]. Copyright 2019 Elsevier.)

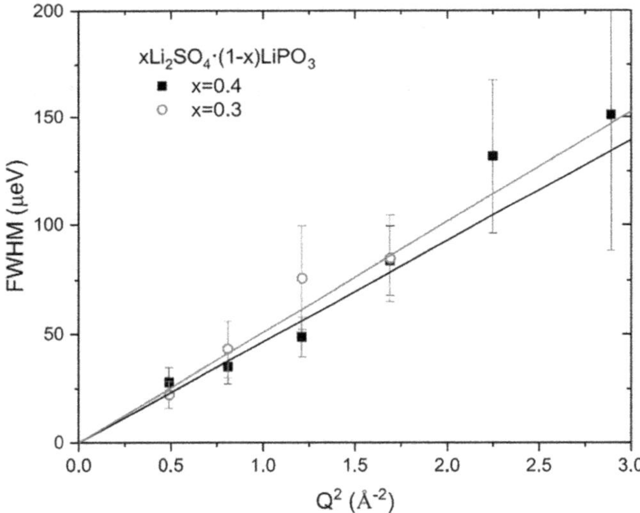

FIGURE 12.50 The quasielastic broadening is plotted vs. Q^2 to demonstrate the quadratic behaviour, indicating isotropic translational diffusion in the length and time scales that are proven in the measurement. (Reprinted with permission from ref. [44]. Copyright 2019 Elsevier.)

This is thought to result from the lack of enough mobile ions at the measurement temperature, which is consistent with the low ionic conductivity of the two samples.

Lithium self-diffusion in garnet-type solid-state electrolyte $Li_5La_3Ta_2O_{12}$ has been investigated by combining QENS experiments and molecular dynamics simulation.[45] QENS spectra with the changes in Q and temperature are shown in

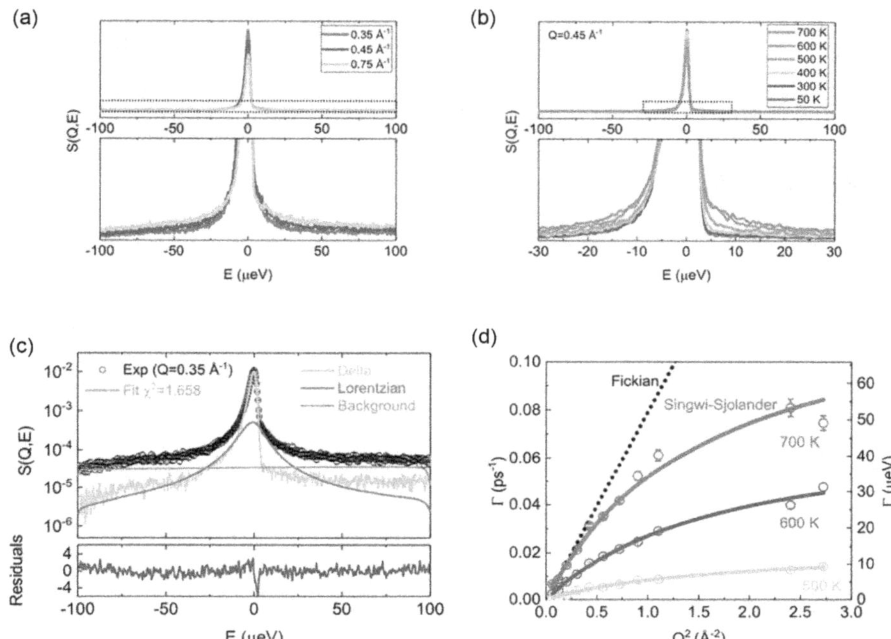

FIGURE 12.51 (a) Dynamic structure factor $S(Q, E)$ from QENS experiments at 700 K for different Q. (b) Dynamic structure factor $S(Q, E)$ from QENS at $Q=0.45\,\text{Å}^{-1}$ for different temperatures. (c) An example of the QENS fit with background, delta and Lorentzian functions convoluted with the resolution function for 700 K and $Q=0.35\,\text{Å}^{-1}$, along with the residuals of the fit. (d) HWHM (Γ) of the Lorentzian as a function of Q^2 at different temperatures. (Reprinted with permission from ref. [45]. Copyright 2017 Elsevier.)

Figure 12.51a and b, respectively. The peaks broaden with increasing Q and temperature above 400 K. The half-width-half-maximum (HWHM) Γ can be obtained by fitting the inelastic features with the Lorentzian function, as shown in Figure 12.51c. The relationship of Γ with Q^2 is shown in Figure 12.51d. The broadening of Γ with Q^2 at low Q appears to be a linear relationship and reaches a horizontal asymptotic line at high Q. The evolution is a characteristic of the jump diffusion process, which can be quantitatively described by the Singwi–Sjölander (SS) jump diffusion model, where the jump time is much shorter than the residence time (τ) of the diffusing nucleus at a site. At low Q, the model reduces to the Fickian model, which exhibits a monotonously linear relationship. In this scenario, $D=\langle r^2\rangle/6\tau$, where $\langle r^2\rangle$ stands for a mean square jump distance. The residence time τ and mean jump length $l=\sqrt{r^2}$ at 700 K can be calculated as 7.5 ± 0.9 ps and $1.96 \pm 0.15\,\text{Å}$, respectively.

Both the relationships derived from the QENS data and calculated MD simulations can be well described by the SS jump diffusion model (Figure 12.52a). The residence time and jump length obtained by fitting the SS model with the two relationships are in good agreement, as shown in Figure 12.52c and d. The mean jump length obtained from the QENS data is ~2.0 Å, which roughly equals the distance from tetrahedral (24d) to octahedral (48 g) sites, as shown in Figure 12.52b. With the

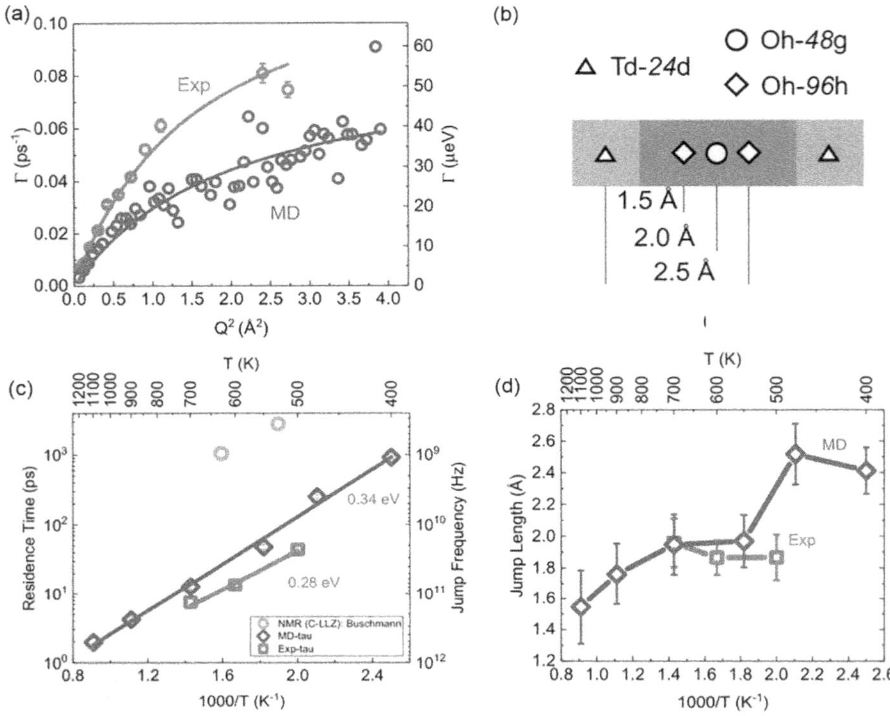

FIGURE 12.52 (a) HWHM (Γ) experimentally measured using QENS and equivalent Γ of Li diffusion derived from MD simulations in $Li_5La_3Ta_2O_{12}$ at 700 K. (b) Schematic of the relation between crystallographic sites for Li within tetrahedral (Td) and octahedral (Oh) cages in $Li_5La_3Ta_2O_{12}$ showing key distances. (c) Mean residence time (τ) of Li diffusing in $Li_5La_3Ta_2O_{12}$ at sites obtained from MD simulation and QENS experiments. (d) Mean jump length of Li diffusing in $Li_5La_3Ta_2O_{12}$ obtained from MD simulation and QENS experiments. (Reprinted with permission from ref. [45]. Copyright 2017 Elsevier.)

increase in temperature to 1,100 K, the jump length decreases to 1.5 Å, corresponding to the distance from tetrahedral (24d) to octahedral (96h) sites. The jump length increases to 2.5 Å at 400 K, corresponding to migration to a farther 96h octahedral site. This result suggests a lithium diffusion mechanism in which jumps occur with larger frequency (smaller residence time) and shorter length at higher temperature, consistent with the delocalization of Li at elevated temperatures.

12.5 X-RAY AND NEUTRON IMAGING

The electrochemical reactions in batteries occur at multiple scales because of the hierarchical structure associated with the chemical composition. Therefore, the microstructure, morphology and distribution of the chemical composition are worth studying to improve battery performance. Imaging methods specialize in the direct visualization of these features, which show powerful capabilities for resolving complicated structure–performance relationships. Next, several X-ray and NI techniques, which have been widely used in the battery field, are briefly introduced in principles and applications.

12.5.1 Contrast Mechanism

To visualize an object in real space, the difference in intensity called "contrast" with the penetration of X-ray needs to be observed. The contrast arises from the intrinsic interactions of X-rays with matter, thus resulting in the great versability of X-ray imaging. Two contrast mechanisms, i.e., absorption contrast and phase contrast, are mainly referred to, where the former results from absorption interactions and the latter arises from scattering processes.

For imaging, absorption is commonly assessed by measuring the total transmitted X-ray intensity. The contrast can be generated because of the difference in penetration thickness. Moreover, the contrast can result from the difference in chemical composition because of the element-dependent absorption cross-section in the fourth order of atomic number, Z^4, at a given energy much larger than the ionization threshold. The contrast for a specific element can be further enhanced by utilizing the absorption edge. If two images are taken at the photon energy right below and above the absorption edge, the difference between the two images displays the spatial distribution of the element. A thorough strategy to utilize the absorption contrast is to collect the entire XANES or NEXAFS spectra by tuning the photon energy. Then, the distribution of oxidation states or even spin states can be obtained at the cost of a longer time for data acquisition. To balance the test time and accuracy, the images can usually be taken at several energy points corresponding to the spectral features. Because of the more significant photoelectric effect in the low energy region, absorption contrast plays a more important role in imaging with soft X-rays.

Phase contrast results from the phase shift of X-rays, which has a direct relationship with the angular deviation during refraction. The refraction process is related to the refraction index δ and the absorption index β, where the complex refraction index can be expressed as $n_r = 1 - \delta - i\beta$. However, δ makes a much larger contribution than β, as demonstrated by the energy-dependent evolution of δ and β of H_2O in Figure 12.53. δ is much greater than β across a very large energy range of X-rays. Moreover, δ decreases more slowly with increasing photon energy than β.

The quantitative relationship of angular deviation with δ is revealed in Figure 12.54 with two simplified models. A homogeneous structure with wedge angle ω is shown on the left. The wedge with a width of Δx is chosen so that both the wavefronts at the left corner and the right corner pass from the material to vacuum. The exit X-ray changes its direction by an amount $\alpha = \frac{\lambda(1+\delta)-\lambda}{\Delta x} = \delta\frac{\lambda}{\Delta x} = \delta \tan\omega$ in passing through the wedge. To the right is a piece of inhomogeneous material with constant thickness. As the density increases toward the right, the outcoming X-ray is refracted by an angle $\alpha = \frac{\lambda\Delta x \frac{\partial\delta(x)}{\partial x}}{\Delta x} = \lambda\frac{\partial\delta(x)}{\partial x}$. The direction correlation of α to

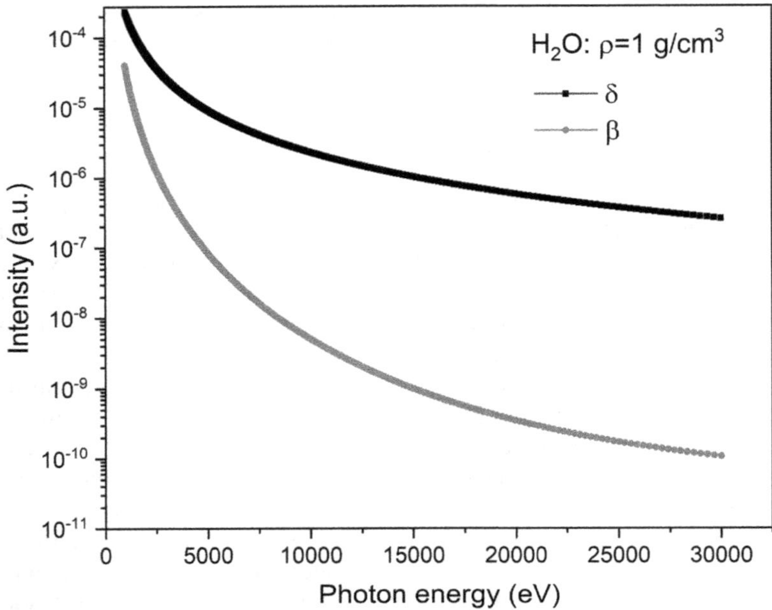

FIGURE 12.53 The energy dependence of the real (δ) and imaginary (β) contributions to the refraction index of water.

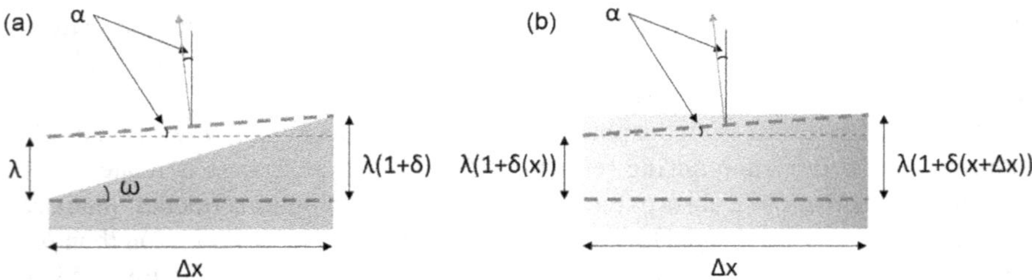

FIGURE 12.54 Relationship of angular deviation with δ when refraction occurs in (a) a wedge-shaped sample of uniform density and (b) a plate where the density linearly increases toward the right.

the dedicative of δ demonstrates the capability of phase contrast imaging in resolving the shape and chemical composition of structural units.

The advantages of phase contrast imaging are shown in Figure 12.55a and b. The attenuation and phase cross-section as a function of photon energy and atomic number are presented, respectively. The phase cross-section is much larger than the attenuation cross-section across a very broad energy range. Moreover, the difference between the two becomes larger with increasing photon energy. Therefore, phase contrast plays a more important role in hard X-ray imaging, which has advantages in measuring thick samples and building in situ experiments. In addition, the phase contrast with hard X-rays is commonly larger than the absorption contrast

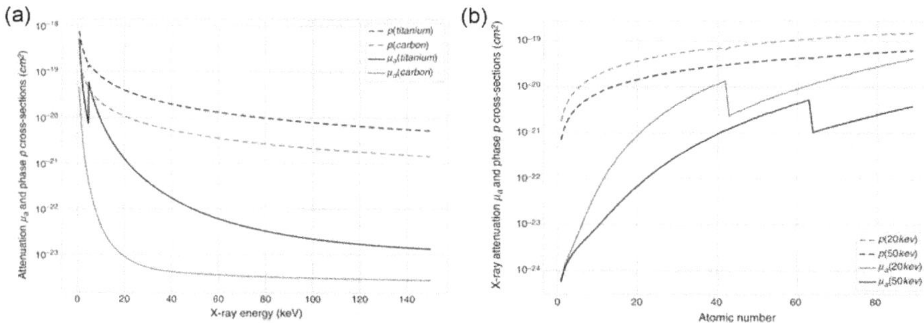

FIGURE 12.55 Phase and attenuation cross-sections as a function of (a) X-ray energy (for carbon and titanium) and (b) atomic number. (Reprinted with permission from ref. [46]. Copyright 2019 Springer Nature Switzerland AG.)

over a large range of atomic numbers, as shown in Figure 12.55b. The difference is particularly substantial for low-Z elements. Therefore, phase contrast imaging with hard X-rays is suitable for materials rich in light elements or mixed with light and heavy elements.

12.5.2 Imaging Methods

The imaging methods can be divided into microscopy and tomography, where the former can provide the distribution of microstructure and chemical composition in two dimensions, and the latter can extend the corresponding information to three dimensions. For microscopy, TXM is one of the major microscopic techniques that has been widely applied to battery studies. TXM methods can be further classified with the data acquisition strategy. The TXM performed without mechanical movements of the sample is named full-field TXM. On the other hand, the images can also be built point by point through a scan of the sample, usually with a focusing beam, which is referred to as STXM. Full-field TXM and STXM are briefly introduced with typical cases. Then, X-ray computed tomography (CT) based on TXM is depicted.

12.5.2.1 TXM

Full-field TXM has a configuration analogous to that of a conventional bright-field optical microscope. A condenser, such as a K-B mirror, zone plate or capillary, is used to shape the X-ray beam hit on the sample, while an area detector, typically a charge-coupled detector (CCD) with a scintillator screen, is used to record the X-ray transmitted through the sample. A Fresnel zone plate can be inserted additionally between the sample and the scintillator screen for the magnification. The total magnification power is approximately 1,000–2,000×, achieving a spatial resolution of 30 nm for hard X-rays and 15 nm for soft X-rays. The advantage of full-field TXM lies in its data acquisition speed, where the exposure time can be shortened to tens of milliseconds. Therefore, the imaging of the entire specimen can be completed quickly by multiple illumination and reimaging with the field of view usually limited to 10–30 μm. Absorption contrast has been the most widely used for full-field

TXM imaging. High-resolution images of biological samples can be obtained with full-field TXM in the soft X-ray range by taking advantage of the absorption contrast between proteins and water. Phase contrast can also be achieved by adding analyser optics or changing the sample–detector distance. This is particularly useful for full-field TXM in the hard X-ray range (hTXM), which is specialized in in situ/operando studies and thus widely used in the battery field. In addition, XANES spectra can be combined with full-field TXM imaging to reveal the distribution of the chemical composition.

In situ hTXM has been used to visualize the reaction heterogeneity of $LiCoO_2$ during cycling.[47] The chemical change in a selected $LiCoO_2$ particle is collected by combining hTXM with XANES of the Co K-edge in Figure 12.56. The pristine particle shows the fully discharged state with pure Co^{3+} coloured green. When charged to 4.6 V at a rate of 1 C, most of the particle turns to a charged state displayed in red. Only approximately half of the reacted particle returns to the pristine state when discharged to 3 V at a rate of 1 C. The recovery portion is demonstrated to be rate-dependent in comparison to the chemical maps in Figure 12.56 (d) for 10 C and 12.56 (e) for 0.2 C. It also suggests the absence of a memory effect. The reaction heterogeneity is thought to result from the nucleation process on delithiation, which is a nonequilibrium process and thus rate-dependent. Instead of the bulk behaviour at the cell scale, this particular active particle can only reflect the rate dependency of the local chemical reaction on cycling, which still offers insight into the fundamental degradation mechanism.

In situ/operando hXAS combined with XANES of the TM K-edge has also been used to study the correlation of the local interfacial environment to the electrochemical performance in solid-state batteries.[48] As shown in Figure 12.57, the changes in

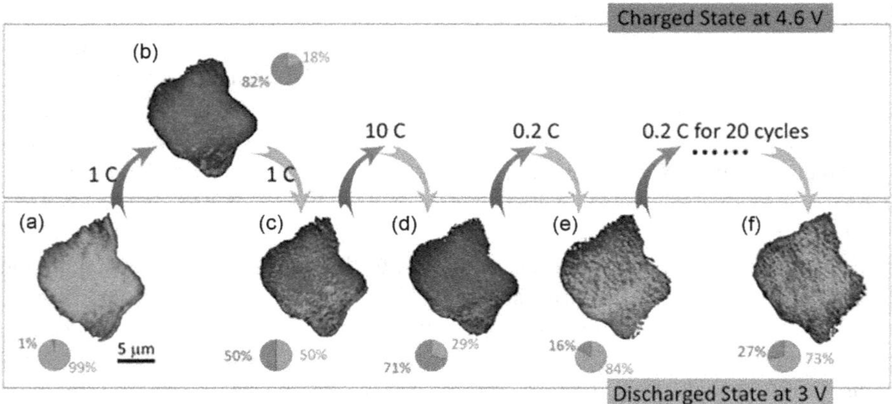

FIGURE 12.56 In situ monitoring of the chemical heterogeneity in a single particle of $LiCoO_2$ up to 20 cycles. The chemical map of the particle at (a) the pristine state and (b) the charged state at 4.6 V. (c–e) Chemical maps at the discharged state at 3 V after the particle went through cycles of different rates at 1C, 10C and 0.2C, respectively. Panel f is the map at the discharged state after 20 cycles at 0.2C. (Reprinted with permission from ref. [47]. Copyright 2017 American Chemical Society.)

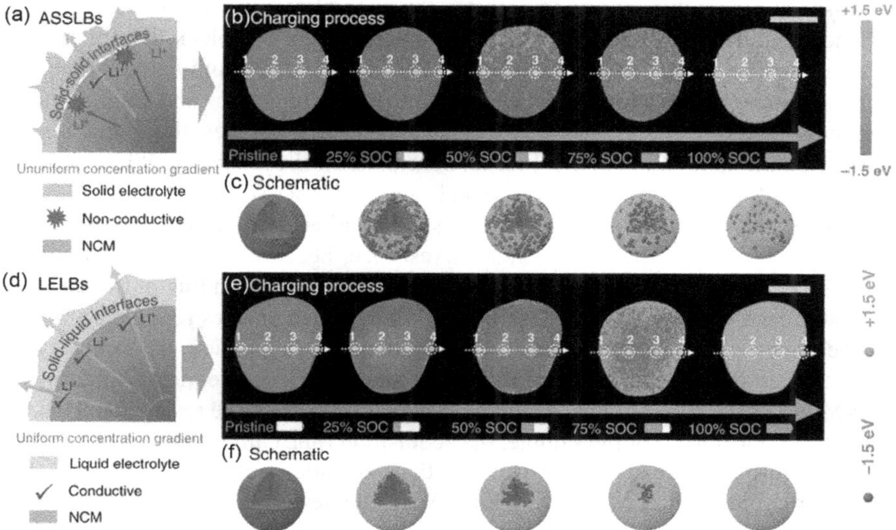

FIGURE 12.57 (a) Schematic of solid–solid interface models and kinetics. (b) TXM-XANES mapping of a single cathode particle as a function of charging time in ASSLBs, and (c) the corresponding schematic diagram to expound the unique solid-state electrochemistry. (d) Schematic of solid–liquid interface models and kinetics. (e) TXM-XANES mapping of a single cathode particle as a function of charging time in LELBs, and (f) the corresponding schematic diagram showing the conventional solid–liquid electrochemistry. Scale bar, 10 μm. (Reprinted with permission from ref. [48]. Copyright 2020 Nature Publishing Group.)

the chemical state of Ni start from the partial particle surface, accompanied by delithiation from the surface. With continuous charge, the Li^+ vacancy gradient moves gradually toward the inner bulk. Although Li^+ is selectively transported through the interfaces because of the discontinuous electrode–electrolyte contact and into the bulk phase in an anisotropic manner, the chemical state presents a uniform distribution across the entire particle in the fully charged state. The equilibrium of the electron and local Li^+ concentrations is thought to compensate for the deficiency due to the physical contact loss at the interfaces. In contrast, the conformal solid–liquid interfaces in liquid electrolyte batteries ensure homogeneous and unrestricted Li^+ transportation at the surface. Therefore, a nearly whole annular reaction region can be formed on the particle surface. During the charge process, the diffusion of lithium vacancies follows a strict "core–shell" delithiation model, and a homogeneous charge distribution finally forms at the fully charged state. Therefore, the local ion concentration equilibrium is demonstrated to be crucial to the persistent performance of solid-state batteries.

12.5.2.2 STXM

The exclusive technical feature of STXM is the placement of a zone plate upstream of the sample. Therefore, the incident X-rays can be demagnified strongly to form a microprobe across which the specimen is raster-scanned in two dimensions. An order-selecting aperture is placed between the zone plate and the sample to remove all undesired diffraction orders, while a beam stop is used to block the direct zero-order light. The spatial resolution of the STXM is primarily dominated by the size of the focusing beam, which is dependent on the focusing optics and configurations and the source properties. For soft X-rays at synchrotron sources, the spatial resolution can reach up to 10 nm with a Fresnel zone plate, while it is below 80 nm for hard X-rays with K-B mirrors, multilayer Laue lenses or refractive lenses.

One of the major advantages of STXM is the simultaneous monitoring of different signals, such as X-rays, electrons and fluorescence, because of the high flux of the focusing beam, which could provide a comprehensive understanding of the samples. Another merit is its tunability of the field of view, which enables rough scanning as an overview and fine scanning for selected small areas. In addition, the samples usually suffer from less beam damage in STXM than in TXM. The major limitation of STXM lies in the speed of data acquisition. Apart from the large restriction from the mechanical movement of the sample, the acquisition time can be affected by the flux and the information detected. For example, the acquisition time can reach up to a few hours for complete fluorescence imaging. With the development of next-generation light sources and advanced focusing optics, the speed of data acquisition will be improved, accompanied by a higher spatial resolution.

In comparison with hard X-rays, STXM with soft X-rays (sSTXM) can fully exploit the capability of absorption imaging because of the enhanced photoelectric effect. Moreover, the direct probe of the TM L-edge enables high sensitivity to TM chemical states. Therefore, sSTXM integrated with a spectroscopic probe plays an irreplaceable role in battery studies despite its performance difficulty in ultrahigh vacuum.

A good example is the application of sSTXM to the origin and hysteresis of compositional spatiodynamics in $LiFePO_4$ particles.[49] As shown in Figure 12.58, the delithiation and lithiation processes in several particles at different C-rates have been compared qualitatively. When lithiated at a high rate of 2C evaluated by electrochemical testing (Figure 12.58a), the composition maps of particles show uniform intercalation, indicating solid-solution reaction behaviour. The quantitative analysis (Figure 12.59a) confirms the small variation in composition across each particle and the continuous increase from $x=0$ to $x=1$ over time. On the other hand, the particles after relaxation for 12 hours (analogous to 0C) displayed sharp phase boundaries between Li-rich and Li-poor regions (Figure 12.58c), where the components can be quantified as $FePO_4$ and $LiFePO_4$ (Figure 12.59b). At intermediate C rates (e.g., 0.2C and 0.6C), the solid-solution regions of fast (de)lithiation and slow (de)lithiation are separated. The continuous change in Li composition occurs separately in domains (Figure 12.59c), which exhibit negligible growth due to the little movement of phase boundaries.

Through the quantification of local current density (Figure 12.59d), the same domains are found to be fast on both lithiation and delithiation. This indicates that the domain structures have little relationship with phase transformations such as random nucleation or spinodal decomposition. The domination of solid Li diffusion in the domain structure can also be ruled out because of the much faster Li diffusion (~ms) than the reaction time (~hours). Therefore, the domain structure is thought to

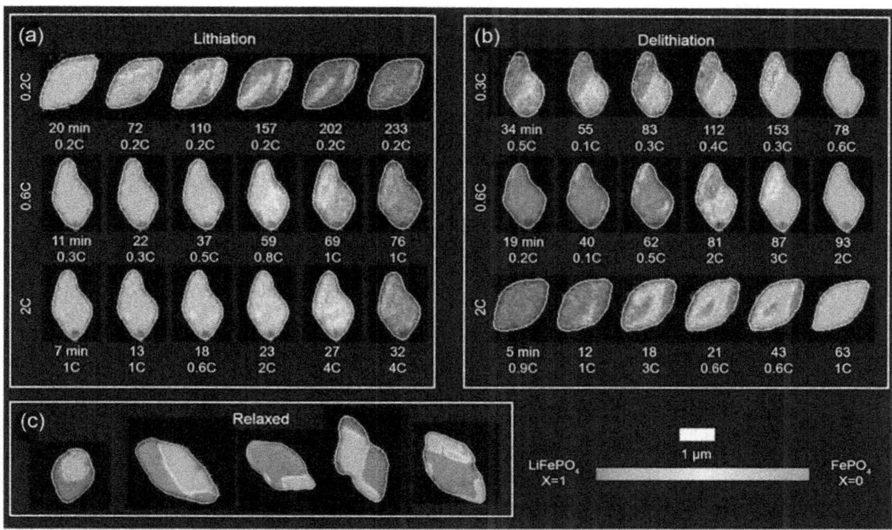

FIGURE 12.58 Representative frames of different particles taken at various (a) lithiation and (b) delithiation rates. (c) Ex situ frames of Li composition for relaxed particles, showing the equilibrium distribution of Li within particles. (Reprinted with permission from ref. [49]. Copyright 2016 American Association for the Advancement of Science.)

arise from the spatial variations in the kinetics at the solid–liquid interfaces, which can be related to the inhomogeneous strain, variations in carbon coating and surface defects induced by cycling.

The reaction kinetics are further studied with the exchange current density j_0 as a function of the Li composition extracted according to the linear relationship between the current and voltage (<120 mV), as shown in Figure 12.60a. The results show that j_0 is strongly dependent on the Li amount, x, which is particularly low for Li-poor and Li-rich end members because of low concentrations of charge carriers. The skewed $j_0 - x$ relationship is thought to be a necessary condition for the solid-solution behaviour above a critical current because of the theoretical prediction of an unstable solid-solution pathway for a symmetric $j_0 - x$ curve. The asymmetric $j_0 - x$ curve can also be responsible for the deviation of solid-solution behaviour between slow and fast domains. As shown in Figure 12.60b, the deviation becomes larger during delithiation because of the larger difference in j_0 between slow and fast domains. In contrast, the deviation decreases upon lithiation. Thus, the solid-solution regions with different Li compositions more readily occur during delithiation than lithiation at a high C-rate (e.g., 2C). This study reveals the spatially varied solid-solution behaviour in $LiFePO_4$, which is conventionally thought to undergo a strict two-phase transformation. The quantification capability of sSTXM is demonstrated through this model study.

12.5.2.3 X-Ray CT

Different from 2D microscopy, CT refers to the 3D spatial distribution constructed by a set of (usually more than 1,000) 2D projection images with algorithms, which are collected at different angles as the sample rotates. Since its construction in the

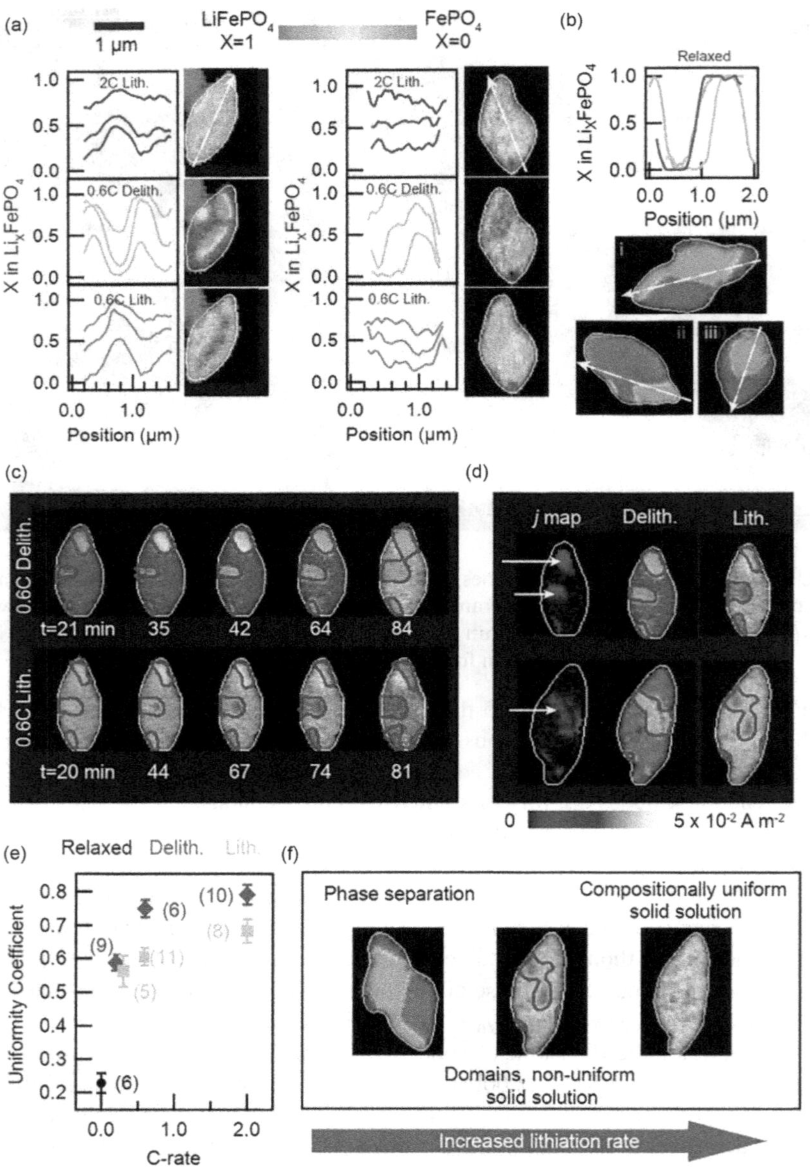

FIGURE 12.59 (a) Line cuts of Li composition (*x*) of the same particles under different cycling conditions. (b) Line cuts of the relaxed, phase-separated particles. (c) The fast domains, outlined in blue, do not substantially grow in size. (d) Current density quantification reveals regions of higher insertion kinetics, calculated from delithiation. The same fast domains are present in both charge and discharge. (e) The uniformity coefficient increases with cycling rate and is consistently higher for lithiation than for delithiation. (f) Scheme of the insertion pathway as a function of the lithiation rate. (Reprinted with permission from ref. [49]. Copyright 2016 American Association for the Advancement of Science.)

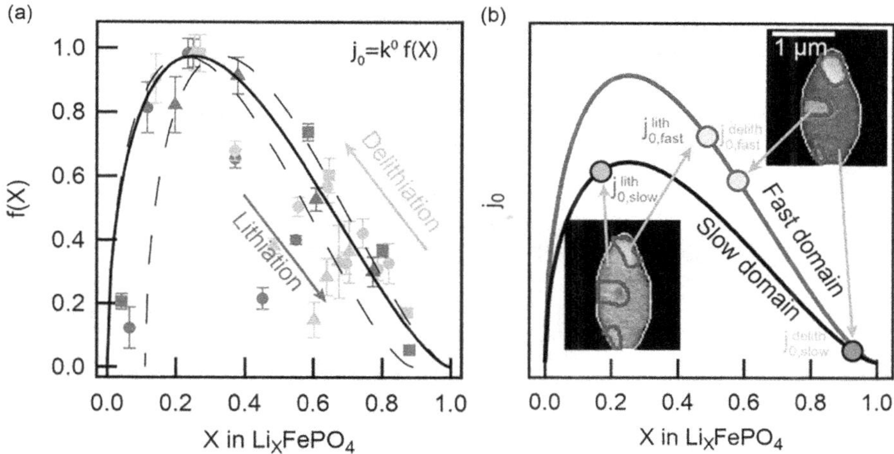

FIGURE 12.60 (a) The measured exchange current density (j_0) varies nonmonotonically with the Li composition *x*. k_0 is ~1×10^{-2} A/m^2. Each marker in each colour represents a different particle. The dashed lines show the shifted $j_0 - x$ curve due to uncertainty arising from the fact that the specific capacity of the particles is less than the theoretical capacity. (b) Because the skewed j_0 peaks at $x \approx 0.25$, the value of j_0 for the fast domains is several times that of j_0 for the slow domains during delithiation, but the two quantities are comparable during lithiation. (Reprinted with permission from ref. [49]. Copyright 2016 American Association for the Advancement of Science.)

1970s by Allan M. Cormack and Godfrey Newbold Hounsfield, CT has been used tremendously in a number of clinical and other research fields. It provides vast opportunities to study the key issues in the battery field, such as the dynamics of 3D morphological changes, when combining synchrotron light sources. CT can be achieved with absorption contrast when the phases in samples have substantially different attenuation coefficients. For materials rich in low-Z elements, such as lithium anodes, phase contrast would be a good choice for CT with hard X-rays. Equipped with different lenses or camera systems, CT is able to nondestructively probe the complex structural and morphologic evolution at multiple scales in batteries for mechanistic studies covering microstructure evolution, phase changes and ionic diffusion.

Absorption contrast CT has been used to visualize and quantify the origins of the reaction and degradation mechanism of SnO.[50] Because of the difference in attenuation coefficients (μ), which is related to the chemical composition and mass density, substantial absorption contrast appears on the reconstructed tomographic image (Figure 12.61a). The yellow–red region corresponds to SnO particles with a high μ of 45 cm^{-1} (Figure 12.61c), while the blue region corresponds to the carbon black, polymer and electrolyte phases filling voids with a low μ of 0.4 cm^{-1} (Figure 12.61d). During lithiation (Figure 12.61b), a low-attenuating (yellow–green) phase starts to form from the surface and progressively penetrates into the particles. This represents the gradual conversion of SnO into Li_2O and Sn mixed clusters, which can be confirmed by the appearance of features at $\mu = 25$ cm^{-1} (Figure 12.61c). The product appears as a nearly annular coating on SnO, indicating the core–shell model of

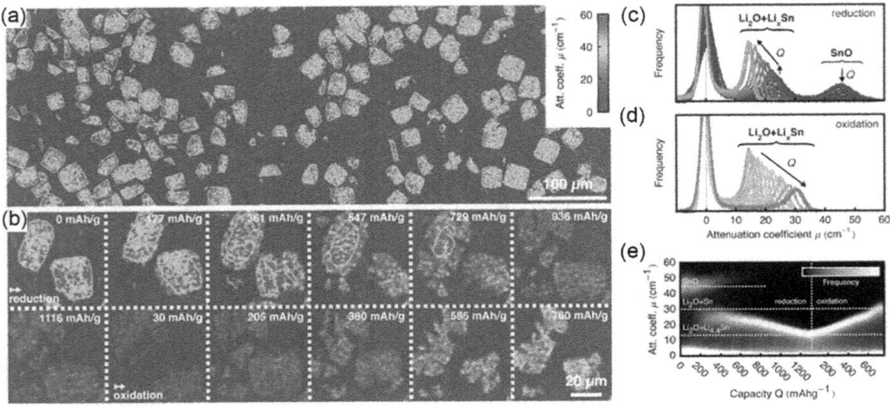

FIGURE 12.61 (a) Unprocessed cross-sectional tomogram showing individual SnO particles in the electrode with high resolution and good contrast against a low-attenuating carbon black, binder and electrolyte phase. (b) A series of cross-sections through two particles demonstrates a core–shell process, volume expansion and particle fracture during the initial reduction and particle redensification during subsequent oxidation. X-ray attenuation coefficient histograms during electrochemical (c) reduction and (d) oxidation. (e) Attenuation coefficient distributions during electrochemical reduction and oxidation. Horizontal dotted white lines indicate theoretical attenuation coefficients for end members of the phase evolution. (Reprinted with permission from ref. [50]. Copyright 2013 American Association for the Advancement of Science.)

the conversion reaction. A closer inspection shows the subtle shifts of μ from 45 to 50 cm^{-1}, which can be visualized by the appearance of a deeper red phase in Figure 12.61b. This is thought to result from the disproportionation reaction of SnO to SnO_2 and Sn. Subsequently, the yellow–green phase homogeneously transforms into the dark green phase (Figure 12.61b), where μ progressively turns into 14 cm^{-1} (Figure 12.61c). This process corresponds to the alloying reaction of Si clusters, where a series of Li_xSn ($x=0.4$, 1, 2.33, 2.5, 2.65, 3.5, 4.4) forms in subsequence. During delithiation, a homogeneous change from dark green to green–yellow can be observed (Figure 12.61b), corresponding to the back-shift of the peak at 14–30 cm^{-1} with decreasing magnitude (Figure 12.61d). This result is consistent with the dealloying of Li_xSn clusters mixed with Li_2O. The deviation of u from the theoretical values for Sn in Li_2O indicates the remanence of Li_xSn. The linear evolution of μ as a function of capacity Q (Figure 12.61e) indicates a steady-state phase transition.

As shown in Figure 12.62a and b, preferential cracks are revealed to form along the grain boundaries in the (001) plane. Moreover, the cracks appear sequentially at the opposite edges. This results from the accumulation and relaxation of stress in the reaction progress. The formation and propagation of the first crack and the progress of the conversion reaction nearby mutually promote each other (Figure 12.62c). However, this can hardly alleviate the stress accumulated at the opposite side with the further reduction of the particle. Therefore, a new crack initializes along a parallel gain boundary at the opposite side. The 3D rendering in Figure 12.62d indicates that the zig-zag morphologic evolution occurs in multiple particles. In this study, CT data can also be used to quantify the volume change in active particles (not shown here).

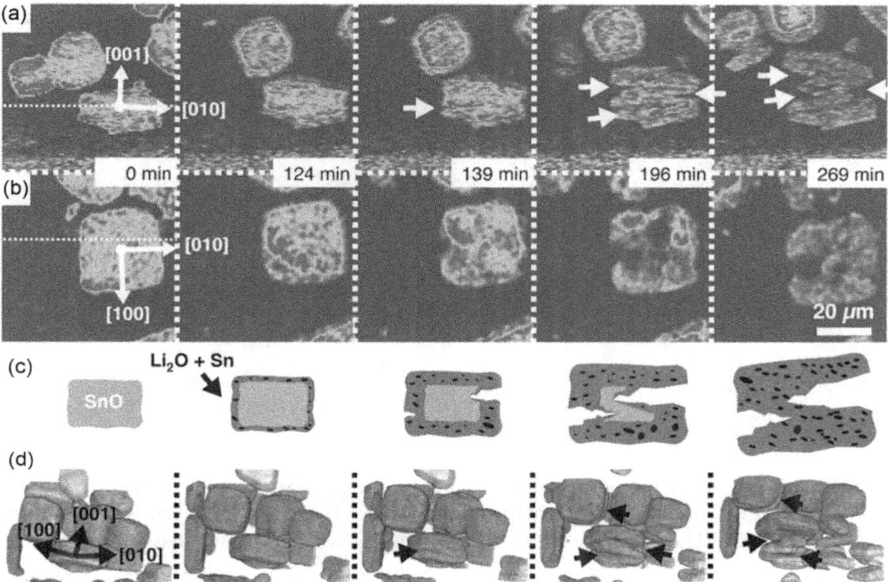

FIGURE 12.62 (a) Coronal and (b) transverse cross-sections through a particle during electrochemical reduction. (c) Schematic of particle phase evolution and crack growth leading to a zig-zag morphology. (d) 3D rendering of subvolume visualizing zig-zag morphology in multiple particles. Black arrows indicate fracture. (Reprinted with permission from ref. [50]. Copyright 2013 American Association for the Advancement of Science.)

Hard X-ray CT in phase contrast mode can be used to monitor the formation of lithium dendrites at the solid-state electrolyte/lithium anode interface.[51] As shown in Figure 12.63, both the slices through typical tomograms and the 3D reconstructed volumes nearby indicate that the formation of dendrites in the electrode is much earlier than that in the electrolyte. The dendrite starts to grow on cycling and propagate into the electrolyte, resulting in the short circuit. Most of the dendrite resides within the electrode instead of the electrolyte. This work suggests that inhibiting the formation of subsurface structures in lithium electrodes might be a more fundamental way to prevent the growth of dendrites.

In addition to the real space imaging methods mentioned above, the other large category is based on the diffraction pattern in reciprocal space. Diffraction imaging can be used to deduce the electron density of samples through the inverse Fourier transform. However, it can hardly be performed under normal circumstances because of the difficulty in phase retrieval. If a finite-sized sample is illuminated with coherent X-rays, the phases can be retrieved, enabling the reconstruction of a real space image by Fourier inversion. The techniques are named coherent X-ray diffraction imaging (xCDI). Based on conventional xCDI, more imaging methods have been developed, including Bragg xCDI, ptychography and holography. The most triggering advantage of diffraction imaging is that the resolution is only limited by the wavelength in principle, which can be as low as 0.1 nm for hard X-rays because of the

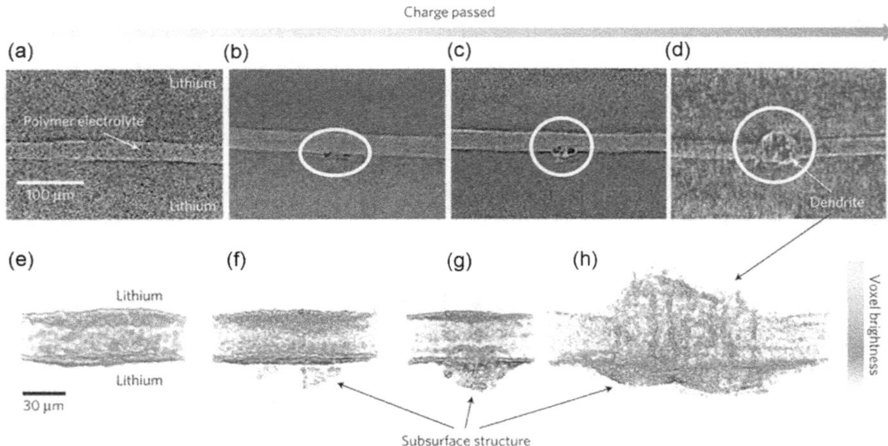

FIGURE 12.63 X-ray tomography slices showing the cross-sections of symmetric lithium cells cycled to various stages with charge amounts of (a) 0 C/cm^2, (b) 9 C/cm^2, (c) 84 C/cm^2 and (d) 296 C/cm^2. The corresponding 3D reconstructed volumes shown in (e–h). (Reprinted with permission from ref. [51]. Copyright 2014 Nature Publishing Group.)

lensless configurations. Their applications in the battery field remain in their infancy and will not be elaborated here. For more discussions on NI, some dedicated reviews are recommended.[52,53]

12.5.2.4 Neutron Imaging

NI can nondestructively resolve the inner structure of materials, which is complementary to X-ray imaging. Neutron radiography (NRg) is one of the major strategies for imaging. Conventional NRg is usually fulfilled with absorption contrast. With the improvement of detectors and neutron sources, other NRgs with various contrast mechanisms can also be developed, such as phase contrast imaging, Bragg-edge imaging, stroboscopic imaging and dark-field imaging. Neutron tomography can be further achieved based on the NRg and computational methods. Thermal neutrons and cold neutrons are typically used for NI, where the former can enable a better contrast because of the stronger interaction with matter. In comparison with X-ray imaging, NI is more suitable for in situ/operando studies because of the larger penetration length of neutrons. Moreover, NI is specialized in probing the distribution of light elements such as hydrogen, lithium and boron. However, the spatial resolution of NI is usually in the range of 10–25 µm, which is much lower than that of X-ray imaging. The most attractive aspect of battery research is that NI could provide direct probing of the changes in lithium concentration because of the activation cross-section of ^{6}Li, which can only be inferred by the changes in the chemical states of heavy elements using X-ray imaging. NI has been regarded as a powerful tool to study the fundamental issues correlated with lithium, such as Li^+ transport and Li^+ distribution.

The mechanical degradation in a commercial Li/MnO_2 primary battery has been revealed by a direct probe of lithium diffusion and electrode wetting by the electrolyte with neutron tomography.[54] As shown in Figure 12.64a and b, the bright region

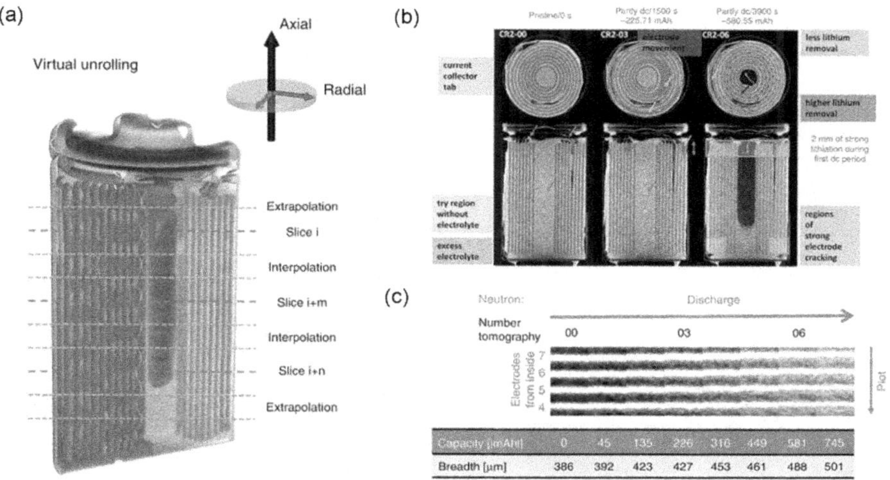

FIGURE 12.64 (a) The reconstructed tomograms from neutron CT along with examples of sections extracted following virtual unrolling of the reconstructions. (b) Orthogonal slices of the neutron tomogram captured during the discharge over a 4.7 Ω resistor, where the lithium electrode and the excess electrolyte in the middle of the cell are clearly visible. Lithium intercalation and electrolyte consumption are observed, as well as electrode cracking and electrolyte consumption. (c) Cut-outs of virtual unrolled multilayer sections of the measured neutron tomograms at different SoCs. The electrode swelling is approximately 30% for a cell discharge to −745.08 mAh (07) and 26.5% to 580.55 mAh (06). (Reproduced with permission from ref. [54]. Copyright 2020 Nature Publishing Group.)

in the neutron tomograms represents lithium, while the dark area corresponds to the cathode with low attenuation. For the pristine cell, excess electrolyte fills the regions in the radical middle of the cell and at the top of the cell (indicated by the pink arrow). After discharging to a capacity of 225.71 mAh in ~1,500 seconds, most changes occur over an ~2 mm region in the upper axial part of the cell. The electrolyte is substantially removed in association with the removal of Li from the anode. The cracks appear at the outer radial electrode windings, which is less pronounced in the radial middle and bottom regions. Furthermore, the outer electrode windings are moved radically outward due to the fast lithium insertion, as displayed by the red arrow. In the further discharging process, most of the excess electrolyte in the radial middle region of the cell moves into the cathode cracks or fills the gaps left with the removal of Li. Strong cracks appear at the bottom and top regions marked by the brown frame. The extent of Li removal can be revealed by the change in anode thickness, which is marked by light blue and dark blue arrows. The resulting volume changes with the progress of discharge can be further evaluated in Figure 12.64c. After discharging to 580.55 mAh, the volume change for the cell can reach up to +26.5%, which is consistent with the values derived from X-ray tomography.

The growth process of Li dendrites with Li flowing from the anode to the cathode has been directly observed by in situ NRg and static tomography.[55] Figure 12.65a shows the 3D reconstructed volumes measured after each charge and discharge step.

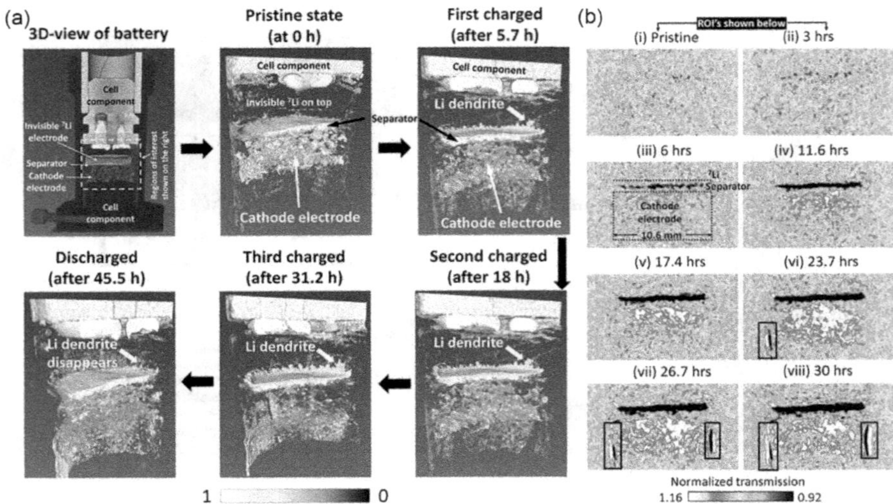

FIGURE 12.65 (a) 3D evolution of the Li distribution in the battery cell at different stages of charging and discharging. The contrast value from 0 to 1 represents an increasing trend of absorption of neutrons. White arrows in the series of images point at the dendritic Li around the disk edge of the ^{7}Li electrode (invisible), growing during charge and vanishing at the end of discharge. (b) 2D evolution of the Li distribution as a function of charging time based on the normalized neutron radiographs. (Reprinted with permission from ref. [55]. Copyright 2019 American Chemical Society.)

Branch-like dendrites start to grow from the edge of the lithium plate, exhibiting progressive growth at different stages of charging. Moreover, the dendrites disappear at the end of discharge. The dynamics of dendrite growth are revealed by real-time 2D radiography, as shown in Figure 12.65b. As shown in (iii–vi), Li migrates to the separator/Li metal interface with increasing time and voltage. The battery eventually becomes shorted with the drop in voltage at 23.7 hours of charging. Meanwhile, depletion of Li across the cathode region can also be observed in (iii–vi), which is displayed as red/white regions. The depletion of Li can be stopped and reversed, as demonstrated by the reduced red/while regions (vi–viii). The change is thought to result from the self-discharge process after battery shorting.

12.6 PERSPECTIVE

In the past few decades, advanced synchrotron X-ray and neutron techniques have been widely applied to the battery field and have greatly promoted understanding of the fundamental science and design of high-performance batteries. These methods have already been effective for exploring long-order structures, local structures and electronic structures because of their resolving capabilities of elements, time or space. However, more novel techniques apart from the abovementioned need to be exploited to provide new insights and strategies for the development of batteries owing to the coupling of multiple processes at multiple time and length scales.

Because of the indispensable role of TMs, the detection of magnetic properties might give more inspiration. In this scenario, spin-resolved X-ray techniques such as magnetic circular dichroism (MCD) and neutron techniques such as magnetic scattering and spin echo spectroscopy have become attractive. For example, strongly frustrated magnetic interactions between spins of TM ions have been revealed by NMS in $LiNi_xMn_yCo_zO_2$ cathode materials, which is thought to be the origin of Li/Ni exchange.[56]

An increasing trend is to use methods with multidimensional sensitivity, which makes the decoupling of complicated electrochemical processes more accessible. In situ measurements and scanning spectroscopic techniques are the two classic methods, which have been mentioned previously. The former can reveal the feature evolution at a certain time scale, while the latter enables the spatial distribution of certain characteristics. In addition, two more methods are worth mentioning here. The first is resonant elastic scattering, which combines chemical and structural detection simultaneously. For hard X-rays, this technique can also be called diffraction anomalous fine structure (DAFS), which can be used to reveal the chemical environment on different sites. For example, the interlayer migration of Ni and Mn in Ni-doped Li_2MnO_3 during the first charge and discharge process has been studied by DAFS, which can be related to cyclability.[57] The second highlighted method is resonant inelastic X-ray scattering with Q-resolved capability (Q-RIXS). For soft X-rays, Q-RIXS can be theoretically used to study low-energy excitations such as phonons, d–d excitation and charge transfer with different correlation lengths, which is closely related to the dynamic process in battery materials. However, Q-RIXS is a very rare technique and has not been applied to battery research as far as we know.

The capabilities of methods with multidimensional sensitivity can be further improved with the development of next-generation light sources, such as diffraction-limited storage rings (DLSRs) and free-electron laser facilities (FELFs), where the generated X-rays have ultrahigh flux, monochromaticity and collimation because of spatial and temporal coherence. Owing to the great temporal structure, the time-resolved capability can be greatly improved with the development of pump probing techniques, which reach ps for DLSR and fs for FELF. The ps resolution lies in the time window of ionic fluctuations and thus is suitable for exploring ionic motions and collective ionic dynamics. The fs resolution can be used to probe electron transfer, which is often the starting point of redox reactions. More techniques have also been developed to take full advantage of the coherent nature of X-rays. A large set of methods is based on coherent X-ray diffraction imaging (xCDI), where the theoretical spatial resolution can only be limited by the wavelength. The nanostrain and lattice displacement in Li-rich and Mn-rich layered oxides have been revealed by the in situ Bragg xCDI, which are assigned as the driving forces for both structure degradation and oxygen loss.[58] Owing to the ultrahigh flux of DLSR or FELF, the time-resolved xCDI images even enable the build-up of movies related to the atom motions. Another experimental technique worth mentioning is X-ray photon correlation spectroscopy (XPCS). It enables probing of the hopping dynamics and structural relaxations under both equilibrium and nonequilibrium conditions from 10^{-3} to 10^3 s. Microscopic properties, including local electrolyte velocities and ion correlations, have been studied by operando XPCS with the Li/PEO-LITFSI/Li configuration, which can be correlated to the macroscopic ion concentration gradients.[59] Analogously, the direct probing of the distribution and diffusion of lithium can be largely enhanced with the development of spallation neutron sources and TOF techniques. Moreover, the development of neutron techniques will make operando studies on commercial batteries more accessible. The combination of synchrotron X-ray techniques and neutron techniques would enable even greater knowledge and productivity on batteries.

ACKNOWLEDGMENTS

This work is supported by the National Key Research and Development Program (2019YFA0405601), the National Natural Science Foundation of China (52130202, U1632269) and the Shanghai Committee of Science and Technology (19ZR1467000).

REFERENCES

[1] Als-Nielsen, J., *Elements of Modern X-Ray Physics*. John Wiley & Sons Ltd, Chichester: 2002; Vol. 55.

[2] Stöhr, J., *NEXAFS Spectroscopy*. Springer Berlin, Heidelberg: 1992.

[3] Liang, L.; Rinaldi, R.; Schober, H., *Neutron Applications in Earth, Energy and Environmental Sciences*. Springer, New York, NY: 2009.

[4] Helmut Fritzsche, J. H., Daniel Fruchart, *Neutron Scattering and Other Nuclear Techniques for Hydrogen in Materials*. Springer, Cham: 2016.

[5] Boothroyd, A. T., *Principles of Neutron Scattering from Condensed Matter*. Oxford University Press: 2020.

[6] Seah, M. P.; Dench, W. A., Quantitative electron spectroscopy of surfaces: A standard data base for electron inelastic mean free paths in solids. *Surf. Interface Anal.* **1979**, *1* (1), 2–11.

[7] Sun, X.; Yang, X. Q.; McBreen, J.; Gao, Y.; Yakovleva, M. V.; Xing, X. K.; Daroux, M. L., New phases and phase transitions observed in over-charged states of $LiCoO_2$-based cathode materials. *J. Power Sources* **2001**, *97–98*, 274–276.

[8] Zhou, Y.-N.; Ma, J.; Hu, E.; Yu, X.; Gu, L.; Nam, K.-W.; Chen, L.; Wang, Z.; Yang, X.-Q., Tuning charge–discharge induced unit cell breathing in layer-structured cathode materials for lithium-ion batteries. *Nat. Comms.* **2014**, *5* (1), 5381.

[9] Padhi, A. K.; Nanjundaswamy, K. S.; Goodenough, J. B., Phospho-olivines as positive-electrode materials for rechargeable lithium batteries. *J. Electrochem. Soc.* **1997**, *144* (4), 1188–1194.

[10] Delmas, C.; Maccario, M.; Croguennec, L.; Le Cras, F.; Weill, F., Lithium deintercalation in $LiFePO_4$ nanoparticles via a domino-cascade model. *Nat. Mater.* **2008**, *7* (8), 665–671.

[11] Orikasa, Y.; Maeda, T.; Koyama, Y.; Murayama, H.; Katsutoshi, F.; Tanida, H.; Arai, H.; Matsubara, E.; Uchimoto, Y.; Ogumi, Z., Direct observation of a metastable crystal phase of Li_xFePO_4 under electrochemical phase transition. *J. Am. Chem. Soc.* **2013**, *135*, 5497–5500.

[12] Liu, H.; Strobridge, F. C.; Borkiewicz, O. J.; Wiaderek, K. M.; Chapman, K. W.; Chupas, P. J.; Grey, C. P., Capturing metastable structures during high-rate cycling of $LiFePO_4$ nanoparticle electrodes. *Science* **2014**, *344* (6191), 1252817.

[13] Zhang, X.; van Hulzen, M.; Singh, D. P.; Brownrigg, A.; Wright, J. P.; van Dijk, N. H.; Wagemaker, M., Direct view on the phase evolution in individual $LiFePO_4$ nanoparticles during Li-ion battery cycling. *Nat. Comms.* **2015**, *6* (1), 8333.

[14] Liu, J.; Kunz, M.; Chen, K.; Tamura, N.; Richardson, T. J., Visualization of charge distribution in a lithium battery electrode. *J. Phys. Chem. Lett.* **2010**, *1* (14), 2120–2123.
[15] Marschilok, A. C.; Bruck, A. M.; Abraham, A.; Stackhouse, C. A.; Takeuchi, K. J.; Takeuchi, E. S.; Croft, M.; Gallaway, J. W., Energy dispersive X-ray diffraction (EDXRD) for operando materials characterization within batteries. *Phys. Chem. Chem. Phys.* **2020**, *22* (37), 20972–20989.
[16] Paxton, W. A.; Zhong, Z.; Tsakalakos, T., Tracking inhomogeneity in high-capacity lithium iron phosphate batteries. *J. Power Sources* **2015**, *275*, 429–434.
[17] Liang, G.; Croft, M. C.; Zhong, Z., Energy dispersive X-ray diffraction profiling of prototype $LiMn_2O_4$-based coin cells. *J. Electrochem. Soc.* **2013**, *160* (8), A1299–A1303.
[18] Kitada, K.; Murayama, H.; Fukuda, K.; Arai, H.; Uchimoto, Y.; Ogumi, Z.; Matsubara, E., Factors determining the packing-limitation of active materials in the composite electrode of lithium-ion batteries. *J. Power Sources* **2016**, *301*, 11–17.
[19] Delacourt, C.; Rodríguez-Carvajal, J.; Schmitt, B.; Tarascon, J.-M.; Masquelier, C., Crystal chemistry of the olivine-type LixFePO4 system ($0 \leqslant x \leqslant 1$) between 25 and 370 °C. *Solid State Sci.* **2005**, *7* (12), 1506–1516.
[20] Nishimura, S.-I.; Kobayashi, G.; Ohoyama, K.; Kanno, R.; Yashima, M.; Yamada, A., Experimental visualization of lithium diffusion in Li_xFePO_4. *Nat. Mater.* **2008**, *7* (9), 707–711.
[21] Rayavarapu, P. R.; Sharma, N.; Peterson, V. K.; Adams, S., Variation in structure and Li^+-ion migration in argyrodite-type Li_6PS_5X (X = Cl, Br, I) solid electrolytes. *J. Solid State Electrochem.* **2012**, *16* (5), 1807–1813.
[22] Rao, R. P.; Gu, W.; Sharma, N.; Peterson, V. K.; Avdeev, M.; Adams, S., In situ neutron diffraction monitoring of $Li_7La_3Zr_2O_{12}$ formation: toward a rational synthesis of garnet solid electrolytes. *Chem. Mater.* **2015**, *27* (8), 2903–2910.
[23] Wiaderek, K. M.; Borkiewicz, O. J.; Pereira, N.; Ilavsky, J.; Amatucci, G. G.; Chupas, P. J.; Chapman, K. W., Mesoscale effects in electrochemical conversion: coupling of chemistry to atomic- and nanoscale structure in iron-based electrodes. *J. Am. Chem. Soc.* **2014**, *136* (17), 6211–6214.
[24] Feng, Z.; Sarnello, E.; Li, T.; Cheng, L., Communication-microscopic view of the ethylene carbonate based lithium-ion battery electrolyte by X-ray scattering. *J. Electrochem. Soc.* **2019**, *166* (2), A47–A49.
[25] Jafta, C. J.; Sun, X.-G.; Veith, G. M.; Jensen, G. V.; Mahurin, S. M.; Paranthaman, M. P.; Dai, S.; Bridges, C. A., Probing microstructure and electrolyte concentration dependent cell chemistry via operando small angle neutron scattering. *Energy Environ. Sci.* **2019**, *12* (6), 1866–1877.
[26] Butala, M. M.; Mayo, M.; Doan-Nguyen, V. V. T.; Lumley, M. A.; Göbel, C.; Wiaderek, K. M.; Borkiewicz, O. J.; Chapman, K. W.; Chupas, P. J.; Balasubramanian, M.; Laurita, G.; Britto, S.; Morris, A. J.; Grey, C. P.; Seshadri, R., Local structure evolution and modes of charge storage in secondary Li-FeS_2 cells. *Chem. Mater.* **2017**, *29* (7), 3070–3082.
[27] Hu, E.; Li, Q.; Wang, X.; Meng, F.; Liu, J.; Zhang, J.-N.; Page, K.; Xu, W.; Gu, L.; Xiao, R.; Li, H.; Huang, X.; Chen, L.; Yang, W.; Yu, X.; Yang, X.-Q., Oxygen-redox reactions in $LiCoO_2$ cathode without O-O bonding during charge-discharge. *Joule* **2021**, *5* (3), 720–736.
[28] Yu, X.; Lyu, Y.; Gu, L.; Wu, H.; Bak, S.-M.; Zhou, Y.; Amine, K.; Ehrlich, S. N.; Li, H.; Nam, K.-W.; Yang, X.-Q., Understanding the rate capability of high-energy-density Li-rich layered $Li_{1.2}Ni_{0.15}Co_{0.1}Mn_{0.55}O_2$ cathode materials. *Adv. Energy Mater.* **2014**, *4* (5), 1300950.
[29] Lin, F.; Liu, Y.; Yu, X.; Cheng, L.; Singer, A.; Shpyrko, O. G.; Xin, H. L.; Tamura, N.; Tian, C.; Weng, T.-C.; Yang, X.-Q.; Meng, Y. S.; Nordlund, D.; Yang, W.; Doeff, M. M., Synchrotron X-ray analytical techniques for studying materials electrochemistry in rechargeable batteries. *Chem. Rev.* **2017**, *117* (21), 13123–13186.

[30] Liu, X.; Liu, J.; Qiao, R.; Yu, Y.; Li, H.; Suo, L.; Hu, Y.-S.; Chuang, Y.-D.; Shu, G.; Chou, F.; Weng, T.-C.; Nordlund, D.; Sokaras, D.; Wang, Y. J.; Lin, H.; Barbiellini, B.; Bansil, A.; Song, X.; Liu, Z.; Yan, S.; Liu, G.; Qiao, S.; Richardson, T. J.; Prendergast, D.; Hussain, Z.; de Groot, F. M. F.; Yang, W., Phase transformation and lithiation effect on electronic structure of Li_xFePO_4: an in-depth study by soft X-ray and simulations. *J. Am. Chem. Soc.* **2012**, *134* (33), 13708–13715.

[31] Liu, X.; Wang, Y. J.; Barbiellini, B.; Hafiz, H.; Basak, S.; Liu, J.; Richardson, T.; Shu, G.; Chou, F.; Weng, T. C.; Nordlund, D.; Sokaras, D.; Moritz, B.; Devereaux, T. P.; Qiao, R.; Chuang, Y. D.; Bansil, A.; Hussain, Z.; Yang, W., Why $LiFePO_4$ is a safe battery electrode: Coulomb repulsion induced electron-state reshuffling upon lithiation. *Phys. Chem. Chem. Phys.* **2015**, *17* (39), 26369–77.

[32] Olalde-Velasco, P.; Jiménez-Mier, J.; Denlinger, J. D.; Hussain, Z.; Yang, W. L., Direct probe of Mott-Hubbard to charge-transfer insulator transition and electronic structure evolution in transition-metal systems. *Phys. Rev. B* **2011**, *83* (24), 241102.

[33] Hong, W. T.; Stoerzinger, K. A.; Moritz, B.; Devereaux, T. P.; Yang, W.; Shao-Horn, Y., Probing $LaMO_3$ metal and oxygen partial density of states using X-ray emission, absorption, and photoelectron spectroscopy. *J. Phys. Chem. C.* **2015**, *119* (4), 2063–2072.

[34] Hong, W. T.; Stoerzinger, K. A.; Lee, Y.-L.; Giordano, L.; Grimaud, A.; Johnson, A. M.; Hwang, J.; Crumlin, E. J.; Yang, W.; Shao-Horn, Y., Charge-transfer-energy-dependent oxygen evolution reaction mechanisms for perovskite oxides. *Energ. Environ. Sci.* **2017**, *10* (10), 2190–2200.

[35] Firouzi, A.; Qiao, R.; Motallebi, S.; Valencia, C. W.; Israel, H. S.; Fujimoto, M.; Wray, L. A.; Chuang, Y.-D.; Yang, W.; Wessells, C. D., Monovalent manganese based anodes and co-solvent electrolyte for stable low-cost high-rate sodium-ion batteries. *Nat. Comms.* **2018**, *9* (1), 861.

[36] Zhuo, Z.; Liu, Y.-S.; Guo, J.; Chuang, Y.-D.; Pan, F.; Yang, W., Full energy range resonant inelastic X-ray scattering of O_2 and CO_2: direct comparison with oxygen redox state in batteries. *J. Phys. Chem. Lett.* **2020**, *11* (7), 2618–2623.

[37] Assat, G.; Iadecola, A.; Delacourt, C.; Dedryvère, R.; Tarascon, J.-M., Decoupling cationic-anionic redox processes in a model Li-rich cathode via operando x-ray absorption spectroscopy. *Chem. Mater.* **2017**, *29* (22), 9714–9724.

[38] Cheng, L.; Crumlin, E. J.; Chen, W.; Qiao, R.; Hou, H.; Franz Lux, S.; Zorba, V.; Russo, R.; Kostecki, R.; Liu, Z.; Persson, K.; Yang, W.; Cabana, J.; Richardson, T.; Chen, G.; Doeff, M., The origin of high electrolyte–electrode interfacial resistances in lithium cells containing garnet type solid electrolytes. *Phys. Chem. Chem. Phys.* **2014**, *16* (34), 18294–18300.

[39] Takamatsu, D.; Koyama, Y.; Orikasa, Y.; Mori, S.; Nakatsutsumi, T.; Hirano, T.; Tanida, H.; Arai, H.; Uchimoto, Y.; Ogumi, Z., First in situ observation of the $LiCoO_2$ electrode/electrolyte interface by total-reflection X-ray absorption spectroscopy. *Angew. Chem. Int. Edit.* **2012**, *51* (46), 11597–11601.

[40] Wang, C.; Gong, Y.; Dai, J.; Zhang, L.; Xie, H.; Pastel, G.; Liu, B.; Wachsman, E.; Wang, H.; Hu, L., In situ neutron depth profiling of lithium metal-garnet interfaces for solid state batteries. *J. Am. Chem. Soc.* **2017**, *139* (40), 14257–14264.

[41] Blagoveshchenskii, N.; Novikov, A.; Savostin, V., Self-diffusion in liquid lithium from coherent quasielastic neutron scattering. *Physica B: Condens. Matter* **2012**, *407*, 4567–4569.

[42] Heitmann, T.; Hester, G.; Mitra, S., Evolution of Boson peak with Li-salt concentration in superionic $xLi_2SO_4 \cdot (1-x)LiPO_3$ glasses. *Physica B Condens. Matter* **2018**, *551*, 315–319.

[43] Muy, S.; Bachman, J. C.; Giordano, L.; Chang, H.-H.; Abernathy, D. L.; Bansal, D.; Delaire, O.; Hori, S.; Kanno, R.; Maglia, F.; Lupart, S.; Lamp, P.; Shao-Horn, Y., Tuning mobility and stability of lithium ion conductors based on lattice dynamics. *Energy Environ. Sci.* **2018**, *11* (4), 850–859.

[44] Heitmann, T.; Hester, G.; Mitra, S.; Calloway, T.; Tyagi, M.; Miskowiec, A.; Diallo, S.; Osti, N.; Mamontov, E., Probing Li ion dynamics in amorphous $xLi_2SO_4 \cdot (1-x)LiPO_3$ by quasielastic neutron scattering. *Solid State Ionics* **2019**, *334*, 95–98.

[45] Klenk, M.; Boeberitz, S.; Dai, J.; Jalarvo, N.; Peterson, V.; Lai, W., Lithium self-diffusion in a model lithium garnet oxide $Li_5La_3Ta_2O_{12}$: a combined quasi-elastic neutron scattering and molecular dynamics study. *Solid State Ionics* **2017**, *312*, 1–7.

[46] Ida, N.; Meyendorf, N., *Handbook of Advanced Nondestructive Evaluation*. Springer, Cham: 2019.

[47] Xu, Y.; Hu, E.; Zhang, K.; Wang, X.; Borzenets, V.; Sun, Z.; Pianetta, P.; Yu, X.; Liu, Y.; Yang, X.-Q.; Li, H., In situ visualization of state-of-charge heterogeneity within a $LiCoO_2$ particle that evolves upon cycling at different rates. *ACS Energy Lett.* **2017**, *2* (5), 1240–1245.

[48] Lou, S.; Liu, Q.; Zhang, F.; Liu, Q.; Yu, Z.; Mu, T.; Zhao, Y.; Borovilas, J.; Chen, Y.; Ge, M.; Xiao, X.; Lee, W.-K.; Yin, G.; Yang, Y.; Sun, X.; Wang, J., Insights into interfacial effect and local lithium-ion transport in polycrystalline cathodes of solid-state batteries. *Nat. Comms.* **2020**, *11* (1), 5700.

[49] Lim, J.; Li, Y.; Alsem Daan, H.; So, H.; Lee Sang, C.; Bai, P.; Cogswell Daniel, A.; Liu, X.; Jin, N.; Yu, Y.-S.; Salmon Norman, J.; Shapiro David, A.; Bazant Martin, Z.; Tyliszczak, T.; Chueh William, C., Origin and hysteresis of lithium compositional spatiodynamics within battery primary particles. *Science* **2016**, *353* (6299), 566–571.

[50] Ebner, M.; Marone, F.; Stampanoni, M.; Wood, V., Visualization and quantification of electrochemical and mechanical degradation in Li ion batteries. *Science* **2013**, *342* (6159), 716–720.

[51] Harry, K. J.; Hallinan, D. T.; Parkinson, D. Y.; MacDowell, A. A.; Balsara, N. P., Detection of subsurface structures underneath dendrites formed on cycled lithium metal electrodes. *Nat. Mater.* **2014**, *13* (1), 69–73.

[52] Kardjilov, N.; Manke, I.; Woracek, R.; Hilger, A.; Banhart, J., Advances in neutron imaging. *Mater. Today* **2018**, *21* (6), 652–672.

[53] Ziesche, R. F.; Kardjilov, N.; Kockelmann, W.; Brett, D. J. L.; Shearing, P. R., Neutron imaging of lithium batteries. *Joule* **2022**, *6* (1), 35–52.

[54] Ziesche, R. F.; Arlt, T.; Finegan, D. P.; Heenan, T. M. M.; Tengattini, A.; Baum, D.; Kardjilov, N.; Markötter, H.; Manke, I.; Kockelmann, W.; Brett, D. J. L.; Shearing, P. R., 4D imaging of lithium-batteries using correlative neutron and X-ray tomography with a virtual unrolling technique. *Nat. Comms.* **2020**, *11* (1), 777.

[55] Song, B.; Dhiman, I.; Carothers, C.; Veith, G.; Liu, J.; Bilheux, H.; Huq, A., Dynamic lithium distribution upon dendrite growth and shorting revealed by operando neutron imaging. *ACS Energy Lett.* **2019**, *4*.

[56] Xiao, Y.; Liu, T.; Liu, J.; He, L.; Chen, J.; Zhang, J.; Luo, P.; Lu, H.; Wang, R.; Zhu, W.; Hu, Z.; Teng, G.; Xin, C.; Zheng, J.; Liang, T.; Wang, F.; Chen, Y.; Huang, Q.; Pan, F.; Chen, H., Insight into the origin of lithium/nickel ions exchange in layered $Li(Ni_xMn_yCo_z)O_2$ cathode materials. *Nano Energy* **2018**, *49*, 77–85.

[57] Komatsu, H.; Minato, T.; Matsunaga, T.; Shimoda, K.; Kawaguchi, T.; Katsutoshi, F.; Nakanishi, K.; Tanida, H.; Kobayashi, S.; Hirayama, T.; Ikuhara, Y.; Arai, H.; Ukyo, Y.; Uchimoto, Y.; Matsubara, E.; Ogumi, Z., Site-selective analysis of nickel substituted Li-rich layered material: migration and role of transition metal at charging and discharging. *J. Phys. Chem. C.* **2018**, *122*.

[58] Liu, T.; Liu, J.; Li, L.; Yu, L.; Diao, J.; Zhou, T.; Li, S.; Dai, A.; Zhao, W.; Xu, S.; Ren, Y.; Wang, L.; Wu, T.; Qi, R.; Xiao, Y.; Zheng, J.; Cha, W.; Harder, R.; Robinson, I.; Wen, J.; Lu, J.; Pan, F.; Amine, K., Origin of structural degradation in Li-rich layered oxide cathode. *Nature* **2022**, *606* (7913), 305–312.

[59] Steinrück, H.-G.; Takacs, C. J.; Kim, H.-K.; Mackanic, D. G.; Holladay, B.; Cao, C.; Narayanan, S.; Dufresne, E. M.; Chushkin, Y.; Ruta, B.; Zontone, F.; Will, J.; Borodin, O.; Sinha, S. K.; Srinivasan, V.; Toney, M. F., Concentration and velocity profiles in a polymeric lithium-ion battery electrolyte. *Energ. Environ. Sci.* **2020**, *13* (11), 4312–4321.

Index

Note: Bold page numbers refer to tables and *italic* page numbers refer to figures.

For Product Safety Concerns and Information please contact our EU
representative GPSR@taylorandfrancis.com
Taylor & Francis Verlag GmbH, Kaufingerstraße 24, 80331 München, Germany

www.ingramcontent.com/pod-product-compliance
Lightning Source LLC
Chambersburg PA
CBHW060424220326
41598CB00021BA/2285

* 9 7 8 1 0 3 2 2 8 9 5 4 0 *